LABORATORY MANUAL FOR EARTH SCIENCE

JESSICA OLNEY
Hillsborough Community College, Florida

ALLAN LUDMAN
Queens College, New York

STEPHEN MARSHAK
University of Illinois

ROBERT RAUBER
University of Illinois

W. W. NORTON & COMPANY
NEW YORK · LONDON

W. W. Norton & Company has been independent since its founding in 1923, when William Warder Norton and Mary D. Herter Norton first published lectures delivered at the People's Institute, the adult education division of New York City's Cooper Union. The firm soon expanded their program beyond the Institute, publishing books by celebrated academics from America and abroad. By mid-century, the two major pillars of Norton's publishing program—trade books and college texts—were firmly established. In the 1950s, the Norton family transferred control of the company to its employees, and today—with a staff of five hundred and hundreds of trade, college, and professional titles published each year—W. W. Norton & Company stands as the largest and oldest publishing house owned wholly by its employees.

Editor: Jake Schindel
Senior Project Editor: Thomas Foley
Production Manager: Eric Pier-Hocking
Editorial Assistant: Mia Davis
Copy Editor: Laura Sewell
Managing Editor, College: Marian Johnson
Managing Editor, College Digital Media: Kim Yi
Digital Media Editor: Michael Jaoui
Associate Media Editor: Arielle Holstein
Digital Media Project Editor: Marcus Van Harpen
Marketing Manager, Geology: Katie Sweeney
Associate Art Director: Lissi Sigillo
Designer: Lisa Buckley Design
Photography Editor: Thomas Persano
Director of College Permissions: Megan Schindel
Permissions Manager: Bethany Salminen
Text Permission Specialist: Joshua Garvin
Composition and page layout: Graphic World, Inc.
Graphic World, Inc. Project Manager: Gary Clark
New Illustrations by Graphic World, Inc.
All Topographic Maps created by Mapping Specialists, Ltd.—Madison, WI
Manufacturing by LSC Communications—Menasha, WI
Permission to use copyrighted material is included in the backmatter of this book.

978-0-393-69712-4

W. W. Norton & Company, Inc., 500 Fifth Avenue, New York, N.Y. 10110
www.wwnorton.com

W. W. Norton & Company Ltd., 15 Carlisle Street, London W1D 3BS

1 2 3 4 5 6 7 8 9 0

CONTENTS

This laboratory manual is based on many years of teaching and coordinating introductory Earth science and geology lab and lecture courses. Those experiences have helped us understand not only how students best learn Earth science principles, but also how to stimulate their engagement with the material to enhance their learning process. Our manual provides (1) an up-to-date, comprehensive background covering the full scope of an introductory Earth science course; (2) step-by-step explanations that facilitate students' understanding of what they're doing, why, and how it connects to broader Earth science concepts; (3) text and exercises that lead students to think like Earth scientists, engaging them in solving real-life problems important to their lives; and (4) the passion and excitement that we still feel after decades as teachers and researchers.

Students often ask us how we maintain this enthusiasm. Our answer is to share with them both the joys and frustrations of facing and solving real-world science problems. You will find many of those types of problems in the following pages—modified for the introductory nature of the course, but still reflecting their challenges and the rewards of solving them. This manual brings that experience directly to the students, *engaging* them in the learning process by *explaining* concepts clearly and providing many avenues for further *exploration*.

Unique Elements

As you read through this manual, you will find a distinctive approach that incorporates many unique elements, distinguishing it from other manuals. These elements include:

Emphasis on inquiry and exploration

We believe that students learn science best by *doing* science, not just by memorizing facts.

"Geotours" Google Earth activities in each lab, authored by M. Scott Wilkerson of DePauw University, allow students to apply the concept knowledge gained through their lab and lecture work to real-life sites and phenomena, just as an Earth scientist would. "What Do You Think?" prompts in each chapter get students to make assessments and decisions on topics relevant to their own lives, such as selecting building codes appropriate for local earthquake hazards (Ch. 2); considering how technological advances that have altered demands for various mineral resources have impacted their community (Ch. 3); and evaluating what unforeseen and unintended results accompany human interactions with natural processes (Ch. 11).

Innovative, hands-on exercises that engage students and provide instructors with choice

Tiered exercises are carefully integrated into the text, leading students to understand concepts for themselves by first reading about the concept and then immediately using what they learned in the accompanying exercise. These unique exercises engage students because they see how important Earth science principles are to our everyday activities. The exercises also engage a variety of modes of student thinking, including foundational concept review; graph creation and interpretation; measurement and calculation; topographic map creation and interpretation; hands-on experimentation in which students collect data and employ the scientific method; and critical thinking and application.

There are more exercises per chapter than can probably be completed in a single lab session. This is done intentionally to provide flexibility for instructors. The complexity and rigor of the exercises increase within each chapter, enabling instructors to assign the exercises that are most appropriate for their student population. This structure also ensures students move beyond simply recording answers to consider the scientific and societal implications of the answers they are generating.

Superb illustration program

Readers have come to expect a superior illustration program in any Norton Earth science text, and this manual does not disappoint. The extensive and highly illustrative photos, line drawings, maps, and DEMs continue the tradition of Stephen Marshak and Robert Rauber's popular *Earth Science* textbook.

Reader-friendly language and layout

Our decades of teaching introductory students help us to identify the concepts that are most difficult for students to understand. The conversational style of this manual and the use of many real-world analogies help to make these difficult concepts clear and enhance student understanding. The crisp, open layout makes the book reader-friendly and accessible for students new to science.

Online Exercises

Nearly every exercise in the manual is available in an online format via Norton Testmaker, at no extra cost to students. These online exercises can also be integrated into your campus learning management system (LMS), with student work reporting to your LMS gradebook. Responses are either auto-graded or require brief responses that can be manually graded. *Note:* Students still need either a print or electronic version of the lab manual for access to figures and background reading. The online exercises can be downloaded at wwnorton.com/instructors.

Pre-Lab Quizzes

Pre-lab quizzes, available as auto-graded assignments or printable worksheets, are also accessible via Norton Testmaker and can be integrated into your LMS. These exercises are designed to ensure that students come to their lab session prepared and having done the assigned reading. The pre-lab quizzes can be downloaded at wwnorton.com/instructors.

Additional Instructor and Student Resources

(available for download at *wwnorton.com/instructors*)
Instructor's Manual. Available in electronic format, the Instructor's Manual contains Word files of the solutions to each exercise, teaching tips for each lab, suggestions for how to set up certain exercises or run them in alternate ways, and a list of all the exercises available in online format.

Electronic Figures. All figures, photographs, charts, maps, and exercise worksheets in the manual are available for instructors to download and incorporate into presentations, handouts, or online courses.

Videos and Animations. Narrative art videos, real-world videos, animations, and interactive simulations, all designed or curated to help students better understand core concepts in Earth science, are available to stream from our site.

Acknowledgments

We are indebted to the talented team at W. W. Norton & Company whose zealous quest for excellence is matched only by their ingenuity in solving layout problems, finding that special photograph, and keeping the project on schedule. Specifically, we'd like to thank our editor Jake Schindel and editorial assistant Mia Davis; the digital media team of Michael Jaoui, Arielle Holstein, Jasvir Singh, and Marcus Van Harpen; project editor Thom Foley; production manager Eric Pier-Hocking; photo editor Tommy Persano; and permissions specialist Josh Garvin. We would also like to thank Matthew Olney for assistance in reviewing exercises and providing photos. Finally, we would like to thank the following reviewers for their input and expertise in making this lab manual the best it can be:

Christine Baack, Tarrant County College
Lanna Bradshaw, Brookhaven College
Melissa Driskell, University of North Alabama
Joseph Garcia, Longwood University
Bryan Gibbs, Richland College
Deborah Hyde, Northeastern State University
Peter Jenkins, Charleston Southern University
Shakira Khan, Broward College
Dominike Merle-Johnson, Montgomery County Community College
David Mrofka, Mt. San Antonio College
Daniel Murphy, Eastfield College
Kuppusamy Panneerselvam, Florida International University
Jeffrey Richardson, Columbus State Community College
Jennifer Sheppard, Moraine Valley Community College
Tawny Tibbits, Broward College
Zhiyong Wang, Troy University

Jessica Olney is associate professor of Earth Sciences at Hillsborough Community College, Ybor City. She holds a B.S. and M.A. from The Ohio State University, and a Ph.D. from Northern Illinois University. She teaches lecture and laboratory courses in Earth Science, Physical Geology, and Oceanography. Her research has ranged from studying ice-rafted debris in ocean sediment cores to igneous petrology and volcanology/geochemistry in Central Utah and Nicaragua. She has been teaching since graduate school, first as a laboratory teaching assistant in the classroom and on field trips (Ohio State and Northern Illinois); as a summer elderhostel lecturer (Northern Illinois); and in 2004, as the Geologist-in-the-Park at Capulin Volcano National Park, New Mexico, where she led interpretive programs and participated in the removal of invasive plant species from the park. Before coming to HCC, she taught at Florida State University and Tallahassee Community College.

Allan Ludman is Professor of Geology in the School of Earth and Environmental Sciences at Queens College, part of the City University of New York. He holds a B.S. from Brooklyn College, an M.A. from Indiana University, and a Ph.D. from the University of Pennsylvania. He has devoted more than five decades to deciphering the evolution of the Northern Appalachian mountain system through field and laboratory studies in Maine and parts of adjacent New Brunswick, Canada. He has more than 50 years of experience teaching introductory geology and has supervised the introductory geology laboratories at Queens College for the past 43 years. Professor Ludman is also the director of GLOBE NY Metro, the southern New York State partner of the GLOBE Program®, an international K-12 science teacher development program, created to promote hands-on, inquiry-based Earth System research for elementary and high-school level teachers and their students. As a field mapper, he compiled geologic maps for over 2,500 square miles in Maine and was a regional compiler for the Bedrock Geologic Map of Maine. In addition to research papers and this manual, he has co-authored *Laboratory Manual for Introductory Geology*.

Stephen Marshak is a Professor Emeritus of Geology at the University of Illinois, Urbana-Champaign, where he also served as Department Head and as the Director of the School of Earth, Society, and Environment. He holds an A.B. from Cornell University, an M.S. from the University of Arizona, and a Ph.D. from Columbia University. Steve's research interests lie in structural geology and tectonics, and he has participated in field projects on a number of continents. Steve loves teaching and has won his college's and university's highest teaching awards. He also received the 2012 Neil Miner Award from the National Association of Geoscience Teachers (NAGT), for "exceptional contributions to the stimulation of interest in the Earth sciences." Steve is a Fellow of the Geological Society of America. In addition to research papers and this manual, he has authored *Earth: Portrait of a Planet*, and *Essentials of Geology*, and has co-authored *Earth Science, Laboratory Manual for Introductory Geology, Earth Structure: An Introduction to Structural Geology and Tectonics*, and *Basic Methods of Structural Geology*.

(continued)

Robert Rauber is a Professor of Atmospheric Sciences at the University of Illinois, Urbana-Champaign, where he served as the Department Head and is currently the Director of the School of Earth, Society, and Environment. He holds a B.S. in Physics and a B.A. in English from the Pennsylvania State University, as well as M.S. and Ph.D. degrees in Atmospheric Science from Colorado State University. In addition to teaching and writing, Bob oversees a research program that focuses on the development and behavior of storms. To carry out this work, Bob flies into hazardous weather in specially equipped airplanes—during these flights, he's so intent on recording data that he usually doesn't notice the bouncing and lightning. Bob has won several campus teaching awards, is a Fellow of the American Meteorological Society (AMS), and served as Publication Commissioner for the AMS. In addition to research papers and this manual, he has authored *Severe and Hazardous Weather: An Introduction to High Impact Meteorology* and *Radar Meteorology, a First Course*, and co-authored *Earth Science*.

Setting the Stage for Learning about the Earth

1

This sunset, over the red cliffs of the Grand Canyon on the Colorado River in Arizona, shows the Earth System at a glance—air, water, and rock all interacting to produce this stunning landscape.

1.1 Introduction

1.1.1 Thinking Like an Earth Scientist

Learning about the Earth is like training to become a detective. Both Earth scientists and detectives need keen powers of observation, curiosity about slight differences, broad scientific understanding, and instruments to analyze samples. And both ask the same questions: What happened? How? When? Why? Much of the logical thinking is the same, but there are big differences between the work of a detective and that of a scientist. A detective's "cold case" may be 30 years old, but "old" to an Earth scientist means hundreds of millions or billions of years. To a detective, a "body" is a human body, but to an Earth scientist, a body may be a mountain range or a continent. Eyewitnesses can help detectives, but for most of the Earth's history there weren't any humans to report to scientists regarding what happened. To study the Earth, scientists must therefore develop different strategies from those of other kinds of investigators. The overall goal of this manual is to help you look at the Earth and think about its mysteries like a scientist.

To help you begin thinking like an Earth scientist, let's start with a typical geologic mystery. Almost 300 years ago, settlers along the coast of Maine built piers (like the modern pier shown in **FIGURE 1.1**) to load and unload ships. Some of these piers are now submerged to a depth of 1 meter (39 inches) below sea level.

Tourists might not think twice about this before heading for a lobster dinner at the local restaurant, but a **geologist** (an Earth scientist who studies our planet with a focus on the materials that compose it, the phenomena that change it, and its long-term history) would want to know what caused the submergence and how rapidly the pier was submerged. How would a geologist go about tackling this problem? Exercise 1.1 will outline the problem and show some of the basic geologic reasoning needed to get answers to the questions raised above. At the same time, this will be your first of many opportunities to see that Earth scientists—including geologists, oceanographers, meteorologists, and astronomers—solve real-world problems affecting real people.

FIGURE 1.1 Subsidence along the coast of Maine. The red circle shows where the subsidence has occurred and submerged the pier.

Name: _____   Section: _____
Course: _____   Date: _____

2020

1720

1 m

1 m

The figure on the left illustrates a pier whose walkway sits 1 meter below the ocean's surface today. Because we weren't there when it was built 300 years ago, we have to make some assumptions—scientists often do this to make estimates. So, let's assume that the pier's walkway was originally built 1 m *above* sea level at high tide, as many are built today (illustrated in the figure on the right), and that submergence occurred at a constant rate. With these assumptions, calculating the rate of submergence for the past 300 years becomes simple arithmetic.

(a) The rate of submergence is the total change in the elevation of the pier (_____ m) divided by the total amount of time involved (_____ yr) and is therefore _____cm/yr. (Remember, 1 meter = 100 centimeters.)

Now consider a problem this equation might solve:

(b) A local restaurant owner is considering the purchase of a pier, whose walkway is 50 cm above the high-water mark, for use in outdoor events. The owner has been advised that piers with walkways less than 30 cm above the high-water mark should be avoided because they can be flooded by storms and very high tides. If submergence continues at the rate you calculated, how many years will pass before the high-water mark is less than 30 cm from the base of the walkway? _____ years

? What Do You Think Now it's time to try really thinking like a scientist.
■ Given your answers to questions (a) and (b), would you recommend that the restaurant owner purchase this pier? In a sentence or two, on a separate sheet of paper, explain why. Then describe another issue that you think the owner should investigate before making a decision.

Congratulations! You've just tackled your first problem. A veteran Earth scientist would also want to explain *why* the piers were submerged. When faced with a problem like this, scientists typically try to come up with as many explanations as possible. For example, which of the following explanations could account for the submergence?

☐ The sea level has risen.

☐ The land has sunk.

☐ Both the sea level and land have risen, but sea level has gone up more.

☐ Both the sea level and land have sunk, but the land has sunk more.

If you thought all four choices might be right (correctly!), you realize that explaining submergence along the Maine coast may be more complicated than it seemed at first. To find the answer, you need more **data**—more observations and/or measurements. One way to obtain more data would be to see if submergence is restricted to just Maine or to the entire east coast of North America, or if it is perhaps worldwide. As it turns out, submergence is observed worldwide, suggesting that the first choice above (sea-level rise) is the most probable explanation, but not necessarily the only one.

With even more data, we could answer questions like, When did submergence happen? Did the sea level rise at a constant rate? Maybe all submergence occurred in the first 100 years after the pier was built and then stopped. Or perhaps it began slowly and then accelerated. Unfortunately, we may not be able to answer all of these questions because, unlike television detectives who always get the bad guys, scientists don't always have enough data and must often live with uncertainty. We still do not have the answers to many questions about the Earth.

1.1.2 The Methods of Science

Scientists (and most people trying to find answers to problems they have identified) follow a logical process that you are probably familiar with—the **scientific method.** You did so instinctively in Exercise 1.1 and will do so many times throughout this course. The scientific method begins with making observations of Earth features or processes—such as, in Exercise 1.1, the observation that a colonial pier is now below sea level. The sequence and cycle of observations and experiments in the scientific method are illustrated schematically in **FIGURE 1.2.**

FIGURE 1.2 **The scientific method.**

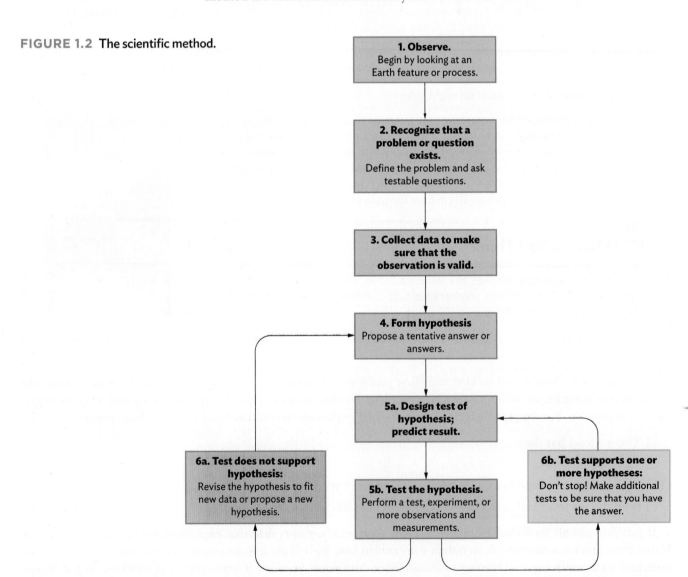

Before moving on, take a few minutes to consider how each step in the process above was represented in the pier investigation. Then see how your breakdown compares to the explanation below.

STEP 1 Observe an Earth feature or process (e.g., the submerged pier).

STEP 2 Recognize that a problem exists and define the problem by asking questions about it. Usually the problem is that we don't understand how what we've observed came to be: What caused the pier in Exercise 1.1 to be submerged? By how much has the pier been submerged? How fast did submergence take place? Was submergence constant or sporadic? We respond with the steps that follow.

STEP 3 Collect more data to (a) confirm that the observation is valid and (b) help us understand what is going on. In Exercise 1.1, for instance, it isn't just one pier being submerged, but many along the Maine coast.

STEP 4 Propose tentative answers to those questions, called **hypotheses** (singular, **hypothesis**). Some versions of the scientific method suggest proposing a single hypothesis, but when we first look at problems, we usually find that more than one hypothesis can explain the phenomenon. We therefore come up with as many hypotheses as we can—a practice called having *multiple working hypotheses.*

STEP 5 Test the hypotheses by getting more data. The new information may support some hypotheses, rule out others, and/or possibly lead to new ones. Some of this testing can be in a classic laboratory experiment, but there are also other types of tests, such as field trips to gain additional information, detailed measurements where there had been only eyeball estimates, and so forth.

STEP 6 Based on the new information, reject or modify those hypotheses that don't fit, continue testing those that do, and propose new ones as needed to meet all the information. If your test supports a hypothesis, continue to perform additional tests to further verify your result.

Continue cycling through steps 4, 5, and 6 as needed until a single hypothesis remains. Then, to be sure, continue testing it. If this hypothesis survives years of further testing, it is then considered to be a **theory**. Nonscientists often don't understand the difference between hypothesis and theory; when they say, "Oh, that's just a theory," they really mean, "That's just a hypothesis"—a *possible* explanation that is not yet proven. A theory has been tested *and* proven. Some theories with which you may be familiar are the theory of evolution, the germ theory of disease, and Einstein's theory of relativity. And during this course you will become very familiar with the plate tectonic theory for how the Earth's major features form and change.

1.2 Studying Matter and Energy in the Earth System

Now that you know *how* Earth scientists study things, let's look at *what* they study. The Earth is a dynamic planet. Unlike the airless, oceanless Moon, which has remained virtually unchanged for billions of years, gases in the Earth's atmosphere and water in its oceans, rivers, and lakes are in constant motion and cause the solid earth beneath them to change rapidly (rapidly in relation to geologic time, that is). Modern scientists envisage an **Earth System** that includes all of the Earth's materials—gases, liquids, solids, and life forms—and the energy that drives their activity. The first step in understanding the Earth System is to understand the nature of its matter and energy and how they interact with one another.

TABLE 1.1 Basic definitions

- An **element** is a substance that cannot be broken down chemically into other substances.
- The smallest piece of an element that still has all the properties of that element is an **atom**.
- Atoms combine with one another chemically to form **compounds**; the smallest possible piece of a compound is called a **molecule**.
- Atoms in compounds are held together by **chemical bonds**.
- A simple **chemical formula** describes the combination of atoms in a compound. For example, the formula H_2O shows that a molecule of water contains two atoms of hydrogen and one of oxygen.

1.2.1 The Nature of Matter

Matter is the "stuff" of which the Universe is made; we use the word matter to refer to any material on and in the Earth, within its atmosphere, or within the broader Universe in which the Earth resides. Geologists, chemists, and physicists have shown that matter consists of 92 naturally occurring elements, and that some of these elements are much more abundant than others. Keep these definitions about the composition of matter in mind as you read further (**TABLE 1.1**).

Matter occurs on the Earth in three states: solid, liquid, or gas. Atoms in *solids*, like minerals and rocks, are held in place by strong chemical bonds. As a result, solids retain their shape over long periods. Bonds in *liquids* are so weak that atoms or molecules move easily, and as a result, liquids adopt the shape of their containers. Atoms or molecules in *gases* are barely held together at all, so a gas expands to fill whatever container it is placed in. Matter changes from one state to another in many geologic processes, as when heat evaporates water to produce water vapor, or when water freezes to form ice, or when lava freezes to become solid rock.

We describe the amount of matter in an object by indicating its **mass** and the amount of space it occupies by specifying its **volume**. The more mass packed into a given volume of matter, the greater the **density** of the matter. You notice density differences every day: It's easier to lift a large box of popcorn than a piece of rock of the same size because the rock is much denser; it has much more mass packed into the same volume and therefore weighs much more. We'll explore density more closely in Section 1.3.2.

1.2.2 Distribution of Matter in the Earth System

Matter is stored in the Earth System in five major spheres, or **reservoirs** (**FIG. 1.3A**). Most *gas* is in the **atmosphere**, a semi-transparent blanket composed of about 78% nitrogen (N_2) and 21% oxygen (O_2), with minor amounts of water vapor (H_2O), carbon dioxide (CO_2), ozone (O_3), and methane (CH_4). Nearly all liquid occurs as water in the **hydrosphere**—the Earth's oceans, rivers, lakes, and groundwater, which is found in cracks and pores beneath the surface. Frozen water makes up the **cryosphere**, which includes snow, thin layers of ice on the surface of lakes or oceans, and huge masses of ice in glaciers and the polar ice caps.

The solid Earth is called the **geosphere**, which geologists divide into concentric layers like those in a hard-boiled egg (**FIG. 1.3B**). The outer layer, the **crust**, is relatively thin, like an eggshell, and consists mostly of rock. Below the crust is the **mantle**, which also consists mostly of different kinds of rock and contains most of

FIGURE 1.3 The Earth System.

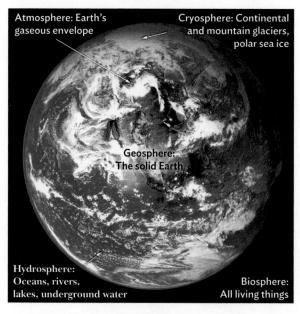

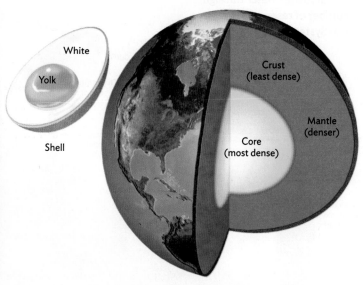

(a) The Earth's major reservoirs of matter.

(b) A simple image of the Earth's internal layering and the hard-boiled egg analogy for its pattern of layers.

the Earth's volume, just as an egg white contains most of an egg's volume. We say these layers are *mostly* rock because about 2% of the crust and mantle has melted to produce liquid material called **magma** (known as **lava** when it erupts on the surface). The central part of the Earth, comparable to the egg yolk, is the **core**. The outer core consists mostly of a liquid alloy of iron and nickel, and the inner core is a solid iron-nickel alloy. Humans have never drilled entirely through the crust; during this course, you will learn how we figured out that our planet is layered, how thick those layers are, and what they are made of.

Continents make up about 30% of the crust and are composed of relatively low-density rocks. The remaining 70% of the crust is covered by the oceans. Oceanic crust is both thinner and denser than the crust under the continents. Three types of solid Earth materials are found at the surface: **bedrock**, a solid aggregate of minerals and rocks attached to the Earth's crust; **sediment**, unattached mineral grains such as boulders, sand, and clay; and **soil**, sediment and rock modified by interactions with the atmosphere, hydrosphere, and organisms so that it can support plant life.

The **biosphere** is the realm of living organisms, extending from a few kilometers below the Earth's surface to a few kilometers above. Earth scientists have learned that organisms, from bacteria to mammals, are important parts of the Earth System because they contribute to many processes. They exchange gases with the atmosphere, absorb and release water, break rock into sediment, and play major roles in converting sediment and rock to soil.

The movement of materials from one reservoir to another is called a **flux** and happens in many Earth processes. For example, rain is a flux in which water moves from the atmosphere to the hydrosphere. Rates of flux depend on the materials, the reservoirs, and the processes involved. In some cases, a material moves among several reservoirs but eventually returns to the first. We call such a path a **cycle**. In this class you will learn about several cycles, including the **rock cycle** (the movement of atoms from one rock type to another) and the **hydrologic** (or **water**) **cycle** (the movement of water from the hydrosphere to and from the other reservoirs) (**FIG. 1.4**). Exercises 1.2 and 1.3 will help you understand the distribution and fluxes of matter.

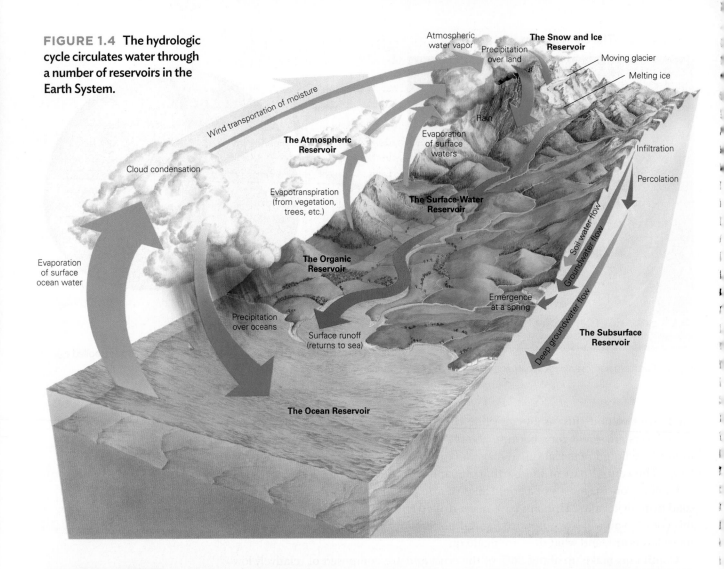

FIGURE 1.4 The hydrologic cycle circulates water through a number of reservoirs in the Earth System.

EXERCISE 1.2 **Reservoirs in the Earth System**

Name: _____ Section: _____
Course: _____ Date: _____

What Earth materials did you encounter in the past 24 hours? List at least 10 in the following table without worrying about the correct scientific terms (for example, "dirt" is okay for now). Place each Earth material in its appropriate reservoir and indicate whether it is a solid (S), liquid (L), or gas (G).

Atmosphere	Hydrosphere	Geosphere	Cryosphere	Biosphere

Name: _____   **Section:** _____

Course: _____   **Date:** _____

You already have an instinctive sense of how water moves from one reservoir to another in the Earth System. Based on your experience with natural phenomena, complete the following table, in which the first column lists several geologic processes. Describe what happens during each process in the second column, using plain language. In the third column, indicate what transfer, if any, has occurred between the major reservoirs of the Earth System. The first process is given as an example.

Process	What happens?	Did matter move from one reservoir to another? If so, from _____ to _____
Sublimation	Solid ice becomes water vapor without melting.	Yes: from cryosphere to atmosphere
Antarctic ice melts		
A puddle evaporates		
A lake freezes		
Plant roots absorb water from the soil		
Clouds form in the sky		
Steam erupts from a volcano		

1.2.3 Energy in the Earth System

Natural disasters in the headlines remind us of how dynamic the Earth is: rivers flood cities and fields; mudslides, lava, and volcanic ash bury villages; earthquakes topple buildings; and hurricanes ravage coastal regions. However, many Earth processes are much slower and less dangerous, like the movement of ocean currents and the almost undetectable creep of soil downhill. All are caused by energy, which acts on matter to change its character, move it, or split it apart.

Energy for the Earth System comes from (1) the Earth's *internal* heat, which melts rock, causes earthquakes, and builds mountains (some of this heat is left over from the formation of the Earth, but some is being produced today by radioactive decay); (2) *external* energy from the Sun, which warms air, rocks, and water on the Earth's surface; and (3) the pull of the Earth's gravity. Heat and gravity, working independently or in combination, drive most Earth processes.

Heat energy is a measure of the degree to which atoms or molecules vibrate in matter—including in solids. When you heat something in an oven, for example, the atoms in the material vibrate faster and move farther apart. Heat energy drives the flux of material from one state of matter to another or from one reservoir of the Earth System to another. For example, heating ice causes **melting** (solid → liquid; cryosphere → hydrosphere) and heating water causes **evaporation** (liquid → gas; hydrosphere → atmosphere). Cooling slows the motion, causing **condensation** (gas → liquid, atmosphere → hydrosphere) or **freezing** (liquid → solid, hydrosphere → cryosphere).

Exercise 1.4 explores evidence for the sources of heat for Earth processes.

EXERCISE 1.4	Sources of Heat

Name: _____ **Section:** _____
Course: _____ **Date:** _____

Some of the heat that affects geologic processes comes from the Sun and some comes from inside the Earth. To begin to understand what role each of these heat sources plays in Earth processes, conduct the simple experiment below by using a heat lamp to represent heat from the Sun, and a hot plate to represent the Earth's internal heat. After you complete the experiment, apply what you observe to the questions that follow.

Your instructor has provided you with sediment, glass containers such as beakers, a heat lamp, a hot plate, and thermometers. (*Note:* If this exercise is available and being conducted online, the setup, materials, and directions may differ. See the instructions in the online lab for how to conduct the exercise.) You will set up your experiment as shown in the figure and described in the procedure below.

Experiment Set-up

Step 1: Fill each container with 500 mL of sediment (sand or soil work best) or the amount given by your instructor.

Step 2: In each container, place one thermometer so that the bulb is just below the surface of the sediment. Insert the other thermometer in each beaker so that its bulb is 5 cm (2 in) below the depth of the first thermometer. Be sure the thermometer does not touch the sides of the container.

Step 3: Record the starting temperatures in °C for all four thermometers in the table below.

Step 4: Turn on the hot plate to a medium-high temperature, such as 6 on a scale of 8 (or 3/4 of the total temperature range on your hot plate). Turn on the heat lamp, making sure that it is not in contact with the thermometers.

Step 5: Record the temperatures from all four thermometers after 5 minutes and 10 minutes. Write these temperatures in the table.

(continued)

EXERCISE 1.4 Sources of Heat *(continued)*

Name: ______________________________ **Section:** ____________

Course: ______________________________ **Date:** ____________

Sources of Heat Data

	Heat Lamp Temperatures (°C)			Hot Plate Temperatures (°C)		
	Start	**5 minutes**	**10 minutes**	**Start**	**5 minutes**	**10 minutes**
Upper thermometer						
Lower thermometer						

(a) Which thermometer recorded a greater temperature change under the heat lamp?

Circle your answer: Upper Thermometer Lower Thermometer

Which thermometer recorded a greater temperature change on the hot plate?

Circle your answer: Upper Thermometer Lower Thermometer

(b) Imagine taking off your shoes on a sandy beach and walking on it on a hot, sunny day. Is the sand likely to be hot or cold? Why?

____

____

____

____

(c) If you were to dig down into the sand a few centimeters, what temperature would you feel? Why?

____

____

____

____

(d) What do these observations suggest about the depth to which heat from the Sun can penetrate the Earth?

____

____

____

____

(e) Based on this conclusion, is the Sun's energy or the Earth's internal heat the cause of melting of rock within the Earth? Explain.

____

____

____

____

Gravity, as Isaac Newton showed more than three centuries ago, is the force of attraction that every object exerts on other objects. The strength of this force depends on the amount of mass in each object and how close the objects are to one another, as expressed by:

$$F_G = G\frac{m_1 \times m_2}{d^2}$$

where F_G = force of gravity; G = a gravitational constant; m_1 and m_2 = masses of the two objects; and d = the distance between them.

The greater the masses and the closer the objects are, the stronger the gravitational attraction between them. The smaller the masses and the farther apart the objects are, the weaker the attraction. The Sun's enormous mass produces a force of gravity sufficient enough to hold the Earth and the other planets in their orbits. The Earth's gravitational force is far less than the Sun's, but it is strong enough to hold the Moon in orbit, hold you on its surface, cause rain or volcanic ash to fall, and enable rivers and glaciers to flow.

The pull of the Earth's gravity produces a force called **pressure**. For example, the pull of gravity on the atmosphere creates a pressure of 1.03 kg/cm² (14.7 lb/in²) at sea level. This means that every square centimeter of surface of the ocean, the land, or your body at sea level is pressed upon by a weight of 1.03 kg. We call this amount of pressure 1 **atmosphere** (**atm**). Scientists commonly specify pressures using a unit called the *bar* (from the Greek *barros*, meaning weight), where 1 bar ≈ 1 atm. Two kinds of pressure play important roles in the hydrosphere and geosphere. **Hydrostatic pressure**, pressure caused by the weight of overlying water, increases as you descend into the ocean and can crush a submarine at great depths. **Lithostatic pressure**, pressure caused by the weight of overlying rock, increases as you go deeper in the geosphere and is great enough in the upper mantle to change the graphite you find in a pencil into diamond. A volume of rock weighs a lot more than the same volume of air or water, so the pull of gravity causes lithostatic pressure to increase much faster than either atmospheric or hydrostatic pressure—so much so that a standard measure of pressure in the Earth is the **kilobar**, equivalent to 1,000 times atmospheric pressure.

1.3 Units for Scientific Measurement

Before we begin to examine the components of the Earth System scientifically, it is important to be familiar with the units used to measure them. We can then examine the challenges of scale that Earth scientists face when studying topics as large as the known Universe and as small as the atoms of which all matter is made.

1.3.1 Units of Length and Distance

People have struggled for thousands of years to describe size in a precise way with widely accepted standard units of measurement. Scientists everywhere, and people in nearly all countries except the United States, use the **metric system** to measure length and distance. The largest metric unit of length is the kilometer (km), which is divided into smaller units: 1 km = 1,000 meters (m); 1 m = 100 centimeters (cm); 1 cm = 10 millimeters (mm). Metric units differ from each other by factors of 10, making it very easy to convert one unit into another. For example, 5 km = 5,000 m = 500,000 cm = 5,000,000 mm. Similarly, 5 mm = 0.5 cm = 0.005 m = 0.000005 km.

The United States uses the U.S. customary system (previously called the English Unit System until Great Britain adopted the metric system) to describe distance. Distances are given in miles (mi), yards (yd), feet (ft), and inches (in), where 1 mi

= 5,280 ft; 1 yd = 3 ft; and 1 ft = 12 in. As scientists, we use metric units in this book, but when appropriate, equivalents are also given (in parentheses). Appendix 1.1, at the end of this chapter, provides basic conversions between U.S. customary and metric units.

Because distances between objects in outer space are so vast, we use separate, larger units to describe them. The distance between the center of the Earth and the center of the Sun—about 150 million km (~93 million mi)—equals one **astronomical unit (AU)**. The distance light travels in a year—9.461 trillion km—equals one **light-year**. These units will be helpful in the astronomy labs.

1.3.2 Other Dimensions, Other Units

Length and distance are just two of the dimensions of the Earth and Universe that you will examine during this course. We still need to look at other units used to describe other aspects of the Earth and of other objects in space: units of time, velocity, temperature, mass, and density.

Time is usually measured in seconds (s), minutes (min), hours (h), days (d), years (yr), centuries (hundreds of years), and millennia (thousands of years). A year is the amount of time it takes for the Earth to complete one orbit around the Sun. Because the Earth is very old, scientists have to use larger units of time to describe the age of an object or process: a thousand years ago (abbreviated **Ka**, for "kilo-annum"), a million years ago (**Ma**, for "mega-annum"), and a billion years ago (**Ga**, for "giga-annum"). The formation of the Earth 4,570,000,000 years ago can thus be expressed as 4.57 Ga, or 4,570 Ma.

Velocity, or the rate of change of the position of an object, is described by units of distance divided by units of time, such as meters per second (m/s), feet per second (ft/s), kilometers per hour (km/h), or miles per hour (mph). You will learn later that geologic materials move at velocities ranging from extremely slow (mm/yr) to extremely fast (km/s).

Temperature—the average speed at which atoms vibrate—is measured in terms of how hot an object is relative to a standard. It is measured in degrees Celsius (°C) in the metric system and degrees Fahrenheit (°F) in the U.S. customary system. The reference standards in both are the freezing and boiling points of water: 0°C and 100°C (32°F and 212°F), respectively. Note that there are 180 Fahrenheit degrees between freezing and boiling, but only 100 Celsius degrees. A change of 1°C is thus 1.8 times larger than a change of 1°F (180°/100°). To convert Fahrenheit to Celsius or vice versa, see Appendix 1.1.

Mass refers to the amount of matter in an object, and **weight** refers to the force with which one object is attracted to another. The weight of an object on the Earth therefore depends not only on its mass but also on the strength of the Earth's gravitational field. Objects that have more mass than others also weigh more on the Earth because of the force of the Earth's gravity. While the mass of an object remains the same whether it is on the Earth or on the Moon, the object *weighs* less on the Moon because of the Moon's weaker gravity.

Grams and kilograms (1 kg = 1,000 g) are the units of mass in the metric system; the U.S. customary system uses pounds and ounces (1 lb = 16 oz). (1 kg = 2.2046 lb and 1 g = 0.0353 oz.)

We saw earlier that the density (δ) of a material is a measure of how much mass is packed into each unit of volume, generally expressed as units of g/cm^3. We instinctively distinguish especially low-density materials like Styrofoam and feathers from low-density materials like water ($\delta = 1$ g/cm^3) and high-density materials like steel ($\delta = 7$ g/cm^3) because the former feel very light *for their sizes* and the latter feel unusually heavy *for their sizes* (**FIG. 1.5**).

FIGURE 1.5 Weights of materials with different densities.

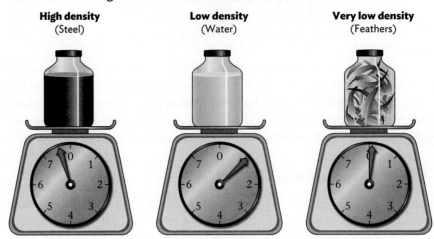

High density
(Steel)

Low density
(Water)

Very low density
(Feathers)

To measure the density of a material, a property useful in studying minerals and rocks, we need to know its mass and volume. Mass is measured with a balance or scale, and volume for regular geometric shapes like cubes, bricks, spheres, or cylinders is calculated from simple formulas. For example, to calculate the volume of a bar of gold, you would multiply its length times its width times its height (**FIG. 1.6A**). But rocks are rarely regular geometric shapes; more typically, they are irregular chunks. To measure the volume of a rock (or other irregular object), submerge it in a graduated cylinder partially filled with water (**FIG. 1.6B**). Measure the volume of water before the rock is added and then with the rock in the cylinder. (Make sure to view the cylinder from eye level when recording the measurement, to avoid over- or underreporting the level.) The rock displaces a volume of water equivalent to its own volume, so simply subtract the initial volume of the water from that of the water plus rock to obtain the volume of the rock. The density of a rock can then be calculated simply from the definition of density:

$$\text{density} = \text{mass} \div \text{volume}.$$

Exercise 1.5 provides practice in determining density.

FIGURE 1.6 Measuring the volume of materials.

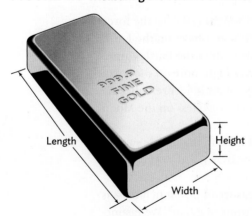

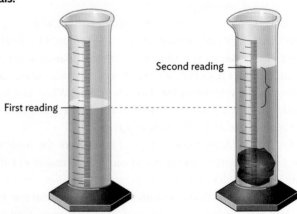

First reading

Second reading

(a) For a rectangular solid, volume = length x width x height.

(b) For an irregular solid, the volume of the solid is the volume of the liquid it displaces in a graduated cylinder, which is the difference between the first and second readings.

Name: _____ Section: _____
Course: _____ Date: _____

(a) Determine the density of a liquid provided by your instructor using a balance, a graduated cylinder, and a container of the liquid. (*Note:* If this exercise is available and being conducted online, the setup, materials, and directions may differ. See the instructions in the online lab for how to conduct the exercise.) Consider how to go about this, and then do it. Make sure to provide the proper units with your answer. The density of the liquid is _____.

(b) Your instructor will give you samples of *granite*, a light-colored rock that makes up a large part of the continental crust, and *basalt*, a dark-colored rock that makes up most oceanic crust and the lower part of the continental crust. (*Note:* If this exercise is available and being conducted online, the setup, materials, and directions may differ. See the instructions in the online lab for how to conduct the exercise.)

Determine their densities and record your answers with units included.
- What is the density of granite? _____ (gm/cm³)
- What is the density of basalt? _____ (gm/cm³)
- If the volume of continental crust is half granite and half basalt, what is its density? _____ (gm/cm³)

1.3.3 Expressing the Earth's "Vital Statistics" with Appropriate Units

Now that you are familiar with some of the units of measurement used in the Earth sciences, we can look at some of our planet's "vital statistics."

- *Place in the Solar System:* The Earth is the third of eight planets from the Sun. It is one of four terrestrial planets, along with Mercury, Venus, and Mars, which are made of rock, in contrast to the Jovian planets (Jupiter, Saturn, Uranus, and Neptune), which are mostly made of methane and ammonia in either gaseous or frozen forms.
- *Revolution:* The Earth takes 1 year (365.25 days or 3.15×10^7 seconds) to complete one orbit around the Sun.
- *Rotation:* The Earth completes one 360° turn (or spin) on its axis in 1 day (24 hours or 86,400 seconds).
- *Shape:* The Earth is almost, but not quite, a sphere. Its rotation produces a slight bulge at the Equator: its equatorial radius of 6,400 km (~4,000 mi) is 21 km (~15 mi) longer than its polar radius. (A circle's or sphere's *radius* is the distance from its center to the perimeter; it is half the *diameter*, the maximum length of a line passing through the center of the circle/sphere.)
- *Temperature:* The Earth's average surface temperature is 15°C (59°F); its core temperature is about 5,000°C (~9,000°F).
- *Highest mountain:* The peak of Mt. Everest is the Earth's highest point, at 8,850 m (29,035 ft) above sea level (and is still growing!).
- *Average ocean depth*: The average depth of the ocean floor is 4,500 m (14,700 ft).
- *Deepest part of ocean floor:* The bottom of the Marianas Trench in the Pacific Ocean is the deepest point on the ocean floor, at 11,033 m (35,198 ft) below sea level.

1.4 The Challenges of Studying a Universe

Problems like submergence along the Maine coast pose challenges to the scientists who are trying to solve them and the students who are trying to learn about the Earth. These challenges require us to:

- understand the many kinds of materials that make up the Earth and how they behave.
- be aware of how energy causes changes at the Earth's surface and beneath it.
- consider features at a wide range of sizes and scales—from the atoms that make up rocks and minerals to the planet as a whole.
- think in *four* dimensions, because processes in Earth science involve not just the three dimensions of space but also *time*.
- realize that some processes occur in seconds, but others take millions or billions of years—so slow that we can detect them only with very sensitive instruments.

The rest of this chapter examines these challenges and how Earth scientists cope with them. You will learn basic terminology and how to use tools of observation and measurement that will be useful throughout this course. Some concepts and terms may be familiar from previous science classes.

1.4.1 Scientific Notation and Orders of Magnitude

When studying the Earth and the Universe in which it lies, we must cope with enormous ranges in scale involving distance (for everything from atoms to sand grains to planets), temperature (below 0°C in the cryosphere and upper atmosphere to more than 1,000°C in some lavas to millions of degrees Celsius in the Sun), and velocity (continents moving at 2 cm/year to tsunamis moving at hundreds of kilometers per hour to light traveling at 299,792 km/second).

Scientists sometimes describe scale in approximate terms, and sometimes more precisely. For example, the terms *mega-scale*, *meso-scale*, and *micro-scale* denote enormous, moderate, and tiny features, respectively, but don't tell exactly how large or small they are, because they depend on a scientist's frame of reference. Thus, *mega-scale* to an astronomer might mean intergalactic distances, but to a geologist it may mean the size of a continent or a mountain range. *Micro-scale* could refer to a sand grain, a bacterium, or an atom. Geologists often express scale in terms that specify the frame of reference, like "outcrop-scale" for a feature in a single roadside exposure of rock (an outcrop), or "hand-specimen-scale" for a rock sample you can hold in your hand.

Scientists can describe objects more precisely as differing in scale by **orders of magnitude**. A feature an order of magnitude larger than another is 10 times larger: a feature one-tenth the size of another is an order of magnitude *smaller*. Something 100 times the size of another is two orders of magnitude larger, and so on. We also use a system of **scientific notation** based on powers of 10 to describe incomprehensibly large or small objects. In scientific notation, 1 is written as 10^0, 10 as 10^1, 100 as 10^2, and so on. Numbers less than 1 are shown by negative exponents: for example, $1/10 = 10^{-1}$, $1/100 = 10^{-2}$, $1/1,000 = 10^{-3}$. A positive exponent tells how many places to move the decimal point to the right of a number, and a negative exponent how many places to the left. Therefore, $3.1 \times 10^2 = 310$; $3.1 \times 10^{-2} = 0.031$.

Scientific notation saves a lot of space in describing very large or very small objects. For example, the 150,000,000-km (93,000,000-mi) distance from the Earth to the Sun becomes 1.5×10^8 km (9.3×10^7 mi), and the diameter of a hydrogen atom (0.0000000001 meters) becomes 1.0×10^{-10} m. The full range of dimensions that scientists must consider spans an incomprehensible 44 orders of magnitude, from the diameter of the particles that make up an atom (about 10^{-18} m across) to the radius of the observable Universe (an estimated 10^{26} m (**TABLE 1.2**).

TABLE 1.2 Orders of magnitude defining lengths in the Universe (in meters)

~2 x 10²⁶	Radius of the observable Universe
2.1 x 10²²	Distance to the nearest galaxy (Andromeda)
9.5 x 10¹⁷	Diameter of the spiral disk of the Milky Way galaxy
1.5 x 10¹¹	Average radius of the Earth's orbit
6.4 x 10⁶	Radius of the Earth
5.1 x 10⁶	East-west length of the United States
8.8 x 10³	Height of Mt. Everest (the Earth's tallest mountain) above sea level
3.8 x 10²	Height of the Empire State Building
9.1 x 10¹	Length of a U.S. football field
2.8 x 10¹	Length of a professional basketball court
1.7 x 10⁰	Average height of an adult human
3.5 x 10⁻²	Length of a standard paperclip
1.0 x 10⁻³	Diameter of a pinhead
6.0 x 10⁻⁴	Diameter of a living cell
2.0 x 10⁻⁶	Diameter of a virus

(Some industries use a variation of scientific notation called *E-notation*, in which the letter "ᴇ" is used to mean "10 to the power of." In this system, 3.1×10^2 would be expressed as $3.1\text{ᴇ}+2$. Likewise, 3.1×10^{-2} would be expressed as $3.1\text{ᴇ}-2$. In this manual, we'll use the form of scientific notation laid out in the previous paragraphs.)

1.4.2 Coping with Issues of Scale

Scientists deal routinely with objects as miniscule as atoms and as enormous as the Appalachian Mountains or the Pacific Ocean. Sometimes we must look at a feature at different scales, as in **FIGURE 1.7**, to understand it fully.

FIGURE 1.7 The White Cliffs of Dover, England, seen at two different scales.

(a) The towering chalk cliffs of Dover, England, stand up to 110 m (350 ft) above the sea.

(b) Microscopic view of the chalk (plankton shells) that the cliffs are made of. The eye of a needle gives an idea of the miniscule sizes of the shells that make up the cliffs. In actuality, about 20 of these plankton shells could fit into the eye of a needle.

One of the challenges we face in studying the Earth is a matter of perspective: to a flea, the dog on which it lives is its entire world; but to a parasite inside the flea, the flea is *its* entire world. For most of our history, humans have had a flea's-eye view of the Earth, unable for many centuries to recognize even the most basic facts about our planet: that it is nearly spherical, not flat; that it isn't the center of the Universe, or even the Solar System; nor that as we sit at a desk, we are actually moving thousands of miles an hour because of the Earth's rotation and orbit around the Sun. Exercise 1.6 will provide some perspective on the matter of perspective. Exercise 1.7 practices scientific notation and gives a sense of how human scales relate to the Earth's basic processes.

EXERCISE 1.6   The Challenge of Perspective and Visualizing Scale

Name: _____   Section: _____
Course: _____   Date: _____

The enormous difference in size between ourselves and our planet gives us a limited perspective on large-scale features and makes understanding major Earth processes challenging. To appreciate this challenge, consider the relative sizes of familiar objects (use Appendix 1.1 for conversions):

1 mm

1 m

Relative sizes of a dog and a flea.

12,800 km

2 m

Relative sizes of the Earth and a tall scientist.

(a) Assume that a flea is 1 mm long and that a dog is 1 m long.
  • To relate this to our U.S. customary system of measurement, how long is the flea in inches?
    _____ inches
  • How many times larger is the dog than the flea? _____
  • How long is the scale bar representing the flea? _____
  • How long would the bar representing the dog have to be if the dog and the flea were shown
    at the same scale? _____

(b) Now, think about the relative dimensions of a scientist and the Earth.
  • How many times larger than the scientist is the Earth?
  • How large would the drawing of the Earth have to be if it were drawn at the same scale as the scientist? Give your
    answer in kilometers and in miles. _____ km _____ miles
  • Based on the relative sizes of flea and dog versus human and the Earth, does a flea have a better understanding of a
    dog than a human has of the Earth? Or vice versa? Explain.

  _____

  _____

  _____

Name: _____   **Section:** _____
Course: _____   **Date:** _____

Complete the following calculations to get a sense of the distances that the Earth travels and the speed at which it moves.

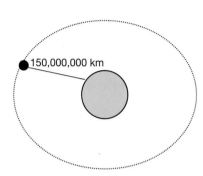

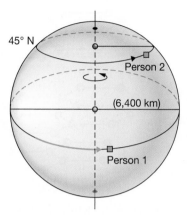

Earth's orbit. The Earth orbits the Sun in an elliptical path. To simplify calculation, here we assume a circular orbit with a radius of 150,000,000 km (93,000,000 mi).

Earth's rotation. The Earth rotates on its axis every 24 hours.

(a) For simplicity, in this exercise we picture the Earth's orbit around the Sun as a circle, with a radius of 150,000,000 km (93 million mi). It takes 1 year for the Earth to orbit the Sun.
  • What distance does the Earth travel during a single complete orbit? Remember that the circumference of a circle is $2\pi r$, and that $\pi = 3.14$. Orbital distance = _____ km (_____ mi).
  • What is the velocity of the Earth as it orbits the Sun? Give your answer in both kilometers per hour and miles per hour. _____ km/h _____ mph
  • Express the above results using scientific notation. _____ km/h _____ mph

(b) Earth rotates on its axis in 1 day. At what velocity is a person standing at the equator in the figure above (Person 1) moving due to our planet's rotation? _____ km/h _____ mph

(c) Now consider a person standing on the Earth's surface at a point halfway between the equator and the geographic pole (Person 2). Is this person moving faster or slower than the person standing at the equator? Explain your answer.

_____

_____

_____

_____

_____

_____

(d) Many components of the Earth System are too small for us to comprehend. For example, atoms are too small to be seen even with the most powerful optical microscope. A sodium atom has an approximate diameter of 2×10^{-8} cm. The diameter of the period at the end of this sentence is approximately 4×10^{-4} cm.
  • If you lined up sodium atoms one next to the other, how many sodium atoms would it take to span the diameter of a period? _____ Express this number in scientific notation: _____
  • How many orders of magnitude larger is the period than the atom? _____

Figure 1.7b showed how we use microscopes to see very small features, and we can use models to scale up even smaller features, like the structures of minerals (**FIG. 1.8A**). Several methods help us deal with very large objects. One is to use models that scale those features down physically (**FIG. 1.8B**). Using models like these, we can re-create the conditions that led to a flood or the ways in which the ground shook during an earthquake, to understand those processes better. We can also use scaled-down diagrams to show changes inside the Earth. Maps, aerial photographs, and satellite images zoom out on surface features of the Earth and other planetary bodies, essentially scaling down the relationships between them to make them more understandable (**FIGS. 1.8C, D, E**).

1.4.3 Geologic Time

Deciphering the Earth's history presents two major challenges: (1) most of the features that record Earth history were formed by processes that occurred long before there were any humans to observe them, and

FIGURE 1.8 Methods for dealing with issues of scale.

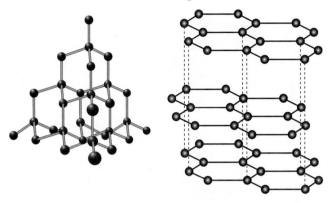

(a) Models showing arrangement of carbon atoms in diamond (left) and graphite (right).

(b) Scale model of St. Paul Harbor, Alaska, used to study water and sediment movement.

(c) Aerial photograph of part of the Grand Canyon.

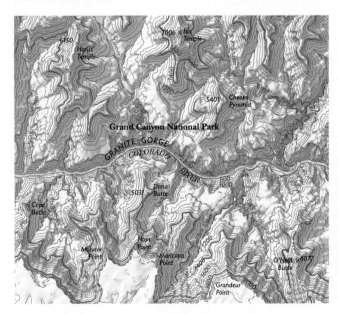

(d) Topographic map showing part of the Grand Canyon.

(e) Digital elevation model showing topography.

FIGURE 1.9 An example of the principle of uniformitarianism.

(a) Asymmetric ripple marks in modern sand on the shore of Cape Cod, Massachusetts.

(b) Asymmetric ripple marks in 145-million-year-old sandstone at Dinosaur Ridge, Colorado.

(2) that history covers an immense span of geologic time, estimated to be 4.56 billion (4,560,000,000, or 4.56×10^9) years. Without witnesses, how can we understand what happened so long ago? The answer came nearly 300 years ago when the Scottish geologist James Hutton noted that some features he observed forming in modern environments looked just like those he observed in rocks. For example, the sand on a beach or dune commonly forms a series of low ridges called ripple marks. You can see identical ripple marks preserved in ancient layers of sandstone, a rock made of sand grains (**FIG. 1.9**). Based on his observations, Hutton proposed the *principle of uniformitarianism*, which states that most ancient geologic features are formed by the same processes as similar modern ones. Stated more succinctly, it tells us that "the present is the key to the past."

As noted above, geologists have amassed a large body of evidence showing that the Earth is about 4.56 billion years old. This enormous span is referred to as **geologic time** (**FIG. 1.10**), and understanding its vast scope is nearly impossible for humans, who generally live less than

FIGURE 1.10 Milestones in geologic time.

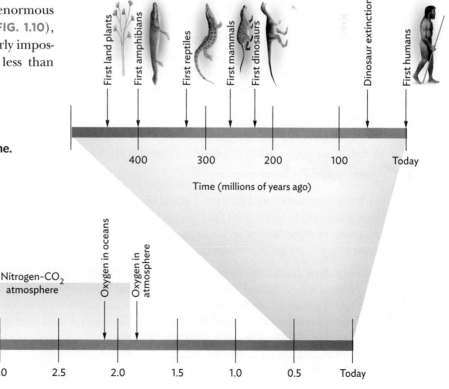

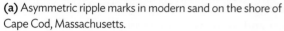

100 years. Our usual frame of reference for time is based on human lifetimes: a war fought two centuries ago happened two lifetimes ago, and 3,000-year-old monuments were built about 30 lifetimes ago.

To help visualize the billions of years of geologic time, it may help to compare its span to a more familiar one: if all of geologic time was condensed into a single year, each second would represent 145 years.

1.5 Rates of Processes

Some geologic, meteorologic, and astronomic processes happen quickly. A meteorite impact takes just a fraction of a second; most earthquakes are over in just a few seconds; and a landslide or explosive volcanic eruption can happen in minutes. But others occur so slowly that it's difficult to recognize that they are happening. It often takes decades before anyone notices the slow downhill creep of soil; layers of mud only an inch thick may require thousands of years to accumulate in the deep ocean; and satellite measurements show that continents are moving at 1 to 15 cm per year. Objects in space can be millions of light-years away, such that the images our space and land-based telescopes record are actually a snapshot of the past, as it takes many years for the light to travel through space to be collected on Earth in an image. A star that has exploded may be breaking news to us on Earth, but the event may have taken place thousands or millions of years ago. Exercise 1.8 investigates the rates of various processes that you will encounter in future labs that take place within the Earth System and in outer space.

EXERCISE 1.8　　**How Long Do Processes in the Earth System Take?**

Name: _____　　Section: _____
Course: _____　　Date: _____

(a) **Rates of uplift and erosion.** The following questions will give you a sense of the rates at which uplift and erosion take place. We will assume that uplift and erosion do not occur at the same time—that mountains are first uplifted, and only then does erosion begin—whereas the two processes actually operate simultaneously.
- If mountains rose by 1 mm/yr, how much higher would they be (in meters) after 1,000 years? _____ m
  10 million years? _____ m
  50 million years? _____ m
- The Himalayas now reach an elevation of 8.8 km, and radiometric dating suggests that their uplift began about 45 million years ago. Assuming a constant rate of uplift, how fast did the Himalayas rise? _____ km/yr _____ m/yr _____ mm/yr
- Evidence shows that there were once Himalaya-scale mountains in northern Canada, in an area now eroded nearly flat. If the Earth were only 6,000 years old, as was once believed, how fast would the rate of erosion have had to be for these mountains to be eroded to sea level in 6,000 years? _____ m/yr _____ mm/yr
- Observations of modern mountain ranges suggest that they erode at rates of 2 mm per 10 years. At this rate, how long would it take to erode the Himalayas down to sea level? _____ years

(b) **Rates of seafloor spreading.** Today the Atlantic Ocean is about 5,700 km wide at the latitude of Boston. At one time, however, there was no Atlantic Ocean because the east coast of the United States and the northwest coast of Africa were joined in a huge supercontinent. The Atlantic Ocean started to form "only" 185,000,000 years ago, as modern North America split from Africa and the two continents slowly drifted apart in a process called *seafloor spreading*.
- Assuming that the rate of seafloor spreading has been constant, at what rate has North America been moving away from Africa? _____ mm/yr _____ km per million years

(continued)

Name: _____   Section: _____
Course: _____   Date: _____

(c) **Rates of light travel.** Light travels at approximately 3.0×10^5 km per second. Light from the Sun, the nearest star, takes approximately 500 seconds (about 8.5 minutes) to reach us here on Earth.
  - Using this information alone, how far is the Sun from the Earth in kilometers? _____ km
  - A light-year is the distance that light travels in one year (9.461×10^{12} km). Our next nearest star, Proxima Centauri, is approximately 4.3 light years away. How far is that in kilometers? _____ km
  - How long does it take light to travel from Proxima Centauri to the Earth in minutes? _____ minutes. In years (Earth years, not light-years)? _____ years.

Since nothing can travel faster than the speed of light, based on our understanding of physics and the known Universe, even if humans could travel at the speed of light, you can see how long a journey it would be to Proxima Centauri!

GEOTOURS EXERCISE 1   Scaling Geologic Time in the Grand Canyon

Name: _____   Section: _____
Course: _____   Date: _____

Exploring Earth Science Using Google Earth

1. Visit **digital.wwnorton.com/labmanualearthsci**
2. Go to the **Geotours** tile to download Google Earth Pro and the accompanying Geotours exercises file.

Check and double-click the Geotour01 folder icon in Google Earth to fly to the Grand Canyon of northern Arizona. Here, erosion by the Colorado River treats visitors to one of the most spectacular exposures of geologic history on the planet (ranging from ~270-million-year-old Kaibab Limestone capping the Grand Canyon's rim to the ~2,000-million-year-old Vishnu Schist paving the Inner Gorge). Right-click on the Bright Angel Trail (red path) and select **Show Elevation Profile** to plot a graph of elevation change (*y*-axis) versus distance along the trail (*x*-axis). *Please note that axis values can vary slightly depending on window size. So students should use the numbers provided for their calculations.*

(a) It is a long, downhill walk (12.6 km) from the youngest rocks (270 Ma) at the rim of the canyon to the oldest rocks (2,000 Ma) at the river. Using this information, estimate how far back in time you would go with each meter along the trail. _____ yr/m

(b) Using the change in elevation from the rim to the river (2,085 m and 734 m from the *y*-axis) and assuming that Hutton's principle of uniformitarianism applies for erosion of the Grand Canyon, use an average downcutting erosion rate of 0.5 mm/yr to calculate a rough estimate for how long it might have taken to carve the Grand Canyon. _____ yrs

(c) What factors might have caused this erosion rate to change over time? _____
_____

Metric-U.S. Customary Conversion Chart

To convert U.S. customary units to metric units	To convert metric units to U.S. customary units
Length or distance	
inches × 2.54 = centimeters feet × 0.3048 = meters yards × 0.9144 = meters miles × 1.6093 = kilometers	centimeters × 0.3937 = inches meters × 3.2808 = feet meters × 1.0936 = yards kilometers × 0.6214 = miles
Area	
in^2 × 6.452 = cm^2 ft^2 × 0.929 = m^2 mi^2 × 2.590 = km^2	cm^2 × 0.1550 = in^2 m^2 × 10.764 = ft^2 km^2 × 0.3861 = mi^2
Volume	
in^3 × 16.3872 = cm^3 ft^3 × 0.02832 = m^3 U.S. gallons × 3.7853 = liters	cm^3 × 0.0610 = in^3 m^3 × 35.314 = ft^3 liters × 0.2642 = U.S. gallons
Mass	
ounces × 28.3495 = grams pounds × 0.45359 = kilograms U.S. (short) tons × 0.907185 = metric tons	grams × 0.03527 = ounces kilograms × 2.20462 = pounds metric tons × 1.10231 = U.S. (short) tons
Density	
lb/ft^3 × 0.01602 = g/cm^3	g/cm^3 × 62.4280 = lb/ft^3
Velocity	
ft/s × 0.3048 = m/s mph × 1.6093 = km/h	m/s × 3.2804 = ft/s km/h × 0.6214 = mph
Temperature	
0.55 × (°F − 32) = °C	(1.8 × °C) + 32 = °F
Pressure	
lb/in^2 × 0.0703 = kg/cm^2 lb/in^2 × 0.06803 = atm lb/in^2 × 0.06895 = bar	kg/cm^2 × 14.2233 = lb/in^2 atm × 14.70 = lb/in^2 bar × 14.504 = lb/in^2
For U.S. customary units	**For metric units**
1 foot (ft) = 12 inches (in) 1 yard (yd) = 3 feet 1 mile (mi) = 5,280 feet	1 centimeter (cm) = 10 millimeters (mm) 1 meter (m) = 100 centimeters 1 kilometer (km) = 1,000 meters 1 milliliter (mL) = 1 cm^3 1 atm = 1,013.25 bars 1 astronomical unit (AU) = 149,597,870,700 m 1 light-year = 9.461 × 10^{12} km 1 parsec = 3 × 10^{16} m = 3.26 light-years

The Way the Earth Works: Examining Plate Tectonics

2

A path through Thingvellir in Iceland, the rift valley at the crest of the Mid-Atlantic Ridge that separates the North American and Eurasian plates.

MATERIALS NEEDED

■ Tracing paper
■ Colored pencils
■ Ruler with divisions in tenths of an inch or millimeters (included in the GeoTools section at the back of this manual)
■ Protractor (included in the GeoTools section at the back of this manual)
■ Calculator or pencil and paper for simple arithmetic

2.1 Introduction

Earthquakes and volcanic eruptions show that the Earth is a dynamic planet with enough energy beneath its surface to cause disasters for those who live atop it. Humans wondered about the causes of these events until the 1960s and 1970s, when geologists developed the **theory of plate tectonics**: a unifying theory that answered geologic questions that had puzzled us for thousands of years. According to this theory, the outer layer of the Earth is made up of separate plates that move with respect to one another and change the Earth's surface as they move. At first it was difficult to accept the concept that the Earth's oceans, continents, and mountains are only temporary features that move and change over time, because the changes are so slow that they could not be detected. Yet, according to plate tectonics theory, planet-wide processes break continents apart, open and close oceans, and build and shrink great mountain chains. Local earthquakes and volcanoes are simply results of the energy released as these processes occur. No one doubts plate tectonics now, because geologists have drawn on conclusive evidence to prove that these processes are happening today and have been operating for billions of years. In this chapter, we will explore the evidence and geologic reasoning that led to plate tectonics theory.

2.2 The Plate Tectonics Theory

Plate tectonics theory is based on many kinds of information about the Earth that you will examine during this course, including the origin and distribution of different rock types, the topography of the continents and ocean basins, and the geographic distribution of earthquakes and volcanic eruptions. The basic concepts of the theory include the following:

■ The Earth's crust and the uppermost part of the layer below it, called the **mantle**, form a relatively rigid outermost layer, called the **lithosphere**, which extends to a depth of 100 to 150 km.
■ The lithosphere rests on the **asthenosphere**, a zone in the upper mantle that, although solid, has such low rigidity that it can flow like soft plastic (**FIG. 2.1**). The asthenosphere acts as a lubricant, permitting the plates above it to move.
■ Continental lithosphere is thicker than oceanic lithosphere because continental crust alone (without the mantle component) is 25 to 70 km thick, whereas oceanic crust is only about 7 km thick.

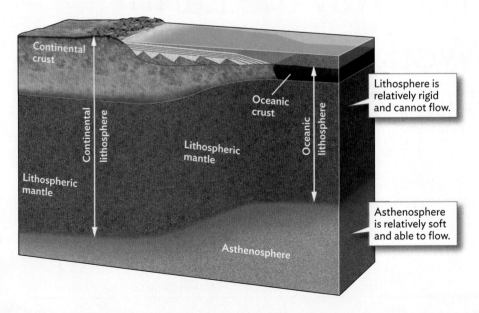

FIGURE 2.1 The lithosphere consists of the crust plus the uppermost mantle, and sits atop the asthenosphere. Lithospheric mantle is thicker beneath continents than beneath oceans.

FIGURE 2.2 The Earth's major lithosphere plates.

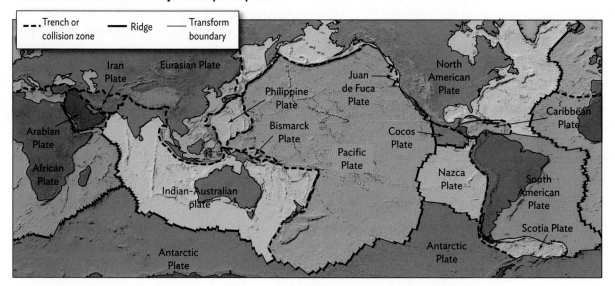

- The lithosphere is not a single shell, but consists of several large pieces called **lithosphere plates**, or simply **plates** (**FIG. 2.2**). There are about 12 major plates that are thousands of kilometers wide and several minor plates that are hundreds of kilometers wide. Plates move relative to one another 1 to 15 cm/yr, roughly the rate at which fingernails grow (at the slower end) or the rate at which your hair grows (at the faster end).

- A place where two plates make contact is called a **plate boundary**. There are three different kinds of plate boundaries, defined by the relative motions of the adjacent plates (**FIG. 2.3**):

1. At a **divergent boundary**, plates move away from one another along the central peaks, or axes, of huge submarine mountain ranges called **mid-ocean ridges**, or oceanic ridges. (These don't always lie in the precise middle of oceans.) Molten material rises from the asthenosphere to form new oceanic lithosphere at the ridge axis. The ocean grows wider through this process, called **seafloor spreading**, as the new lithosphere moves outward from the axis to the flanks of the ridge.

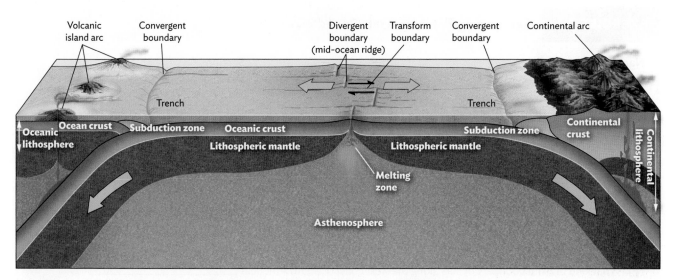

FIGURE 2.3 **Cross section showing activity at convergent, divergent, and transform boundaries.** At divergent boundaries (mid-ocean ridges), new lithosphere is created; at convergent boundaries (subduction zones and areas of continental collision), old lithosphere is destroyed; and at transform boundaries, lithosphere is neither created nor destroyed.

Continental rifts are places where continental lithosphere is stretched and pulled apart in the process of breaking apart at a new divergent boundary. If rifting is "successful," a continent splits into two pieces separated by a new oceanic plate, which gradually widens by seafloor spreading.

2. At a **convergent boundary**, two plates move toward each other, and the oceanic lithosphere of one plate (the subducting plate) sinks into the mantle below the other (the overriding plate), forming a **subduction zone**. The lithosphere of the overriding plate may be oceanic or continental. The boundary between the two plates is a deep-sea **trench**. At depths of 100 to 150 km, gases (mostly steam) released from the heated subducting plate rise into the lower part of the lithosphere. These gases help melt the mantle component of the lithosphere, and the resulting magma rises to the surface to produce volcanoes as either a **volcanic island arc**, where the overriding plate is made of oceanic lithosphere (as in the islands of Japan), or as a **continental arc**, where the overriding plate is made of continental lithosphere (as in the Andes of South America).

Continental crust cannot be subducted because its lower density makes it too buoyant to sink into the mantle. When subduction completely consumes an oceanic plate between two continents, **continental collision** occurs, forming a **collisional mountain belt** like the Himalayas, Alps, or Appalachians. Folding during the collision thickens the crust to the extent that the thickest continental crust is found in these mountains.

3. At a **transform boundary**, two plates slide past each other along a vertical zone of fracturing called a **transform fault**. Most transform faults break ocean ridges into segments. A few transform faults, however, such as the San Andreas fault in California, the Alpine fault in New Zealand, and the Great Anatolian fault in Turkey, cut through continental plates.

■ In the **tectonic cycle**, new oceanic lithosphere is created at the mid-ocean ridges, moves away from the ridges during seafloor spreading, and returns to the mantle in subduction zones. However, oceanic lithosphere is neither created nor destroyed at transform faults, where movement is almost entirely horizontal.

EXERCISE 2.1 **Recognizing Plates and Plate Boundaries**

Name: _____ Section: _____
Course: _____ Date: _____

Using Figure 2.2 as a reference, answer the following questions:

(a) What is the name of the plate on which the contiguous United States and Alaska reside? _____

(b) Where is the eastern edge of this plate? _____

(c) Does this plate consist of continental lithosphere, oceanic lithosphere, or both? _____

(d) On what plate is Hawaii located? _____

(e) Does this plate consist of continental lithosphere, oceanic lithosphere, or both? _____

(f) Where and how does the lithosphere of the Atlantic Ocean form? _____

(g) What kind of plate boundary occurs along the west coast of South America? _____

(h) Is the west coast of Africa a plate boundary? Explain. _____

2.3 Early Evidence for Plate Tectonics

The simple problem of scale and the low rate of plate movement delayed the discovery of plate tectonics until the late 1960s: lithospheric plates are so big, and move so slowly, that we didn't realize they were moving at all. Today there is no question that they move, because global positioning satellites and sensitive instruments can measure their directions and rates of movement. In the rest of this chapter we will look at some of the evidence that led geologists and other Earth scientists to accept the hypothesis that plates move, and then see how we can deduce the nature and rates of processes at the three types of plate boundaries.

2.3.1 Evidence from the Fit of the Continents

As far back as 500 years ago, mapmakers drawing the coastlines of South America and Africa noted that the two continents looked as if they might have fit together once in a larger continent. Those foolish enough to say it out loud were ridiculed, but today this fit is considered one of the most obvious lines of evidence for plate tectonics (**FIG. 2.4**). Note that the true edge of a continent is *not* the shoreline, but rather the edge of the continental shelf: the shallow water shown in light blue in Figure 2.4a.

FIGURE 2.4 **The fit of South America and Africa.**

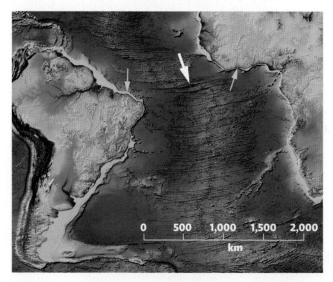

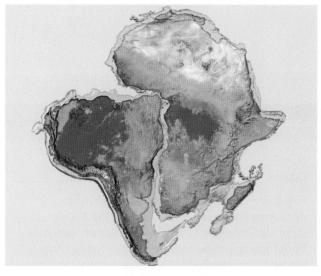

(a) The geography of the two continents and the South Atlantic Ocean today.

(b) The geography of the two continents about 180 million years ago.

2.3.2 Evidence from Reconstructing Paleoclimate Zones

The Earth's climate zones are distributed symmetrically about the equator today (**FIG. 2.5**), and the tropical, temperate, and polar zones each support animals and plants unique to that environment. For example, walruses live in Alaska, but palm trees don't, and large coral reefs grow in the Caribbean Sea, but not in the Arctic Ocean. Climate zones also produce different sediments that give rise to distinctive rock types: regional sand dunes and salt deposits form in arid regions, debris deposited by continental glaciers accumulates in polar regions, and coal-producing plants grow in temperate and tropical forests.

However, geologists have found many rocks and fossils far from the modern climate zones where we would expect them to form. For example, 390-million-year-old

FIGURE 2.5 The Earth's major climate zones.

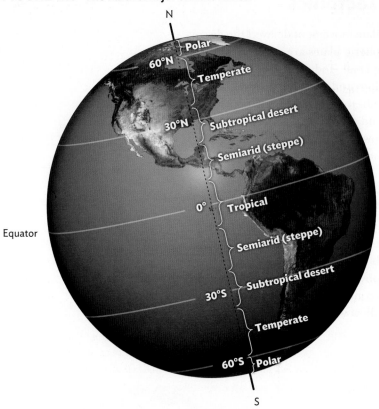

(390-Ma) limestone containing reef-building organisms crops out along the entire Appalachian Mountain chain, far north of where reefs exist today; thick 420-Ma layers of salt underlie humid, temperate Michigan, Ohio, and New York; and deposits from 260-Ma to 280-Ma continental glaciers are found in tropical Africa, South America, India, and Madagascar.

According to the principle of uniformitarianism, these ancient reefs, salt deposits, and glacial deposits should have formed in climate zones similar to those where they form today. Either uniformitarianism doesn't apply to these phenomena, or the landmasses where these deposits are found have moved from their original climate zone into another. Observations like these led Alfred Wegener, a German meteorologist, to suggest in the 1920s that the continents had moved—a process he called *continental drift*. Modern geologists interpret these anomalies as the result of plate motion, as continents change their positions on the globe by moving apart during seafloor spreading or by coming together as subduction closes an ocean.

EXERCISE 2.2 | **Interpretations of Past Plate Locations Based on Paleoclimate Indicators**

Name: _____   Section: _____
Course: _____   Date: _____

Fossil corals about 370 million years old have been discovered along much of the length of the Appalachian Mountain system in the United States—as shown by the red stars on the map to the right. Today corals are generally found in tropical and subtropical climate zones, between 30° N and 30° S latitudes, and rarely as far north as the northern part of the Appalachians.

(a) What does the presence of these coral fossils suggest about the location of North America 370 million years ago? _____

_____

_____

_____

(continued)

Name: _____   **Section:** _____
Course: _____   **Date:** _____

(b) The east coast of the United States today is oriented nearly straight north-south.
  • Do you think the orientation of the east coast was the same 370 million years ago? Why or why not?

_____

_____

  • When they were alive, were the corals closer to or farther from the equator than the fossils are today?

_____

_____

The map below shows where geologists have found evidence of continental-scale glaciation that occurred at about 260 million years ago on all of the southern continents (South America, Africa, Australia, and India), along with the directions in which the ice flowed (arrows). The movement of ice in a continental glacier, like the one covering most of Antarctica today, is outward in all directions from a central snowfield.

(c) On a separate sheet of paper, trace the continents (which are outlined with a heavy black line) with their glaciated areas (outlined in blue). Cut out the continents and reconstruct the continental glacier by fitting the continents together and considering the ice-flow directions shown by the arrows. Mark the location of the central snowfield on the map below with a colored pencil.

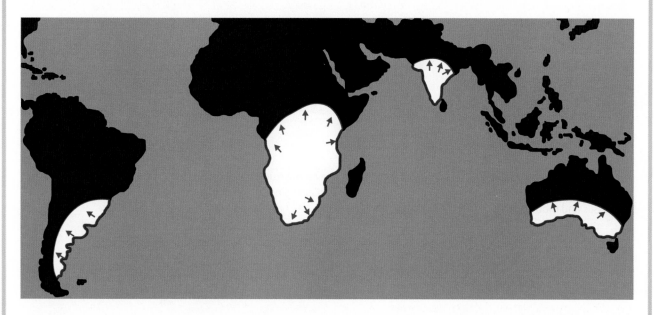

Let's further test the idea of the existence of a unified continent (called Pangaea) by assembling the present-day continents based on their rock and fossil evidence. Use the puzzle pieces located at the back of this manual in the Tools section.

  • First, choose a color for each item in the legend (type of fossil and mountain belt) and color in the key. Then color in the corresponding areas on the continent with the same colors you chose for your key.
  • Cut out the continents along their borders. These pieces represent the probable shape of the continents following the breakup of Pangaea.
  • Place the continents in the empty space on the next page and match them up using the fossil and mountain evidence. They may not fit perfectly together.
  • If your instructor has asked you to include your reconstruction (puzzle) here, glue or tape the assembled pieces and the legend in the space on the next page or on another piece of paper.

(*continued*)

Name: _____ Section: _____
Course: _____ Date: _____

(d) Does your reconstruction of the continents agree with the reconstruction based on glaciers from question (c) above and the past existence of the supercontinent Pangaea? Explain using evidence from the reconstruction (locations of fossils and mountain ranges and fit of the continents).

_____

_____

_____

_____

2.3.3 Geographic Distribution of Active Volcanoes and Earthquakes

By the mid-1800s, scientists realized that the Earth's active volcanoes are not distributed randomly. Most *subaerial volcanoes* (those that protrude into the air) are concentrated in narrow belts near the edges of continents, like the chain called the Ring of Fire surrounding the Pacific Ocean, whereas the centers of most continents host no volcanoes (**FIG. 2.6**). In the late 1900s, as technological advances provided new insights into the nature of the ocean floor, we learned that *submarine volcanoes* (underwater volcanoes) are found in every major ocean basin.

More clues soon came from seismologists (geologists who study earthquakes). Records showing the locations of earthquakes worldwide revealed a concentration along linear belts (**FIG. 2.7**). Although this pattern was more complex than that for volcanoes, particularly in the deep ocean and in continental interiors, the two

FIGURE 2.6 Worldwide distribution of active subaerial volcanoes. The red triangles are volcanoes and the orange line outlines the Pacific Ring of Fire.

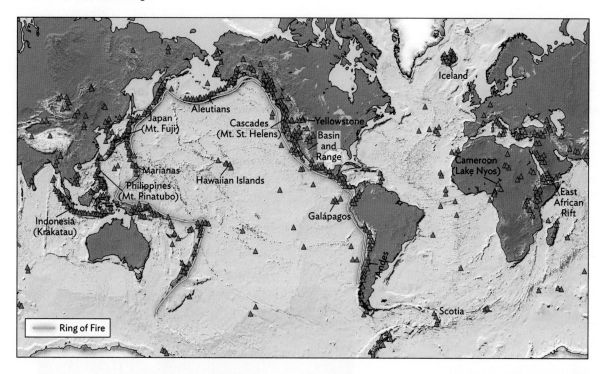

FIGURE 2.7 Worldwide distribution of earthquakes, 1960–1980.

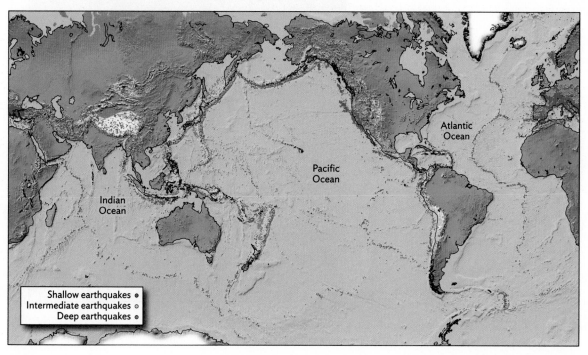

patterns demonstrated important similarities. Something unique was happening along the volcanic chains and where earthquakes occurred, but geologists couldn't agree on what that was. In Exercise 2.3 you will follow the reasoning geologists used to build the basic framework of plate tectonics theory.

EXERCISE 2.3 | **Putting the Early Evidence Together**

Name: _____ Section: _____
Course: _____ Date: _____

Examine the patterns shown in Figures 2.6 and 2.7 and answer the following questions to understand the initial reasoning underlying plate tectonics theory.

(a) Where are the volcanic and earthquake patterns most similar?

_____

(b) In what parts of the world are there abundant earthquakes but not (apparently) active volcanoes?

_____

(c) Focus on the patterns in the middle of the Atlantic Ocean. Only a few volcanoes are shown along the Mid-Atlantic Ridge, but earthquakes seem to occur all along it. Volcanic activity does, in fact, occur all along the ridge. Why are only a few volcanoes shown?

_____

_____

(d) The dots in the figure at right indicate the locations of earthquakes at depth within the Earth's interior, resulting from an interaction between two plates.
• What is the region where these plates meet called? _____
• What kind of plate boundary does this represent? _____

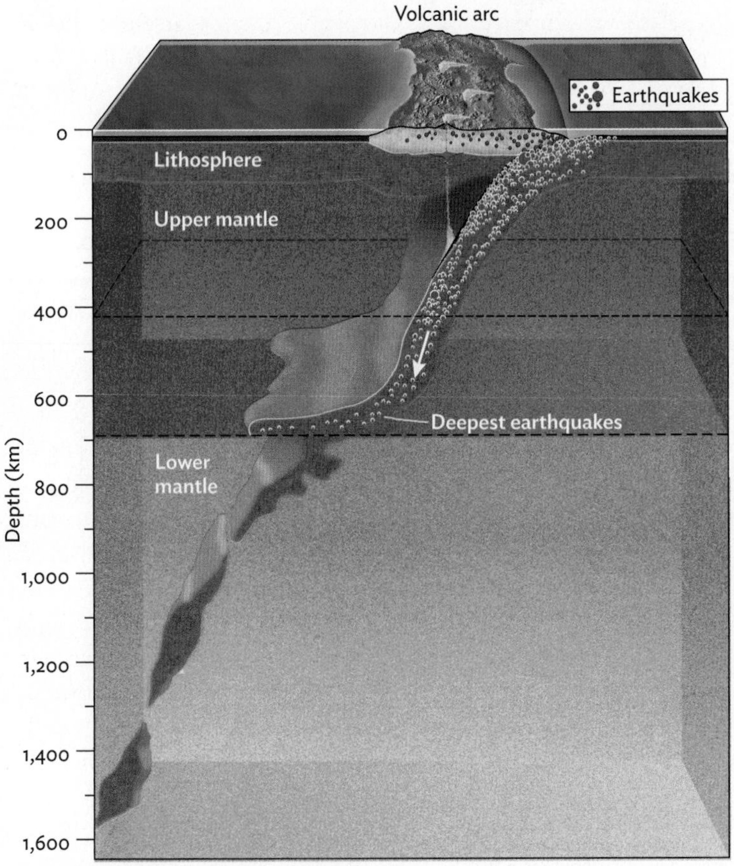

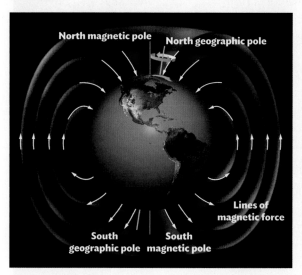
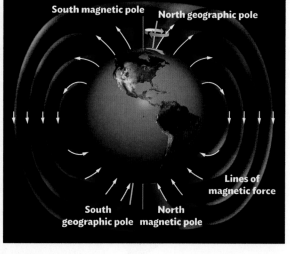

(a) Normal polarity.　　　　**(b)** Reversed polarity.

2.4 Modern Evidence for Plate Tectonics

The geographic fit of continents and paleoclimate evidence convinced some geologists that plate tectonics was a reasonable hypothesis, but more information was needed to convince the rest. That evidence came from an improved understanding of the Earth's magnetic field, the ability to date ocean-floor rocks, careful examination of earthquake waves, and direct measurements of plate motion using global positioning satellites and other exciting new technologies. The full body of evidence has converted nearly all doubters to ardent supporters.

2.4.1 Evidence for Seafloor Spreading: Oceanic Magnetic Anomalies

The Earth has a magnetic field that can be thought of as having "north" and "south" poles like a bar magnet (**FIG. 2.8A**). Navigational compasses are aligned by magnetic lines of force that emanate from one magnetic pole and re-enter the Earth at the other. This magnetic field has been known for centuries, but two discoveries in the mid-20th century provided new insights into how the field works and, soon afterward, the evidence that confirmed plate tectonics theory.

First, geologists learned that when grains of magnetite or hematite—minerals containing magnetic materials—crystallize, their magnetic fields align parallel to the lines of force of the Earth's magnetic field. Some rocks that contain magnetite or hematite therefore preserve a weak record of the Earth's ancient magnetic field—a record called **paleomagnetism**. Then, geologists learned that the Earth's magnetic field reverses polarity at irregular intervals, so that what is now the north magnetic pole becomes the south magnetic pole, and vice versa. During periods of **normal polarity**, the field is the same as it is today, but during periods of **reversed polarity**, a compass needle that points to today's north magnetic pole would swing around and point south (**FIG. 2.8B**). By finding and determining the polarities of rocks throughout the world, geologists have dated magnetic reversals back millions of years. **FIGURE 2.9** shows the last 4 million years of the paleomagnetic record.

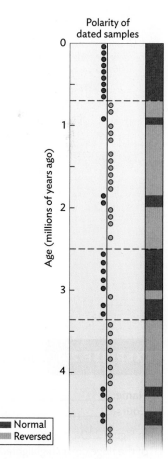

FIGURE 2.9 **Magnetic reversals of the past 4 million years.**

FIGURE 2.10 Magnetic anomaly stripes in the Atlantic and Pacific oceans.

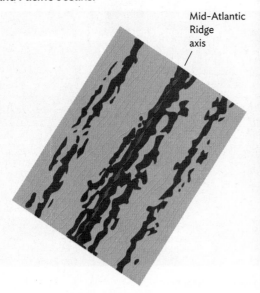

Mid-Atlantic Ridge axis

Dark bands = normal polarity
Light bands = reversed polarity

(a) The Mid-Atlantic Ridge southwest of Iceland.

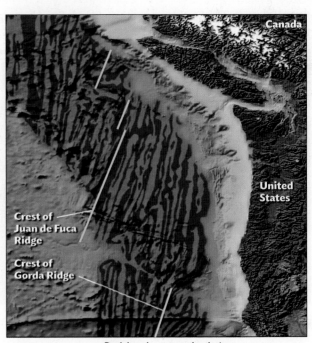

Canada

United States

Crest of Juan de Fuca Ridge

Crest of Gorda Ridge

Dark bands = normal polarity
Light bands = reversed polarity

(b) The Juan de Fuca and Gorda Ridges in the North Pacific off the state of Washington and the province of British Columbia.

Then, in the late 1960s, researchers discovered that the history of normal and reversed polarities is also recorded in the seafloor, which produces parallel linear magnetic belts in which the magnetic field is either anomalously (unusually) stronger than the Earth's average magnetic field, or anomalously weaker than the Earth's average magnetic field. These belts are therefore called **magnetic anomaly stripes** (**FIG. 2.10**). Every area where the paleomagnetism was recorded for a time in which the Earth's magnetic field had the same polarity as today's, displayed an anomalously stronger magnetic field (a positive anomaly); every area where the paleomagnetism was recorded for a time when the Earth's magnetic field was the reverse of today's, yielded a weaker field (negative anomaly).

This pattern of marine magnetic stripes has now been found at every mid-ocean ridge, proving that the magnetic reversals are truly worldwide events. Earth scientists determined that as lava erupting at mid-ocean ridges cools, it records the magnetic field polarity present at that time. If the Earth's polarity reverses, new lava will adopt the new polarity, as shown in Figure 2.10. Exercise 2.4 shows the reasoning by which geologists connected marine magnetic anomaly stripes in the oceans with plate tectonics theory.

EXERCISE 2.4	Interpreting Magnetic Anomaly Stripes at Mid-Ocean Ridges

Name: _____   Section: _____
Course: _____   Date: _____

In Figure 2.10a, compare the orientation of the magnetic anomaly stripes near the Mid-Atlantic Ridge with the orientation of the ridge crest (highlighted by the red line). In Figure 2.10b, do the same for the Juan de Fuca Ridge (shown in yellow) and its associated magnetic anomaly stripes.

(continued)

Name: _____ Section: _____

Course: _____ Date: _____

(a) Are the individual anomalies oriented randomly? Are they parallel to the ridge crests? Oblique (at an angle) to the ridge crests?

_____

_____

(b) Explain how the process of seafloor spreading can produce these orientations and relationships.

_____

_____

_____

(c) Some magnetic anomaly stripes are wider than others. Knowing what you do about seafloor spreading and magnetic reversals, suggest an explanation.

_____

_____

2.4.2 Direct Measurement of Plate Motion

Skeptics can no longer argue that the Earth's major features are fixed in place. Satellite instruments, which can measure these Earth features with precision not even dreamed of 10 years ago, make it possible to measure the directions and rates of plate motion. Data for the major plates are shown in **FIGURE 2.11**, in which the length of each black arrow indicates the rate of plate motion, relative to the adjacent plate, caused by seafloor spreading. We will see later how geologists were able to deduce the same information using other data.

FIGURE 2.11 Rates of motion of the Earth's major plates (in centimeters per year) measured by satellite instruments.

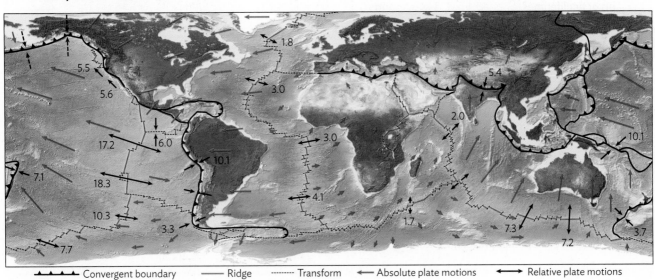

△△△ Convergent boundary —— Ridge -------- Transform ← Absolute plate motions ←→ Relative plate motions

2.5 Processes at Plate Boundaries Revealed by Earth Features

2.5.1 Seafloor Spreading

The next few exercises examine how geologists can deduce details of plate boundary geometry, the rates of plate motion involved, and their histories. Let's start with the information that we can gather about seafloor spreading at divergent boundaries. Your results in Exercises 2.5 and 2.6 will dramatically demonstrate how slowly the South Atlantic Ocean seafloor is spreading and why plate tectonics met with widespread disbelief initially. They also reinforce the importance of understanding the vast expanse of geologic time discussed in Chapter 1. Even extremely slow processes can have large effects given enough time to operate!

EXERCISE 2.5　**Estimating Seafloor Spreading Rates**

Name: _____　Section: _____

Course: _____　Date: _____

The South Atlantic Ocean formed by seafloor spreading at the Mid-Atlantic Ridge. Geologists can get a rough estimate of the spreading rate (i.e., the rate at which South America and Africa are moving apart) by measuring the distance between the two continents in a direction parallel to the transform faults and determining the time over which the spreading occurred.

(a) In Figure 2.4a, the two points on Africa and South America (indicated by the yellow arrows) were once together. Measure the distance that they have been separated, along the transform fault indicated by the white arrow.

The oldest rocks in the South Atlantic Ocean, immediately adjacent to the African and South American continental shelves, are 120,000,000 years old.

(b) Calculate the average rate of seafloor spreading for the South Atlantic Ocean over the last 120 million years. Remember, to calculate rate, you divide the width of the ocean by the amount of time it took to reach that distance.

_____ km/million years

_____ km/yr

_____ cm/yr (*Hint*: there are 100,000 cm in a kilometer)

_____ mm/yr

(c) Assuming someone born today lives to the age of 100, how much wider will the South Atlantic Ocean become during his or her lifetime? _____ cm

---

EXERCISE 2.6　**Comparing Seafloor Spreading Rates of Different Mid-Ocean Ridges**

Name: _____　Section: _____

Course: _____　Date: _____

Magnetic reversals are found worldwide, so magnetic anomaly stripes should be the same width in every ocean *if the rate of seafloor spreading were the same at all mid-ocean ridges*. As you investigated earlier, if a particular anomaly stripe is wider in one ocean than in another, however, the difference must result from faster spreading. The figure on the next page shows simplified magnetic anomaly stripes from the South Atlantic and South Pacific oceans, the ages of the rocks in the stripes, and their distance from the spreading center (the red line).

(continued)

Name: _____   Section: _____

Course: _____   Date: _____

For simplicity, only the most *recent* 40 million years of data are shown for the two oceans, and we will estimate the spreading rate only for that time span.

(a) Provide the width of the South Atlantic Ocean, as given in the nearer figure on the right. _____ km

(b) Estimate the average rate at which the South Atlantic Ocean has been opening over the 40 million years for which data are provided. _____ cm/yr _____ km per million years

(c) Now look at the data for the South Pacific Ocean and its spreading center, the East Pacific Rise. Considering the width of this ocean, is the spreading rate the same, greater, or less than that of the South Atlantic?

_____

(d) Now get the details. Provide the width of the South Pacific Ocean, as given in the farther figure on the right. _____ km

(e) What is the spreading rate of the East Pacific Rise? _____ cm/yr _____ km per million years

These spreading rates are typical of the range measured throughout the world's oceans and represent "fast spreaders" and "slow spreaders."

(f) Based on your answer in (c) above, how does the different rate of spreading affect the width of the seafloor anomalies (wide versus narrow)? Explain.

_____
_____
_____
_____

Map view of magnetic anomaly stripes in two oceans.

2.5.2 Transform Faults

While most transform faults occur in the oceans, some transform faults also cut continental lithosphere. The most famous of these faults constitutes the San Andreas fault system of California, which has caused major damage and loss of life over the past 100 years. The San Andreas fault *system* (shown in Exercise 2.7) is not a single fault, but rather a zone several miles wide containing numerous active transform faults. It extends for more than 1,000 km, connecting segments of the Juan de Fuca Ridge and Cascade Trench at its northern end to an unnamed mid-ocean ridge segment in the Gulf of California to the south.

Name: _____   Section: _____

Course: _____   Date: _____

The more we know about the history of a continental transform fault that lies close to heavily populated regions, the better we can prepare for its next pulse of activity. Geologists try to find out how long a continental transform fault has been active, how much it has offset features on the plates it separates, and how fast it has moved in the past.

Geologists estimate that the San Andreas fault system, mapped below, has been active for about 20 million years. This exercise shows how geologic markers cut by transform faults enable us to measure the amount and rate of motion along a fault.

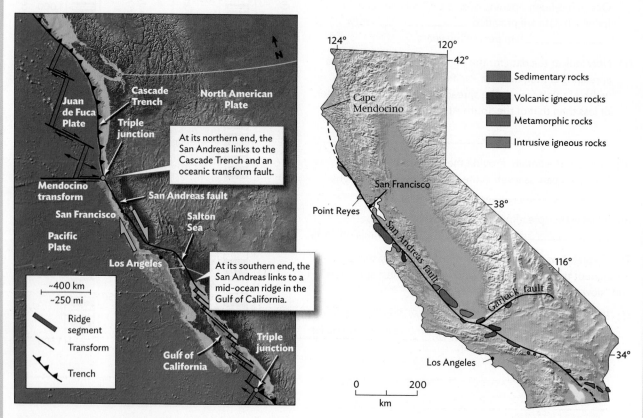

Simplified geologic setting of the San Andreas fault system.

Amount of fault movement indicated by offset bodies of identical rock.

Before tackling the San Andreas system, let's practice on a much simpler area (see the map on the next page). Field geologists have mapped an active continental transform fault (the orange line) for several hundred kilometers. A 50-million-year-old (50-Ma) body of granite and a 30-Ma vertical dike have been offset by the fault as shown.

(a) Draw arrows on opposite sides of the fault to show the direction in which the two plates moved relative to each other.

(b) Measure the amount of offset of the 50-Ma granite body. _____ km

(c) The geologists have proved that the faulting began almost immediately after the granite formed and continues today. Assuming that the plates moved at a constant rate for the past 50 million years, calculate the rate of offset. _____ cm/yr

(d) Now do the same for the 30-Ma dike. Offset: _____ km.
Assuming a constant rate of movement for the past 30 million years, calculate the rate of offset. _____ km/yr

(continued)

Name: _____   **Section:** _____

Course: _____   **Date:** _____

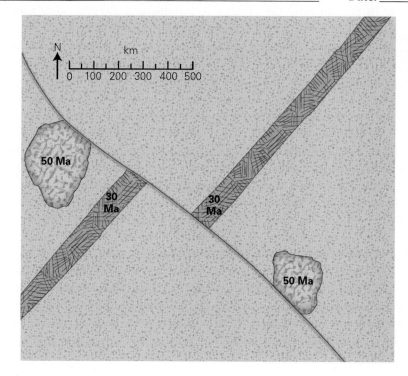

Geologic markers displaced by a continental transform fault.

(e) Compare the two rates. Has faulting taken place at a constant rate, or has the rate increased or decreased over time? Explain in as much detail as possible.

_____

_____

Now let's look at the San Andreas fault system again in the maps at the beginning of this exercise.

(f) San Francisco and Los Angeles are on opposite sides of the San Andreas fault. If the fault is undergoing offset at the rates you calculated in questions (c) and (d) above, how many years will it take for the two cities to be located directly opposite each other? _____

? What Do You Think Because of the earthquake threats from the San Andreas fault, the city of San Francisco has rigorous building codes requiring that buildings be designed to withstand earthquakes. On the other side of the United States, however, the building codes are less rigorous. An earthquake in Virginia in August 2011 was felt along most of the east coast, prompting questions about whether existing building codes should be changed to meet the more rigorous San Francisco standards. But meeting the San Francisco standards makes construction a lot more expensive in San Francisco than it is in New York. Imagine you had to make a recommendation to the New York City Council about whether it should or should not adopt San Francisco building codes. Using what you've learned in this chapter, on a separate sheet of paper, explain what your recommendation would be, and why.

2.5.3 Hot Spots and Hot-Spot Tracks

Paleoclimate anomalies show that plates have moved, magnetic anomaly stripes record the rate of seafloor spreading, and satellites measure the rates and directions of plate motion today. But how can we determine whether a plate has always moved at its current rate? Or whether it has changed direction over time? The answers come from the study of hot-spot volcanic island chains. **FIGURE 2.12** shows how hot-spot island chains form. Each volcano in the chain forms at a **hot spot**: an area of unusual volcanic activity not associated with processes at plate boundaries. The cause of this activity is still controversial, but many geologists propose that hot spots form above a narrowly focused source of heat called a **mantle plume**: a column of very hot rock that rises by slow plastic flow from deep in the mantle.

When the plume reaches the base of the lithosphere, it melts the lithosphere rock and produces magma that rises to the surface, erupts, and builds a volcano. The plume is thought to be relatively motionless. If the plate above it moves, the volcano is carried away from its magma source and becomes extinct. A new volcano then forms above the hot spot until it, too, is carried away from the hot spot. Over millions of years, a chain of volcanic islands forms, the youngest at the hot spot, the oldest farthest from it. As the volcanoes cool, they become denser, subside, and are eroded by streams and ocean waves. Eventually, old volcanoes sink below the ocean surface, forming **seamounts**. The chain of islands and seamounts traces plate motion above the hot spot, just as footprints track the movement of animals. The Hawaiian Islands, for example, are the youngest volcanoes in the Hawaiian–Emperor seamount chain. Most of the older volcanoes in the chain are seamounts detected by underwater oceanographic surveys. The Hawaiian–Emperor seamount

FIGURE 2.12 Origin of hot-spot island chains and seamounts.

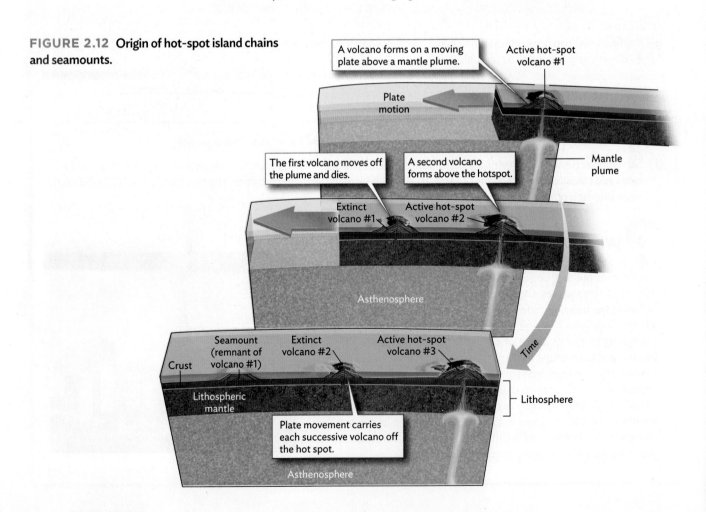

chain tracks the motion of the Pacific Plate and lets us interpret its motion over a longer time span than that recorded by the Hawaiian Islands alone. In Exercises 2.8 and 2.9, you will examine both the Hawaiian Islands (as hotspot volcanic islands) and the Hawaiian–Emperor seamount chain to determine the history of Pacific Plate movement.

<table>
<tr><td>**EXERCISE 2.8**</td><td>**Determining Ancient Plate Direction: Footprints of a Moving Plate**</td></tr>
</table>

Name: _____ Section: _____

Course: _____ Date: _____

The Hawaiian Islands, located in the Pacific Ocean far from the nearest mid-ocean ridge, are an excellent example of hot-spot volcanic islands (see the figure below). Volcanoes on Kauai, Oahu, and Maui haven't erupted for millions of years, but the island of Hawaii hosts five huge volcanoes, one of which (Kilauea) has been active continuously for more than 30 years. In addition, a new volcano, already named Loihi, is growing on the Pacific Ocean floor just southeast of Kilauea. As the Pacific Plate continues to move, Kilauea will become extinct and Loihi will become the primary active volcano.

(a) Where is the Hawaiian hot-spot plume located today relative to the Hawaiian Islands? Explain your reasoning.

_____

_____

(b) Draw a line connecting the volcanic center (highlighted in red) on Maui to Kilauea (the currently active volcano on Hawaii). Connect the volcanic centers of Maui to Molokai, Molokai to Oahu, and Oahu to Kauai as well.

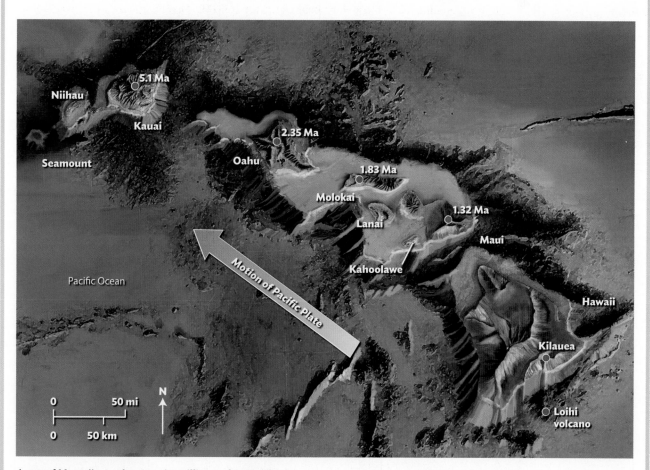

Ages of Hawaiian volcanoes in millions of years (Ma).

(continued)

Name: _____   Section: _____
Course: _____   Date: _____

(c) Geologists use the *azimuth method* to describe direction precisely. In the azimuth method, north is 0°, east is 090°, south is 180°, and west is 270°, as shown in the following figure. The green arrow points to 052° (northeast). To practice using the azimuth method, estimate and then measure the directions shown by arrows A through D, using your circular protractor located in the GeoTools section at the back of this manual.

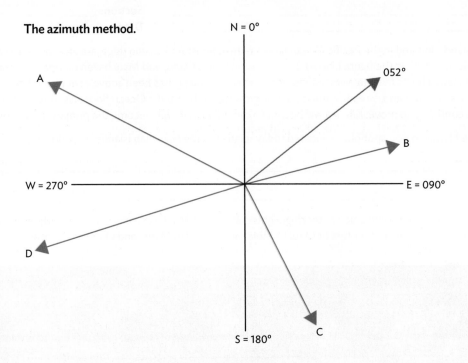

The azimuth method.

(d) Measure the direction and distance between the volcanic centers of the Hawaiian Islands using a ruler, a protractor, and the map scale. Calculate the rate of plate motion (distance between volcanoes divided by the time interval between eruption ages) and fill in the table below. Express the rates in millimeters per year.

	Distance between volcanic centers (km)	Number of years of plate motion	Rate of plate motion (mm/yr)	Azimuth direction of plate motion (e.g., 325°)
Hawaii to Maui				
Maui to Molokai				
Molokai to Oahu				
Oahu to Kauai				

Name: _____   Section: _____
Course: _____   Date: _____

The oldest volcano of the Hawaiian–Emperor seamount chain was once directly above the hot spot, but is now in the northern Pacific, thousands of kilometers away (see the figure below). Seamount ages show that the hot spot has been active for a long time and reveal the direction and rate at which the Pacific Plate has moved.

(a) What evidence is there that the Pacific Plate has not always moved in the same direction?

_____

_____

_____

_____

(b) How many years ago did the Pacific Plate change direction? Explain your reasoning.

_____

_____

_____

Bathymetry (measurements of water depths) of the Pacific Ocean floor, showing the ages of volcanoes in the Hawaiian–Emperor seamount chain.

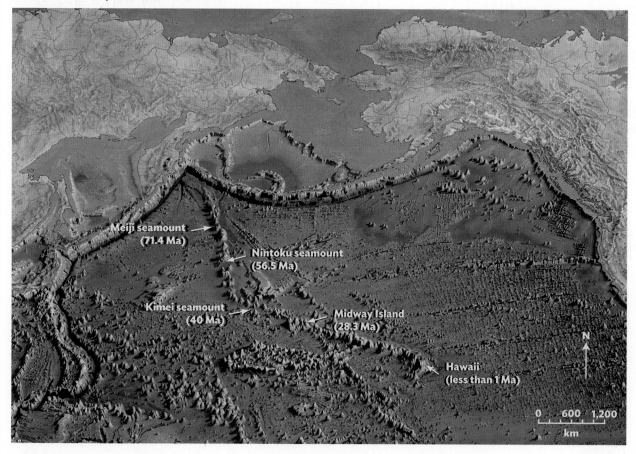

(continued)

EXERCISE 2.9 Long-Term Movement of the Pacific Plate *(continued)*

Name: ________________________ **Section:** ____________

Course: ________________________ **Date:** ____________

(c) Based on the age data on the bathymetric map, in what azimuth direction did the Pacific Plate move originally?

____

(d) How far has the Meiji seamount moved from the hot spot? Explain your reasoning.

____

____

____

____

(e) At what rate has the Pacific Plate moved?

- based on data from the Hawaii-Midway segment?____________ km per million years
- based on data from the Hawaii-Kimei segment?____________ km per million years
- based on data from the Kimei-Meiji segment? ____________ km per million years

(f) Has the Meiji seamount moved at a constant rate? Explain your reasoning.

____

____

____

____

(g) Per data in the map on the previous page, in what direction is the Pacific Plate moving today? Explain your reasoning.

____

____

____

____

(h) Assuming that the directions of motion depicted in the map on the previous page were to continue, what would be the eventual fate of the Meiji seamount? Explain in as much detail as possible.

____

____

____

____

____

____

Name: _____   Section: _____

Course: _____   Date: _____

Exploring Earth Science Using Google Earth

1. Visit digital.wwnorton.com/labmanualearthsci

2. Go to the Geotours tile to download Google Earth Pro and the accompanying Geotours exercises file.

Check and double-click the Geotour02 folder icon in Google Earth to fly to a position over southern Idaho. This view highlights the locations and ages of the main calderas that formed due to the passage of the North American Plate over the Yellowstone Hot Spot during the past 16.5 Ma *(note that you can select the folder and use the transparency slider at the bottom of the Places panel to make items in this folder semi-transparent).*

(a) Using the caldera ages, in what general direction (N, S, E, W, NE, SE, SW, NW) is the North American Plate moving over the stationary hot spot?

_____

_____

_____

(b) Click on the **Ruler tool** icon in the top menu bar and use the **Path** tab to draw a path connecting all the calderas' center dots. Measure this distance (cm) along the hot-spot trail, and calculate the average velocity (cm per year) of the North American Plate using 16.5 Ma and 0.63 Ma as the ends of the path. _____ cm/yr

(c) Note that the last three calderas have ages of 2.0 Ma, 1.3 Ma, and 0.63 Ma. What pattern do you see in these ages, and what might this suggest about the timing of the next major eruption?

_____

_____

(d) Before becoming overly concerned with your answer to question (c), look at the eruptive ages for the older calderas. Does this same pattern hold over the entire hot-spot trail? What does this information tell you regarding the prediction of future eruptions for the Yellowstone Hot Spot?

_____

_____

_____

(e) Suppose the North American Plate began moving due south. After 5–10 million years (given the current rate of plate motion), what state likely would be directly over the hot spot?

_____

_____

_____

Minerals

3

Giant gypsum crystals in the Cave of the Crystals, Chihuahua, Mexico.

3.1 Introduction

This chapter begins our study of the materials of which the Earth is made. It starts by examining the different kinds of materials in the geosphere, then focuses on minerals, the basic building blocks of most of the Earth. You will learn what makes minerals different from other substances, become familiar with their physical properties, and use those properties to identify common minerals. Minerals are important to Earth scientists because they record the conditions and processes of ancient Earth history and to society because many are natural resources that contribute trillions of dollars annually to the U.S. and global economies.

3.2 Classifying Earth Materials

Imagine that an octopus is swimming in the ocean when a container falls off a freighter overhead, breaks up, and spills its entire cargo of sneakers, sandals, flip-flops, shoes, moccasins, and boots into the sea. The octopus is curious about these objects and wants to learn about them, but it doesn't have heels or toes, doesn't walk, has eight legs, doesn't understand "left" and "right," and doesn't wear clothes. How would it begin to study these totally alien objects? A first step might be to sort, or classify, the objects into subgroups of objects sharing similar features. But what kind of system would it use? One possibility might be to separate items that are mostly enclosed (shoes, boots, sneakers, moccasins) from those that are open (sandals, flip-flops). Another might be to separate objects made of leather from those made of cloth, or brown objects from black objects, or big ones from small ones. There are many ways to classify footwear, some of which might lead our octopus to a deeper understanding of the reasons for these differences.

Seventeenth-century scientists faced a similar task when they began the modern study of Earth materials. They started by describing and classifying these materials because classification reveals similarities and differences between things that lead to an understanding of the processes that formed them. Biologists classify organisms, art historians classify paintings, and geologists classify Earth materials. Exercise 3.1 leads you through the thought processes involved in developing a classification scheme for Earth materials.

3.3 What Is a Mineral and What Isn't?

Most people know that the Earth is made of minerals and rocks, but don't know the difference between them. The words *mineral* and *rock* have specific meanings to geologists, often somewhat different from those used in everyday language. For example, what a dietitian calls a mineral would be considered an element by geologists. Geologists define a **mineral** as a naturally occurring, homogeneous solid formed by geologic processes, which has an ordered internal arrangement of atoms, ions, or molecules and has a composition that is defined by a chemical formula. Historically, there was also a requirement that minerals be inorganic—not produced by an animal or plant. Today this requirement is dropped by many geologists in recognition that the hard parts of bones and teeth are identical to the mineral apatite, many clamshells are identical to the mineral calcite, and many microscopic creatures build shells from the mineral quartz.

Let's examine the definition more closely:

- *Naturally occurring* means that a mineral forms by natural Earth processes. Thus, human-made materials, such as steel and plastic, are not minerals.
- *Homogeneous* means that a piece of a mineral contains the same pure material throughout.

Name: _____   Section: _____
Course: _____   Date: _____

(a) Examine the specimens of Earth materials provided by your instructor. Group them into categories you believe are justified by your observations, and explain the criteria you used to set up the groups. *Note:* If taking this lab online, and directed by your instructor, use the specimen images provided by your instructor in lieu of mineral specimens.

Group	Defining criteria for each group	Specimens in group

(b) Compare your results with those of others in the class. Did you use the same criteria? Are your classmates' specimens in the same groups as yours?

_____

_____

(c) What does your comparison tell you about the process of classification?

_____

_____

_____

- *Solid* means that minerals retain their shape indefinitely under normal conditions. Therefore, liquids such as water and oil, and gases such as air and propane, aren't minerals.
- *Formed by geologic processes* traditionally implied processes such as solidification or precipitation, which did not involve living organisms; however, as noted above, many geologists now consider solid, crystalline materials produced by organisms to be minerals.
- *An ordered internal arrangement of atoms, or ions* is an important characteristic that separates minerals from substances that might fit all other parts of the definition. Atoms in minerals occupy fixed positions in a grid called a *crystal structure*. Solids in which atoms occur in random positions, rather than being locked into a crystal structure, are called *glasses*.
- *Definable chemical composition* means that the elements present in a mineral and their proportions can be expressed by a simple chemical formula—for example, the formula for quartz is SiO_2 (one silicon atom for every two oxygen atoms), and the formula for calcite is $CaCO_3$ (one calcium atom and one carbon atom for every three oxygen atoms)—or by one that is more complex—such as that of the mineral muscovite, which is $KAl_2(AlSi_3O_{10})(OH)_2$.

When a mineral grows without interference from other minerals, it will develop smooth, flat surfaces and a symmetric geometric shape that we call a **crystal**. When a mineral forms in an environment where other minerals interfere with its growth, it will have an irregular shape but will still have the appropriate internal (crystal) structure for that mineral. An irregular or broken piece of a mineral is called a **grain**, and a single piece of a mineral, either crystal or grain, is called a **specimen**.

In the geosphere, most minerals occur as parts of rocks. It is important to know the difference between a mineral specimen and a rock. A **rock** is *a coherent, naturally occurring, solid consisting of an aggregate of mineral grains, pieces of older rocks, or a mass of natural glass.* ("Coherent," in this context, means that a rock holds together as a solid mass, rather than as a loose pile of grains.) Some rocks, such as granite, contain grains of several different minerals; others, such as rock salt, are made of many grains of a single mineral. Still others are made of fragments of previously existing rock that are cemented together. And a few kinds of rock are natural glasses, cooled so rapidly from a molten state that their atoms did not have time to form the grid-like crystal structures required for minerals. Exercise 3.2 helps you practice this terminology with specimens provided by your instructor.

EXERCISE 3.2   **Is It a Mineral or a Rock?**

Name: _____   Section: _____
Course: _____   Date: _____

(a) Based on the definitions of *mineral* and *rock,* determine which specimens used in Exercise 3.1 are minerals, which are rocks, and which, if any, are neither minerals nor rocks. Write the specimen numbers in the appropriate columns in the following table. *Note:* If taking this lab online, and directed by your instructor, use the specimen images provided by your instructor in lieu of mineral specimens.

Minerals	Rocks	Other

(b) Choose one of the rock specimens and look at it carefully. How many different minerals are there in this rock?

_____

(c) How do you know? What clues did you use to distinguish a mineral from its neighbors?

_____

_____

(d) Describe up to four of the minerals in your own words.
Mineral 1:

_____

_____

_____

(continued)

Name: _____   Section: _____
Course: _____   Date: _____

Mineral 2:

_____

_____

_____

Mineral 3:

_____

_____

_____

Mineral 4:

_____

_____

_____

3.4 Physical Properties of Minerals

Mineralogists (geologists who specialize in the study of minerals) have named more than 4,000 minerals that differ from one another in chemical composition and crystal structure. These characteristics determine a mineral's physical properties, which include how it looks (color and luster), breaks, feels, smells, and even tastes. Some minerals are colorless and nearly transparent; others are opaque, dark-colored, and shiny. Some are hard, others soft. Some form needle-like crystals, others blocky cubes. You instinctively used some of these physical properties in Exercise 3.2 to decide the number of minerals in your rocks and then to describe them. In this section we will discuss the major physical properties of minerals so that you can use them to identify common minerals—in class, at home, or while on vacation.

3.4.1 Diagnostic versus Ambiguous Properties

Geologists use physical properties to identify minerals much as detectives use physical descriptions to identify suspects. And as with people, some physical properties of minerals are *diagnostic properties*—they immediately help identify an unknown mineral or rule it out as a possibility. Other properties are *ambiguous properties* because they may vary in different specimens of the same mineral. For example, color is a notoriously ambiguous property in many minerals (**FIG. 3.1**). Size doesn't really matter either; a large specimen of quartz has the same properties as a small one. Exercise 3.3 shows how diagnostic and ambiguous properties affect everyday life.

3.4.2 Luster

One of the first things we notice about a mineral is its **luster**: the way light interacts with its surface. For mineral identification, we distinguish minerals that have a *metallic* luster from those that are *nonmetallic*. Something with a metallic luster is shiny and opaque, like an untarnished piece of metal. Materials with a nonmetallic luster are said to look earthy (dull and powdery), glassy, waxy, silky, or pearly—all terms relating their luster to familiar materials. Luster is a diagnostic property for many minerals, but be careful: some minerals may tarnish, and their metallic luster may be dulled.

3.4.3 Color

The **color** of a mineral is controlled by how the different wavelengths of visible light are absorbed or reflected by the mineral's atoms. Color is generally a diagnostic property for minerals with a metallic luster as well as for some with a nonmetallic luster. But specimens of some nonmetallic minerals, like the fluorite in Figure 3.1, have such a wide range of colors that they were once thought to be different minerals. We now know that the colors are caused by impurities. For example, rose quartz contains a very small amount of titanium.

3.4.4 Streak

The **streak** of a mineral is the color of its powder. We don't crush minerals to identify them, but instead get a small amount of powder by rubbing them against an unglazed porcelain streak plate. Streak and color are the same for most minerals, but for those for which they are not the same, the difference is an important diagnostic property. A mineral's color may vary widely, as we saw in Figure 3.1, but its streak is generally similar for all specimens regardless of their color, as seen in **FIGURE 3.2**.

FIGURE 3.2 Color and streak.

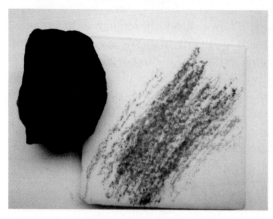 

(a) Red hematite has a reddish-brown streak.

(b) But this dark, metallic-looking specular hematite also has a reddish-brown streak.

EXERCISE 3.3 **People Have Diagnostic Properties Too**

Name: _____   Section: _____

Course: _____   Date: _____

Your father has asked you to pick up his old college roommate at the airport. You've never met him, but your father gave you a yearbook photo and described what he looked like 30 years ago: height, weight, hair color, beard, eye color. Which of these features would still be diagnostic today? Which, considering the passage of time, might be ambiguous? What other properties might also be diagnostic? Indicate in the following chart which properties, considering the passage of time, might be diagnostic and which might be ambiguous. Then suggest two others that would be diagnostic despite the years that have passed.

Property	Diagnostic (explain)	Ambiguous (explain)
Height		
Weight		
Hair color		
Beard		
Eye color		
Others		

3.4.5 Hardness

The **hardness** of a mineral is a measure of how easily it can scratch or be scratched by other substances. A 19th-century mineralogist, Friedrich Mohs, created a mineral hardness scale, using ten familiar minerals that we still use today. He assigned a hardness of 10 to the hardest mineral and a hardness of 1 to the softest (**TABLE 3.1**). This scale is a *relative* one, meaning that a mineral can scratch those lower in the scale but cannot scratch those that are higher. It is not an absolute scale, in which diamond would be 10 times harder than talc and corundum would be 3 times harder than calcite. The hardness of common materials, such as the testing materials listed in **FIGURE 3.3**, can also be described using the Mohs hardness scale. To determine the

TABLE 3.1 Mohs hardness scale and its relationship to common testing materials

Mineral	Mohs hardness number (H)	Mohs hardness of testing materials	
Diamond	10		HARD
Corundum	9		HARD
Topaz	8	Streak plate (6.5–7)	HARD
Quartz	7	Streak plate (6.5–7)	HARD
Orthoclase	6		MODERATE
Apatite	5	Common nail or pocket knife (5.0–5.5) / Window glass; steel-cut nail (5.5)	MODERATE
Fluorite	4		MODERATE
Calcite	3	U.S. penny (3.0)	SOFT
Gypsum	2	Fingernail (2.5)	SOFT
Talc	1		SOFT

EXERCISE 3.4   **Constructing and Using a Relative Hardness Scale**

Name: _____   Section: _____

Course: _____   Date: _____

Your instructor will tell you which specimens from your mineral set to use for this exercise. Arrange them in order of increasing hardness by seeing which can scratch the others and which are most easily scratched.

Softest ————————————————————▶ Hardest

Specimen no.: _____   _____   _____   _____

Now use the testing materials listed in Table 3.1 to determine the Mohs hardness of minerals in your set.

Mohs hardness: _____   _____   _____   _____

FIGURE 3.3 **Testing for hardness.**

(a) A fingernail (H = 2.5) can scratch gypsum (H = 2.0), but not calcite (H = 3.0).

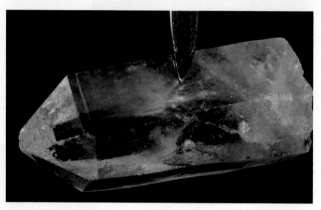

(b) A knife blade (H = 5.0) can scratch fluorite (H = 4.0), but not quartz (H = 7.0).

hardness of a mineral, see which of these materials it can scratch and which ones can scratch it. In some cases, it may only be possible to give a range of hardness— e.g., softer than a streak plate and harder than a nail.

3.4.6 Crystal Habit

The crystal shapes found in the mineral kingdom range from simple cubes with 6 faces to complex 12-, 24-, or 48-sided (or more) crystals (**FIG. 3.4**). Some crystals are flat like a knife blade; others are needle-like. Each mineral has its own diagnostic **crystal habit**, a preferred crystal shape that forms when it grows unimpeded by other grains. For example, the habit of halite is a cube, and that of garnet is a 12-sided crystal with sides of equal length. The habit of quartz is elongate hexagonal crystals topped by a 6-sided pyramid. Remember that crystal growth requires very special conditions. As a result, most mineral specimens are irregular grains; few display their characteristic crystal habit.

3.4.7 Breakage

Some minerals break along one or more smooth planes, others along curved surfaces, and still others in irregular shapes. The way a mineral breaks is controlled by whether there are zones of weak bonds in its structure. Instead of breaking the minerals in your sets with a hammer, you can examine specimens using a microscope or

FIGURE 3.4 Crystals of some common minerals with diagrams of their crystal shapes.

(a) Cubes of pyrite.

(b) A dodecahedral (twelve-sided) garnet crystal.

(c) Slender, prismatic tourmaline crystals.

(d) Prismatic quartz crystals.

(e) Potassium feldspar.

(f) Bladed kyanite crystals.

(g) Scalenohedral calcite crystals.

(h) Needle-like crystals of natrolite.

FIGURE 3.5 Types of fracture in minerals.

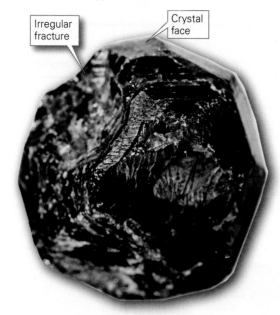

Irregular fracture

Crystal face

(a) Irregular fracture in garnet.

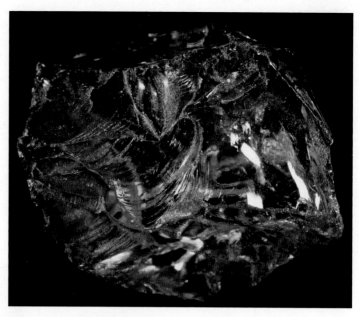

(b) Conchoidal fracture in quartz.

magnifying glass to see how they have already been broken. Two kinds of breakage are important: *fracture* and *cleavage*.

Fracture occurs when there are no zones of particularly weak bonding within a mineral. When such a mineral breaks, either irregular (irregular fracture) or curved (conchoidal fracture) surfaces form (**FIG. 3.5**). Conchoidal fracture surfaces are common in thick glass and in minerals whose bond strength is nearly equal in all directions (e.g., quartz).

Cleavage occurs when the bonds holding atoms together are weaker in some directions than in others. The mineral breaks along these zones of weakness, producing flat, relatively smooth surfaces. Some minerals have a single zone of weakness, but others may have two, three, four, or six (**FIG. 3.6**). If there is more than one zone of weakness, a mineral cleaves in more than one direction. In describing cleavage, we note how many directions there are and the angles between them. Two different minerals might each have two directions of cleavage, but those directions might be different. For example, amphiboles and pyroxenes (two important groups of minerals) are similar in most other properties and each have two directions of cleavage, but amphiboles cleave at 56° and 124° whereas pyroxenes cleave at 87° and 93°.

Note in Figure 3.6 that there may be many cleavage surfaces, but several are parallel to one another, as shown for halite. All of these parallel surfaces define a single cleavage direction. To observe a mineral's cleavage, hold it up to the light and rotate it. Parallel cleavage surfaces reflect light at the same time, making different *cleavage directions* easy to see.

Both crystal faces and cleavage surfaces are smooth, flat planes and might be mistaken for each other. If you can see many small, parallel faces, these are cleavage faces because crystal faces are not repeated. In addition, breakage occurs after a crystal has grown, so cleavage or fracture surfaces generally look less tarnished or altered than crystal faces. Exercise 3.5 will help you recognize the difference between cleavage and fracture.

FIGURE 3.6 Cleavage in common minerals. Some minerals cleave in four or six directions, but it is rare to see all of them in a single specimen.

Types of cleavage	Diagram	Visual	Examples
1 direction	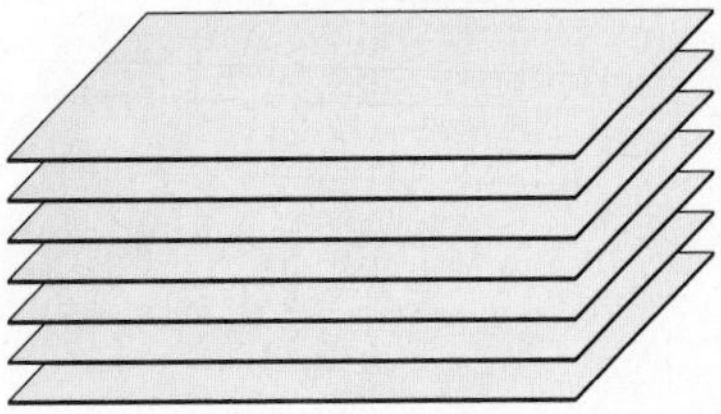		Muscovite
2 directions at 90°	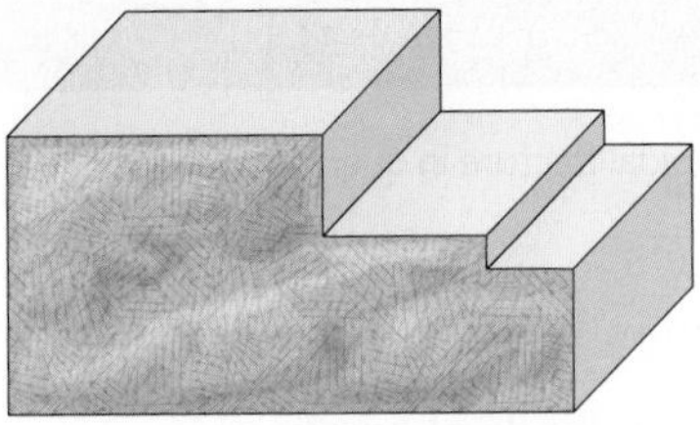	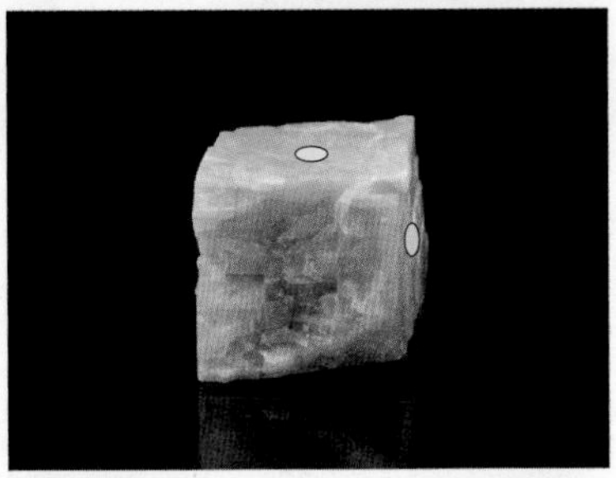	Potassium feldspar
2 directions not at 90°	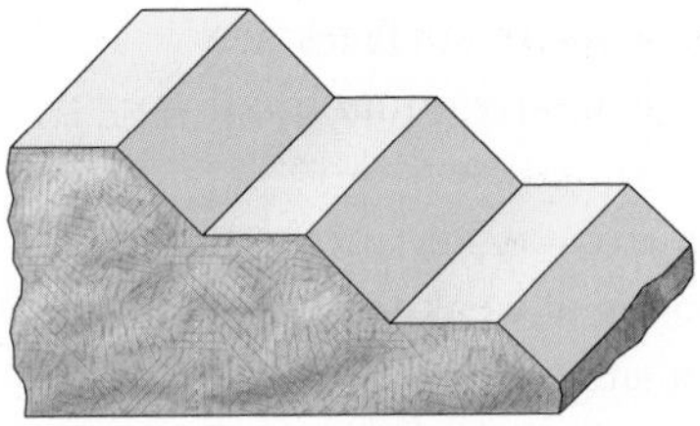		Amphibole
3 directions at 90°	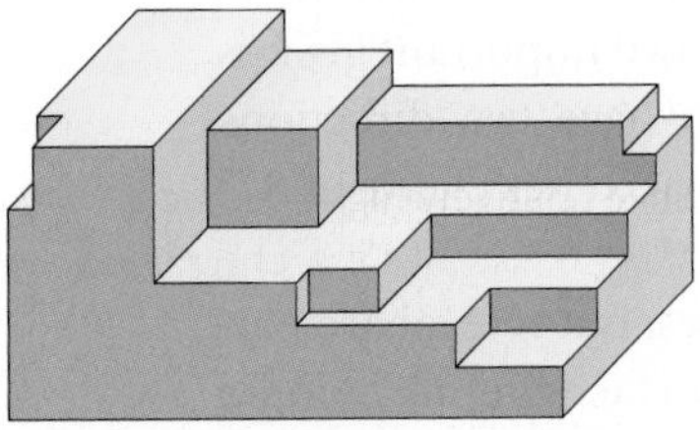		Halite
3 directions not at 90°	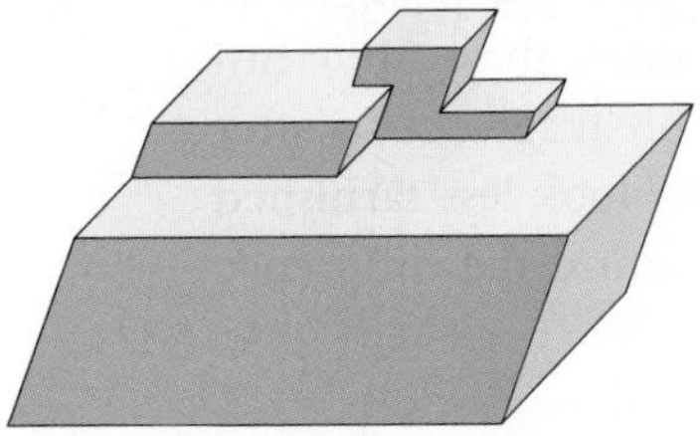	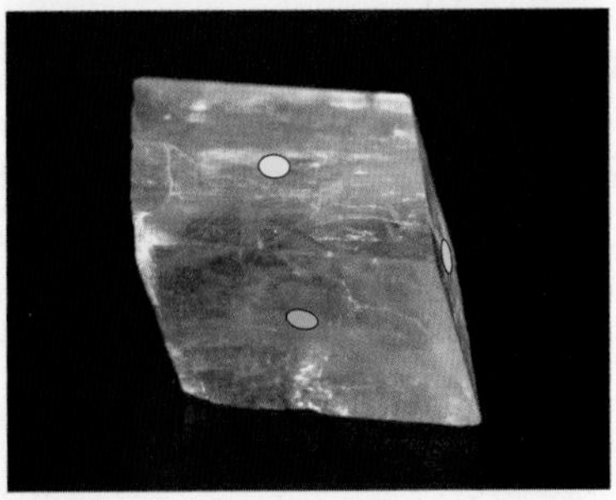	Calcite

Recognizing Breakage in Minerals

Name: _____   Section: _____
Course: _____   Date: _____

Examine the specimens indicated by your instructor. Which have cleaved and which have fractured? For those with cleavage, indicate the number of cleavage directions and the angles between them. (*Note:* If taking this lab online, and directed by your instructor, use the specimen images provided by your instructor in lieu of mineral specimens.)

_____

_____

_____

_____

3.4.8 Specific Gravity

The **specific gravity** (**SpG**) of a mineral is a comparison of its density with the density of water. The density of pure water is 1 g/cm³; so, if a mineral has a density of 4.68 g/cm³, its specific gravity is 4.68. Specific gravity is calculated by dividing the density of the mineral by the density of water, so it has no units because the units cancel in the calculation:

$$\text{SpG} = \frac{\text{Density of mineral}}{\text{Density of water}} = \frac{4.68 \text{ g/cm}^3}{1.00 \text{ g/cm}^3} = 4.68$$

You can measure specific gravity by calculating the density of a specimen (density = mass ÷ volume). But geologists generally *estimate* specific gravity by *hefting* a specimen and determining whether it seems heavy or light. To compare the specific gravities of two minerals, pick up similar-sized specimens to get a general feeling for their densities. You will feel the difference—just as you would feel the difference between a box of Styrofoam packing material and a box of marbles. In Exercise 3.6 you will practice estimating specific gravity by heft, and then measure it precisely.

Heft and Specific Gravity

Name: _____   Section: _____
Course: _____   Date: _____

Separate the minerals provided by your instructor into those that have relatively high specific gravity and those that have relatively low specific gravity by hefting (holding) them. (*Note:* If taking this lab online, and directed by your instructor, use the specimen images provided by your instructor in lieu of mineral specimens.)

(a) What luster do most of the high-specific gravity minerals have?_____ In general, minerals with this luster have higher specific gravity than minerals with other lusters.

(b) To become familiar with the range of specific gravity in common minerals, select the densest and the least dense of your specimens based on their heft. Calculate their densities by weighing them, measuring their volume by submerging them in a graduated cylinder partly filled with water (see Figure 1.6), and dividing mass (g) by volume (cm³).

   SpG: High-density specimen: _____ g/cm³ Low-density specimen: _____ g/cm³

(c) This technique would not work with halite (rock salt). Why not?

_____

3.4.9 Magnetism

A few minerals are attracted to a magnet or act like a magnet and attract metallic objects such as nails or paper clips. The most common example is, appropriately, called magnetite. Because so few minerals are magnetic, magnetism is a very useful diagnostic property.

3.4.10 Feel

Some minerals feel greasy or slippery when you rub your fingers over them. They feel greasy because their chemical bonds are so weak in one direction that the pressure of your fingers is enough to break them and slide planes of atoms past one another. Talc and graphite are common examples.

3.4.11 Taste

Yes, geologists sometimes taste minerals! Taste is a *chemical* property, determined by the presence of certain elements. The most common example is halite (common salt), which tastes salty because it contains the chloride ion (Cl^-). **Do not taste minerals in your set unless instructed to do so!** We taste minerals only after we have narrowed the possibilities down to a few for which taste would be the diagnostic property. Why not taste every mineral? Because some taste bitter (such as sylvite, KCl), some are poisonous, and some might already be covered in other students' germs!

3.4.12 Odor

Geologists use all of their senses to identify minerals. A few minerals, and the streak of a few others, have a distinctive odor. For example, the streak of minerals containing sulfur smells like rotten eggs, and the streak of some arsenic minerals smells like garlic.

3.4.13 Reaction with Dilute Hydrochloric Acid

Many minerals containing the carbonate ion (CO_3^{2-}) effervesce (fizz) when dilute hydrochloric acid is dropped on them. The acid frees carbon dioxide from the mineral, and the bubbles of gas escaping through the liquid acid produce the fizz.

3.4.14 Tenacity

Tenacity refers to how materials respond to being pushed, pulled, bent, or sheared. Most adjectives used to describe tenacity are probably familiar: *malleable* materials can be bent or hammered into a new shape; *ductile* materials can be pulled into wires; *brittle* materials shatter when hit hard; and *flexible* materials bend. Flexibility is a diagnostic property for some minerals. After being bent, thin sheets of *elastic* minerals return to their original unbent shape, but sheets of flexible minerals retain the new shape.

3.5 Identifying Mineral Specimens

You are now ready to use these physical properties to identify minerals. There are more than 4,000 minerals, but only 30 occur commonly, and an even smaller number make up most of the Earth's crust—the part of the geosphere with which we are most familiar. Identifying unknown minerals is easiest if you follow the systematic approach used by geologists.

STEP 1 Assemble the equipment available in most geology classrooms to study minerals:

- A glass plate, penny, and knife or steel nail to test hardness
- A ceramic streak plate to test streak and hardness
- A magnifying glass, hand lens, or microscope to help determine cleavage
- Dilute hydrochloric acid to identify carbonate minerals
- A magnet to identify magnetic minerals

STEP 2 Observe or measure each mineral specimen's physical properties. Profile the properties on a standardized data sheet like those at the end of the chapter, as in the example in **TABLE 3.2.**

STEP 3 Eliminate from consideration all minerals that do not have the properties you have recorded. This can be done systematically by using a flowchart (**FIG. 3.7**) that asks key questions in a logical sequence, so that each answer eliminates entire groups of minerals until only a few remain (one, if you're lucky). Or you can use a determinative table in which each column answers the same questions, like those at the branches of a flowchart. Appendices 3.1 and 3.2 at the end of this chapter provide flowcharts and determinative tables. Experiment to find out which tool works best for you. Follow these steps to complete Exercise 3.7.

TABLE 3.2 Sample profile of a mineral's physical properties

Specimen number	Luster	Color	Hardness	Breakage	Other diagnostic properties
1	Metallic	Dark gray	Less than a fingernail	Excellent cleavage in one direction	Leaves a mark on a sheet of paper

FIGURE 3.7 Flowchart showing the steps used to identify minerals.

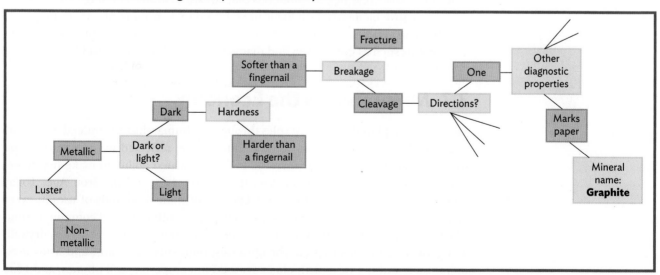

Note: For simplicity, only the path for the unknown mineral is shown.

Name: _____ Section: _____

Course: _____ Date: _____

Your instructor will provide a set of minerals to be identified. (*Note:* If taking this lab online, and directed by your instructor, use the specimen images provided by your instructor in lieu of mineral specimens.) Observe or measure the properties of each specimen and record the profiles of their properties in the data sheets at the end of the chapter. Then either use the flowcharts (Appendix 3.1) or determinative tables (Appendix 3.2) at the end of the chapter to identify each mineral. If there is more than one possibility, look at Appendix 3.3 for additional information.

3.6 Mineral Classification

The minerals in your set were chosen because they are important rock-forming minerals that make up much of the geosphere, are economically valuable resources, or illustrate the physical properties used to study minerals. Geologists classify all minerals into a small number of groups based on their chemical composition. These groups include:

- *silicates,* such as quartz, feldspars, amphiboles, and pyroxenes, which contain silicon and oxygen. Silicates are divided into *ferromagnesian* minerals, which contain iron and magnesium, and *nonferromagnesian* minerals, which do not contain those elements.
- *oxides,* such as magnetite (Fe_3O_4) and hematite (Fe_2O_3), in which a cation (positively charged ion) is bonded to oxygen anions (negatively charged ions).
- *sulfides,* such as pyrite (FeS_2) and sphalerite (ZnS), in which a cation is bonded to sulfur anions.
- *sulfates,* such as gypsum ($CaSO_4 \cdot 2H_2O$), in which calcium cations are bonded to the sulfate anion (SO_4^{2-}).
- *halides,* such as halite (NaCl) and fluorite (CaF_2), in which cations are bonded to halogen anions (elements like fluorine and chlorine in the second column from the right in the periodic table).
- *carbonates,* such as calcite ($CaCO_3$) and dolomite ($CaMg[CO_3]_2$), containing the carbonate anion (CO_3^{2-}).
- *native elements,* which are minerals that consist of atoms of a single element. The native elements most likely to be found in your mineral sets are graphite (carbon), sulfur, and copper. You are unlikely to find more valuable native elements such as gold, silver, and diamond (which is carbon, just like graphite).

3.7 Minerals and the Economy

Minerals have played important roles throughout human history. Indeed, mileposts in the evolution of technology after the Stone Age, reflected in the names of the subsequent eras, indicate our increasingly sophisticated ability to make new materials from minerals: the Copper Age, the Bronze Age, and the Iron Age. Today minerals are more valuable than ever because of their thousands of uses in every area of manufacturing, construction, agriculture, health care, and communication. What would our lives be like without the steel, aluminum, copper, concrete, drywall, and glass used in construction; the abrasives, pigments, colorants, and acids used in industry; and the rare-earth elements used in computers, cell phones, and LED lighting? The U.S. Geological Survey estimates that in 2018, the value of mineral

Name: _____   Section: _____
Course: _____   Date: _____

In the following table, indicate what physical property you think would make a mineral appropriate for the use indicated, and name a mineral from your set that could be used for this purpose. In some instances, more than one mineral will meet the requirements and more than one property is required.

Economic use	Physical property or properties needed	Mineral(s)
Abrasives (e.g., sandpaper)		
Old-time window materials before glass was widely available		
Modern windowpanes		
Writing on paper		
Lubricant for locks		
After-bath powder		
Bright eye shadow		
Pigment for paints		
Navigation with a compass		

mining and processing amounted to a little more than 3 *trillion* dollars, almost 15% of the national economy.

3.7.1 What Makes Minerals Valuable?

Geologists, engineers, and financiers classify useful minerals as **ore minerals** (those from which we extract metals) or **industrial minerals** (those useful in manufacturing processes, health care, agriculture, and the arts, but not as sources of metals). A mineral's chemical composition and physical properties determine how it can be used. For the ore minerals, composition is the critical factor: we separate the desired metal from the other elements in those minerals to extract iron, copper, zinc, lead, aluminum, titanium, chromium, and so forth. Conversely, physical properties such as hardness, color, luster, and specific gravity are responsible for the usefulness of many industrial minerals.

3.7.2 Economic Mineral Deposits

Few minerals form alone. Most are combined with others in rocks, so it takes time, energy, and a lot of money to separate them. For a mineral deposit to be economically valuable, there must be:

- a sufficient amount of ore containing the desired mineral or element (no one will mine a mere handful of iron ore).
- the ability to mine that ore at a cost low enough to make a profit.
- the ability to separate out and ship the desired mineral or element at a cost low enough to make a profit.

The minerals and elements we most rely on are typically present in small amounts in the crust, in quantities far too small to mine profitably. Economically viable mineral deposits are found where the rock-forming processes that we will explore in Chapter 4 concentrate the desired elements in amounts far beyond their average abundances. Such deposits are rare and are not distributed equally across the globe. Locations of selected mineral deposits in the United States are shown in **FIGURE 3.8**, and the economic values of 2018 production are shown in **FIGURE 3.9**. (To learn more about the status of the mineral commodity industry, you can examine the annual summaries published by the United States Geological Survey, available online at https://minerals.usgs.gov/minerals/pubs/mcs/.)

FIGURE 3.8 Locations of selected mineral deposits in the contiguous United States.

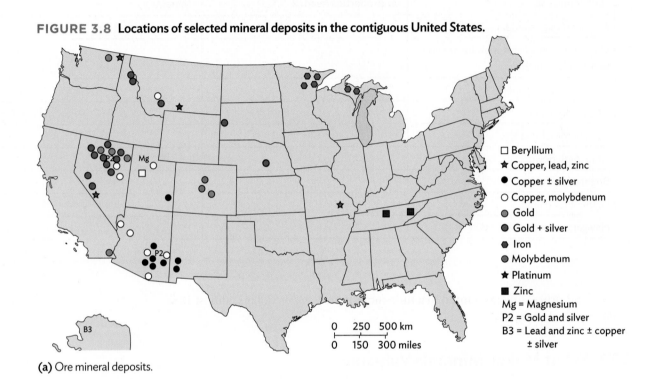

(a) Ore mineral deposits.

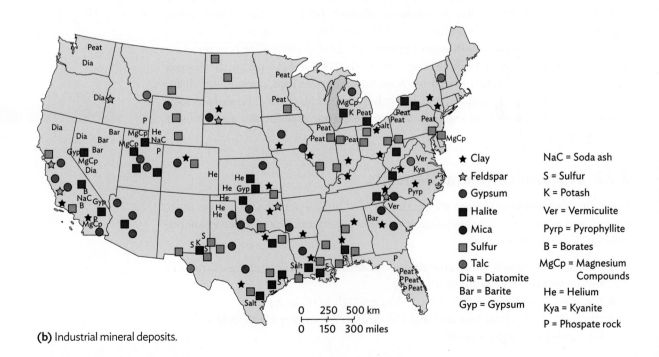

(b) Industrial mineral deposits.

FIGURE 3.9 Value of U. S. mineral mining, 2018.

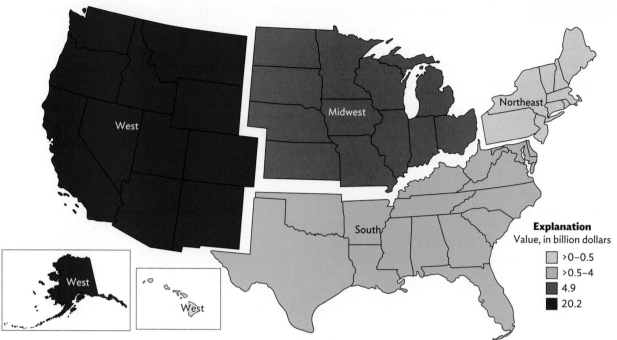

(a) The value of ore metals produced in 2018 per U.S. region (in billions of dollars).

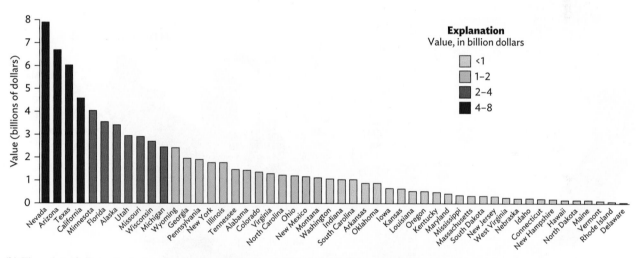

(b) The value of all non-fuel minerals per state (in billions of dollars).

EXERCISE 3.9   **Mineral Resources in Your State**

Name: _____   Section: _____
Course: _____   Date: _____

Answer the following questions using Figures 3.8 and 3.9.

(a) What minerals are being mined in the state in which you are taking this course?

_____

_____

_____

(continued)

Name: _____   Section: _____
Course: _____   Date: _____

(b) What region of the United States supplies most of the metallic ores?

_____

_____

_____

(c) Compare the value of all of the states' non-fuel mineral resources with the values for gross state product (GSP) shown in the following map.

What percentage of your state's GSP does mining provide? _____

For which state is mining the greatest contributor to GSP? _____

The smallest? _____

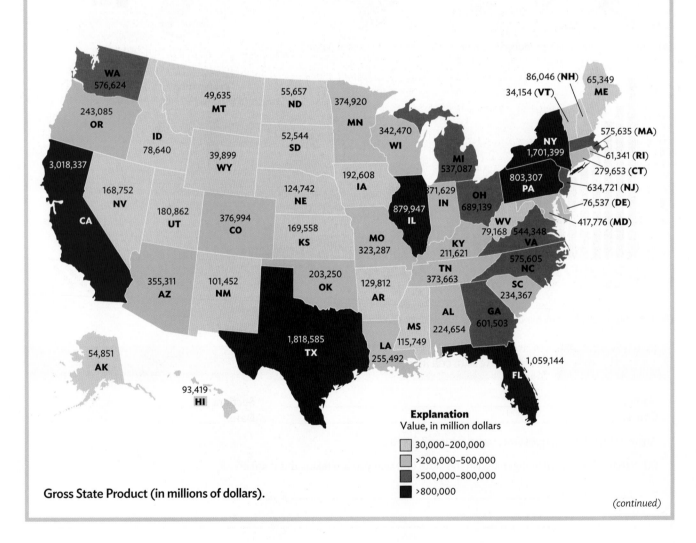

Gross State Product (in millions of dollars).

Explanation
Value, in million dollars

- 30,000–200,000
- >200,000–500,000
- >500,000–800,000
- >800,000

(continued)

Name: _____   Section: _____

Course: _____   Date: _____

? What Do You Think It took ancestral humans thousands of years to progress from the Stone Age to the Copper Age, and thousands more to reach the Iron Age. The rate of technological advance today is astonishingly faster than anything that those ancestors could have imagined, and even faster than what we thought possible only 50 years ago. Electrification of cities happened only a little more than 100 years ago, as did the beginnings of air travel. Interplanetary exploration was science fiction in the 1960s and 70s. The "two-way wrist radio" introduced in the Dick Tracy comic strip in 1946 was considered to be wildly imaginative fiction—but the cell phone and Apple Watch are now almost universal.

As technology advances, some materials that once had no commercial value have become important resources, and some that once were important resources now have no significant economic value. It is hard to predict what the next "new" resource will be, but it is possible to understand from

recent experiences the impact of some of these changes. Among recent industrial changes, aluminum studs are replacing 2" x 4" lumber as the preferred construction material in many modern buildings; copying machines and chemically treated paper have made carbon paper obsolete (have you seen any lately?); and incandescent (and soon fluorescent) light bulbs will soon become as extinct as the dodo or *Tyrannosaurus*. Graphite fiber is replacing metals in airplane and automobile construction.

Consumers adjust rapidly to these changes, but local economies do not, with thousands losing their jobs as entire industries become obsolete. Has your community been affected negatively by or benefited from changes in the demand for natural resources? How has it tried to adjust? What difference has it made in your life? Answer on a separate sheet of paper.

For information on what materials your state produces, visit: https://www.usgs.gov/centers/nmic/state-minerals-statistics-and-information.

Name: _____   Section: _____

Course: _____   Date: _____

Exploring Earth Science Using Google Earth

1. Visit **digital.wwnorton.com/labmanualearthsci**
2. Go to the **Geotours** tile to download Google Earth Pro and the accompanying Geotours exercises file.

Expand the Geotour03 folder in Google Earth by clicking the triangle to the left of the folder icon. Check and double-click the Banded Iron Formation icon to fly to a position south of Ishpeming, Michigan. Here, the placemark shows a large quarry where Fe-rich minerals are being extracted from the ground as a source of iron. In this area, metallic gray Fe-rich minerals are commonly interbedded with jasper (red microcrystalline quartz) to form the beautiful (and often folded) banded iron formation shown in the placemark photo.

(a) Dark gray Fe-rich minerals in banded iron formations typically are either magnetite (Fe_3O_4) or specular hematite (Fe_2O_3). To differentiate between the two, recall that color typically isn't diagnostic for minerals, but that streak on unglazed porcelain tends to be relatively consistent for a given mineral regardless of its color. Specular hematite, like other forms of hematite, produces a prominent red-brown streak, whereas magnetite typically exhibits a dark gray/black streak. Check and double-click the zoomable photo labeled Streak. Given the color of the streak in the zoomable photo, which dark gray Fe-rich mineral is present in the banded iron formation? _____

Minerals are incredibly valuable in our day-to-day lives. However, their extraction often comes at a price to our environment (use Google Earth to tour the open pits and waste spoil piles that define the quarry).

(b) Check and double-click the Pit Floor and Pit Rim placemarks. In order to get a sense of scale of these landscape scars, pause your cursor over each placemark and note their elevations (**elev** is given in meters along the bottom right of the Google Earth viewer window). Which range best describes the depth of this pit (i.e., their difference in elevation)? Less than 100 m? 100–200 m? 200–300 m? 300–400 m? More than 400 m? _____ m

(c) Fly back out to a regional perspective. What other ecological impact of mining is prominent in this area?

_____

_____

In these appendices, we present two ways of identifying or determining minerals. First, in Appendix 3.1, we provide flowcharts for students who may be more visually oriented. Then, in Appendix 3.2, we present the same information in the more standard determinative tables. Because both students and geologists think and work in different ways, we felt it was important to provide both options. Use whichever works best for you!

Mineral Identification Flowcharts

A. Minerals with Metallic Luster

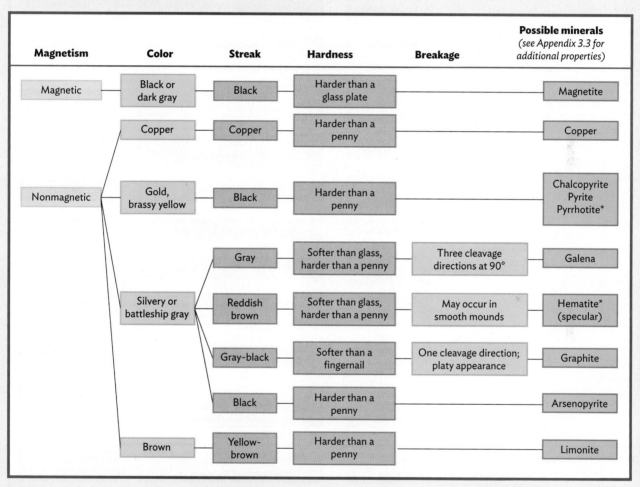

* Pyrrhotite and hematite are sometimes weakly magnetic.

Mineral Identification Flowcharts

B. Minerals with Nonmetallic Luster, Dark Colored

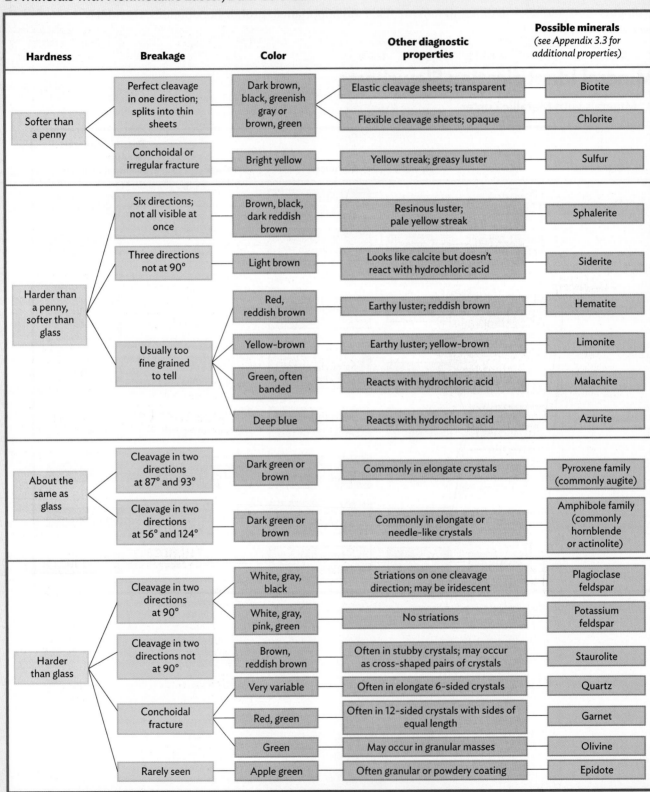

Mineral Identification Flowcharts

C. Minerals with Nonmetallic Luster, Light Colored

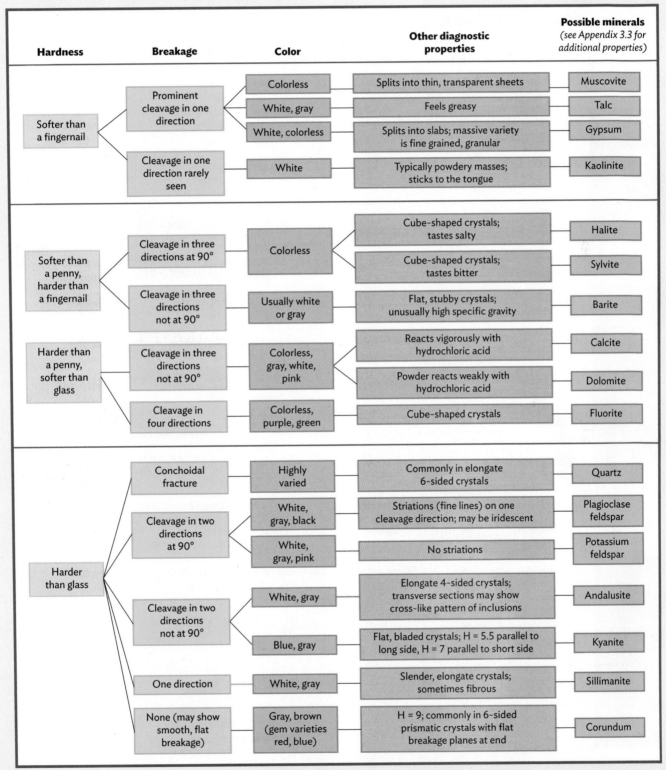

Hardness	Breakage	Color	Other diagnostic properties	Possible minerals (see Appendix 3.3 for additional properties)
Softer than a fingernail	Prominent cleavage in one direction	Colorless	Splits into thin, transparent sheets	Muscovite
		White, gray	Feels greasy	Talc
		White, colorless	Splits into slabs; massive variety is fine grained, granular	Gypsum
	Cleavage in one direction rarely seen	White	Typically powdery masses; sticks to the tongue	Kaolinite
Softer than a penny, harder than a fingernail	Cleavage in three directions at 90°	Colorless	Cube-shaped crystals; tastes salty	Halite
			Cube-shaped crystals; tastes bitter	Sylvite
	Cleavage in three directions not at 90°	Usually white or gray	Flat, stubby crystals; unusually high specific gravity	Barite
Harder than a penny, softer than glass	Cleavage in three directions not at 90°	Colorless, gray, white, pink	Reacts vigorously with hydrochloric acid	Calcite
			Powder reacts weakly with hydrochloric acid	Dolomite
	Cleavage in four directions	Colorless, purple, green	Cube-shaped crystals	Fluorite
Harder than glass	Conchoidal fracture	Highly varied	Commonly in elongate 6-sided crystals	Quartz
	Cleavage in two directions at 90°	White, gray, black	Striations (fine lines) on one cleavage direction; may be iridescent	Plagioclase feldspar
		White, gray, pink	No striations	Potassium feldspar
	Cleavage in two directions not at 90°	White, gray	Elongate 4-sided crystals; transverse sections may show cross-like pattern of inclusions	Andalusite
		Blue, gray	Flat, bladed crystals; H = 5.5 parallel to long side, H = 7 parallel to short side	Kyanite
	One direction	White, gray	Slender, elongate crystals; sometimes fibrous	Sillimanite
	None (may show smooth, flat breakage)	Gray, brown (gem varieties red, blue)	H = 9; commonly in 6-sided prismatic crystals with flat breakage planes at end	Corundum

Determinative Tables for Systematic Mineral Identification

Sequence of questions: Luster? Approximate hardness? Streak? Breakage? Color? Other?

TABLE 1 Minerals with metallic luster

			(a) Hardness less than 2.5 (softer than a fingernail)		
Streak	**Cleavage or fracture**	**H**	**Color**	**Other diagnostic properties**	**Mineral name (composition)**
Black	Perfect cleavage in one direction	1	Dark gray-black	Greasy feel; leaves a mark on paper; SpG = 2.23	Graphite C
Yellow-brown	Difficult to see	—	Yellow-brown	Very rarely in masses with metallic luster; more commonly dull, earthy; SpG = 3.6–4	Limonite $FeO(OH) \cdot nH_2O$
			(b) Hardness between 2.5 and 5.5 (harder than a fingernail; softer than glass)		
Streak	**Cleavage or fracture**	**H**	**Color**	**Other diagnostic properties**	**Mineral name (composition)**
Gray	Three directions at 90° angles	2.5	Lead gray	Commonly in cubic crystals; SpG = 7.4–7.6	Galena PbS
Black	Rarely seen	3	Bronze-brown when fresh	Commonly with purplish, iridescent tarnish; SpG = 5.06–5.08	Bornite Cu_5FeS_4
	Rarely seen	3.5–4	Brassy yellow	Often tarnished; similar to pyrite but not in cubes	Chalcopyrite $CuFeS_2$
	Rarely seen	4	Brown-bronze	Slightly magnetic; SpG = 4.62	Pyrrhotite $Fe_{1-x}S$
Copper-red	Rarely seen	2.5–3	Copper	Often in branching masses; SpG = 8.9	Copper (Cu)
			(c) Hardness greater than 5.5 (harder than glass; cannot be scratched by a knife)		
Streak	**Cleavage or fracture**	**H**	**Color**	**Other diagnostic properties**	**Mineral name (composition)**
Black	Conchoidal fracture	6–6.5	Brassy yellow	Commonly in 12-sided crystals or cubes with striated faces; SpG = 5.02	Pyrite FeS_2
	Rarely seen	6	Iron black	Strongly magnetic; SpG = 5.18	Magnetite Fe_3O_4
	Rarely seen	5.5–6	Silver white	Streak smells like garlic because of arsenic; SpG = 6.07	Arsenopyrite FeAsS
Reddish brown	Rarely seen	5.5–6.5	Black, red	Black variety is metallic; red variety is more common and has nonmetallic, earthy luster	Hematite Fe_2O_3

Determinative Tables for Systematic Mineral Identification

TABLE 2 Minerals with nonmetallic luster

(a) Hardness less than 2.5 (softer than a fingernail)					
Streak	**Cleavage or fracture**	**H**	**Color**	**Other diagnostic properties**	**Mineral name (composition)**
Yellow	Conchoidal or uneven fracture	1.5–2.5	Bright yellow	Resinous luster; SpG = 2.05–2.09	**Sulfur** S
White or colorless	Perfect cleavage in one direction	2–2.5	Colorless, light tan, yellow	Can be peeled into transparent, elastic sheets; SpG = 2.76–2.88	**Muscovite** $KAl_2(AlSi_3O_{10})(OH)_2$
	Perfect cleavage in one direction	1	Green, gray, white	Greasy feel; may occur in irregular masses (soapstone); SpG = 2.7–2.8	**Talc** $Mg_3Si_4O_{10}(OH)_2$
	Perfect cleavage in one direction; may show two other directions not at 90°	2	Colorless, white, gray	Occurs in clear crystals or gray or white, earthy masses (alabaster); SpG = 2.32	**Gypsum** $CaSO_4 \cdot 2H_2O$
	Three directions at 90°	2	Colorless, white	Cubic crystals like halite but has very bitter taste; SpG = 1.99	**Sylvite** KCl
	Perfect in one direction, but rarely seen	2–2.5	White	Usually in dull, powdery masses that stick to the tongue; SpG = 2.6	**Kaolinite** $Al_2Si_2O_5(OH)_4$
	Perfect in one direction, but not always visible	2–5	Green, white	Platy and fibrous (asbestos) varieties; greasy luster; SpG = 2.5–2.6	**Serpentine** $Mg_3Si_2O_5(OH)_4$
	—	—	White, brown, gray	Not really a mineral; rock often made of small, spherical particles containing several clay minerals; SpG = 2–2.55	**Bauxite** Mixture of aluminum hydroxides
Brown or green	Perfect cleavage in one direction	2.5–3	Brown, black, green	Can be peeled into thin, elastic sheets; SpG = 2.8–3.2	**Biotite** $K(Fe,Mg)_3AlSi_3O_{10}(OH)_2$
	Perfect cleavage in one direction	2–2.5	Green, dark green	A mica-like mineral, but sheets are flexible, not elastic; SpG = 2.6–3.3	**Chlorite** Complex Fe-Mg sheet silicate
(b) Hardness between 2.5 and 5.5 (harder than a fingernail; softer than glass)					
Streak	**Cleavage or fracture**	**H**	**Color**	**Other diagnostic properties**	**Mineral name (composition)**
Green	—	3.5–4	Bright green	Occurs in globular or elongate masses; reacts with HCl; SpG = 3.9–4.03	**Malachite** $Cu_2CO_3(OH)_2$

Determinative Tables for Systematic Mineral Identification

TABLE 2 Minerals with nonmetallic luster (*continued*)

Streak	Cleavage or fracture	H	Color	Other diagnostic properties	Mineral name (composition)
Blue	—	3.5	Intense blue	Often in platy crystals or spherical masses; reacts with HCl; SpG = 3.77	**Azurite** $Cu_3(CO_3)_2(OH)_2$
Reddish brown	—	—	Reddish brown	Usually in earthy masses; also occurs as black, metallic crystals; SpG = 5.5–6.5	**Hematite** Fe_2O_3
Yellow-brown	—	—	Brown, tan	Earthy, powdery masses and coatings on other minerals; SpG = 3.6–4	**Limonite** $FeO(OH) \cdot nH_2O$
	Three directions not at 90°	3.5–4	Light to dark brown	Often in rhombic crystals; reacts with hot HCl; SpG = 3.96	**Siderite** $FeCO_3$
	Six directions, few of which are usually visible	3.5	Brown, white, yellow, black, colorless	Resinous luster; SpG = 3.9–4.1	**Sphalerite** ZnS
White or colorless	Three directions at 90°	2.5	Colorless, white	Cubic crystals or massive (rock salt); salty taste; SpG = 2.5	**Halite** $NaCl$
	Three directions not at 90°	3	Varied; usually white or colorless	Rhombic or elongated crystals; reacts with HCl; SpG = 2.71	**Calcite** $CaCO_3$
	Three directions not at 90°	3.5–4	Varied; commonly white or pink	Rhombic crystals; *powder* reacts with HCl but crystals may not; SpG = 2.85	**Dolomite** $CaMg(CO_3)_2$
	Three directions at 90°	3–3.5	Colorless, white	SpG = 4.5 (unusually high for a nonmetallic mineral)	**Barite** $BaSO_4$
	Four directions	4	Colorless, purple, yellow, blue, green	Often in cubic crystals; SpG = 3.18	**Fluorite** CaF_2
	One direction, poor	5	Usually green or brown	Elongate 6-sided crystals; may be purple, blue, colorless; SpG = 3.15–3.20	**Apatite** $Ca_5(PO_4)_3(OH,Cl,F)$

(c) Hardness between 5.5 and 9 (harder than glass or a knife; softer than a streak plate)

Streak	Cleavage or fracture	H	Color	Other diagnostic properties	Mineral name (composition)
	Conchoidal fracture	7	Red, green, brown, black, colorless, pink, orange	12-sided crystals with sides of equal length; SpG = 3.5–4.3	**Garnet family** Complex Ca, Fe, Mg, Al, Cr, Mn silicate
	Two directions at 90°	6	Colorless, salmon, gray, green, white	Stubby prismatic crystals; three polymorphs: orthoclase, microcline, sanidine; may show exsolution lamellae; SpG = 2.54–2.62	**Potassium feldspar** $KAlSi_3O_8$

Determinative Tables for Systematic Mineral Identification

TABLE 2 **Minerals with nonmetallic luster** *(continued)*

Streak	Cleavage or fracture	H	Color	Other diagnostic properties	Mineral name (composition)
White or colorless	Two directions at 90°	6	Colorless, white, gray, black	Striations (fine lines) on one of the two cleavage directions; solid solution between Na (albite) and Ca (anorthite) plagioclase; SpG = 2.62–2.76	**Plagioclase feldspar** $CaAl_2Si_2O_8$, $NaAlSi_3O_8$
	Conchoidal fracture	7	Colorless, pink, purple, gray, black, green, yellow	Elongate 6-sided crystals; SpG = 2.65	**Quartz** SiO_2
	Rarely seen	7.5	Gray, white, brown	Elongate 4-sided crystals; SpG = 3.16–3.23	**Andalusite** Al_2SiO_5
	One direction	6–7	White, rarely green	Long, slender crystals, often fibrous; SpG = 3.23	**Sillimanite** Al_2SiO_5
	One direction	5 *and* 7	Blue, blue-gray to white	Bladed crystals; *two hardnesses*: H = 5 parallel to long direction of crystal, H = 7 across the long direction	**Kyanite** Al_2SiO_5
Colorless to light green	Conchoidal fracture	6.5–7	Most commonly green	Stubby crystals and granular masses; solid solution between Fe (fayalite) and Mg (forsterite)	**Olivine family** Fe_2SiO_4, Mg_2SiO_4
	Two directions at 56° and 124°	5–6	Dark green to black	An amphibole with elongate crystals; SpG = 3–3.4	**Hornblende** Complex double-chain silicate with Ca, Na, Fe, Mg
	Two directions at 56° and 124°	5–6	Pale to dark green	An amphibole with elongate crystals; SpG = 3–3.3	**Actinolite** Double-chain silicate with Ca, Fe, Mg
	Two directions at 87° and 93°	5–6	Dark green to black	A pyroxene with elongate crystals; SpG = 3.2–3.3	**Augite** Single-chain silicate with Ca, Na, Mg, Fe, Al
	One direction perfect, one poor; not at 90°	6–7	Apple green to black	Elongate crystals and fine-grained masses; SpG = 3.25–3.45	**Epidote** Complex twin silicate with Ca, Al, Fe, Mg
No streak; mineral scratches streak plate	One direction, imperfect	7.5–8	Blue-green, yellow, pink, white; gem variety: emerald (green)	6-sided crystals with flat ends; SpG = 2.65–2.8;	**Beryl** $Be_3Al_2Si_6O_{18}$
	—	7–7.5	Dark brown, reddish brown, brownish black	Stubby or cross-shaped crystals; SpG = 3.65–3.75	**Staurolite** Hydrous Fe, Al silicate
	No cleavage	9	Gray, light brown; gem varieties red (ruby), blue (sapphire)	6-sided crystals with flat ends	**Corundum** Al_2O_3

Common Minerals and Their Properties

Mineral	Additional diagnostic properties and occurrences
Actinolite	Elongate green crystals; cleavage at 56° and 124°; H = 5.5–6. An amphibole found in metamorphic rocks.
Amphibole*	Stubby rod-shaped crystals common in igneous rocks; slender crystals common in metamorphic rocks; two cleavage directions at 56° and 124°.
Andalusite	Elongate gray crystals with rectangular cross sections.
Apatite	H = 5; pale to dark green, brown, white; white streak; 6-sided crystals.
Augite	H = 5.5–6; green to black rod-shaped crystals; cleavage at 87° and 93°. A pyroxene common in igneous rocks.
Azurite	Deep blue; reacts with HCl. Copper will plate out on a steel nail dipped into a drop of HCl and placed on this mineral.
Barite	H = 3–3.5; SpG is unusually high for a nonmetallic mineral.
Bauxite	Gray-brown earthy *rock* commonly containing spherical masses of clay minerals and mineraloids.
Beryl	6-sided crystals; H = 7.5–8.
Biotite	Dark-colored mica; one perfect cleavage into flexible sheets.
Bornite	High SpG; iridescent coating on surface gives it "peacock ore" nickname.
Calcite	Reacts with HCl. Produces double image from text viewed through transparent cleavage fragments.
Chalcopyrite	Similar to pyrite, but softer and typically has iridescent tarnish.
Chlorite*	Similar to biotite, but does not break into thin, flexible sheets; forms in metamorphic rocks.
Copper	Copper-red color and high specific gravity are diagnostic.
Corundum	6-sided prismatic crystals with flat ends; hardness of 9 is diagnostic. Most lab specimens have dull luster and are gray or brown.
Dolomite	Similar to calcite, but only weak or no reaction with HCl placed on a grain of the mineral; *powder* reacts strongly. Slightly curved rhombohedral crystals.
Epidote	Small crystals and thin, granular coatings form in some metamorphic rocks and [by alteration of] some igneous rocks.
Fluorite	Commonly in cube-shaped crystals with four cleavage directions cutting corners of the cubes.
Galena	Commonly in cube-shaped crystals with three perfect cleavages at 90°.
Garnet*	Most commonly dark red; 12- or 24-sided crystals in metamorphic rocks.
Graphite	Greasy feel; leaves a mark on paper.
Gypsum	Two varieties: *selenite* is colorless and nearly transparent with perfect cleavage; *alabaster* is a rock—an aggregate of grains with an earthy luster.
Halite	Cubic crystals and taste are diagnostic.
Hematite	Two varieties: most common is reddish brown masses with earthy luster; rare variety is black crystals with metallic luster.

Common Minerals and Their Properties *(continued)*

Mineral	Additional diagnostic properties and occurrences
Hornblende	Dark green to black amphibole; two cleavages at 56° and 124°.
Kaolinite	Earthy, powdery white to gray masses; sticks to tongue.
Kyanite	Bladed blue or blue-gray crystals; H = 5 parallel to blade, H = 7 across blade.
Limonite	Earthy, yellow-brown masses, sometimes powdery; forms by the "rusting" (oxidation) of iron-bearing minerals.
Magnetite	Gray-black; H = 6; magnetic.
Malachite	Bright green. Copper will plate out on a steel nail dipped into a drop of HCl and placed on this mineral.
Muscovite	A colorless mica; one perfect cleavage; peels into flexible sheets.
Olivine	Commonly as aggregates of green granular crystals.
Plagioclase feldspar*	Play of colors and striations distinguish plagioclase feldspars from potassium feldspars, which do not show these properties.
Potassium feldspar*	
Pyrite	Brassy gold color, hardness, and black streak are diagnostic. Cubic crystals with striations on their faces or in 12-sided crystals with 5-sided faces.
Pyrrhotite	Brownish bronze color; black streak; may be slightly magnetic.
Pyroxene*	Two cleavages at 87° and 93°; major constituent of mafic and ultramafic igneous rocks.
Quartz	Wide range of colors; 6-sided crystal shape, high hardness (7) and conchoidal fracture are diagnostic.
Serpentine	Dull white, gray, or green masses; sometimes fibrous (asbestos).
Siderite	Three cleavages not at 90°; looks like brown calcite; powder may react to HCl.
Sillimanite	Gray, white, brown; slender crystals, sometimes needle-like.
Sphalerite	Wide variety of colors (including colorless); distinctive pale yellow streak.
Staurolite	Reddish brown to dark brown stubby crystals in metamorphic rocks.
Sulfur	Bright yellow with yellow streak; greasy luster.
Sylvite	Looks like halite but has very bitter taste.
Talc	Greasy feel; H = 1.

*Indicates mineral family.

MINERAL PROFILE DATA SHEET   Name _____

Sample	Luster	Hardness	Streak	Color	Cleavage or fracture (describe)	Other properties	Mineral name and composition

MINERAL PROFILE DATA SHEET

Name _____

Sample	Luster	Hardness	Streak	Color	Cleavage or fracture (describe)	Other properties	Mineral name and composition

MINERAL PROFILE DATA SHEET   Name _____

Sample	Luster	Hardness	Streak	Color	Cleavage or fracture (describe)	Other properties	Mineral name and composition

MINERAL PROFILE DATA SHEET

Name _____

Sample	Luster	Hardness	Streak	Color	Cleavage or fracture (describe)	Other properties	Mineral name and composition

MINERAL PROFILE DATA SHEET

Name _____

Sample	Luster	Hardness	Streak	Color	Cleavage or fracture (describe)	Other properties	Mineral name and composition

Rocks and
the Rock Cycle

4

The three classes of rocks, from left to right: sedimentary (conglomerate), metamorphic (gneiss), and igneous (granite).

LEARNING OBJECTIVES

- Describe the differences between minerals and rocks
- Identify the three main classes of rock: igneous, sedimentary, metamorphic
- Describe the different processes that form rocks
- Explain how the grains in a rock act as records of geologic processes
- Use textures and mineral content to determine whether rocks are igneous, sedimentary, or metamorphic
- Describe the rock cycle and how rocks are recycled to make new ones

MATERIALS NEEDED

- An assortment of igneous, sedimentary, and metamorphic rocks
- Supplies to make artificial igneous, sedimentary, and metamorphic rocks: forceps or tongs for moving hot petri dishes, hot plate, sugar, thymol, sodium acetate in a dropper bottle, sand grains, glass petri dishes, calcium hydroxide solution, straws, modeling clay (e.g., Play-Doh or Silly Putty), two beakers, seawater or homemade saltwater, plastic coffee stirring rods, plastic chips
- Magnifying glass or hand lens

4.1 Introduction

A rock is an aggregate of mineral grains, fragments of previously existing rock, or a mass of natural glass. Rocks form in several ways—by cooling and solidification of a melt, cementation of loose grains, precipitation from water solutions, or changes that happen underground in response to increasing temperature and pressure. Geologists and other Earth scientists use two principal characteristics of rocks to evaluate how they formed and to interpret aspects of the Earth's history: their composition (the identity of minerals or glass that make up a rock) and their texture (the sizes and shapes of grains, and the ways that grains are arranged, oriented, and held together). In this chapter, you will first be introduced to the three basic classes of rock, then learn how each of them forms by making your own "rocks." These simulations show that rock-forming processes leave textural evidence that enables us to identify which class a rock belongs to and, in some instances, the conditions under which it formed. The ultimate goal is to combine your skills in mineral identification with this new textural knowledge to correctly group rock samples into the three classes. On our dynamic planet, rocks don't survive forever. The Earth is the original recycler, reusing materials from one rock to form new ones. This section will also introduce you to the rock cycle, the complex interaction among rock-forming processes by which this recycling takes place.

4.2 The Three Classes of Rocks

Geologists struggled for centuries with the question of how to classify rocks. They finally concluded that rocks can best be classified on the basis of how they formed, and using this standard, we now group all rocks into three categories: igneous, sedimentary, and metamorphic.

- **Igneous rocks** form through the cooling and solidification of molten rock, which is created by melting of pre-existing rock in the mantle or lower crust. We refer to molten rock below the Earth's surface as *magma* and to molten rock that has erupted onto the surface as *lava*. Some volcanoes erupt explosively, blasting rock fragments into the air, and when these fragments fall back to the Earth, coalesce, and solidify, the resulting rock is also considered to be igneous.
- **Sedimentary rocks** form at or near the surface of the Earth in two basic ways: (1) when grains of pre-existing rock accumulate, are buried, and then are cemented together by minerals that precipitate from groundwater in the pore spaces between the grains; and (2) when minerals precipitate out of water near the Earth's surface, either directly or through the life function of an organism, and either form a solid mass or are cemented together later. The materials that become incorporated in sedimentary rocks form when pre-existing rocks are broken down into pieces by processes involving interactions with air, water, and living organisms. These interactions are called *weathering*. Some weathering simply involves the physical fragmentation of rock. Other kinds of weathering involve chemical reactions that produce new minerals, most notably clay. The products of weathering can be transported (eroded) by water, ice, or wind to the site where they are deposited, buried, and transformed into new rock.
- **Metamorphic rocks** form when pre-existing rock is subjected to physical and chemical conditions within the Earth different from those under which the original rock formed: increased pressure and temperature and/or shearing at elevated temperatures. For example, when buried very deeply, rock is warmed

to high temperatures and squeezed by high pressure. The texture and/or mineral content of the pre-existing rock changes in response to the new conditions, but the rock remains a coherent solid and does not break into smaller pieces as in sedimentary rock, or melt as in igneous rocks.

4.2.1 The Rock Cycle

An igneous rock exposed at the surface of the Earth will not last forever. Minerals that crystallized from magma to form the rock can be broken apart at the surface by weathering, undergo erosion (the grinding effects of moving ice, water, or air), be transported by streams, be deposited in the ocean, and be cemented together to form a sedimentary rock. These same minerals, now part of a sedimentary rock, can later be buried so deeply beneath other sedimentary rocks that they are heated and squeezed to form a metamorphic rock with new minerals. Erosion may eventually expose the metamorphic rock at the Earth's surface, where it can be weathered to form new sediment and eventually a different sedimentary rock. Or the metamorphic rock may be buried so deeply that it melts to produce magma and eventually becomes a new igneous rock with still different minerals. This movement of material from one rock type to another over geologic time is called the rock cycle (**FIG. 4.1**). The rock cycle involves the reuse of mineral grains or the breakdown of minerals into their constituent atoms and reuse of those atoms to make other minerals. Rocks formed at each step of the rock cycle look very different from their predecessors because of the different processes by which they formed. As Figure 4.1 shows, each rock type is capable of being transformed directly into any of the other rock types.

4.3 Igneous Rocks

Every rock has a story to tell. The story of an igneous rock begins when rock in the lower crust or upper mantle melts to form molten material called **magma**, which rises up through the crust. Some magma reaches the surface, where it flows as **lava**. Other magma explodes upward through cracks or vents in the Earth, spewing tiny particles of **volcanic ash** into the air. Igneous rock that forms from solidified lava or ash is called **extrusive** igneous rock because it comes out (extrudes) onto the Earth's surface.

FIGURE 4.1 The rock cycle.

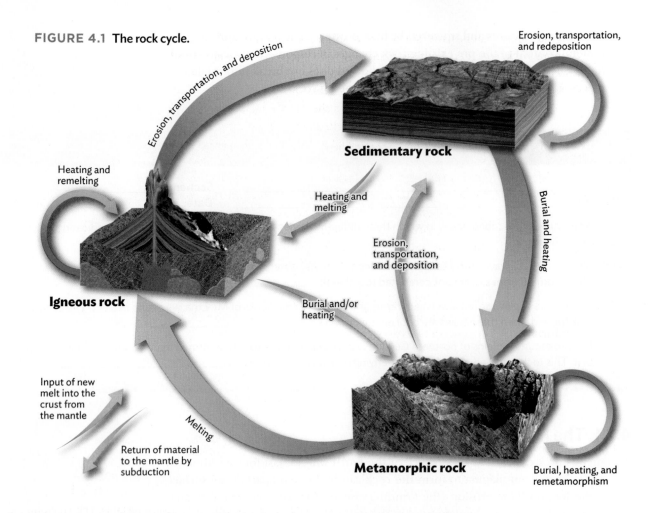

Erosion, transportation, and deposition

Erosion, transportation, and redeposition

Sedimentary rock

Heating and remelting

Heating and melting

Erosion, transportation, and deposition

Burial and heating

Igneous rock

Burial and/or heating

Input of new melt into the crust from the mantle

Return of material to the mantle by subduction

Melting

Metamorphic rock

Burial, heating, and remetamorphism

FIGURE 4.2 Intrusive and extrusive igneous rock bodies. The intrusive bodies occur below the Earth's surface, while the extrusive bodies occur above ground.

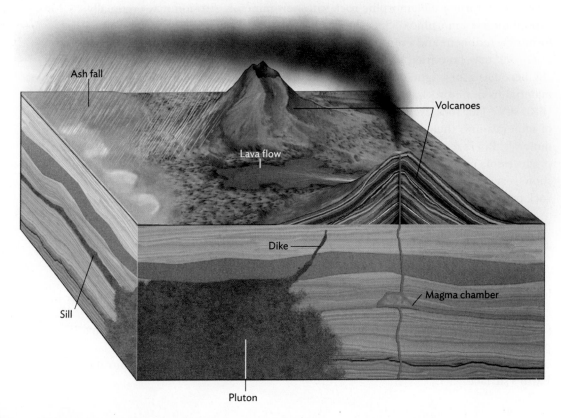

Ash fall

Volcanoes

Lava flow

Dike

Magma chamber

Sill

Pluton

Some magma never reaches the surface and solidifies underground to form **intrusive** igneous rock, so called because it squeezes into (intrudes) the surrounding rocks. Bodies of intrusive igneous rock come in many shapes. Massive blobs are called **plutons**, and the largest of these—called **batholiths**—are generally composed of several smaller plutons. Other intrusions form thin sheets that cut across layering in the **host rock** (the rock into which the magma intruded). These intrusions are called **dikes**, and those that form thin sheets parallel to the layers of host rock are called **sills** (FIG. 4.2).

When looking at an igneous rock, geologists ask several questions: Where did the rock cool (was it intrusive or extrusive)? Where in the Earth did the rock's parent magma form? In what plate-tectonic setting—ocean ridge, mid-continent, subduction zone, hot spot—did it form? Most of these questions can be answered with three simple observations: the rock's grain size, color, and specific gravity. In this section of the chapter you will learn how to answer these questions and how to identify common types of igneous rock. Few new skills are needed—just observe carefully and apply scientific reasoning.

Exercise 4.2 shows how easy the process is. The following subsections provide the additional information you need to interpret the history of an igneous rock.

EXERCISE 4.2 **A First Look at Igneous Rocks**

Name: _____   Section: _____
Course: _____   Date: _____

There are many ways to classify igneous rocks, but for now let's use three easily observable criteria: grain size, color, and specific gravity. Using the set of igneous rocks provided by your instructor, first separate the specimens into coarse- and fine-grained categories. Record the specimen number in the appropriate column in the following table. Then regroup the specimens using color and using specific gravity. (*Note:* If taking this lab online, and directed by your instructor, use the specimen images provided by your instructor in lieu of rock specimens.)

Grain size		Color		Specific gravity (heft)	
Coarse*	Fine**	Light-colored	Dark-colored	Relatively high	Relatively low

* larger than 5 mm; grains are large enough so you can identify the minerals or grains present
** smaller than 0.1 mm; individual grains are too small to be seen with the naked eye or identified with a hand lens

(a) Which two properties seem to be related to each other? Which is not related to the other two?

_____

_____

_____

(b) Based on your answer to question (a), if you know the color of an igneous rock sample, would examining its grain size or specific gravity be more useful for narrowing down the rock's identity? Why?

_____

_____

_____

4.3.1 Origin of Magmas: Where and Why Do Rocks and Minerals Melt?

Some science-fiction movies and novels suggest that the rigid outer shell of the Earth floats on a sea of magma. In fact, most of the crust and mantle is solid rock. Melting occurs only in certain places, generally by one of the following three processes:

1. *Addition of heat:* Iron in a blast furnace melts when enough heat is added to break the bonds between atoms, and some magmas form the same way in continental and oceanic crust. Some very hot magmas that rise into the crust from the mantle bring along enough heat to start melting the crust, much as hot fudge melts the ice cream it is poured on. This process occurs in subduction zones and above mantle plumes.

2. *Decreasing pressure:* Stretching of the lithosphere at a divergent plate boundary lowers the confining pressure on the mantle below. The amount of heat in the mantle cannot by itself overcome the bonds forming solid rock *and* the confining pressure. But once the pressure decreases, the heat already present in the mantle is enough to overcome the bonds and cause melting.

3. *Addition of fluids:* Some minerals in subducted oceanic crust contain water in their structures. When the subducted crust reaches a depth of about 150 km, water and other fluids are released from these minerals and rise into the asthenosphere, where they cause melting. The presence of water lowers the melting point of the rock, so no additional heat is needed.

4.3.2 Cooling Histories of Igneous Rocks and Their Textures

Imagine that you are looking at an outcrop of ancient igneous rock. That rock might have formed from volcanic debris blasted into the air, lava that frothed out of a volcano or flowed smoothly across the ground, magma that cooled just below the surface, or magma that solidified many kilometers below the surface. But millions of years have passed since the rock formed, and if a volcano was involved, it has long since been eroded away. If the rock is intrusive, kilometers of overlying rock must have been removed to expose it at the surface. How can you determine which of these possibilities is the right one?

The key to understanding the cooling history of an igneous rock is its **texture**: the size, shape, and arrangement of its grains. The texture of an igneous rock formed by the settling of ash and other volcanic fragments is very different from that of a rock formed by magma cooling deep underground or by lava cooling at the surface.

Specimens of igneous rock composed of interlocking grains—whether those grains are large enough to be identified or too small to be identified using only your eyes—are called **crystalline** rocks. Those that are shiny and contain no grains are called **glasses**. Sponge-like masses are said to be **porous**, and those that appear to have pieces cemented together are said to be **fragmental**.

Each of these textures indicates a unique cooling history—if you understand how to "read" the textural information. Exercise 4.3 will help you start this process.

4.3.2a Crystalline Igneous Textures Grain size is the key to understanding the cooling history of most igneous rocks that solidified underground or on the surface (you will experiment with these processes in Exercise 4.4). When magma or lava begins to cool, small crystal seeds form (a process called *nucleation*), and crystals grow outward from each seed until they interfere with one another. The result is the three-dimensional interlocking texture found in most igneous rocks. But why do some igneous rocks have coarser (larger) grains than others?

A First Look at the Textures of Igneous Rocks

Name: _____   Section: _____
Course: _____   Date: _____

We look first at igneous rocks that have the same chemical composition so that their different textures could only be caused by the different ways in which they cooled. For example, the rocks in Figure 4.3 contain the same minerals and thus have similar compositions. Separate the light-colored igneous rocks in your set. Describe their textures, paying careful attention to the sizes and shapes of the grains and the relationships among adjacent grains. Use everyday language—the appropriate geologic terms will be introduced later. (*Note:* If taking this lab online, and directed by your instructor, use the specimen images provided by your instructor in lieu of rock specimens.)

Specimen	Textural Description

Crystals grow as ions migrate through magma to crystal seeds, so anything that assists ionic migration increases grain size in an igneous rock. **Cooling rate** is the most important factor controlling grain size. The slower a magma cools, the more time ions have to migrate, and the larger crystals can grow; the faster it cools, the less time there is, and the smaller the grains will be. Another factor is a magma's **viscosity** (its resistance to flow). The *less* viscous a magma is (i.e., the more *fluid*), the easier it is for ions to migrate through it, and the larger the crystals can become.

Differences in cooling rate and viscosity cause some igneous rocks to have fine grains (an *aphanitic* texture), and others to have coarse grains (sometimes called a *phaneritic* texture). Extremely coarse-grained igneous rock is said to be *pegmatitic*. **FIGURE 4.3** illustrates the difference between pegmatitic, coarse, and fine grains. Some igneous rocks have grains of two different sizes, one much larger than the other (**FIG. 4.4**). These rocks are said to have a *porphyritic* texture. The larger grains are called *phenocrysts* and the smaller grains are called, collectively, the rock's *groundmass*.

FIGURE 4.3 Grain sizes in light-colored igneous rock.

(a) Very coarse-grained (pegmatitic crystals).

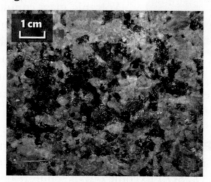

(b) Coarse-grained (phaneritic crystals).

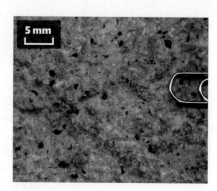

(c) Fine-grained (aphanitic crystals; most grains are too small to see with naked eye).

FIGURE 4.4 Rocks with a porphyritic texture have two different grain sizes.

Phenocrysts

(a) Large light-colored plagioclase phenocrysts in fine-grained groundmass.

(b) Microscopic view of feldspar, pyroxene, and olivine phenocrysts in very fine-grained groundmass.

Phenocrysts   Small groundmass grains

Phenocrysts   Coarse groundmass grains

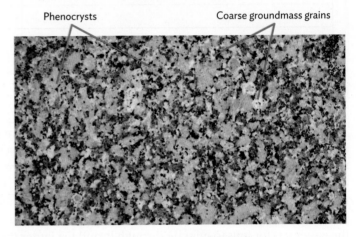

(c) Dark-colored rock with fine-grained groundmass.

(d) Light-colored rock with coarse-grained groundmass.

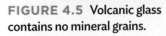

FIGURE 4.5 Volcanic glass contains no mineral grains.

4.3.2b Glassy Igneous Textures Some magmas cool so fast or are so viscous that crystal seeds can't form. Instead, atoms in the melt are frozen in place, arranged haphazardly rather than in the orderly arrangement required for mineral grains. Because haphazard atomic arrangement is also found in window glass, we call a rock that has cooled this quickly **volcanic glass**, and it is said to have a **glassy texture** (**FIG. 4.5**). Like thick glass bottles, volcanic glass fractures conchoidally.

Volcanic glass commonly looks black, but impurities may cause it to be red-brown or streaked. Most volcanic glass forms at the Earth's surface when lava exposed to air or water cools very quickly, but some may form just below the surface in the throat of a volcano. In addition, much of the ash blasted into the air during explosive eruptions is made of volcanic glass, formed when tiny particles of magma freeze instantly in the air.

4.3.2c Porous (Vesicular) Textures As magma rises toward the surface, the pressure on it decreases. This decrease in pressure allows gases dissolved in the magma (H_2O, CO_2, SO_2) to come out of solution and form bubbles, like those that appear when you open a can of soda. If bubbles form just as the lava solidifies, their shapes are preserved in the lava. The material between the bubbles may be fine-grained

Name: _____   Section: _____

Course: _____   Date: _____

In this exercise, you will conduct simple experiments to better understand rock textures.

(a) Crystalline igneous rock. Partially fill (1/2 inch) a glass petri dish with the thymol powder provided by your instructor. (*Note:* If this exercise is available and being conducted online, the setup, materials, and directions may differ. See the instructions in the online lab for how to conduct the exercise.) Heat on a hot plate until the material melts completely. Carefully remove the dish from the hot plate using forceps (or tongs) and allow the liquid to cool. Observe the crystallization process closely. You may have to add a crystal seed if crystallization does not begin in a few minutes. Congratulations—you have just made a "magma" and then an "igneous rock" with an interlocking crystalline texture.

• Describe the crystallization process. How and where did individual grains grow, and how did they eventually join with neighboring grains?

_____

_____

_____

Sketch and describe the texture of the cooled "rock."

Sketch: Describe: _____

_____

_____

_____

_____

_____

_____

_____

This crystalline texture formed when crystals in the cooling "magma" grew until they interfered with one another and eventually interlocked to form a cohesive solid. Crystals grow only when the magma cools slowly enough for atoms to have time to fit into a crystal lattice.

(b) Glassy igneous texture. Add sugar ($\frac{1}{2}$ inch) to another petri dish, melt it on a hot plate, and allow it to cool.

Describe the cooling process and sketch the resulting texture.

Sketch: Describe: _____

_____

_____

_____

_____

_____

_____

This glassy texture is typical of igneous rocks that cool so quickly that atoms do not have time to arrange into crystalline lattices. In some cooling lavas, bubbles like those you saw may be preserved as the lava freezes.

FIGURE 4.6 Porous (vesicular) igneous textures.

(a) Dark-colored porous igneous rock.

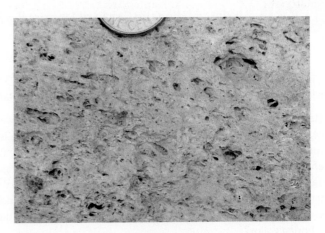

(b) Light-colored porous igneous rock.

FIGURE 4.7 Fragmental (pyroclastic) igneous rocks in a range of grain sizes and textures.

(a) Coarse volcanic bombs in a finer-grained pyroclastic matrix at Kilauea volcano in Hawaii.

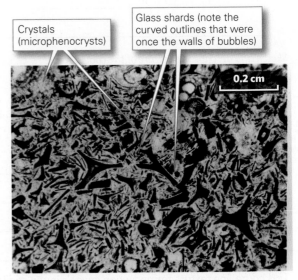

Crystals (microphenocrysts)

Glass shards (note the curved outlines that were once the walls of bubbles)

0.2 cm

(b) Photomicrograph of glass shards (black) welded together with a few tiny crystals.

Volcanic rock fragments

3.5 cm

(c) Hand sample with volcanic rock fragments in a very fine-grained brown ash.

crystalline material (**FIG. 4.6A**) or volcanic glass (**FIG. 4.6B**). These rocks are said to have a porous or vesicular texture, and the individual bubbles are called vesicles. "Lava rock" used in outdoor grills and pumice used to smooth wood and remove calluses are common and useful examples of porous igneous rocks.

4.3.2d Fragmental Textures Violent volcanic eruptions can blast enormous amounts of material into the air. This material is collectively called pyroclastic debris (from the Greek *pyro*, meaning fire, and *clast*, meaning broken), and is said to have a fragmental or pyroclastic texture. Pyroclastic debris includes volcanic glass, large blocks or bombs erupted as liquid and cooled as fine-grained crystalline rock (**FIG. 4.7A**), crystals formed in the magma before eruption (**FIG. 4.7B**), tiny ash particles, and pieces of rock broken from the walls of the volcanic vent or ripped from the ground surface during an eruption (**FIG. 4.7C**).

You are now able to read the story recorded by igneous textures and can do so without identifying a single mineral or naming the rocks. **TABLE 4.1** summarizes the origin of common igneous textures and the terms used by geologists to describe them. In Exercise 4.5 you will sharpen your skills by interpreting the history of igneous rocks provided by your instructor.

4.3.3 Four Major Compositional Groups

In Exercise 4.2, you saw that some of the igneous rocks in your set are relatively dark-colored and that others are light-colored. You also saw that some have high specific gravities, others relatively low specific gravities. A rock's color and specific gravity are controlled mostly by its minerals, which are, in turn, determined by its chemical composition. Oxygen and silicon are by far the two most abundant elements in the lithosphere, so it is not surprising that nearly all igneous rocks are composed primarily of *silicate minerals*, such as quartz, feldspars, pyroxenes, amphiboles, micas, and olivine.

TABLE 4.1 Interpreting igneous rock textures

Texture	Pegmatitic	Coarse-grained (phaneritic)	Fine-grained (aphanitic)	Porphyritic	Glassy	Porous (vesicular)	Fragmental (pyroclastic)
Description	Very large grains (>2.5 cm)	Individual grains are visible with the naked eye	Individual grains cannot be seen without magnification	A few large grains (phenocrysts) set in a finer-grained groundmass	Smooth, shiny; looks like glass; no mineral grains present	Spongy; filled with large or small holes	Mineral grains, rock fragments, and glass shards welded together
Interpretation	Very slow cooling or cooling from an extremely fluid magma (usually the latter)	Slow cooling; generally *intrusive*	Rapid cooling; generally *extrusive*	Two cooling rates: slow at first to form the phenocrysts, and then more rapid to form the groundmass	Extremely rapid cooling; generally *extrusive*	Rapid cooling accompanied by the release of gases	Explosive eruption of ash and rock into the air
Example							

Name: _____   Section: _____

Course: _____   Date: _____

Examine the igneous rocks provided by your instructor. Apply what you have learned about the origins of igneous textures to fill in the "cooling history" column in the study sheets at the end of the chapter. Use the following questions as a guide to your interpretation. (*Note:* If taking this lab online, and directed by your instructor, use the specimen images provided by your instructor in lieu of rock specimens.)

- Which specimens cooled quickly? Which cooled slowly?
- Which specimens cooled very rapidly at the Earth's surface (i.e., are extrusive)?
- Which specimens cooled slowly beneath the surface (i.e., are intrusive)?
- Which specimens cooled from a magma rich in gases?
- Which specimens experienced more than one cooling rate?

There are many different kinds of igneous rock, but they all fit into one of four compositional groups—felsic, intermediate, mafic, and ultramafic—defined by how much silicon and oxygen (*silica*) they contain and by which other elements are most abundant (**TABLE 4.2**).

- **Felsic** igneous rocks (from *fel*dspar and *si*lica) have the most silica and the least iron and magnesium. They contain abundant potassium feldspar (also called orthoclase) and sodium-rich plagioclase feldspar, commonly quartz, and only sparse ferromagnesian minerals—usually biotite or hornblende. Like their most abundant minerals, felsic rocks are light-colored and have low specific gravities.
- **Intermediate** igneous rocks have chemical compositions, colors, specific gravities, and mineral assemblages between those of felsic and mafic rocks: plagioclase feldspar with nearly equal amounts of calcium and sodium, both amphibole and pyroxene, and only rarely quartz.
- **Mafic** igneous rocks (from *ma*gnesium and the Latin *ferrum*, meaning iron) have much less silica, potassium, and sodium than felsic rocks and much more calcium, iron, and magnesium. Their dominant minerals—calcium plagioclase, pyroxene, and olivine—are dark green or black and have higher specific gravities than minerals in felsic rocks. Even fine-grained mafic rocks can therefore be recognized by their dark color and relatively high specific gravity.
- **Ultramafic** igneous rocks have the least silica and the most iron and magnesium, but very little aluminum, potassium, sodium, or calcium. As a result, they contain mostly ferromagnesian minerals such as olivine and pyroxene and very little, if any, plagioclase. Ultramafic rocks are very dark- colored and have the highest specific gravities of the igneous rocks.

TABLE 4.2 The four major compositional groups of igneous rocks

Igneous rock group	Approximate % silica (SiO2) by weight	Other major elements	Most abundant minerals
Felsic	>66	Aluminum (Al), potassium (K), sodium (Na)	K-feldspar, Na-plagioclase, quartz
Intermediate	~52–66	Al, Na, calcium (Ca), iron (Fe), magnesium (Mg)	Ca-Na-plagioclase, amphibole, pyroxene
Mafic	~45–52	Al, Ca, Mg, Fe	Ca-plagioclase, pyroxene, olivine
Ultramafic	<45	Mg, Fe	Olivine, pyroxenes

4.3.4 Identifying Igneous Rocks

The name of an igneous rock is based on its mineral content *and* texture (**FIG. 4.8**). Each of the four compositional groups of igneous rock contains rock types with coarse, fine, porphyritic, glassy, porous, and fragmental textures. Rocks in a given column of Fig. 4.8 may have exactly the same minerals but look so different that they are given different names. For example, granite and rhyolite are both felsic, but look

FIGURE 4.8 Classification of igneous rocks. To use the chart, determine the color/composition of an igneous rock sample by estimating the relative amounts of the light- and dark-colored minerals it contains, using the images as a guide. Then, go down the column under that color/composition and stop at the texture description given on the left that matches your sample. The name in the box where the color and texture descriptions meet is the name of the rock sample.

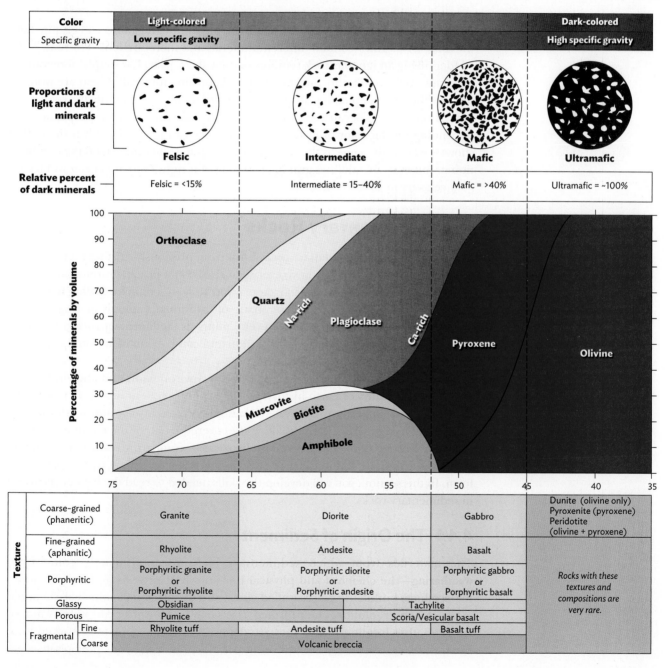

Name: _____   Section: _____

Course: _____   Date: _____

Name the specimens in your igneous rock set by following the steps below. (*Note:* If taking this lab online, and directed by your instructor, use the specimen images provided by your instructor in lieu of rock specimens.)

Step 1: Place the rock in the correct column in Figure 4.8a by noting its color and heft and estimating mineral abundances if its grains are coarse enough.

Step 2: Determine the texture of the rock and note which row in Figure 4.8b it corresponds to.

Step 3: The name of the rock can be found in the box at the intersection of the column and row.

Record the names of the rocks on the study sheets at the end of the chapter.

different because of their different textures. And although gabbro and granite have the same *texture*, they contain different minerals and therefore look different.

Identifying an igneous rock requires no new skills, just a few simple observations and your ability to identify the common rock-forming minerals. If you are wondering why there aren't pictures of all the rock types to help you identify the specimens in your rock set, it's because granite may be gray, red, white, or even purple or black depending on the color of its feldspars. A picture can thus help only if the rock it shows is exactly the same as the rock in your set. But if you understand the combination of minerals that defines granite, you will get it right every time. You will have a chance to try in Exercise 4.6.

4.4 Sedimentary Rocks

Like igneous rocks, sedimentary rocks form from previously existing rocks—but by very different processes. If you've ever seen shells on a beach, gravel along a riverbank, mud in a swamp, or sand in a desert, you've seen sediment. **Sediment** consists of loose grains (**clasts**) that may be made of individual minerals or fragments of previously existing rock containing several minerals. **Sedimentary rock** forms at or near the Earth's surface by the cementation and compaction of accumulated layers of these different kinds of sediment.

Sedimentary rocks preserve a record of past environments and ancient life and thus record the story of Earth history—once you know how to read it. For example, the fact that a sedimentary rock called limestone occurs throughout North America means that in the past, a warm, shallow sea covered much of the continent. Geologists have learned to read the record in the rocks by comparing features in sedimentary rocks with those found today in environments where distinct types of sediment form. In this section, you will develop the skills needed to read the history recorded in sedimentary rocks.

4.4.1 The Origin of Sediment

The materials from which sedimentary rocks form ultimately come from **weathering**—the chemical and physical breakdown of pre-existing rock. Weathering produces both loose pieces, called clasts, and **dissolved ions**, which are charged atoms in a water solution. Once formed, clasts may be transported by water, wind, or ice to another location, where they are deposited and accumulate. Dissolved ions, meanwhile, enter streams and groundwater. Some of these ions precipitate from groundwater in the spaces between clasts, forming a **cement** that holds the clasts together. Other ions get carried to lakes or seas, where organisms extract them to form shells. Finally, some dissolved ions precipitate directly from water to form layers of new sedimentary minerals. In some environments, sediment may also include organic material, the carbon-containing compounds that remain when plants, animals, and microorganisms die.

4.4.2 The Basic Classes of Sedimentary Rocks

Geologists recognize many kinds of sedimentary rocks, each with a name based on the nature of the material that the rock contains and the process by which it formed. They use a simple classification scheme that groups all sedimentary rocks into four major classes—clastic, chemical, biochemical, and organic—based on the origin of the particles of which they are made. In this section, we will first examine these classes and then introduce a more comprehensive classification scheme that includes chemical and mineral composition.

4.4.2a Clastic Sedimentary Rock **Clastic sedimentary rocks** form from clasts (mineral grains or rock fragments) produced by physical weathering of previously existing rocks. Clastic rocks have a characteristic texture in which discrete grains are held together by a chemical cement or by a very fine-grained clastic matrix. Of the most common clastic sedimentary rocks, the majority are derived from silicate rocks and thus contain clasts composed of silicate minerals.

Formation of clastic sedimentary rock involves the following processes (**FIG. 4.9A**):

1. **Weathering:** The process of weathering reduces solid bedrock to loose clasts, also known as *detritus*.
2. **Erosion:** Erosion occurs as moving water (streams and waves), moving air (wind), and moving ice (glaciers) pluck or pick up the clasts.
3. **Transportation:** Moving water, wind, and ice carry clasts away from their source.

FIGURE 4.9 **The five steps in clastic sedimentary rock formation.**

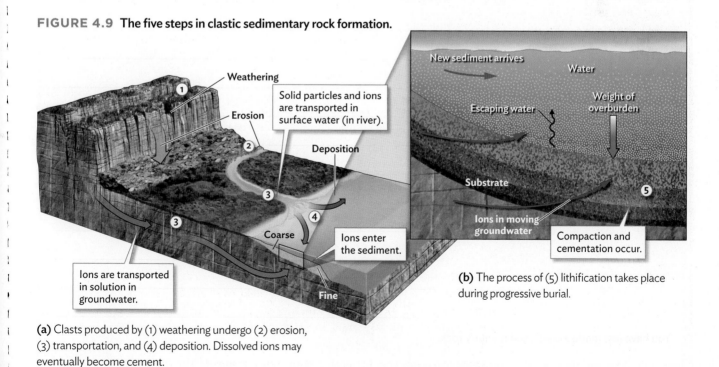

(a) Clasts produced by (1) weathering undergo (2) erosion, (3) transportation, and (4) deposition. Dissolved ions may eventually become cement.

(b) The process of (5) lithification takes place during progressive burial.

4. **Deposition:** When moving water or wind slows, or when ice melts, clasts settle out and accumulate. The places where this occurs (such as the land surface, the seafloor, or a riverbed) are called **depositional environments**.

5. **Lithification:** This process transforms loose sediment into solid rock through compaction and cementation (**FIG. 4.9B**). Over time, accumulations of clasts are buried. When this happens, the weight of overlying sediment squeezes out air and water, fitting the clasts more tightly together. This process is called **compaction**. As ion-rich groundwater passes through the compacted sediment, minerals precipitate and bind, or "glue," the clasts together. This process is called **cementation**, and the mineral glue between clasts is called **cement**. Quartz and calcite are the most common cements in sedimentary rock, along with limonite and hematite. In Exercise 4.7, you'll investigate the formation of sedimentary rock by creating your own.

EXERCISE 4.7 **Understanding the Origin of Clastic Sedimentary Rock Textures**

Name: _____ Section: _____
Course: _____ Date: _____

Cover the bottom of a petri dish with a mixture of sand grains and small pebbles. Add a small amount of the liquid provided by your instructor and allow the mixture to sit until the liquid has evaporated. Now turn the dish upside down. (*Note:* If this exercise is available and being conducted online, the setup, materials, and directions may differ. See the instructions in the online lab for how to conduct the exercise.)
  • Why don't the grains fall out of the dish?

_____

_____

Examine the "rock" with your hand lens. Sketch and describe the texture.

Sketch: Describe: _____

                                 _____

                                 _____

                                 _____

                                 _____

                                 _____

                                 _____

                                 _____

                                 _____

                                 _____

                                 _____

You have just made a clastic sedimentary rock.

In a final process, the application of pressure and the circulation of fluids over time may gradually change the characteristics of a sedimentary rock (e.g., grain size and composition of cement). Any chemical, physical, or biological process that alters the rock after it is formed is called **diagenesis**. It is important to note that diagenesis occurs at relatively low temperatures—below those at which the changes are considered to be metamorphism.

The names of clastic sedimentary rock types are based primarily on the size of their clasts (**FIG. 4.10**). From coarsest to finest, clastic sedimentary rocks (and their grain sizes) include the following:

FIGURE 4.10 Types of clastic sedimentary rocks.

(a) Conglomerate.

(b) Breccia.

(c) Sandstone.

(d) Siltstone.

(e) Shale.

- **Conglomerate and breccia** consist of pebbles and/or cobbles, often in a matrix of sand and finer-grained material. Clasts in conglomerate are rounded (have no sharp corners), whereas those in breccia are angular (have sharp corners). Modifiers are sometimes added to these names when one kind of clast dominates (e.g., quartz conglomerate or limestone breccia).

- **Sandstone** applies to any clastic sedimentary rock composed predominantly of sand-sized grains.

- **Siltstone** consists of silt-sized clasts.

- **Shale/mudstone** consist mostly of clay-sized particles, invisible even with a hand lens.

4.4.2b Chemical Sedimentary Rock **Chemical sedimentary rocks** form when minerals precipitate directly from a water solution that has become oversaturated. Ions that can no longer be dissolved bond together to form solid mineral grains, which often display a crystalline texture as they grow and interlock with one another. These crystals either settle out of the solution or grow outward from the walls of the container holding the solution. In Exercise 4.8 you can simulate the chemical precipitation process in the lab to see how the textures of these rocks develop.

Groundwater, oceans, and saline lakes all contain significant quantities of dissolved ions and are therefore sources of chemical sedimentary rocks. Precipitation to form chemical sedimentary rocks happens in many environments, including (1) hot springs, where warm groundwater seeps out at the Earth's surface and cools; (2) cave walls, where groundwater seeps out, evaporates, and releases CO_2; (3) the floors of saline lakes or restricted seas, where saltwater evaporates; (4) within sedimentary rocks, when reactions with groundwater replace the original minerals with new ones; and (5) on the deep-sea floor, where the shells of plankton dissolve to form a gel-like layer that then crystallizes.

EXERCISE 4.8	Simulating Chemical Sedimentary Textures

Name: _____   Section: _____
Course: _____   Date: _____

(a) Place a beaker with seawater (or homemade saltwater) on a hot plate and heat it gently until the water evaporates. Partially fill a second beaker with a clear, concentrated solution of calcium hydroxide [$Ca(OH)_2$]. Using a straw, blow *gently* into the solution until you notice a change.

Describe what happened in each demonstration and sketch the resulting textures.

(i) Evaporated seawater description   Sketch:

_____

_____

_____

(ii) $Ca(OH)_2$ solution description   Sketch:

_____

_____

_____

_____

_____

_____

(b) Compare the texture in (i) with that of a granite, a rock formed by cooling of a melt. Describe and explain the similarities in texture.

_____

_____

_____

_____

FIGURE 4.11 Textures of chemical sedimentary rocks.

(a) Coarse, interlocking halite grains in rock salt.

(b) Cryptocrystalline silica grains in chert. Note the almost glassy appearance and conchoidal fracture.

The composition of a chemical sedimentary rock depends on the composition of the solution it was derived from—some chemical sedimentary rocks consist of salts (e.g., halite, gypsum), whereas others consist of silica or carbonate. In some chemical sedimentary rocks, the grains are large enough to see (**FIG. 4.11A**). But in others, the grains are so small that the rock looks almost like porcelain or obsidian (**FIG. 4.11B**). Such rocks are called **cryptocrystalline rocks**, from the Latin *crypta–*, meaning hidden. Geologists distinguish among different types of chemical sedimentary rocks primarily by their composition:

■ **Evaporites** are chemical sedimentary rocks, composed of crystals formed when saltwater evaporates, and typically occur in thick deposits. The most common examples of evaporites are rock salt and rock gypsum (alabaster).

■ **Oolitic limestone** is a unique form of limestone in which calcite precipitates onto small mineral grains; as the grains roll back and forth, the calcite coating thickens, producing bead-shaped grains that are then lithified.

■ **Travertine** (*chemical limestone*) Travertine is a special kind of limestone composed of calcite ($CaCO_3$) formed by chemical precipitation from groundwater that has seeped out of the ground either in hot- or cold-water springs or from the walls of caves.

■ **Dolostone**, like limestone, is a carbonate rock, but it differs from limestone in that it contains a significant amount of dolomite, a mineral with equal amounts of calcium and magnesium [$CaMg(CO_3)_2$].

■ **Chert** is composed of very fine-grained silica. One kind of chert, called biochemical chert, is formed from microscopic plankton shells, as we will see shortly.

4.4.2c Biochemical and Organic Sedimentary Rocks While alive, some organisms build shells by extracting dissolved ions from their environment. For example, clams, oysters, and some types of algae and plankton use calcium, carbon, silicon, and oxygen ions dissolved in water to produce shells made of calcite or quartz. Plants extract CO_2 from the air to produce the cellulose of leaves and wood. And all organisms produce *organic chemicals*, meaning chemicals containing rings or chains of carbon atoms bonded to hydrogen, nitrogen, oxygen, and other elements. When these organisms die, the materials that compose their shells or cells can accumulate, just as inorganic clasts do. Calcite and quartz shells are quite durable—after all, they're composed of relatively hard minerals compared with the organism's soft organic tissues, which commonly decompose and oxidize. But in special depositional settings where there is relatively little oxygen and the organisms are buried rapidly, organic chemicals can also be preserved in rocks.

Rocks composed primarily of the hard shells of once-living organisms are classified as **biochemical sedimentary rocks**, indicating that living organisms extracted the dissolved ions. Rocks containing significant amounts of carbon-rich soft tissues are called **organic sedimentary rocks**.

The most common biochemical sedimentary rocks include the limestones and chert:

- **Limestone** is a general class of sedimentary rock composed of calcite. Biochemical limestone, which forms from calcite-containing shells, commonly tends to form chunky blocks, light gray to dark bluish gray in color. Geologists recognize distinct subcategories of biochemical limestone by their texture:

    - **Fossiliferous limestone** contains abundant visible fossils in a matrix of fossil fragments and other grains (**FIG. 4.12A**).

    - **Micrite** (or **micritic limestone**) is very fine-grained limestone formed from the lithification of carbonate mud. The mud may be made up of the tiny spines of sponges or the shells of algae or bacteria, or it may form after burial beneath sediment via diagenesis.

FIGURE 4.12 Types of biochemical and organic sedimentary rocks.

(a) Fossiliferous limestone.

(b) Chalk.

(c) Coquina.

(d) Bituminous coal.

- **Chalk** is a soft white limestone composed of the shells of plankton (**FIG. 4.12B**).

- **Coquina** consists of a mass of shells that are only poorly cemented together and have undergone minimal diagenesis (**FIG. 4.12C**).

- **Biochemical chert** forms when shells of silica-secreting plankton accumulate on the seafloor and partially dissolve to form a very fine-grained gel, which then solidifies. We use the word *biochemical* to distinguish this chert from *replacement* chert, which forms by diagenesis in previously lithified limestone.

Of the organic sedimentary rocks, there are two:

- **Coal**, composed primarily of carbon, is derived from plant material (wood, leaves) that was buried and underwent diagenesis. Geologists distinguish three *ranks* of coal based on the proportion of carbon they contain: lignite (50%) and bituminous (85%) coals (**FIG. 4.12D**), which are sedimentary, and anthracite (95%), a metamorphic rock.

- **Oil shale**, like regular shale, is composed mostly of clay, but unlike regular shale, oil shale contains a significant amount of kerogen (a waxy organic chemical derived from fats in the bodies of plankton and algae).

4.4.3 Identifying Sedimentary Rocks

Now that you are familiar with the nature and origins of sediment, the four sedimentary rock classes, and some of the most common sedimentary rock types, you are ready to identify and interpret the histories of sedimentary rocks in the laboratory and the field.

Geologists approach sedimentary rock identification systematically, following the steps outlined below. As with mineral identification, we offer two different visual reference tools to help you: (1) **TABLE 4.3** presents different classes of sedimentary rocks, with their basic characteristics, clues for identification, and their names, and (2) **FIGURE 4.13** presents the same information in a flowchart format. The colors used to distinguish the different classes of sedimentary rock are the same in both schemes.

TABLE 4.3 Classification of common sedimentary rocks

(A) Clastic sedimentary rocks (silicate mineral grains or rock fragments held together by cement)			
Composition	**Texture or clasts (grain size in mm)**	**Clues to identification**	**Rock name**
Usually silicate minerals	Boulders, cobbles, pebbles (>2)	Clasts are bigger than peas; you can measure clasts with a ruler.	Conglomerate (if clasts are rounded); Breccia (if clasts are angular)
Usually silicate minerals	Sand (0.063 – 2)	Grains are easily visible and identifiable, but too small to measure with a ruler.	Sandstone
Usually silicate minerals	Silt (0.004 – 0.062)	Grains are visible but too small to identify; feels gritty.	Siltstone
Usually silicate minerals	Clay (<0.004)	Individual grains are not visible; composed dominantly of clay.	Mudstone (breaks into blocky pieces); Shale (breaks into thin plates)

*Lithic clasts are sand-sized rock fragments.

(continued)

TABLE 4.3 Classification of common sedimentary rocks (*continued*)

(B) Chemical sedimentary rocks (composed of grains that precipitated from a water solution)

Composition	Texture or clasts (grain size in mm)	Clues to identification	Rock name
Halite	Crystal (generally >2)	Clear to gray, with visible interlocking crystals; tastes salty.	Rock salt
Gypsum	Crystal (generally >2)	Clear to whitish–pale gray or pinkish; soft—can be scratched with a fingernail.	Rock gypsum
Calcite	Grains appear like tiny balls	Very fine grained; HCl test yields a vigorous fizz.	Oolitic limestone
Quartz	Grains not visible	Won't be scratched by a nail or knife; grains are too small to see; tends to be porcelainous; fractures conchoidally.	Chert – Jasper (reddish) – Flint (black)
Dolomite	Grains not visible	Grayish to rusty tan; scratched-off powder has a moderate reaction with HCl.	Dolostone

(C) Biochemical sedimentary rocks (composed of minerals originally extracted by organisms to form shells)

Composition	Texture or clasts (grain size in mm)	Clues to identification	Rock name
Calcite	Visible shells or a few shells in very fine grains (generally 0.004 to 2); some crystal appearance	Commonly grayish, but may be white, yellow, or pink; vigorously reacts with HCl; some examples may consist of shell fragments cemented together and have a clastic texture; some have recrystallized to form a crystalline texture.	Limestone (general name) – Fossiliferous limestone – Micrite (very fine-grained; grains aren't visible)
Calcite	Visible shells (>2)	A weakly cemented mixture of shells.	Coquina (a type of limestone)
Calcite	Grains not visible	Whitish; can be used to write on slate. Vigorously reacts with HCl.	Chalk (a type of limestone)

(D) Organic sedimentary rocks (containing organic chemicals derived from the bodies of organisms)

Composition	Texture or clasts (grain size in mm)	Clues to identification	Rock name
Clay and kerogen	Grains not visible	Dark gray to black; may have an oily smell; may burn; can't see grains.	Oil shale
Carbon (± clay and quartz)	Grains not visible	Black; may have a subtle bedding; typically breaks into blocks; may contain plant fossils.	Coal – Lignite coal (50% carbon; fairly soft) – Bituminous coal (85% carbon; medium hard, dull) – Anthracite coal (95% carbon; quite hard; shiny)

STEP 1 Examine the rock's texture. A simple textural observation will be easy for sedimentary rocks with grains coarse enough to see with the naked eye or a hand lens. Does the specimen consist of clasts, cemented-together grains, intergrown crystals, or fossils? If your specimen has grains too small to be seen even with a hand lens, go directly to step 2. Otherwise follow step (a), (b), or (c) depending on the texture of your specimen. Use Table 4.3 or Figure 4.13 to help you make your determination.

a) If the rock is clastic, determine clast *size* and *roundness*. First, use the clast size chart tool at the back of the manual to find the correct name for the clast size. Then, use the other clues for identification in Table 4.3 or Figure 4.13 to find the rock name. Remember that larger, rounded clasts define a conglomerate and angular clasts a breccia. If the clasts are smaller (the size of sand or silt), you are looking at a form of sandstone, siltstone, shale, or mudstone. Use the specifics in Table 4.3 or Figure 4.13 to choose the correct one.

b) If the rock is crystalline, locate the crystalline rocks in Table 4.3 or Figure 4.13, and use the observation skills and tests you have learned for hardness, cleavage, luster, HCl reaction, and if necessary, taste, to identify the minerals. Then apply the appropriate rock name from the table or flow chart.

c) If the rock is made entirely of cemented fossils and fossil fragments, it is coquina. If a few fossils are set in a fine-grained matrix, use HCl to determine if it is a fossiliferous limestone (it will fizz strongly) or mudstone (it likely won't).

FIGURE 4.13 Flow chart for identifying sedimentary rocks.

STEP 1: Examine the rock's texture (if individual grains can be seen).

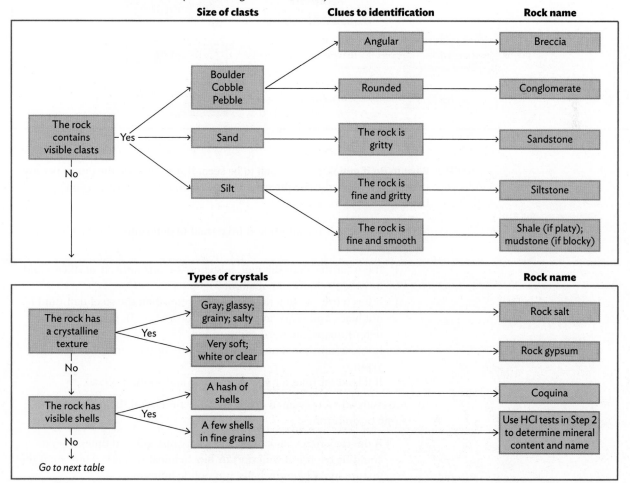

FIGURE 4.13 Flow chart for identifying sedimentary rocks. (*continued*)

STEP 2: If the grains are too small to see, use physical (a) and HCl (b) tests to determine rock type.

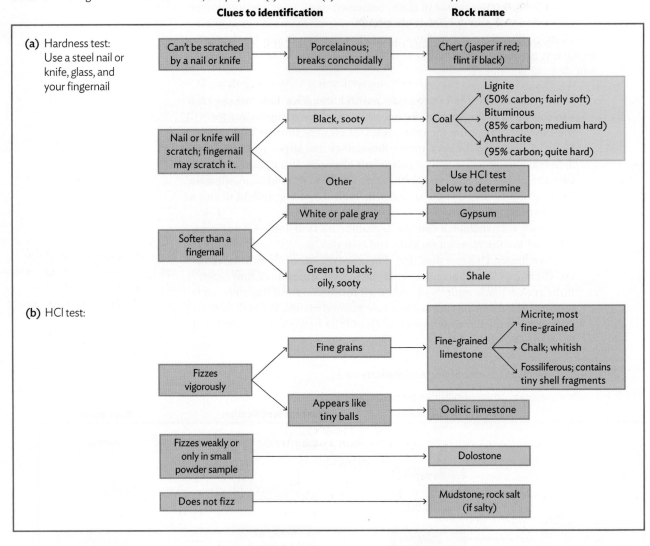

STEP 2 **What to do if grains are too small to be seen.** If you can't see the grains because they are too small, use the physical (a) and chemical HCl tests (b) you learned in Chapter 3.

a) Use a steel nail (or knife) and fingernail to determine the rock's hardness.

i. If it is harder than the nail, it is chert—made entirely of silica—and has a Mohs hardness of about 7.

ii. If it is harder than a fingernail but softer than the steel nail (and it's not black and sooty, meaning it's a form of coal), go to step 2b to determine its chemical composition.

iii. If it is softer than a fingernail and white or pale gray, it is rock gypsum.

iv. If it is softer than a fingernail, black, and sooty, it is coal.

b) Carefully add a very small drop of HCl to find out if the rock is a limestone or dolostone.

i. If the rock fizzes vigorously, it is a very fine-grained limestone. It could be chemical (micrite) or biochemical (chalk). However, if the texture looks like tiny balls, it is oolitic limestone.

    ii. If the rock fizzes weakly, scratch it again and put a drop of acid on the powder. If it reacts more strongly, it is a very fine-grained dolostone (dolomicrite).

    iii. If the rock doesn't fizz, go to step 2c.

  c) Moisten your finger, rub it against the rock, and taste. If it tastes salty, the specimen is rock salt.

Exercise 4.9 will help you practice identifying rock samples by using these steps. Remember that you can look to either the table or the chart for help. Both present the same information, so use whichever style with which you are more comfortable.

EXERCISE 4.9   **Identifying Sedimentary Rock Samples**

Name: _____   Section: _____

Course: _____   Date: _____

Examine the rock samples in your sedimentary rock set. Read Section 4.4.3 closely to remind yourself of the steps. (You can use the sediment grain size scale tool at the back of the manual to help assess clast sizes.) Fill in the study sheets at the end of the chapter to identify each sample. Keep these samples and your study sheets until you have finished the chapter. At that point, you will be able to add an interpretation of the histories of the rocks. (*Note:* If taking this lab online, and directed by your instructor, use the specimen images provided by your instructor in lieu of rock specimens.)

4.5 Metamorphic Rocks

Biologists use the term *metamorphosis* (from the Greek *meta*, meaning change, and *morph*, meaning form) to describe what happens when a caterpillar turns into a butterfly. Geologists use the similar term **metamorphism** for the processes of change that a rock undergoes when exposed to physical and chemical conditions significantly different from those under which it first formed. The original rock, called the **protolith** (or **parent rock**), may be any kind of igneous, sedimentary, or older metamorphic rock. Logically, the end product of metamorphism is a **metamorphic rock**. Metamorphism differs from sedimentary and igneous rock-forming processes because it is entirely a *solid-state phenomenon*. This means that the original rock remains in a solid state without breaking apart or melting. By definition, metamorphism occurs at temperatures higher than those involved in sedimentary rock diagenesis and lower than those at which melting takes place.

Metamorphic rocks form at depths below those at which sedimentary rocks generally form and above those where melting generally occurs, so they may yield important clues about parts of the Earth for which neither igneous nor sedimentary rocks can provide information. For example, what happens to rocks as the opposite sides of the San Andreas fault grind past one another? Or to rocks along the west coast of South America as the Nazca Plate is subducted beneath the continent? Or to rocks deep below Mt. Everest as India collides with Asia? How did rocks in California change when they were intruded by the granitic magma that today forms the enormous Sierra Nevada batholith?

This section shows how we can answer these questions and unravel the geologic history of rocks that have been changed—sometimes dramatically—from their original forms. We'll begin by describing metamorphic changes and the agents that cause them, and then examine metamorphic textures and minerals for clues to the type of metamorphism that occurred and the temperature and pressure conditions under which it took place.

4.5.1 What Changes During Metamorphism?

Nearly all characteristics of the protolith may change during metamorphism, including its texture, its mineral composition, and even its chemical composition. In some cases, only one of these properties may change, but in others, two or all three of them may change so much that the metamorphic rock looks nothing like its protolith.

4.5.1a Changes in Texture A rock's texture includes the size, shape, and relationships of its grains. These properties may change during metamorphism by any of three processes:

1. *Recrystallization* (change in grain size and shape): Mineral grains may regrow to form new grains of the same mineral, changing in shape, size, or both.
2. *Pressure solution* (change in grain shape and relationships): If water is present during metamorphism, some minerals may partially dissolve and the ions may re-precipitate, changing the shapes and relationships among grains.
3. *Alignment of grains* (change in grain orientation): In most instances, minerals in the protolith are oriented randomly, but during certain kinds of metamorphism, they become aligned parallel to one another. The resulting textures, called *foliation* and *lineation*, are described later in this section.

4.5.1b Changes in Mineralogy The original protolith minerals may not be able to survive the new temperature and pressure conditions and may therefore change *in the solid state* to new metamorphic minerals that are more stable under those new conditions.

Sometimes a rock changes only slightly during metamorphism, so that most of its original sedimentary, igneous, or prior metamorphic characteristics are still recognizable. In many cases, however, the changes are so drastic that a metamorphic rock can look as different from its protolith as a butterfly does from a caterpillar (**FIG. 4.14**). If a rock changes only a little during metamorphism, we say it has undergone *low-grade* metamorphism; if it has changed drastically, it has experienced *high-grade* metamorphism. Note that the same protolith in Figure 4.14a.i and b.ii produces very different metmorphic rocks depending on the intensity of metamorphism and whether textural or mineralogic changes occurred.

4.5.2 The Agents of Metamorphism

Geologists refer to the heat, pressure, hydrothermal fluids, and stress that cause metamorphic changes as **agents of metamorphism**. Each of these agents acting alone can cause metamorphism, but in most geologic settings two or more act simultaneously. In this section, we will examine how the first three of these agents of metamorphism cause change and observe their effects on protoliths. We discuss the effects of stress in the next section, when we examine metamorphic rock textures.

4.5.2a The Effect of Heat When rocks are heated, either by an intrusive magma or during burial by sedimentation, mountain building, or subduction, their ions vibrate more rapidly. The chemical bonds holding those ions in mineral structures stretch, and some begin to break. The freed ions migrate slowly through the solid rock by a process called **diffusion**, much as atoms of food coloring spread when dripped into a glass of water. Eventually, the wandering ions bond with other ions to produce metamorphic minerals that are stable under the higher temperatures. Heat is thus the major cause of changes and chemical reactions that replace protolith minerals with metamorphic minerals. Diffusion is much slower in solid rock

FIGURE 4.14 Examples of changes from protolith (left) to metamorphic rock (right).

(a) Changes in texture.

　(i) Very fine-grained calcite in micritic limestone recrystallizes to form coarse-grained calcite in marble.

　(ii) Coarse quartz and feldspar grains in granite are smeared in a fault zone to very fine-grained ultramylonite.

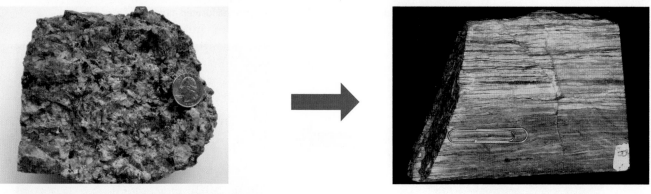

　(iii) Rounded clasts in conglomerate are flattened in metamorphosed conglomerate.

(b) Changes in mineralogy.

　(i) Clay minerals in mudstone react with quartz to form biotite and garnet in gneiss.

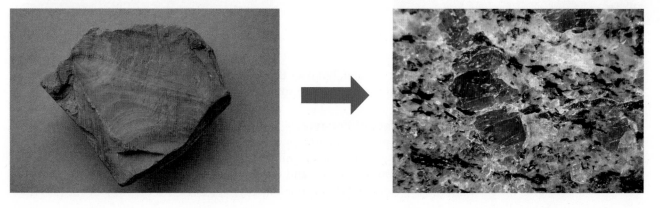

FIGURE 4.14 Examples of changes from protolith (left) to metamorphic rock (right) (continued).
(ii) Dolomite and quartz in micrite react to produce actinolite (green; an amphibole) in marble.

(c) Multiple changes.
(i) Tiny, randomly oriented clay grains and quartz in mudstone react to form coarse-grained foliated muscovite flakes with large garnet crystals in schist.

(ii) Randomly oriented plagioclase gray and pyroxene crystals in basalt react to form aligned amphibole crystals (hornblende) and a different plagioclase in amphibolite. This alignment of rod-shaped crystals is called *lineation*.

than in magma or water, but over thousands or millions of years, new metamorphic minerals can grow to impressive sizes (**FIG. 4.15**).

4.5.2b The Effect of Pressure If you pull an inflated balloon underwater, it is affected by the pressure caused by the surrounding water. The balloon responds to this pressure by getting smaller while retaining its shape. Rocks behave the same way when they are buried deeper and deeper in the Earth. Pressure causes the ions in minerals to be jammed closer together (the opposite of the effect of heat). Minerals

FIGURE 4.15 Growth of metamorphic minerals.

(a) A large, nearly perfect garnet crystal.

(b) Andalusite crystals.

formed close to the surface have relatively open structures compatible with their low-pressure environment. Under high pressure, those structures collapse, and the ions are rearranged in more densely packed structures. Bury graphite, for example, deeply enough in the mantle and it undergoes a change to diamond.

Both temperature and pressure increase with depth, so rocks must respond to these agents simultaneously. The *geothermal* gradient varies, but is typically 30°C per kilometer, whereas the *geobaric* gradient is about 0.33 kilobars (kbar) per kilometer (1 kilobar = 1,000 times standard atmospheric pressure). Thus, temperature–pressure conditions 10 km below the surface would be about 300°C and 3.3 kbar. At 25 km, the temperature would be about 750°C and pressure about 8.3 kbar. Therefore, as a rock is buried deeper in the Earth, it must adjust to progressively increasing temperatures and pressures. An assemblage of metamorphic minerals that formed at 10 km would change to those that form under more intense conditions if it were buried more deeply at 25 km.

4.5.2c The Effect of Hydrothermal Fluids In many subsurface environments, rocks contain hot liquids, gases, and, at the highest temperatures, fluids that have properties of both gases and liquids. These fluids are composed mainly of H_2O and CO_2 and are collectively called **hydrothermal fluids** (from *hydro*, meaning water, and *thermal*, meaning hot). Some of these fluids are given off from magma when it solidifies, others form when H_2O or CO_2 is released from protolith minerals as they react to form metamorphic minerals, and still others are produced when groundwater percolates deep into the crust.

Hydrothermal fluids act as a catalyst, speeding up metamorphic reactions by greatly increasing the diffusion of ions. They can carry large amounts of ions and may change the basic chemistry of a protolith by dissolving some elements and/or depositing others as they pass through the rock. The change in protolith chemistry through interaction with hydrothermal fluids is called **metasomatism**.

4.5.3 Metamorphic Rock Classification

Recognizing some rocks as metamorphic can be challenging because they may preserve some of the original igneous or sedimentary characteristics of their protoliths. Once you know a rock is metamorphic, a few simple observations can reveal aspects of its history, such as what agents of metamorphism were involved and how intense those agents were.

Name: _____   Section: _____
Course: _____   Date: _____

The minerals andalusite, sillimanite, and kyanite are polymorphs of aluminosilicate (Al_2SiO_5) that differ in their internal structures and in the conditions under which they are stable. The aluminosilicate *phase diagram* on the right, which uses different colors to delineate the stability zones of these polymorphs, provides insight into how metamorphic minerals respond to changes in temperature and pressure conditions. Examine the phase diagram and answer the following questions:

(a) Indicate the temperature-pressure (T-P) conditions and estimated depths for rocks A, B, and C in the following table. Assume average geothermal (30°C per kilometer) and geobaric gradients (0.33 kbar per kilometer).

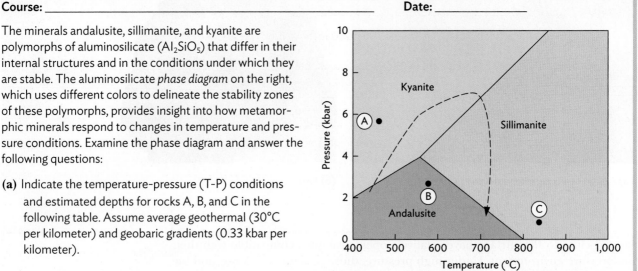

Phase diagram showing stability zones for three aluminosilicate ($Al_2 Sio_5$) polymorphs.

Rock	Polymorph	T (°C)	Depth (km) based on geothermal gradient	P (kbar)	Depth (km) based on geobaric gradient
A					
B					
C					

(b) What does the discrepancy between depth estimates based on geothermal and geobaric gradients suggest about metamorphism?

_____

_____

_____

(c) What mineral changes would occur, and at what temperature and pressure, if
• Rock A was heated to 800°C without a change in pressure?

_____

_____

_____

• Rock A experienced a drop in pressure to 1.5 kbar without a change in temperature?

_____

_____

_____

_____

(continued)

Name: _____   Section: _____
Course: _____   Date: _____

• Rock B was heated to 700°C and pressure on it was increased to 10 kbar?

_____

_____

_____

• Rock C was cooled to 500°C without a change in pressure?

_____

_____

_____

(d) In some cases, it is possible to detect how rocks have changed position in the crust during metamorphism. Suppose that Rock X has experienced a complex tectonic history in which T-P conditions changed over time as shown by the dashed arrow in the diagram.
  • Describe the sequence of aluminosilicate phase changes in Rock X and the T-P conditions and depths at which they occurred.

_____

_____

_____

_____

_____

Texture and mineralogy are the keys to recognizing that a rock is metamorphic. In some cases, texture or mineralogy alone will point to metamorphism, whereas in others it is necessary to consider both. As with igneous rocks, we first note a rock's most visible property—its texture—and then identify the minerals it contains.

4.5.3a Textural Classes of Metamorphic Rocks The pressure rocks experience during burial is equal from all directions. During plate-tectonic activity, however, the forces acting on rocks are greater in some directions than in others, subjecting the rocks to **stress**. Stress can change the shapes of grains in a protolith, like the flattened metaconglomerate clasts in Figure 4.14a.iii. It can also change protolith grains from random orientations to *preferred orientations*; thus, the flattened metaconglomerate clasts in Figure 4.14a.iii are all flattened in the same direction. Preferred orientation most commonly develops when new metamorphic minerals crystallize under stress, producing **foliation**, in which platy minerals like mica flakes are aligned parallel to one another, as in the muscovite in Figure 4.14c.i, or **lineation**, in which rod-shaped crystals lie parallel to one another, as in the amphibolite in Figure 4.14c.ii. Exercise 4.11 will help you visualize how stress causes preferred orientation to develop in a rock.

Name: _____   Section: _____

Course: _____   Date: _____

Take a handful of small plastic chips and push them with random orientation into a mass of modeling clay. Then flatten the clay with a book.

Sketch and describe the orientation of the plastic chips in the "rock." Are they randomly oriented, as they were originally? Is their alignment related to the pressure you applied?

Sketch: Describe: _____

_____

_____

_____

_____

_____

Parallel alignment of platy minerals, called *foliation*, is found in metamorphic rocks that have been strongly squeezed, as at a convergent boundary between colliding lithosphere plates.

• What real minerals would you expect to behave the way the plastic chips did?

_____

_____

So to determine if a rock is metamorphic, you can start by observing its texture:

■ **Foliated or lineated textures:** If a rock is foliated or lineated (see Fig. 4.14c), it must be metamorphic, because stress causes these preferred mineral orientations, and metamorphic rock is the only class of rock in whose formation stress plays a role.

■ **Gneissic banding:** The alternating bands of light and dark minerals in gneiss (**FIG. 4.16**) are characteristics of high-grade metamorphism. They are different *texturally* from the depositional layering of sedimentary and fragmental igneous rocks in that (1) the grains are not cemented clasts or compacted pyroclastic debris, but rather interlock with one another; and in that (2) one set of bands is typically foliated or lineated (as in Fig. 4.16a, in which the dark bands are strongly foliated biotite flakes). They also differ *mineralogically* in that the grains in gneiss are interlocking grains of metamorphic minerals.

■ **Nonfoliated textures:** Some metamorphic rocks contain interlocking grains with no preferred orientation. For some, this is because stress was not one of the agents of their metamorphism, so neither protolith grains nor new minerals could be aligned. For others, stress was indeed a metamorphic agent, but their grains were neither platy nor rod-like and were therefore unable to show preferred alignment (**FIG. 4.17**). You've seen that interlocking grains also occur in igneous rocks and in crystalline sedimentary rocks, so an interlocking texture alone cannot prove or disprove a metamorphic origin. For that, it is necessary to identify the minerals the rock contains.

FIGURE 4.16 Two examples of gneissic banding in which light and dark minerals have been separated.

(a) Pink potassic feldspar layers (unfoliated) separated by foliated biotite-rich layers.

(b) Similar to (a), but light layers are quartz and gray-white feldspar.

FIGURE 4.17 Photomicrographs of rocks with nonfoliated textures showing metamorphic changes from their protoliths.

(a) Interlocking calcite grains in a very fine-grained micritic limestone protolith.

(b) Photomicrograph of large calcite grains in marble.

(c) Clasts and cement in this sandstone protolith can be distinguished easily.

(d) Interlocking grains in quartzite. Quartz clasts and quartz cement have recrystallized and can no longer be distinguished from each other.

4.5.3b Compositional Classes of Metamorphic Rocks Some minerals form almost exclusively during metamorphism, so their presence immediately identifies a rock as metamorphic; unfortunately, most of these minerals are not usually found in student mineral sets. Metamorphic minerals, such as muscovite and biotite, not only form during metamorphism, but can also crystallize in igneous rocks. Similarly, garnet can be found in small amounts as clasts in sedimentary rocks. The critical clue is *how much* muscovite, biotite, or garnet a rock contains. Rocks containing large amounts of—or, in some cases, made nearly entirely of—muscovite or biotite must be metamorphic.

4.5.4 Identifying Metamorphic Rocks

TABLE 4.4 presents a classification scheme for metamorphic rocks based on texture and mineralogy, summarizing the process we just described. To identify a metamorphic rock, first determine whether it is foliated or nonfoliated (column 1), and then select its mineral content and grain size; the appropriate rock name will be found in column 4. Check the description of each rock type below for examples and additional details.

TABLE 4.4 Metamorphic rock classification

	Texture	*Dominant minerals	Grain size	Rock name	Grade	Comments
Foliated	Aligned minerals	Micas, chlorite	Very fine	Slate	Low	Breaks along smooth, flat surfaces (slaty cleavage)
			Fine	Phyllite	Low	Breaks along wavy, shimmering surfaces; micas barely visible
			Coarse	*Schist	High	Strong alignment of visible mica flakes; often with garnet and other minerals
	Layered and aligned	†Alternating light and dark layers	Medium to coarse	Gneiss	High	Light and dark minerals separated into layers; typically, one of these is foliated and the other nonfoliated
	Aligned deformed grains	Quartz, rock fragments	Coarse	Metaconglomerate	Low to high	Foliation defined by elongate, flattened and stretched clasts
		Quartz, rock fragments	Fine to very fine	Mylonite	High	Grains "smeared" in fault zones by dynamic metamorphism
Nonfoliated	Interlocking grains with sides of equal length and no preferred orientation	*Calcite, dolomite	Medium to coarse	*Marble	Medium to high	Recrystallized grains interlock and get larger with increasing metamorphic grade
		Quartz	Fine to medium	Quartzite	Medium to high	Recrystallized grains interlock and get larger with increasing metamorphic grade
		Quartz, feldspars	Fine to medium	Metasandstone	Medium to high	Matrix clay minerals typically produce interstitial micas
		*Quartz, feldspars, micas	Fine to medium	Hornfels	Medium to high	Forms at contacts with igneous rocks; may contain coarse index minerals
		Dominant composition is carbon	Nearly glassy	Anthracite coal	Medium to high	Shiny black surface; resembles obsidian

† Additional metamorphic index minerals may also be present. such as garnet, staurolite, sillimanite, kyanite, andalusite, and actinolite.
ˡ Mineralogy of gneiss depends on its chemical composition.

The following descriptions of the more common metamorphic rock types you are likely to find in your classroom are given in the order in which those rock types appear in Table 4.4. The first three types—slate, phyllite, and schist—represent progressively increasing grades of metamorphic change for the same protolith.

Slate: A low-grade metamorphic rock formed from shale or mudstone, composed mostly of clay minerals. Individual grains are too small to be seen with a hand lens. Foliation is revealed by the tendency of slate to break along smooth, closely spaced planes (a pattern called *slaty cleavage*, which makes the rock useful for roofing and paving walkways). Slates are commonly gray, black, green, or red.

Phyllite: A rock that forms from shale or mudstone under higher-grade metamorphic conditions than slate as clay minerals react to form fine-grained flakes of mica and the platy mineral chlorite. Individual mica or chlorite flakes may be barely visible with a hand lens, but their foliation typically produces a silky sheen regardless of their size.

Schist: A medium- to coarse-grained rock that forms from phyllite with increasing metamorphic grade, characterized by well-developed foliation and mica flakes that are easily seen with the naked eye (see Fig. 4.14c.i). The most common mineral components of schist are muscovite and biotite; temperatures at this grade are typically too high for chlorite to survive. Additional minerals such as garnet, kyanite, or sillimanite may form, depending on protolith composition and temperature.

Gneiss: A coarse-grained high-grade metamorphic rock in which light- and dark-colored minerals have been segregated by diffusion into parallel bands. Either light or dark bands are typically foliated. Coarse-grained metamorphic index minerals—most commonly garnet—may be present. Gneisses may form from several different protoliths, contain different minerals, and therefore have very different appearances.

Marble: A medium- to coarse-grained, nonfoliated rock formed by metamorphism of limestone or dolostone (see Fig. 4.14a.i). Common varieties are white, pink, gray, or beige. Impurities in the protolith (clay, quartz, feldspar) may react to form unique metamorphic minerals.

Quartzite: A fine- to medium-grained, nonfoliated, very tough rock formed by metamorphism of silica-cemented quartz sandstone. Typically white or gray, but green and red varieties are also common (**FIG. 4.18**). Outlines of individual clasts may be visible (as in Fig. 4.18b), depending on original grain size and degree of recrystallization, but at high grades of metamorphism, quartz cement and quartz clasts recrystallize and can no longer be distinguished.

FIGURE 4.18 Varieties of quartzite.

(a) Pink quartzite.

(b) The original clast outlines are visible in this raw specimen of quartzite.

(c) Crystalline quartzite (ortho-quartzite).

FIGURE 4.19 Anthracite.

Anthracite: A black, shiny rock formed by metamorphism of bituminous coal. Anthracite, which is also called "hard coal," commonly exhibits conchoidal fracture and may resemble obsidian (**FIG. 4.19**), but is softer and, because it is nearly pure carbon, less dense.

4.6 Rocks in Everyday Life

Most people understand that minerals have economic value, if only because of the gemstones we see in jewelry. Similarly, the economic importance of rocks should be obvious because the evidence is present in some of the most famous buildings, monuments, and sculptures in the world (**FIG. 4.20**), to say nothing of the kitchen countertops in our homes. Rocks are also used in many less prominent ways. Limestone, for example, is processed to make lime, which is the principal constituent of cement and concrete as well as a common additive used to reduce soil acidity. Fine-grained igneous rocks broken into pieces a few inches across form the stable roadbeds for the railroad tracks that cross many continents. Even the intermediate products of the rock cycle are valuable, especially the physically weathered material not yet converted to clastic sedimentary rock. In some states, sand and gravel deposits are the most valuable geologic resources. Exercise 4.13 explores the properties that make rock useful.

Name: _____   **Section:** _____
Course: _____   **Date:** _____

(a) Just for fun, how many of the objects in Figure 4.20 can you identify?

Figure 4.20a: i _____ ii _____ iii _____

Figure 4.20b: i _____ ii _____ iii _____

Figure 4.20c: i _____ ii _____ iii _____

(b) Granite has been used throughout the ages for buildings (Fig. 4.20c, ii). What properties do you think make granite useful for construction?

_____

_____

(c) Marble (Fig. 4.20b, ii) and alabaster (Fig. 4.20a, ii), have physical properties much the same as those of the minerals of which they are made: calcite and gypsum, respectively. Which of these properties have made these rocks favorites of sculptors for centuries? Explain your reasoning.

_____

_____

(d) Some sculptures, however, are made of igneous rock, like the granite in Figure 4.20c, i and iii. Granite is made mostly of quartz, potassium feldspar, and plagioclase feldspar. How do the minerals in granite affect the ease with which sculptors can shape the stone?

_____

_____

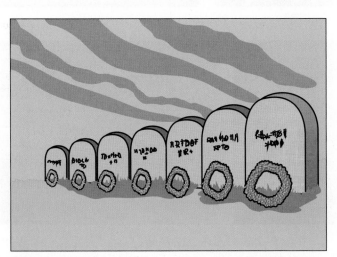

? What Do You Think A veterans' group has contacted you about the best stone to use for memorials honoring the men and women who have served our country. They want to use local stone whenever possible, and they have found nearby sources of marble and granite. Assuming that the memorials will be built near your institution, what factors should you consider before making a recommendation? (Consider the ease of carving and polishing, the response of the materials to weathering, etc.) What factors would you consider if you lived in a very wet and humid climate? What factors would you consider if you lived in a hot and arid climate? (*Hint 1: Consider the diagnostic qualities of the minerals composing each rock. Hint 2: The kind of acid that impacts certain minerals can form in the atmosphere when carbon dioxide mixes with water.*) Answer on a separate sheet of paper.

FIGURE 4.20 Examples of sedimentary, metamorphic, and igneous rock used in buildings, monuments, and sculpture.

i

ii

iii

(a) Sedimentary rocks.

i

ii

iii

(b) Metamorphic rocks.

i

ii

iii

(c) Igneous rocks.

Name: _____ Section: _____
Course: _____ Date: _____

Exploring Earth Science Using Google Earth

1. Visit **digital.wwnorton.com/labmanualearthsci**
2. Go to the **Geotours** tile to download Google Earth Pro and the accompanying Geotours exercises file.

The rock cycle provides an important framework that depicts relationships between various geologic processes and the resulting products. That is, observing a geologic process allows you to predict the product. Conversely, given a specific product, you can often infer the formative geologic process.

Expand the Geotour04 folder in Google Earth by clicking the triangle to the left of the folder icon. Inside the folder are three placemarks that will direct you to locations where distinct geologic processes are operative. Use this information to identify the operative geologic process (i.e., *melting/crystallization, heat/pressure/ burial, or weathering/erosion/ deposition/lithification*) and to predict the geologic product (i.e., *rock class*).

(a) Check and double-click the Argentina placemark to fly to South America to a landscape dotted with numerous cone-shaped features. Focus on the dark black rock and use landscape details to infer the process of how it formed and to interpret the rock class to which it belongs.

_____ process

_____ rock class

(b) Check and double-click the Canada placemark to fly to northern Quebec. Here, the lakes serve to highlight two distinct trends of folded rock layers (~N-S and ~SW-NE). These rocks are part of the Precambrian Canadian Shield that forms the core of the North American craton. Given the "texture" of the landscape, interpret the process(es) that were likely operative to form these foliations. To what rock class do these rocks likely belong?

_____ process

_____ rock class

(c) Check and double-click the Peru placemark to fly to the Atacama Desert. Here, a large landform appears to have diverted the stream flowing through this region. Interpret the operative geologic processes and the likely rock class for rocks that make up this landform.

_____ process

_____ rock class

Name _____

IGNEOUS ROCKS STUDY SHEET

Sample number	Texture	Minerals present (approximate %)	Name of rock	Cooling history; source of magma; tectonic setting

IGNEOUS ROCKS STUDY SHEET

Sample number	Texture	Minerals present (approximate %)	Name of rock	Cooling history; source of magma; tectonic setting

SEDIMENTARY ROCKS STUDY SHEET

Name _____

Sample	Texture (grain size, shape, sorting)	Components			Minerals/rock fragments present (approximate %)	Name of sedimentary rock	Rock history (transporting agent, depositional environment, etc.)
		Clastic	Chemical	Biochemical/ Organic			

SEDIMENTARY ROCKS STUDY SHEET

Name _____

Sample	Texture (grain size, shape, sorting)	Components			Minerals/rock fragments present (approximate %)	Name of sedimentary rock	Rock history (transporting agent, depositional environment, etc.)
		Clastic	Chemical	Biochemical/ Organic			

METAMORPHIC ROCKS STUDY SHEET   Name _____

Sample	Minerals present	Texture (grain orientation, size, shape)	Type of metamorphism	Metamorphic grade (low, medium, or high grade)	Rock name	Possible protolith

METAMORPHIC ROCKS STUDY SHEET   Name _____

Sample	Minerals present	Texture (grain orientation, size, shape)	Type of metamorphism	Metamorphic grade (low, medium, or high grade)	Rock name	Possible protolith

Volcanoes and Volcanic Hazards

5

Lava from Kilauea engulfing a housing subdivision on the island of Hawaii in the summer of 2018.

- Identify the kinds of volcanoes and explain why they have different shapes
- Describe the products of volcanic eruptions and how they relate to volcano type
- Assess the various volcanic hazards and their impacts on society

- Five samples of liquids of varying viscosity in containers that can be securely covered
- Hot plate
- Beaker or saucepan
- Water to heat (or hot water from tap)
- Ice
- Thermometer and stopwatch
- Graph paper
- Colored pencils
- Poster board or stiff cardboard
- Round particles in three different sizes (e.g., different-size round particles such as barley, small white beans, and pinto beans; or sand, pea gravel, and gravel)
- Small containers for holding round particles
- 4 × 6 index cards
- Protractor

5.1 Introduction

As we saw in Chapter 4, when magma reaches the Earth's surface, it spews out in an event called a volcanic eruption. As magma erupts, it may spread across the land as a glowing **lava flow**, or it may be blasted by gases into the air and fall back to the Earth as **pyroclastic debris**—solid particles ranging from tiny grains of ash to blocks several feet across (**FIG. 5.1**). Eruptions can be beneficial to human society by creating new land. For example, the state of Hawaii and many island nations such as Japan, the Philippines, and Indonesia are hotspot volcanoes or volcanic island arcs. In addition, the lava and pyroclastic material that created (and are still creating) these islands form extremely fertile soils. But lava and pyroclastic material can also be extremely dangerous, burying cities and killing thousands of people.

To understand these hazards, geologists study the deposits of past eruptions, which provide important clues about the different kinds of volcanoes that exist, the types of eruptions that produce them, and where they occur. We'll explore these questions in this lab. Then you can use this knowledge to consider the risk of volcanoes locally, and how a large eruption might affect how and where you live.

5.2 Volcanic Eruptions

5.2.1 Eruption Styles

Some volcanoes produce lavas that flow down the volcano's flanks (slopes), such as in Hawaii. Others, like Mt. St. Helens in Washington State, explode catastrophically, sending material high into the atmosphere and hurtling it hundreds of miles per hour down the volcano's flanks and into the surrounding area, blanketing the landscape with ash and pyroclastic materials. As we can witness from current and past volcanoes, not only do they behave differently from one another, but successive eruptions from the same volcano can also vary. In this section, we study the different ways in which volcanoes erupt—called **eruptive styles**. Later in the section we will examine the causes of these differences.

When volcanoes release only flows or fountains of lava, geologists refer to this style of eruption as **effusive**, or even "quiet" (**FIG. 5.2A**). This style occurs with very hot, fluid lavas that are mafic in composition—like those in Hawaii, Iceland, or the Columbia River Flood Basalts. These lavas are "generally very fluid," and so the

FIGURE 5.1 Volcanic eruptions can take different styles and produce a variety of products.

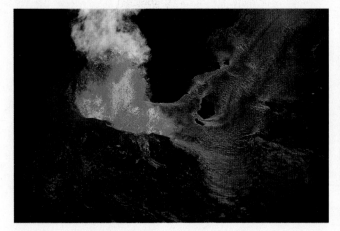

(a) A 2018 effusive eruption of Kilauea volcano in Hawaii spews a lava flow.

(b) An explosive eruption of Mt. Pinatubo in 1991 blasts pyroclastic debris into the air.

FIGURE 5.2 Examples of effusive eruptions.

(a) An effusive eruption on Hawaii. Lava fountains out of a small crater and collects in a fast-moving flow.

(b) An example of a lava lake. Note that a thin layer of dark new rock has formed over the molten lava.

gases that are dissolved in the magma are able to escape easily, leaving less gas to push the lava out quickly. Once extruded, the lava may pool within the crater that surrounds the vent, producing a *lava lake* (**FIG. 5.2B**). Effusive eruptions in which lava fountaining occurs, spurting the lava into the air, have more gas than those that flow or spill out.

In contrast, magmas that are thicker trap the gases that are trying to escape, building up gas pressure inside the magma chamber. Eventually, the gas pressure is so great that it fragments the lava as it erupts at the surface, blasting out clouds of ash and avalanches of debris in an **explosive eruption** (**FIG. 5.3**). The force is often so great in these eruptions that part of the volcano itself is blasted away, leaving a **crater** or **caldera** at the summit.

FIGURE 5.3 Examples of explosive eruptions.

Convective plume

Ash umbrella

Wind

Stratospheric haze

Falling ash and lapilli

Rising convective column

Eruption jet (gas-thrust region)

Collapsing column

Pyroclastic flow (density current)

Ash cloud

Pyroclastic flow

(a) A large explosive cloud contains several components.

(b) The pyroclastic flow and ash cloud from an explosive eruption on Mt. Etna, Italy.

While these are the two broad styles of eruptions that volcanoes may exhibit, geologists often label the degree of explosiveness of a given eruption after a well-known volcano that has shown similar eruption characteristics: Strombolian, Vulcanian, Vesuvian, Pelean, and Hawaiian, for example. We'll revisit these specific types of volcanoes later in this chapter.

5.2.2 Products of Volcanic Eruptions

When you hear a report of a volcano erupting, most likely you picture an explosive eruption with the classic image of a steep-sided, cone-shaped volcano and a large column of gas and ash spewing out the top. However, scientists have long known that not only do volcanoes erupt differently, but the style and products from single eruptions of individual volcanoes may change. In this section, you will examine the products that may be produced by volcanic eruptions, which are dependent on the style of eruption.

5.2.2a Lava Both effusive and explosive eruptions can be accompanied by lava flows. Mafic lavas produced by effusive eruptions flow over long distances (**FIG. 5.4A**);

FIGURE 5.4 **Examples of lava flows.**

(a) Mafic lava flows, like this one in Hawaii, have low viscosity and can flow far and fast.

(b) Pahoehoe developing on a new lava flow in Hawaii.

(c) The rubbly surface of an a'a' flow in Sunset Crater, Arizona.

(d) Felsic lava flows have high viscosity and may build into a lava dome, like this one that sits in the middle of the Crater Lake caldera in Oregon.

but due to their fluid nature, they do not form thick layers until there have been many successive eruptions. As the lavas cool, gas bubbles may escape so that the top layer has a vesicular texture, or it may form the smooth, ropey texture of **pahoehoe** (**FIG. 5.4B**). If the surface solidifies while the lava underneath is still moving, the top layer breaks up, transforming into the blocky lava flow called **a'a** (**FIG. 5.4C**).

In the case of explosive eruptions, the intermediate- to felsic-composition lava is thicker and sluggish, such that it does not travel far from the volcano, forming bulkier individual flows than those produced by mafic lavas (**FIG. 5.4D**).

5.2.2b Volcanic Glass As you saw in Chapter 4, lava cools rapidly when it comes into contact with water or colder air. This rapid cooling does not give the atoms time to form ordered crystalline structures—the lava's internal structure is *amorphous*. The result is volcanic glass, known as obsidian (**FIG. 5.5A**). Fountaining of lava can also form volcanic glass as the droplets of lava that are thrown into the air are rapidly cooled as they fall to the ground. Scientists refer to these glass beads as Pele's tears after the Hawaiian volcano goddess (**FIG. 5.5B**). If the glass cools in strands, then it is called Pele's hair (**FIG. 5.5C**). When hot mixtures of gas and volcanic debris are deposited, the overlying weight of rock compresses glass fragments, forming lenses of volcanic glass, like those in **FIGURE 5.5D** at Bandelier National Monument, New Mexico.

FIGURE 5.5 Examples of volcanic glass.

(a) Obsidian.

(b) Pele's tears.

(c) Pele's hair.

Compressed glass fragments

(d) Compressed glass fragments in a volcanic tuff.

5.2.2c Vesicular Rocks Gas bubbles escaping from erupting and cooling lava leave behind empty spaces (vesicles) giving rise to rocks with vesicular textures. You saw these textures when you identified igneous rocks in Chapter 4. In order to form this texture, there must be enough gas remaining in the magma when it erupts at the surface. **FIGURE 5.6A** shows *pumice*—the felsic igneous rock with vesicular texture. Its high silica content combined with the trapping of air by its numerous small, connected vesicles allows pumice to float in water. Reports from passing ships of large rafts of pumice (**FIG. 5.6B**) in parts of the world's oceans have sometimes been the only indication that remote volcanoes have erupted! *Scoria*—the mafic counterpart to pumice—has fewer large, unconnected vesicles (**FIG. 5.6C-D**). Its high iron and magnesium content means that scoria has a higher density than pumice and sinks. You might find this rock used as landscaping material around your campus or town—it is often bagged and sold at stores as "lava rock."

FIGURE 5.6 Examples of vesicular rocks.

(a) A hand sample of pumice.

(b) A pumice raft in Bequ Lagoon, Fiji.

(c) Scoria deposits in Hverfjall crater in northern Iceland.

(d) A hand sample of scoria.

FIGURE 5.7 Examples of pyroclastic rocks.

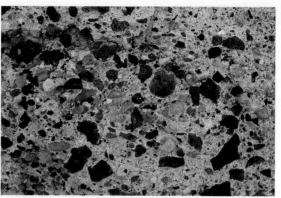

(a) Examples of volcanic tuff (left) and breccia (right).

(b) Cliff dwellings in the Bandelier Tuff, Bandelier National Monument, New Mexico.

5.2.2d Pyroclastic Rocks During explosive volcanic eruptions, glass, pyroclastic fragments, and pieces of rock carried to the surface by the magma are erupted and deposited as pyroclastic rocks, or simply *pyroclasts*. The more violent the explosion, the smaller the particles that form. Pyroclasts are often so hot when they are deposited that they are welded together to form rocks with a pyroclastic or fragmental texture. If the pyroclasts are smaller than 4 mm (0.2 in) in diameter, the rock is called tuff; a rock with larger particles is called volcanic breccia (**FIG. 5.7A**). The degree of welding depends on the temperature of the material when it is deposited, so that some pyroclastic rocks are crumbly while others are extremely hard. In places such as Bandelier National Monument in New Mexico, these deposits form cliffs into which people carved cave dwellings (**FIG. 5.7B**).

In Exercise 5.1, you will identify volcanic deposits to determine eruption style.

Name: _____   Section: _____
Course: _____   Date: _____

The products of volcanic eruptions are highly varied and their formation is controlled by the properties of magmas and eruption styles that you have investigated in this section. Study the following figures and answer the questions that follow each.

A recent lava flow over a road in Hawaii.

(a) What type of lava flow is shown in the photo above?

_____

Based on the color of the image, what do you think it is composed of? Explain.

_____

_____

What style of eruption occurred to produce this feature? Explain your choice.

_____

_____

_____

Samples of pumice (left) and scoria (right).

(continued)

Name: _____ Section: _____
Course: _____ Date: _____

(b) Study the photos of pumice and scoria. What is the rock texture illustrated?

_____

Compare the size of the holes in each sample. Why do you think the holes are smaller in the sample of pumice than in the scoria sample?

_____

_____

_____

What style of eruption produced these rocks? How do you know?

_____

_____

_____

An electron photomicrograph of ash.

(c) The photo above has the smallest-sized fragments that occur in volcanic eruptions. What does this small size indicate about the style of eruption that occurred?

_____

_____

_____

How did gas content play a role in the formation of this deposit?

_____

_____

_____

5.3 Determining Volcanic Activity: Magma Viscosity

In Section 5.2.1 we observed the difference in eruptive style between more fluid and thicker magmas. Molten rock is liquid, and like all liquids it can flow easily or sluggishly, depending on its **viscosity** (resistance to flow). Viscosity originates from the internal friction generated by the chemical bonds within the liquid. High-viscosity magmas have more internal friction between molecules and as a result can flow only slowly. In contrast, low-viscosity magmas have less internal friction and therefore flow rapidly.

In this section, you will investigate how the viscosity of magmas is controlled by three factors: composition, temperature, and dissolved gas content. Understanding the way that magmas and lavas flow is important to geologists who try to predict volcanic eruptions and to those living in actively volcanic areas.

5.3.1 Temperature Effects on Viscosity

Magmas come in a range of temperatures, typically from 700°C to 1300°C (1300°F–2400°F). As liquids get hotter, their molecules move more freely as they have less *surface tension* (the cohesive forces between liquid molecules), and less internal friction. Therefore, hotter magmas have lower viscosities and flow more rapidly, whereas cooler magmas have higher viscosities and flow more slowly.

5.3.2 Magma Composition and Viscosity

When determining the viscosity of a given magma, the most important compositional factor is the amount of *silica* (SiO_2) present. Within magmas—well before crystallization takes place—silicon and oxygen bond together to form the silicon-oxygen tetrahedron molecules $(SiO_2)^+$ that are the building blocks of the silicate minerals found in igneous rocks (**FIG. 5.8**). These tetrahedra bond to each other, attracted by electrical forces between the molecules. As more and more bonds are formed, they begin to create a "network" of silica tretrahedra that increases the friction between all other atoms within the magma. This friction makes the magma more viscous. **TABLE 5.1** shows the silica content of the common magma types. Since mafic magmas have lower silica content, they have lower viscosities; intermediate and felsic magma have higher silica content, and therefore have higher viscosities.

In Exercise 5.2, we will work with the movement of liquids, simulating the movement of magma and lava, and investigate how both temperature and composition of liquids affect their viscosities.

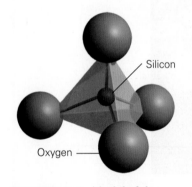

FIGURE 5.8 Model of the silicon-oxygen tetrahedron.

Table 5.1 **Silica content of common magma types**

Silica content	Magma type
~50%	Mafic
~60%	Intermediate
~65-70%	Felsic

Name: _____   Section: _____

Course: _____   Date: _____

In this exercise, you will work with five liquids of varying compositions, provided by your instructor: liquids A, B, C, D, and E. (*Note:* If this exercise is available and being conducted online, the setup, materials, and directions may differ. See the instructions in the online lab for how to conduct the exercise.)

(a) First take a look at the samples. Can you tell that they are different liquids? Explain your reasoning:

_____

_____

_____

Now invert a couple of the containers (first be sure they are tightly sealed!) and note the air bubbles that rise through the liquids. You may see that the bubbles rise faster in some of the containers than others. While inverted, mark a line at a point a bit below where the top of the liquid sits on each container (if not already done for you). Now return the containers right-side up. In this experiment, you will be asked to record the seconds it takes for the bubble in each liquid to rise to the line you marked in the table below. You will be recording this time at three different temperatures to investigate how temperature affects the viscosity of different fluids.

Viscosity of unknown liquids data sheet

	Time required for air bubble to rise to the top of the container (seconds)		
Sample	Ice-water bath (___°C)	Room temperature (___°C)	Hot-water bath (___°C)
A			
B			
C			
D			
E			

Using a thermometer, measure the room temperature in your lab and record it in the data table above. Now, using a stopwatch, time how long it takes for the bubble in sample A to rise to the line you marked on the container when you invert it. Record this time in the "Room temperature" column in the table above. Repeat for samples B through E.

(b) Based on your room-temperature results, is there a liquid that appears the most viscous? If so, which one? Explain your choice:

_____

_____

_____

Least viscous? Explain your choice: _____

_____

_____

To see the effect of temperature on viscosity, prepare an ice-water bath. When its temperature is at or near 0°C (32°F), record this temperature in the data table. Place sample A in the ice-water bath—but do not submerge it—and let it cool for 2 minutes. Once the 2 minutes are up, remove sample A from the ice-water bath and measure the time required for the bubble to rise to the line, just as you did earlier. Record this time in the table in the "Ice-water bath" column. Perform the same steps for samples B through E.

(continued)

Name: _____   Section: _____
Course: _____   Date: _____

(c) At the ice-water temperature, which sample is most viscous? _____ Has the viscosity of the liquids changed because of temperature? _____ Which sample is least viscous? _____ As far as you can measure it, has the viscosity of this liquid changed? _____

Now, let's experiment with another change in temperature. Follow your instructor's directions to either place your samples in a hot-water bath or hold them under hot running water in a sink. Keep the samples in the hot water for at least 2 minutes. Measure the temperature of the hot water and record it in the table.

Once the samples have warmed, measure the time it takes for the air bubbles to rise to the line marked on each sample, exactly as you did before. Record the results for samples A through E in the "Hot-water bath" column in the table.

(d) In the hot-water bath, which liquid sample is most viscous? _____

Recalling that chemical bonds affect all properties of materials in any state, why do you think this liquid has the highest viscosity of the set?

_____

_____

_____

(e) Compare the times for the most viscous samples in the ice-water bath and the hot-water bath. How many more seconds does it take the air bubble to rise to the top at the ice-water temperature than at the hot-water temperature? _____

Which sample, if any, has the least measurable change in viscosity at all three temperatures? _____

Make a line graph of your results on the graph paper provided at the end of this chapter. Label the axes with values for seconds and temperatures. For each sample, plot three marks for the three data points, using a different color for each sample. Connect the data points with a line of the same color. Be sure to include a legend for the colors and their corresponding samples, and give the graph a title.

You just observed that liquids of different compositions have varied viscosities. Therefore, it should not be surprising that magmas of different compositions have different viscosities. As we discussed earlier, mafic magmas (those with low silica content) are "runnier" than felsic magmas (those with high silica content). Additionally, when erupted, mafic magmas are hotter than felsic magmas.

(f) Judging by the graph you just made, which samples would have viscosities similar to mafic magmas?

_____

Similar to felsic magmas? _____

5.3.3 Gas content and Viscosity

While temperature and composition have predictable effects on magma viscosity, a more unpredictable variable is the amount of dissolved gas. The gases (mainly water vapor, carbon dioxide, and minor amounts of sulfur, chlorine, and fluorine) are dissolved in the magmas while the magma is under pressure from the surrounding rocks—this pressure is called the *confining pressure*. An example of an everyday liquid under confining pressure is the carbonated water used in soft drinks. Carbonated water is water containing dissolved carbon dioxide gas (CO_2); in order to dissolve the CO_2, it must be bottled under pressure. As the cap is removed from the carbonated beverage, the confining pressure is released and the inherent pressure of the CO_2 gas (vapor pressure) allows the gas to be released as bubbles. Similarly, as magma rises toward the surface, the confining pressure decreases and tiny gas bubbles form. At the beginning, the bubbles prevent the linking of silica tetrahedra, keeping the viscosity of the magma low. However, as more bubbles form and gas escapes from the magma, the viscosity of the remaining liquid begins to increase as the silica tetrahedra begin to form links.

Experiments have shown that the overall viscosity of a magma depends on both the amount of gas bubbles the liquid contains and how those bubbles are distributed within the magma. These gases are the most important driving force behind the styles of eruptions that occur and their eruptive products. The ease with which the gas escapes from the magma controls the vapor pressure and in turn how explosive an eruption will be. As we see in the next section, the explosive nature of an eruption determines the volcanic landforms that it creates.

5.4 Volcanic Landforms and Types of Volcanoes

When asked to describe volcanic eruptions, most people immediately think of the mountains we call volcanoes, built from within the Earth by lava and/or pyroclastic material. It is first important to note that not all eruptions build volcanoes. For example, highly fluid basalt lavas that flowed from long fissures now underlie broad, flat areas such as the Columbia Plateau in Washington, Oregon, and Idaho (**FIG. 5.9**).

However, most eruptions *do* build volcanoes, which fall into four types based on whether they are made of lava, pyroclastic material, or a unique combination of

FIGURE 5.9 Aerial view of part of the Columbia Plateau basalt flows.

both. Two types are made almost entirely of lava. **Shield volcanoes** (FIG. 5.10A) are formed by eruptions of highly fluid lava. The lava's fluidity produces broad structures with gentle slopes—the largest volcanoes on Earth (and Mars) are shield volcanoes. Because ketchup or maple syrup has a consistency similar to that of highly fluid lava, you can simulate the formation of a shield volcano by pouring one of them onto a flat surface.

Lava domes (FIG. 5.10B) are much smaller than shield volcanoes, often forming in the craters of other types of volcanoes. As their name implies, they too are made of lava, but in this case the lava is extremely viscous, more like toothpaste or cookie dough than ketchup. As a result, the lava can pile up to form steeper slopes than those of shield volcanoes. To simulate formation of a lava dome, squeeze a blob of toothpaste onto a flat surface.

Eruptions of lava domes and shield volcanoes are the least violent of all volcanic eruptions, but even they can pose serious hazards. Lava from Hawaiian shield volcanoes, such as the eruption of Kilauea in 2018 (see Figure 5.1a), has buried homes, forests, and fields, displacing people and blocking roads at a cost of millions of dollars.

Cinder cones (FIG. 5.10C) are made mostly from relatively fine-grained pyroclastic material. They are formed partially from *above*, when ash erupted from the

FIGURE 5.10 The four major types of volcanoes.

(a) A shield volcano (fluid lava): Mauna Loa, Hawaii.

(b) A lava dome (viscous lava) at the Soufriere Hills volcano in Montserrat, Caribbean.

(c) A cinder cone (pyroclastic material): Sunset Crater, Arizona.

(d) A stratovolcano (pyroclastic material plus lava): Mt. Mayon, Philippines.

volcano's vent falls to the ground, like sugar being poured into a pile on a plate. Eruptions that build cinder cones are relatively peaceful, and the resulting volcanoes are much smaller than shield volcanoes, with moderate slopes.

Stratovolcanoes (**FIG. 5.10D**) include some of the most famous volcanoes on Earth: Vesuvius, Mt. Etna, Denali, Mt. Rainier, Mt. St. Helens, and Mt. Fuji, to name only a few. This type of volcano has by far the most violent eruptions and poses the greatest danger to humans. Stratovolcanoes are made of large amounts of both lava and pyroclastic material. They have the steepest slopes of all volcanoes because the pyroclastic material is so hot when it accumulates that the particles are welded together, so the pile is not flattened by the pull of gravity. Stratovolcanoes are most commonly found in subduction zones, and their explosive eruptions are caused by large amounts of steam produced by melting associated with the subducting plate.

EXERCISE 5.3 | **Comparing and Contrasting Stratovolcanoes, Shield Volcanoes, and Cinder Cones**

Name: _____ Section: _____
Course: _____ Date: _____

There are two ways in which we can observe the form of a volcano based on its deposits: directly by experiment and indirectly by extrapolation. In each case, you can determine the volcano's angle of repose—how steep a slope can be before it fails, or collapses. The angle of repose is controlled by particle shape and size, which in turn is determined by the erupted lava's composition, viscosity, and eruption style. First, we will investigate by means of a simple experiment.

(a) Gather the materials provided by your instructor. (*Note:* If this exercise is available and being conducted online, the setup, materials, and directions may differ. See the instructions in the online lab for how to conduct the exercise.) Lay the large piece of cardboard or poster board on your table. Pour about one cup of one size of particle onto the board and try to make as steep a cone as possible. Then, take the protractor and measure the angle of the slope of the cone from the horizontal base (see the figure below for reference). It might make it easier to see and measure the angle if you take another piece of cardboard and place it behind the cone when you measure the slope. Record the slope in the data table below. Using a metric ruler, determine the average diameter of the particles you used. Record this in the data table. Repeat this process for the other two particle sizes.

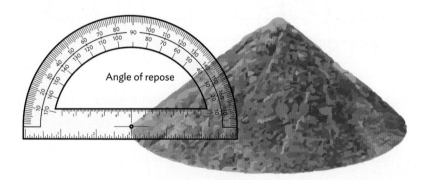

Angle of repose

Kind of particle	Average particle size (mm)	Angle of repose (degrees)

(continued)

Name: _____   Section: _____

Course: _____   Date: _____

(b) Make a bar graph of particle size versus angle of repose. Use a different color for each kind of particle and include a legend. Label the axes.

(c) Based on your graph, describe how the angle of repose changes with particle size:

_____

_____

_____

(d) Let's apply your results to one type of volcanic landform: a cinder cone. The slope at the top of a cinder cone tends to be steeper than near its base. Explain this difference in slope in terms of particle size at the summit versus the base.

_____

_____

_____

(continued)

EXERCISE 5.3 Comparing and Contrasting Stratovolcanoes, Shield Volcanoes, and Cinder Cones *(continued)*

Name: ______________________ Section: ____________
Course: ______________________ Date: ____________

Now let's determine the characteristics of a volcano indirectly, by measuring the slopes from images of volcanoes, such as those shown below. For each image, use a protractor to measure the angle of repose of the slope at two heights: (1) about a third of the way down from the top, and (2) near the base. Record your measurements in the table below.

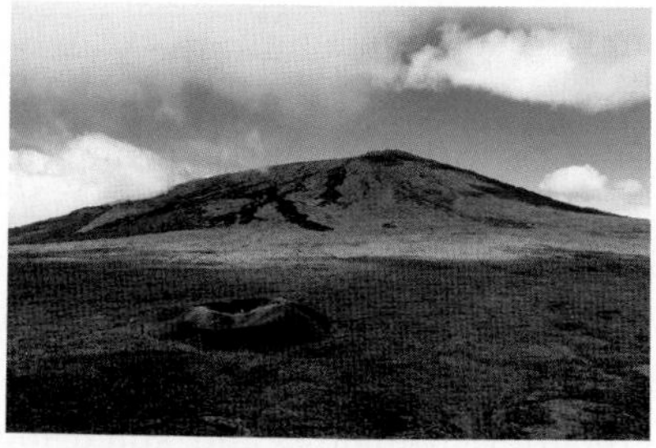

A shield volcano: Mauna Kea Volcano in Hawaii.

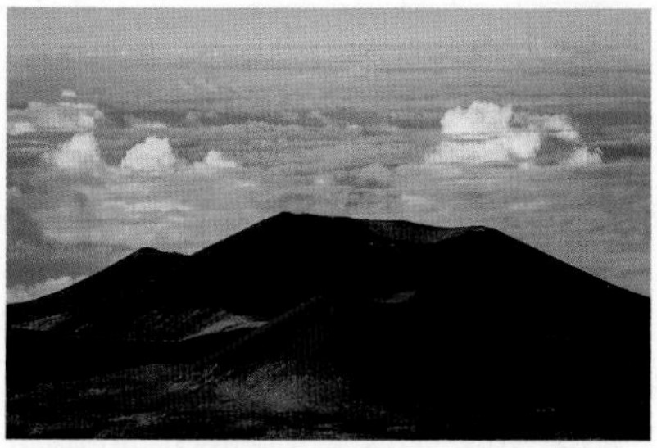

Cinder cones atop Mauna Kea Volcano in Hawaii

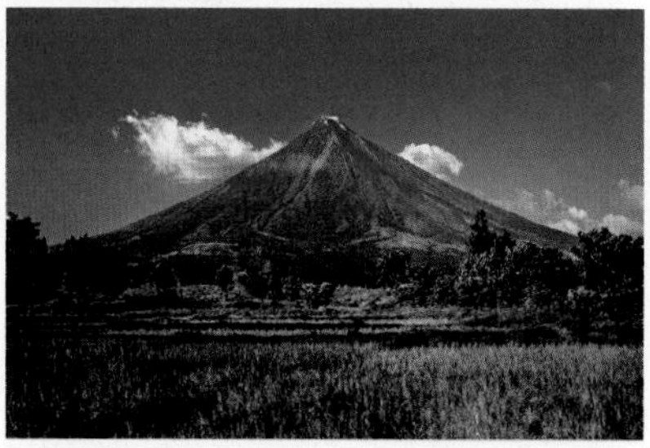

A stratovolcano: Mt. Mayon Volcano in the province of Albay, Bicol, Philippines.

	Angle of repose near top (degrees)	Angle of repose near base (degrees)
Shield volcano		
Cinder cone volcano		
Stratovolcano		

(e) From this information alone, do you think the lava that flows from each volcano is largely silica-rich (felsic) or silica-poor (mafic)? High-temperature lava or low-temperature lava? Explain.

____

____

____

____

____

(f) For each volcano you measured, what types of eruptions would you expect—gentle or explosive? Support your choices with evidence.

____

____

____

____

____

____

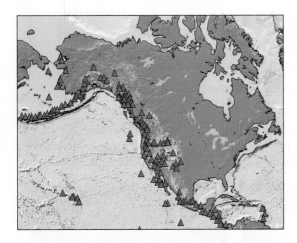

FIGURE 5.11 **Distribution of volcanoes in North America.**

5.5 Volcanic Hazards

The hazards posed by volcanoes may seem limited to only a few areas of the Earth. However, with a rapidly increasing population and the expansion of urban areas, more people are living dangerously close to volcanoes today. Indeed, hundreds of millions of people literally live *on* volcanoes, including citizens of some of the world's most populous nations; more than half a billion people combined live on volcanoes in Indonesia (the fourth most populous country), Japan (the tenth), and the Philippines (the thirteenth) alone. Millions more people in South America, Italy, Africa, and, yes, the United States (**FIG. 5.11**) also live on or close to active volcanoes. All must cope with the potential impact of volcanic eruptions.

The power of a major stratovolcano eruption is, in the true meaning of the word, awesome. The 1883 eruption of Krakatau in Indonesia was heard 4,800 kilometers (3,000 miles) away (greater than the distance between Los Angeles and New York!), created 30-meter (100-foot) tsunamis, and killed more than 35,000 people. The 1912 eruption of Novarupta in Alaska is described by the U.S. Geological Survey as the loudest sound of the twentieth century, dwarfing all

FIGURE 5.12 **Impacts of stratovolcano eruptions.**

(a) Trees from a forest scattered like pick-up sticks by the shock wave of an explosive eruption.

(b) Plaster casts of bodies of Roman citizens of Pompeii, preserved in the very fine-grained ash that buried them.

(c) Pyroclastic flow from a stratovolcano.

(d) A volcano-generated mudflow (lahar) destroys homes near a stratovolcano.

nuclear bomb tests. It should come as no surprise that a stratovolcano eruption in or near a major population center would be a huge natural disaster.

To understand the severity of the hazard, we need to understand the many different ways the eruption of a stratovolcano can affect the local area and areas as far as thousands of miles away (**FIG. 5.12**):

- *The explosive blast:* The sound effects reported above are an indication of the energy that can be released suddenly when a stratovolcano erupts explosively. When Mt. St. Helens erupted in 1980, the blast devastated areas up to 24 kilometers (15 miles) from the volcano, tossing entire forests around like pick-up sticks (Fig. 5.12a). If the blast is underwater, it may trigger devastating tsunamis.

- *Ash fall:* Enormous volumes of ash may be blasted into the air and cover the ground hundreds of kilometers from the volcano (**FIG. 5.13**). Entire villages may disappear, buried beneath many meters of ash that fall quietly but relentlessly (see Fig. 5.12b). The area affected by ash fall from the 1980 Mt. St. Helens eruption is impressive (red zone in Figure 5.13), but pales in comparison with those from older eruptions. Some "super eruptions" have global impacts. In 1815, the Mt. Tambora (Indonesia) eruption put so much ash into the atmosphere that the summer of 1816 never happened. Temperatures were far below normal, crops failed, and food shortages were widespread. Modern technology leads to new problems associated with volcanic ash. An ash cloud from the 2010 eruption of Eyjafjallajökull (go ahead—try to pronounce it) in Iceland disrupted air traffic between North America and Europe for weeks because trans-Atlantic planes had to be diverted around it to avoid damaging their engines.

- *Pyroclastic flows* (sometimes called *nuées ardentes*, French for "burning clouds"): Red-hot, glowing clouds of ash, steam, and toxic gases erupted from stratovolcanoes roll downslope and speed across the surrounding countryside at velocities measured between 600 and 800 kilometers per hour (400–500 mph) (Fig. 5.12c). A nuée ardente from the 1902 eruption of Mt. Pelée, on the Caribbean island of Martinique,

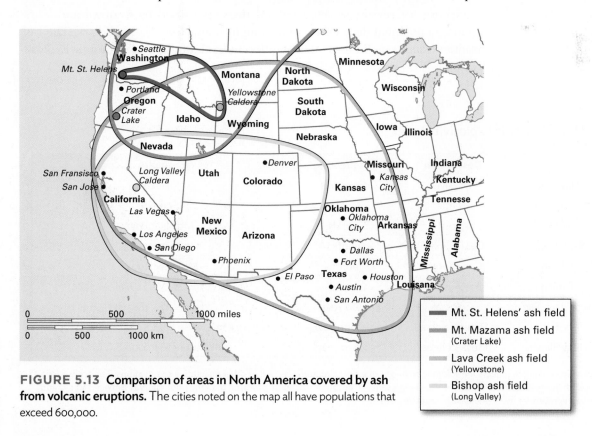

FIGURE 5.13 Comparison of areas in North America covered by ash from volcanic eruptions. The cities noted on the map all have populations that exceed 600,000.

FIGURE 5.14 St. Pierre, Martinique, before (left) and after (right) the 1902 nuée ardente.

flattened the capital city of St. Pierre in minutes (**FIG. 5.14**), bent 6-meter-high (20-foot) iron gates like pretzels, and killed nearly all of the 50,000 people who had fled to the city for safety.

■ *Lahars* (volcanic mudslides): Tiny ash particles shot into the atmosphere act as seeds for rain, so heavy downpours commonly accompany stratovolcano eruptions. The rain mixes with loose ash, forming a slurry that flows downhill with enormous power, dams rivers, and can bury local communities (Fig. 5.12d). Lahars can also occur in volcanic areas even when no eruptions are taking place. Heavy seasonal rains or the melting of glaciers on the summits of volcanoes can also trigger deadly lahars.

Exercise 5.4 reviews selected natural disasters caused by stratovolcanoes and explores a potential disaster in the making in Yellowstone National Park. The case histories examine volcanoes rated at different levels of the USGS Volcanic Explosivity Index (VEI) shown in **TABLE 5.2**.

TABLE 5.2 USGS Volcanic Explosivity Index

VEI	Type	Description	Frequency	Examples
0	Hawaiian	Effusive	Continuous	Kilauea
1	Strombolian	Gentle	Daily	Stromboli
2	Vulcanian	Explosive	Biweekly	Mt. Sinabung
3	Sub-Plinian	Catastrophic	3 months	Mt. Lassen, 1915
4	Plinian	Cataclysmic	18 months	Mt. Pelée, 1902
5	Plinian	Paroxysmic	12 years	Vesuvius, 79 C.E.; Mt. St. Helens, 1980
6	Plinian–Ultra-Plinian	Colossal	50–100 years	Krakatau, 1883
7	Ultra-Plinian	Super-colossal	500–1,000 years	Mt. Mazama, 5600 B.C.E.; Thera, 1630 B.C.E.
8	Ultra-Plinian	Mega-colossal	>50,000 years	Yellowstone, 640,000 B.C.E.

Note: The VEI for a given eruption reflects the volume and height of erupted material, as well as other qualitative observations. The scale is logarithmic, meaning each interval between VEI-2 and VEI-8 represents an eruption 10× more powerful than the prior interval. Between VEI-0 and VEI-2, each interval represents a 100×-increase in eruptive power.

Name: _____ Section: _____

Course: _____ Date: _____

I. Vesuvius, Italy [VEI 5]: Vesuvius erupted in 79 C.E., burying an estimated 16,000 people beneath 20 to 60 feet of ash. The towns of Pompeii and Herculaneum were wiped from the map, their locations lost to the world until the 1700s. The eruption was witnessed by the Roman admiral Pliny (who died on a ship in the harbor), and this type of stratovolcano eruption is named *Plinian* in his honor. The very fine-grained ash preserved the shapes of residents, animals, and household objects so well that archeologists have been able to reconstruct minute details of Roman life in the 1st century C.E., such as the size and shape of loaves of bread (see Fig. 5.12b).

Vesuvius remains highly active today, having undergone 53 "significant" eruptions since 79 C.E. and hundreds of small ones. None were as explosive or damaging as the 79 C.E. event. Most involved lava with varying amounts of ash fall, pyroclastic flows, and lahars, and in 1944 lava flowed into the suburbs of the modern city of Naples. Increasing earthquake and steam vent activity since 2012 suggest that magma is again rising toward the surface, and thus may be precursors to the next eruption. Whether it will be as explosive as the 79 C.E. eruption or not, its effects on Naples could be severe. Despite ample evidence of the volcano's activity, the NASA image of Vesuvius below shows that the city has expanded far beyond its boundaries in Roman days. Today, nearly 4 million people occupy a metropolitan area that is encroaching on the slopes of the volcano.

(a) The lateral blast from the 1980 eruption of Mt. St. Helens extended outward approximately 15 miles from the crater. Vesuvius and Mt. St. Helens have the same VEI rating, so a similar event at Vesuvius is certainly possible. Using the distance scale on the Naples image, outline a 15-mile blast radius around Vesuvius.

(b) Based on this blast radius, how much of the Naples metropolitan area do you think is endangered and should be evacuated?

_____

_____

_____

(c) Based on the past history of Vesuvius, what volcanic hazards besides an explosive blast would probably develop?

_____

_____

_____

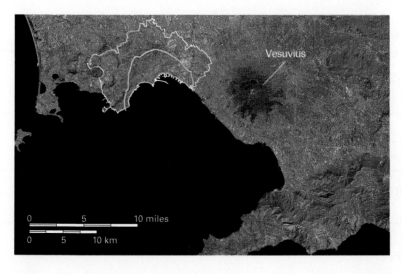

False-color satellite image showing Vesuvius and Naples. Naples' current metropolitan area is outlined in white. The city boundaries during Roman times is outlined in yellow.

(continued)

Name: _____ Section: _____
Course: _____ Date: _____

II. Cascade Range, western North America [VEI 3-7]: The Cascade Range is a chain of stratovolcanoes that extends from Northern California to southern British Columbia. The range is being formed by the subduction of a small plate within the Pacific Ocean (the Juan de Fuca Plate) beneath North America. Its prominent peaks include Mts. Rainier, Baker, Hood, Shasta, Lassen (VEI = 3), and St. Helens (VEI = 5), as well as Mt. Mazama, the apparently extinct stratovolcano whose eruption produced what is now Crater Lake (VEI = 7).

Mt. St. Helens has been by far the most active of these peaks, and it and Mt. Lassen are the only peaks to have erupted since 1900. In 2004, 24 years after its explosive 1980 eruption, Mt. St. Helens erupted again, this time quietly, exuding mostly viscous lava. In 2017, around the anniversary of the 1980 eruption, geologists detected several swarms of small earthquakes beneath the mountain, similar to those that had preceded that eruption. Preliminary analysis suggests that the earthquakes indicate magma rising from depth, but an imminent eruption is not anticipated.

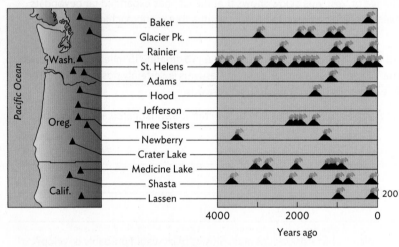

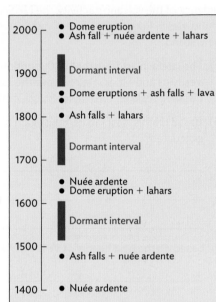

Locations of major Cascade Range stratovolcanoes and their eruptive histories over the past 4,000 years (left). Eruptive history of Mt. St. Helens since 1400 C.E. (right)

Using the right-hand figure (above) of the timeline and nature of significant eruptions at Mt. St. Helens as a reference, answer the following questions:

(d) What is the average number of years that the volcano has remained dormant between eruptions during this period?

_____

(e) Does the active/dormant cycle appear to be regular or erratic? Explain.

_____

_____

_____

(f) Suggest a geologic explanation for this pattern. Explain.

_____

_____

(continued)

Name: _____ Section: _____
Course: _____ Date: _____

(g) Based on the average dormant interval following significant eruptions, predict when the next significant eruption is likely to occur.

_____

(h) What factors make such predictions difficult and sometimes inaccurate? How do the 2004 eruption and 2017 earthquake swarms affect your prediction?

_____

_____

_____

_____

_____

_____

_____

? **What Do You Think** Every time Old Faithful spews steam into the air in Yellowstone National Park, it reminds us that there is hot magma beneath the park capable of producing an eruption. This deceptively peaceful volcanic activity is not the first to occur in the Yellowstone area. About 640,000 years ago, one of the most violent eruptions ever detected (VEI = 8, the maximum) blasted ash over much of the central and southwestern United States, blanketing the area shown in Figure 5.13 and dwarfing the areas affected by ash from other major eruptions.

Comparison of the Yellowstone eruption with the effects of Mt. St. Helens gives us an idea of the difference between VEI 5 and VEI 8 eruptions. We cannot predict when or if a VEI 8 eruption will happen at Yellowstone again, but if it did, what would be its effect on the areas directly impacted by the ash? Which cities and states would suffer the most?

Consider effects on the local economy, health, food supplies, and transportation networks. And although not all of the United States would be directly affected by the ash fall or explosion, what do you think the effects would be on the rest of the country? Answer on a separate sheet of paper.

Name: _____   Section: _____

Course: _____   Date: _____

Exploring Earth Science Using Google Earth

1. Visit **digital.wwnorton.com/labmanualearthsci**
2. Go to the **Geotours** tile to download Google Earth Pro and the accompanying Geotours exercises file.

Expand the Geotour05 folder in Google Earth by clicking the triangle to the left of the folder icon. The folder contains placemarks keyed to questions that investigate the differences between the three basic types of volcanoes: shield, stratovolcano, and cinder cone.

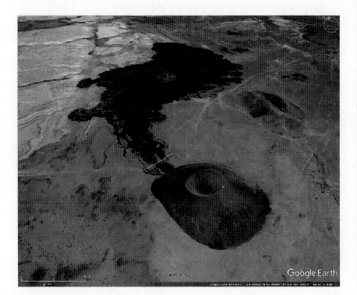

(a) Check and double-click the S P Mountain placemark to fly to a volcano in Arizona. This relatively small volcano is comprised of mostly pyroclastic material with a late-stage basaltic lava flow erupting from the base of the volcano. Using descriptions from the volcanic landforms section, what type of volcano is S P Mountain?

_____

(b) Check and double-click the Payun Matru placemark to fly to a volcano in Argentina. This large volcano is dotted with numerous, small parasitic volcanoes and is covered by many dark, fluid lava flows. The relatively gentle slopes also are characteristic of this type of volcano. What type of volcano is Payun Matru? _____

(c) Check and double-click the Tao-Rusyr placemark to fly to a volcano in the Kuril Islands of Russia. This volcano resembles Crater Lake NP in the United States in that it violently erupted to form a caldera in which a large lake now resides, with a resurgent volcano building up out of the lake (note that the volcano lies within a volcanic arc above a subduction zone). What type of volcano is Tao-Rusyr? _____

(d) Check and double-click the Parasitic Volcano placemark to fly back to Payun Matru in Argentina. Given your knowledge of volcanic landforms, what type of volcano are these small parasitic volcanoes?

_____

Given their composition, can you hypothesize why several of these volcanoes are asymmetric to the east?

_____

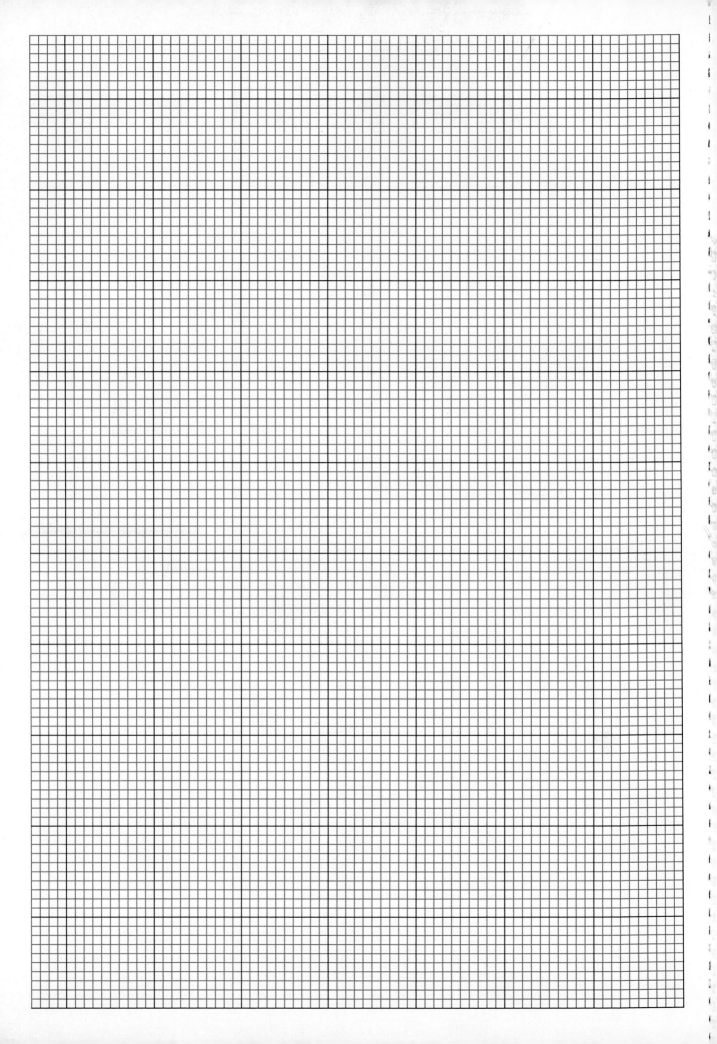

Interpreting Geologic Structures on Block Diagrams, Geologic Maps, and Cross Sections

6

Stress applied to rock produces deformation, such as the recumbent folds and thrust fault seen in this image of a coastal cliff in Cornwall, United Kingdom.

- Define common geologic structures such as folds and faults
- Visualize structures in three dimensions using block diagrams, maps, and cross sections
- Identify folded and faulted rock based on landscape features
- Interpret the geologic structure of an area from a geologic map

MATERIALS NEEDED

- Colored pencils
- A fine-tipped black pen
- Tracing paper
- A pair of scissors and tape
- An azimuth protractor (included in the GeoTools section at the back of this manual)
- A straightedge

6.1 Introduction

The Earth is a dynamic place! Over time, lithosphere plates move relative to one another: at convergent boundaries, one oceanic plate sinks into the mantle beneath another plate, or continents collide; at rifts, a continental plate stretches and may break apart; at a mid-ocean ridge, two oceanic plates move away from each other; and at a transform boundary, two plates slip sideways past each other. All these processes generate stress that acts on the rocks in the crust. In familiar terms, stress refers to any of the following (**FIG. 6.1**): **pressure**, which is equal squeezing from all sides (Fig. 6.1a); **compression**, which is squeezing or squashing in a specific direction (indicated by the inward-pointing arrows in Fig. 6.1b); **tension**, which is stretching or pulling apart (indicated by the outward-pointing arrows in Fig. 6.1c); or **shear**, which happens when one part of a material moves laterally, or sideways, relative to another part (indicated by adjacent arrows pointing in opposite but parallel directions in Fig. 6.1d).

The application of stress to rock produces **deformation**, which includes many phenomena:

- the displacement of rocks on sliding surfaces called **faults**
- the bending or warping of layers to produce arch-like or trough-like shapes called **folds**
- the overall change in the shape of a rock body by thickening or thinning, often producing a texture called **foliation**, caused by the alignment of platy or elongate minerals

The products of deformation, such as faults, folds, and foliation textures, are called **geologic structures**. Some geologic structures are very small and can be seen in their entirety within a single hand specimen. Typically, however, geologic structures in the Earth's crust are large enough to affect the orientation and geometry of rock layers, which in turn may control the pattern of erosion and, therefore, the shape of the land surface. Geologists and other Earth scientists represent the shapes and configurations of geologic structures in the crust with the aid of three kinds of diagrams. A **block diagram** is a three-dimensional representation of a region of the crust that depicts the configuration of structures on the ground (the map surface) as well as on one or two vertical slices into the ground (cross-sectional surfaces).

FIGURE 6.1 Kinds of stress.

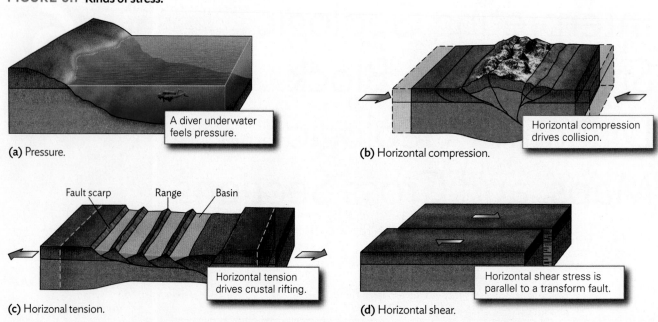

(a) Pressure.

A diver underwater feels pressure.

(b) Horizontal compression.

Horizontal compression drives collision.

(c) Horizontal tension.

Fault scarp   Range   Basin

Horizontal tension drives crustal rifting.

(d) Horizontal shear.

Horizontal shear stress is parallel to a transform fault.

A **geologic map** represents the Earth's surface as it would appear looking straight down from above, showing the boundaries between rock units and where structures intersect the Earth's surface. A **cross section** represents the configuration of structures as seen in a vertical slice through the Earth. **FIGURE 6.2** shows how these different representations depict Sheep Mountain in Wyoming.

The purpose of this chapter is to help you understand geologic structures and to visualize those structures and other geologic features by examining block diagrams, cross sections, and geologic maps. In addition, this chapter will help you to see how geologic structures help to shape the landscape. Geologic structures can be very complex, and in this chapter we can work with only the simplest examples. Again, our main goal here is to help you develop the skill of visualizing geologic features in three dimensions.

FIGURE 6.2 Geology of Sheep Mountain in Wyoming.

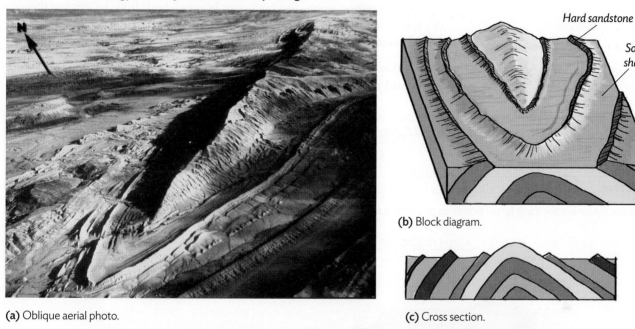

(a) Oblique aerial photo.

(b) Block diagram.

(c) Cross section.

(d) Geologic map. The color bands represent rock units. The map corresponds to the photo in part (a)—though it is rotated approximately 43° clockwise.

6.2 Beginning with the Basics: Contacts and Attitude

6.2.1 Geologic Contacts and Geologic Formations

When you look at Figure 6.2d, you see patterns of lines. What do these lines represent? Each line is the trace of a **contact**, the boundary between two geologic units. In this context, a **trace** is simply the line representing the intersection of a planar feature with the plane of a map or cross section. Geologists recognize several types of contacts:

- An **intrusive contact** is the boundary surface of an intrusive igneous body.
- A **conformable contact** is the boundary between successive beds, sedimentary formations, or volcanic extrusions in a continuous stratigraphic sequence.
- An **unconformable contact** (or **unconformity**) occurs where a period of erosion and/or nondeposition has interrupted deposition. It can often be detected by noting where the boundary between beds is no longer parallel (we'll look at unconformities more closely later in this chapter).
- A **fault contact** is where two units are juxtaposed across a fault.

Throughout this chapter, you'll gain experience interpreting contacts, but to be sure you understand these definitions from the start, complete Exercise 6.1.

EXERCISE 6.1	The Basic Types of Contacts

Name: _____   Section: _____
Course: _____   Date: _____

In the figure below, each arrow points to one of the four basic kinds of contacts. Add the labels.

Paleozoic sedimentary strata

Cretaceous granite

Precambrian gneiss

Note: Geologic Time units (Paleozoic, Cretaceous, Precambrian) are included in the labels in the figure above, but are not relevant to the exercise. Geologic time is discussed in Lab 8.

6.2.2 Describing the Orientation of Layers: Strike and Dip

You can efficiently convey information about the orientation, or **attitude**, of any planar geologic feature, such as a bed or a contact, by providing two numbers. The first number, called the **strike**, is the compass direction of a horizontal line drawn on the surface of the feature (**FIG. 6.3A**). We can give an approximate indication of strike by saying "the bed strikes northeast," or we can be very exact by saying "the bed has a strike of 45° east of north," meaning that there is a 45° angle between the strike line and due north, as measured in a horizontal plane. The second number, called the **dip**, is the angle of tilt or the angle of slope of the bed, measured relative to a horizontal surface. A horizontal bed has a dip of 00°, and a vertical bed has a dip of 90°. A bed dipping 15° has a gentle dip, and a bed dipping 60° has a steep dip. The direction of dip is perpendicular to the direction of strike (**FIG. 6.3B, C**). Because strike represents a line with two ends, a strike line actually trends in two directions: a strike line that trends north must also trend south; one that trends northwest must also trend southeast. How do we pick which direction a strike line trends? By convention, strike is read relative to north, so you will generally only see strikes described as angles east or west of north, or due north, east, or west (**FIG. 6.3D**). Thus, the beds on the left side of Figure 6.3a strike 38° east of north. Also by convention, beds that dip directly to the north are considered to strike west, while those that dip directly to the south strike east.

Geologists write the strike as a three-digit number, based on the division of the compass dial into 360 degrees, or azimuths (**FIG. 6.3E**). A strike of 000° (or 360°) means the bed strikes due north; a strike of 045° means that the strike line trends 45° east of north (i.e., northeast); a strike of 090° is 90° east of north (i.e., due east); and a strike of 320° is 40° west of north. Writing strike as a three-digit number is referred to as **azimuth notation**. Because strike is measured relative to north, the allowable azimuth values are 000° to 090° for a northeast- to east-trending strike line and 270°

FIGURE 6.3 Strike and dip show the orientation of planar structures.

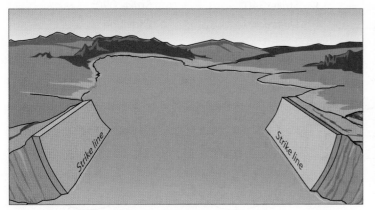

(a) A strike is the intersection of a horizontal plane with the bed surface. Strike lines for two sets of beds oriented in opposite directions are shown here.

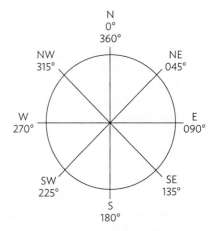

(b) Strike line and dip direction shown for tilted beds at Turners Falls in Massachusetts.

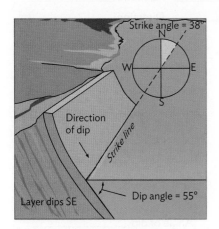

(c) This detail of the beds on the left side of part (a) shows their strike and dip, which are used to describe the orientation, or attitude, of these beds.

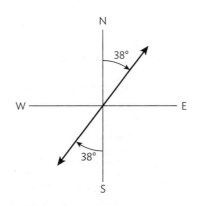

(d) The strike line in part (c) can be described as trending 38° northeast or 38° southwest. By convention, geologists will say the strike is N 38°E or 038°.

(e) Strike can also be specified by azimuths (compass angles between 000°and 360°). In azimuth notation, the strike line of N 38°E is written as 038° in part (c).

to 360° for a northwest- to west-trending strike line. We write the dip as a two-digit number (an angle between 00° and 90°) followed by a general direction. Let's consider an example: if a bed has an attitude of 045°/60° NW, we mean that it strikes northeast and dips steeply northwest. A bed with an attitude of 053°/72° SE strikes approximately northeast and dips steeply to the southeast. Exercise 6.2 will give you some practice measuring strike and dip.

EXERCISE 6.2	Determining Strike and Dip

Name: _____   Section: _____
Course: _____   Date: _____

(a) Use the azimuth protractor in the back of the manual to draw the indicated strike lines on each compass below. Translate each azimuth reading into a direction (e.g., northeast).

Strike: 060°   Strike: 340°   Strike: 090°
Direction: _____   Direction: _____   Direction: _____

(b) Use the protractor to measure the strikes below. Give the strike direction in azimuth notation and describe the direction (e.g., northeast).

(Here the bed dips north.)
Azimuth: _____   Azimuth: _____   Azimuth: _____
Direction: _____   Direction: _____   Direction: _____

(c) Use the protractor to measure the angle of dip for the two beds below. The dashed lines indicate the level of the horizontal. The shaded surfaces are the surfaces of the beds. Using the strikes given beneath the diagram for each bed, indicate the rough direction of dip (e.g., west northwest) for each bed. Remember, the direction of dip is perpendicular to the direction of strike.

Strike: 0°   Strike: 290°
Angle of dip: _____   Angle of dip: _____
Direction of dip: _____   Direction of dip: _____

6.3 Working with Block Diagrams

We start our consideration of how to depict geologic features on a sheet of paper by considering block diagrams, which represent a three-dimensional chunk of the Earth's crust using the artist's concept of perspective (**FIG. 6.4A**). Typically, geologists draw blocks so that the top surface and two side surfaces are visible. The top surface is called the **map view**, and the side surfaces are **cross-section views**. In the real world, the map view would display the topography of the land surface, but for the sake of simplification, our drawings portray the top surface as a flat plane. In this section, we introduce a variety of structures as they appear on block diagrams.

6.3.1 Block Diagrams of Flat-Lying and Dipping Strata

The magic of a block diagram is that it allows you to visualize rock units underground as well as at the surface (**FIG. 6.4A**). For example, **FIGURE 6.4B** shows three horizontal layers of strata. If the surface of the block is smooth and parallel to the layers, you can see only the top layer in the map view; the layers underground are visible only in the cross-section views. But if a canyon erodes into the strata, you can see the layers on the walls of the canyon, too (**FIG. 6.4C**). Now, imagine what happens if the layers are tilted during deformation so that they have a dip. **FIGURE 6.4D** shows the result if the layers dip to the east. In the front cross-section face, we can see the dip. Because of the dip, the layers intersect the map-view surface, so the contacts between layers now appear as lines (the traces of the contacts) on the map-view surface. Note that, in this example, the beds strike due north, so their traces on the map surface trend due north. Also note that, in the case of tilted strata, the true dip angle appears in a cross-section face only if the face is oriented perpendicular to the strike. On the right-side face in Figure 6.4d, the beds look horizontal because the face is parallel to the strike. Practice drawing tilted strata in Exercise 6.3.

FIGURE 6.4 Block diagrams.

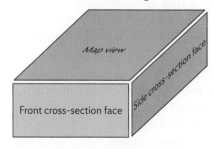

(a) Construction of a block diagram.

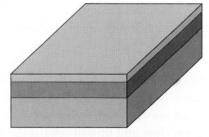

(b) A block diagram of three horizontal layers of strata represented by different colors.

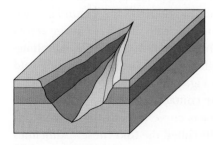

(c) A canyon cut into horizontal strata.

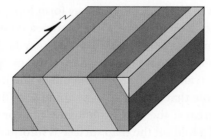

(d) A block diagram of east-dipping strata.

Name: _____   Section: _____
Course: _____   Date: _____

(a) On the block diagram template to the right, sketch what a sequence of three layers would look like if their contacts had traces that trended north-south on the map view and dipped to the west at about 45°.

(b) On the block diagram template to the right, sketch what a sequence of three layers would look like if the traces had an east-west trend and dipped south at about 45°.

(c) On the block diagram template to the right, sketch what a sequence of three layers would look like if the traces had a northeast-southwest trend and the layers dipped to the southeast at about 45°. (*Hint:* This is a bit trickier, because tilt appears in both cross-section faces.)

6.3.2 Block Diagrams of Simple Folds

When rocks are deformed and geologic structures develop, the geometry of layers depicted on a block diagram becomes more complicated. If deformation causes rock layers to bend and have a curve, we say that a **fold** has developed. Geologists distinguish between two general shapes of folds: an **anticline** is an arch-like fold whose layers dip away from the crest, whereas a **syncline** is a trough-like fold whose layers dip toward the base of the trough (**FIG. 6.5A,B**). Anticlines arch layers of rocks upward; synclines do the opposite—their strata bow downward. The side of a fold is a fold **limb**, and the line that separates the two limbs (i.e., the line along which curvature is greatest) is the fold **hinge**. We can represent the hinge with a dashed line and associated arrows on the map: the arrows point outward from the hinge on an anticline and inward toward the hinge on a syncline. On a block diagram of the folds, we see several layers exposed (**FIG. 6.5C**). Note that the same set of layers appears on both sides of the hinge. If the folded strata include a bed that is resistant to erosion, the bed may form topographic ridges at the ground surface (**FIG. 6.5D**). Note that the layers are repeated symmetrically, in mirror image across the fold hinges. The hinge of a fold may be horizontal, producing a **nonplunging fold** (**FIG. 6.6A**), or it may have a tilt, or "plunge," producing a **plunging fold** (**FIG. 6.6B**); an arrowhead on the hinge line in Fig. 6.6b indicates the direction of plunge. Note that if the fold is nonplunging, the contacts are parallel to the hinge trace, whereas if the fold is plunging, the contacts curve around the hinge—this portion of a fold on the map surface is informally called the fold "nose." Anticlines plunge toward their noses, synclines away from their noses. Curving ridges may

FIGURE 6.5 The basic types of folds.

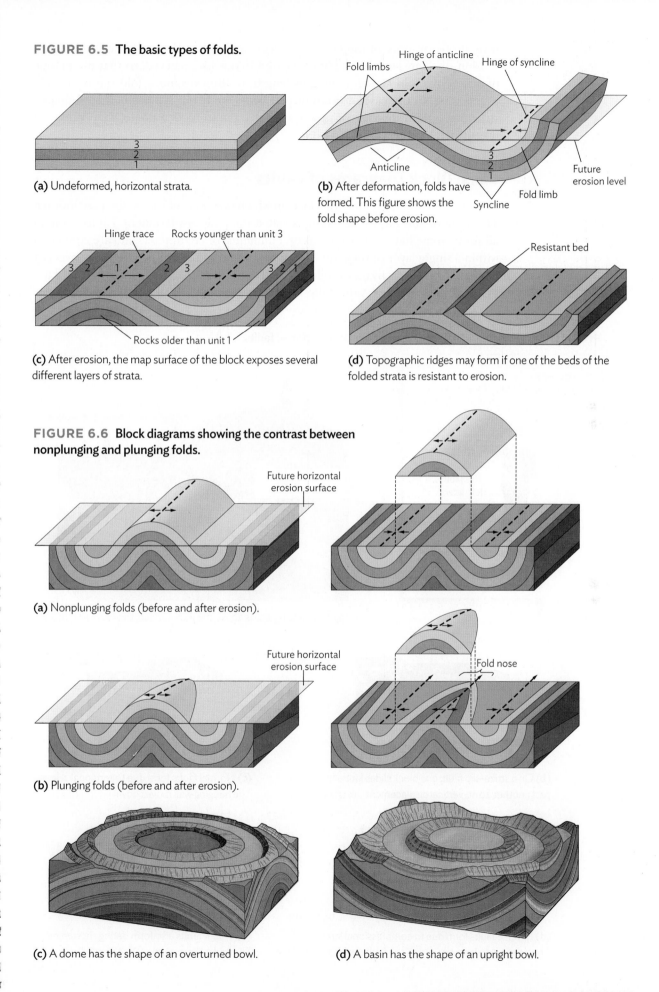

(a) Undeformed, horizontal strata.

Fold limbs

Hinge of anticline

Hinge of syncline

Anticline

Syncline

Fold limb

Future erosion level

(b) After deformation, folds have formed. This figure shows the fold shape before erosion.

Hinge trace

Rocks younger than unit 3

Rocks older than unit 1

(c) After erosion, the map surface of the block exposes several different layers of strata.

Resistant bed

(d) Topographic ridges may form if one of the beds of the folded strata is resistant to erosion.

FIGURE 6.6 Block diagrams showing the contrast between nonplunging and plunging folds.

Future horizontal erosion surface

(a) Nonplunging folds (before and after erosion).

Future horizontal erosion surface

Fold nose

(b) Plunging folds (before and after erosion).

(c) A dome has the shape of an overturned bowl.

(d) A basin has the shape of an upright bowl.

form if one or more of the beds that occur in the folded sequence are resistant to erosion. In some situations, the hinge of a fold is itself curved, so that the plunge direction of a fold changes along its length. In the extreme, a fold can be as wide as it is long. In the case of downwarped beds, the result is a **basin**, a bowl-shaped structure, and in the case of upwarped beds, the result is a **dome**, shaped like an overturned bowl (**FIG. 6.6C,D**).

6.3.3 Block Diagrams of Faults

As we noted earlier, a fault is a surface on which one body of rock slides past another. The amount and nature of sliding is called the fault **displacement**. Faults come in all sizes—some have displacements of millimeters or centimeters and are contained within a single layer of rock; others are larger and offset contacts between layers or between formations by many miles. Not all faults have the same dip: some faults are nearly vertical, whereas others dip at moderate or shallow angles. If the fault is not

FIGURE 6.7 Hanging wall, footwall, and the classification of faults.

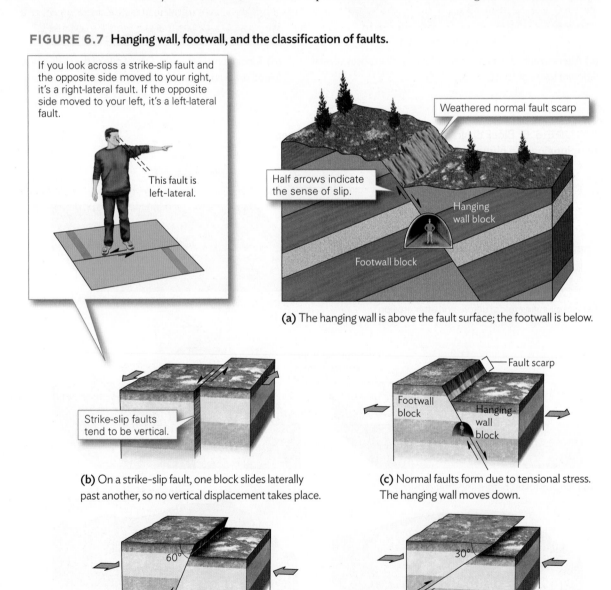

If you look across a strike-slip fault and the opposite side moved to your right, it's a right-lateral fault. If the opposite side moved to your left, it's a left-lateral fault.

This fault is left-lateral.

Weathered normal fault scarp

Half arrows indicate the sense of slip.

Hanging wall block

Footwall block

(a) The hanging wall is above the fault surface; the footwall is below.

Strike-slip faults tend to be vertical.

(b) On a strike-slip fault, one block slides laterally past another, so no vertical displacement takes place.

Fault scarp

Footwall block

Hanging-wall block

(c) Normal faults form due to tensional stress. The hanging wall moves down.

60°

(d) Reverse faults form due to compressional stress. The hanging wall moves up, and the fault is steep.

30°

(e) Thrust faults also form during shortening. The fault's dip is gentle (less than 30°).

vertical, the block of rock above the fault surface is the **hanging wall**, and the block of rock below is the **footwall** (**FIG. 6.7A**). Geologists distinguish among different kinds of faults by the direction of displacement. **Strike-slip faults** tend to be nearly vertical, and the displacement on these faults is horizontal, parallel to the *strike* of the fault (**FIG. 6.7B**). On **dip-slip faults**, the displacement is parallel to the *dip* direction of the fault; if the hanging wall block moves down-dip, it's a **normal fault** (**FIG 6.7C**), and if it moves up-dip, it's a **reverse fault** (**FIG. 6.7D**). If a reverse fault has a gentle dip (less than about 30°) or curves at depth to attain a gentle dip, then geologists generally refer to it as a **thrust fault** (**FIG. 6.7E**). Reverse and thrust faults form in response to compression, and normal faults form in response to tension.

You can recognize faulting, even if the fault surface itself is not visible (due to soil or vegetation cover), if you find a boundary along which contacts terminate abruptly (**FIG. 6.8**). The configuration that you find depends on both the attitude of the fault and the attitude of the layers, as you will see in Exercise 6.4.

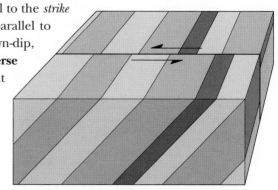

FIGURE 6.8 Example of the consequences of faulting. In this case, displacement on a strike-slip fault causes strata to terminate abruptly.

EXERCISE 6.4	Faulted Strata on a Block Diagram

Name: _____   Section: _____

Course: _____   Date: _____

Answer the following questions referring to Blocks 1 and 2 below.

(a) Block 1 shows a vertical fault cutting across a nonplunging syncline. Complete the block diagram by adding arrows to show the direction of displacement across the fault and by adding colored bands for the appropriate stratigraphic units in the blank areas. What type of fault is it?

_____

(b) Block 2 shows a dip-slip fault. Is this a normal or reverse fault? _____

(c) As you walk from west to east across the map surface of Block 2, you cross layer 3 more than once. Explain how the faulting caused this.

_____

_____

(d) The red line on the front cross-section face of Block 2 represents a drill hole. Does the drill hole cut through the complete stratigraphic section, or do you see repetition or loss of some layers? Explain.

_____

_____

Block 1

Block 2

6.3.4 Block Diagrams of Unconformities

An unconformity is a contact that represents a period of nondeposition and/or erosion, as we noted earlier. Geologists recognize three different kinds of unconformities. (1) At a **disconformity**, bedding above and below the unconformity is mostly parallel, but there is a significant time gap between the age of the strata below and the age of the strata above. These can be the most difficult to spot without knowing the ages of the strata; one clue is if the layers are all sedimentary and are not quite parallel (due to erosion that has occurred between times of deposition). (2) At a **nonconformity**, strata are deposited on an eroded "basement" of intrusive igneous and/or metamorphic rock. (3) At an **angular unconformity**, the orientation of the beds above the unconformity is not the same as that of the beds below. Exercise 6.5 gives you a chance to distinguish among these three types.

EXERCISE 6.5 | **Interpreting Unconformities on a Block Diagram**

Name: _____   Section: _____
Course: _____   Date: _____

(a) In the space provided below each block in the figure that follows, indicate what type of unconformity is shown.

Block diagrams of unconformities.

Block 1: _____   Block 2: _____

Block 3: _____   Block 4: _____

(b) The sedimentary rocks in Block 2 were deposited as horizontal layers. What has happened to the layers? Was the unconformity originally horizontal or tilted? Explain.

_____

_____

(c) In Block 3, in which direction are the post-unconformity strata dipping?

_____

(d) In Block 4, a gray area at the north edge of the block appears on the map surface. What geologic observation(s) could prove that the contact between the gray area and the sedimentary beds is an unconformity and not an intrusive contact?

_____

_____

6.3.5 Block Diagrams of Igneous Intrusions

An igneous intrusion forms when molten rock (magma) pushes into, or "intrudes," pre-existing rock. Geologists distinguish between two general types of igneous intrusions. (1) **Tabular intrusions** have roughly parallel margins; these intrusions include **dikes**, which cut across pre-existing layering, and **sills**, which are parallel to pre-existing layering. (2) **Plutons** are irregularly shaped, blob-shaped, or bulb-shaped intrusions. On a block diagram, you can generally distinguish among different types of intrusions based on their relationship to adjacent layering. To see how, try Exercise 6.6. In Exercise 6.7, you will complete your own block diagrams.

EXERCISE 6.6   **Interpreting Intrusions on a Block Diagram**

Name: _____   Section: _____
Course: _____   Date: _____

(a) Blocks 1 and 2 (below) show sedimentary beds and intrusions. Match the type of intrusion to the appropriate letter on the block.

Pluton   Block 1: _____   Block 2: _____

Dike   Block 1: _____   Block 2: _____

Sill   Block 1: _____   Block 2: _____

(b) The principle of **cross-cutting relationships** states that if one rock cuts across another, it must be the younger rock (see Chapter 8 for more on cross-cutting relationships). Using common sense to interpret cross-cutting relationships, list the sequence of intrusions for each block. If you can't determine an answer from the information shown, write "Can't be determined."

	Oldest	Middle	Youngest
Block 1:	_____	_____	_____
Block 2:	_____	_____	_____

(c) Analyses indicate the unlabeled intrusion in the front cross-section face of Block 2 is part of the same body as Intrusion A exposed on the top surface. Explain why you can't *see* the connection between the map-view exposure and the subsurface cross section on this block.

_____

_____

_____

Block diagrams of igneous intrusions.

Block 1   Block 2

Name: _____   Section: _____
Course: _____   Date: _____

Four cutout block diagrams are provided at the end of this manual for additional practice and to help you visualize struc-tures in three dimensions. Cut and fold the diagrams as indicated, and use tape to hold the tabs together to make three-dimensional block diagrams.

(a) Complete the blank cross-section faces for Block 1 and describe the structure present. Does the block show horizontal or tilted strata? Folds? Faults?

_____

_____

(b) Complete the map view and blank cross-section faces for Blocks 2 and 3. Compare and contrast the structures in these two blocks.

_____

_____

_____

(c) Complete the map view and cross-section faces for Block 4. Be sure to add arrows showing the direction of slip. Describe the nature of the faulting. Is it dip-slip? If so, is it normal or reverse? Is it strike-slip? If so, is it left-lateral or right-lateral? Explain your reasoning.

_____

_____

_____

6.4 Geologic Maps and Cross Sections

6.4.1 Introducing Geologic Maps and Map Symbols

Now that you've become comfortable reading and interpreting block diagrams, we can focus more closely on how to interpret geology on the map-view surface. A map that shows the positions of contacts, the distribution of rock units, the orientation of layers, the positions of faults and folds, and other geologic data is called a geo-logic map (**FIG. 6.9A**). Contacts between rock units are shown by lines (traces), and the units themselves are highlighted by patterns and/or colors and symbols that indicate their ages. The orientation of beds, faults, and foliations, as well as the position of fold hinges, can be represented by strike and dip symbols. The map's explanation (or legend) defines all the symbols, abbreviations, and colors on the map. **FIGURE 6.9B**, a geologic map of the Bull Creek quadrangle in Wyoming,

FIGURE 6.9 Geologic maps.

(a) A geologic map is the top surface of a block diagram. It shows the pattern of geologic units and structures as you would see them by looking straight down from above.

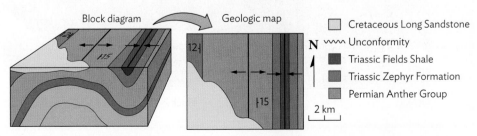

Block diagram   Geologic map

☐ Cretaceous Long Sandstone
〜〜〜 Unconformity
■ Triassic Fields Shale
■ Triassic Zephyr Formation
■ Permian Anther Group

N

2 km

FIGURE 6.9 **Geologic maps** (*continued*).

M.L. Schroeder, 1976

Rock Units

Qal	Alluvium
Qc	Colluvium
Qt	Talus
Qf	Alluvial fan deposits
Qs	Slump blocks

Ql — Mostly composite slumps and earthflows derived from rocks of several formations

Qlt	Tr Landslide debris
Qlh	Th Landslide debris
Qls	KS Landslide debris
Qg	Glacial deposits
Tcu	Camp Davis Formation

Tb	Porphyritic basalt
Tr	Redbeds
Th	Hoback Formation
Ks	Sedimentary rocks, Undivided
Kb	Bear River Formation
Kg	Gannett Group, Undifferentiated
Jsp	Stump and Preuss Sandstones
Jt	Twin Creek Limestone
Jtn	Nugget Sandstone
Tc	Chugwater Formation
Td	Dinwoody Formation
Pp	Phosphoria Formation
PPMw	Wells Formation and Associated Rocks
Mm	Madison Group, Undivided
Dd	Darby Formation

Q = Quaternary; T = Tertiary;
K = Cretaceous; J = Jurassic;
TR = Triassic; P = Permian;
P = Pennsylvanian;
M = Mississippian

Structural symbols

Strike and dip   Thrust fault

High-angle fault

Anticline   Syncline

0 _____ 1 mile

0 _____ 1 km

N   Contour interval = 40'

(b) Geologic map of the Bull Creek quadrangle, Teton and Sublette Counties, in Wyoming.

illustrates the components of a geologic map. All geologic maps should have a scale, north arrow, and explanation.

Let's begin our discussion of geologic maps by considering the various features that can be portrayed on these maps. (You will practice mapping these features in Exercise 6.8.)

- *Rock units:* Geologic maps show the different rock units in an area. These units may be bodies of intrusive igneous rock, layers of volcanic rock, sequences of sedimentary rock, or complexes of metamorphic rock. The most common unit of sedimentary and/or volcanic rock is a stratigraphic formation, as noted earlier. A formation is commonly named for a place where it is well exposed. A formation may consist entirely of beds of a single rock type (e.g., the Twin Creek Limestone consists only of limestone), or it may contain beds of several different rock types (e.g., Chugwater Formation contains sandstone, shale, gypsum, and limestone). Typically, geologic maps use patterns, shadings of gray, or colors to indicate the area in which a given unit occurs. On geologic maps produced in North America, an abbreviation for the map unit may also appear within the area occupied by the formation. This abbreviation generally has two parts: the first part represents the formation's age, in capital letters; the second part, in lowercase letters, represents the formation's name. For example, J indicates rocks of Jurassic age (Jt = Twin Creek Limestone, Jsp = Stump and Preuss Sandstones), and JT indicates Jurassic or Triassic age (JT$_R$c = Chugwater Formation, JT$_R$d = Dinwoody Formation). For further explanation of geologic time periods, see Lab 8 and specifically Figure 8.7, the geologic time scale.

- *Contacts:* Different kinds of contacts are generally shown with different types of lines. For example, a conformable or intrusive contact is a thin line, a fault contact is a thicker line, and an unconformity may be a slightly jagged or wavy line. In general, a visible contact is a solid line, whereas a covered contact (buried by sediment or vegetation) is a dashed line (see dashed line in Fig. 6.9b).

- *Strike and dip:* On maps produced in North America, geologists use a symbol to represent the strike and dip of a layer. The symbol consists of a line segment drawn exactly parallel to the direction of strike and a short tick mark drawn perpendicular to the strike and pointing in the direction of dip (**FIG. 6.10A, B**). A number written

FIGURE 6.10 Indicating features on geologic maps.

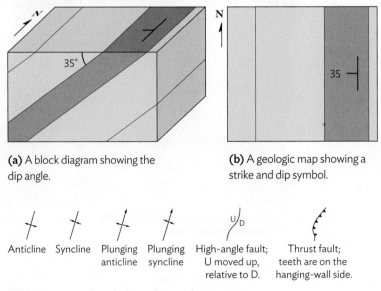

(a) A block diagram showing the dip angle.

(b) A geologic map showing a strike and dip symbol.

Anticline   Syncline   Plunging anticline   Plunging syncline   High-angle fault; U moved up, relative to D.   Thrust fault; teeth are on the hanging-wall side.

(c) Basic structural symbols used on geologic maps.

next to the tick mark indicates the angle of dip. (It is not necessary to write a number indicating the strike angle because that is automatically represented by the map trend of the strike line.).

- *Other structural symbols:* The explanation also includes symbols representing the traces of folds and faults. **FIGURE 6.10C** illustrates some of these symbols.

EXERCISE 6.8   **Using Symbols to Indicate Structures on a Geologic Map**

Name: _____   Section: _____
Course: _____   Date: _____

(a) On the blank map below of a region with no hills or valleys, draw the appropriate strike and dip symbol next to the appropriate point. To do this, you must use a protractor and measure the angle between the north direction (the side edge of the map) and the strike angle. Then, look at the direction of dip so that you put the dip tick on the correct side of the strike symbol. These points are on a contact between two formations:

   A: 045°/30° SE

   B: 280°/10° SW

   C: 350°/25° W

(b) Based on the strike and dip symbols you show, draw a line representing the contact that passes through these points. Remember, the line needs to be parallel to the strike symbol.

_____

_____

(c) What is the structure shown by the structural symbols?

_____

_____

Blank map.

6.4.2 Constructing Cross Sections

We've seen that a cross section represents a vertical slice through the crust of the Earth. Thus, the sides of a block diagram are cross sections. If you start with a block diagram, you can construct the structure in the cross-section faces simply by drawing lines representing the contacts so that they connect to the contact traces in the map view—the strike and dip data on the map tell you what angle the contact makes, relative to horizontal, and you use a protractor to draw the correct angle. If a fold occurs on the map surface, it generally also appears in the cross section.

So far, we've worked with data depicted on block diagrams. Now let's consider the more common challenge of producing a cross section from a geologic map. This takes a couple of extra steps—to see how to do it, refer to **FIGURE 6.11**.

On the left side of Figure 6.11a, you see a simple geologic map. The cross section is a vertical plane inserted into the ground along the line of section. The line of section (X–X') is the line on the map view along which you want to produce the cross section. Take a strip of paper and align it with your line of section. First mark the beginning and ending points of the section on your cross-section paper, so that if you need to lift the paper at any point, you can be sure you're lining it up in the same location when you place it back down on the section. Then mark the points where contacts cross the strip of paper (Step 1). Transfer these points to the cross-section face. Using a protractor, make a little tick mark indicating the dip of the contact; use the strike and dip symbol closest to each contact to provide this angle (Step 2). Next, in the subsurface, sketch in lines that conform to the positions of the contacts and the dip angles (Step 1; in Fig. 6.11a, the contacts dip gently to the

FIGURE 6.11 Constructing a cross section.

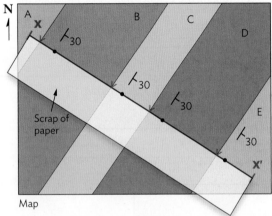

Map

Step 1: Mark data locations on the cross-section paper, including the beginning and ending points.

(a) Example with gently dipping beds.

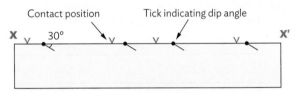

Step 2: Identify contact positions. Add dip marks at correct angles.

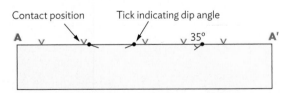

Step 3: Draw contacts so they obey location and dip data.

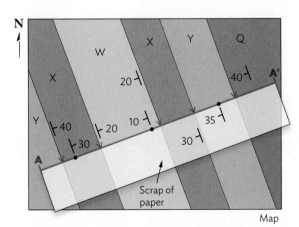

Map

Step 1: Mark data locations on the cross-section paper.

(b) Example with folded beds (syncline).

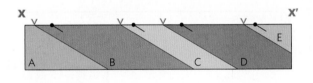

Step 2: Identify contact positions. Add dip marks at correct angles.

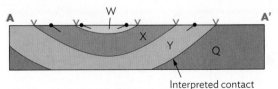

Step 3: Draw contacts so they obey location and dip data.

southeast). In Figure 6.11b, the contacts curve underground to define a syncline. Unless there is a reason to think otherwise, the layers should have constant thickness. Note that, because of this constraint, Layer Q in Figure 6.11b appears in the lower left corner of the cross section; it would come to the surface to the west of the map area. Construct your own cross sections in Exercise 6.9.

EXERCISE 6.9   **Constructing a Cross-Section View of Block Diagram**

Name: _____   Section: _____
Course: _____   Date: _____

(a) The map surface of the block diagram in the figure below provides strikes and dips of the layers shown. From this information, show the layers with their proper angles in the front and side cross-section faces. Note that the strike of the layers is perpendicular to the front face of the block.

Block diagram with strikes and dips of the layers.

What kind of structure is shown?

_____

(b) Complete the map view and cross-section views of the block diagrams below by showing a sequence of sedimentary rocks with the indicated orientations. Show at least three layers in each block and plot the strike and dip symbol on the map surface.

Block A: 090°/40° S

Block B: 000°/60° E

(c) Complete the cross-section views of the block below.

Block diagram for creating cross-section views.

6.4.3 Basic Geologic Map Patterns

Geologic maps can get pretty complex, especially where structures are complicated or where topography is rugged. But by applying what you have learned so far about block diagrams, you can start to interpret them. To make things simple, we begin with some very easy maps of areas that have no topography (i.e., the ground surface is flat), as in the block diagrams that you've worked with. Exercise 6.10 challenges you to look at a map and imagine the three-dimensional structure it represents. Keep in mind that sedimentary and extrusive igneous rocks are commonly deposited in horizontal layers, with the youngest layer at the top of the pile and the oldest at the bottom.

Earlier, in the context of discussing block diagrams, we distinguished between nonplunging folds and plunging folds. We can recognize these folds on geologic maps simply by the pattern of color bands representing formations—on a map, the formation contacts of nonplunging folds trend parallel to the hinge trace, whereas those of plunging folds curve around the hinge trace so that we can see the fold nose. Furthermore, we can distinguish between anticlines and synclines by the age relationships of the color bands—strata get progressively younger away from the hinge of an anticline and progressively older away from the hinge of a syncline. If the hinge isn't shown on the map, you can draw it in where the reversal of age takes place. See **FIGURE 6.12** for an example of a map and cross section of plunging folds.

FIGURE 6.12 Patterns of plunging folds.

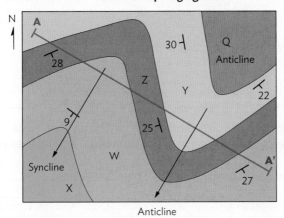

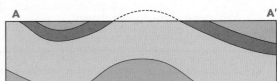

(b) Cross section of folds along the A-A' line.

(a) Map view of plunging anticline and syncline. Hinge traces of the folds are indicated by labeled lines, with arrows indicating the direction of plunge.

EXERCISE 6.10 **Interpreting Simple Geologic Maps**

Name: _____   Section: _____
Course: _____   Date: _____

For each of the following geologic maps, identify the structure or geologic features portrayed. Does the map show a fault, fold, tilted strata, dike, pluton, unconformity, or some combination? To answer these questions, you may need to refer to the block diagrams presented earlier in the chapter. Remember—think in three dimensions! With a little practice, geologists learn to recognize the basic patterns quickly. (*Note:* The geologic periods in the maps below, in order from oldest to youngest, are the Ordovician, Silurian, Devonian, Triassic, Jurassic, Cretaceous, and Tertiary.)

(continued)

Name: _____   **Section:** _____
Course: _____   **Date:** _____

Map A

Map B

Map C

Jrt	Jurassic Tinta Formation
Tr	Triassic Jones Shale
Dt	Devonian Tella Formation
Db	Devonian Bouser Formation
Dn	Devonian Norfolk Shale
Sh	Silurian Hallo Formation

Tb	Tertiary basalt
Kg	Cretaceous granite
Da	Devonian Alsen Formation
Db	Devonian Becraft Limestone
Dn	Devonian Norfolk Shale
Sc	Silurian Cligfell Formation

Ka	Cretaceous Altoona Formation
Kb	Cretaceous Barrell Formation
Dp	Devonian Potomoo Limestone
Sw	Silurian Wala Shale
Si	Silurian Jack Formation
Ot	Ordovician Trent Formation

Note: In legends like these, geologists put the youngest unit at the top, and the units get progressively older as you move to the bottom.

(a) Describe the geologic features of Map A.

i. _____-aged strata are overlain at an unconformity by _____-aged strata.
What type of unconformity is it? _____

ii. The Triassic and Jurassic strata dip in a _____ direction at a _____ angle.

iii. The Silurian and Devonian strata dip in a _____ direction at a _____ angle.

(b) Describe the geologic features of Map B

i. _____-aged strata are folded into a(n) _____. The hinge of the
fold trends _____ . Do the folded layers plunge? _____ Are the folded
layers symmetric? _____ If not, which side is steeper? _____ .

ii. A _____ intrudes the western portion of the folded layers (give age, rock, and type of the intrusion).

iii. A _____ intrudes both the folded layers and the intrusion described in question ii (give age, rock, and type of the intrusion).

(c) Describe the geologic features of Map C.

i. A _____-trending fault cuts strata that strike in a _____ direction and dip to the _____ .

ii. The fault is either a _____ fault or a dip-slip fault in which the (N,S,E,W) block moved
_____ relative to the (N,S,E,W) block. Without seeing a cross section of the fault, it is not
possible to determine if the fault is normal or _____ .

iii. _____-aged strata are overlain at an unconformity by _____-aged strata.
What type of unconformity is it? _____

(d) If you walk from left to right (west to east) along the southern edge of Map A, are you walking "up-section" (i.e., into rocks of progressively younger age) or "down-section" (i.e., into rocks of progressively older age)? _____

6.5 Structures Revealed in Landscapes

Unless you live in the Great Plains or along the Gulf Coast of the United States, you know that landscapes tend not to be as flat as the tops of the idealized block diagrams that we've worked with so far. In many cases, the distribution of rock units controls the details of the landscape, so erosion may cause structures to stand out in the landscape, especially in drier climates. For example, resistant rocks form cliffs, whereas nonresistant rocks form gentler slopes. Thus, a cliff exposing alternating resistant and nonresistant rocks develops a **stair-step profile** (**FIG. 6.13A**). Where strata are tilted, resistant rocks form topographic ridges, whereas nonresistant rocks tend to form valleys (**FIG. 6.13B**). Generally, the ridges are asymmetric—a dip slope

FIGURE 6.13 Structural control of topography.

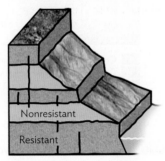

(a) Profile of a stair-step canyon.

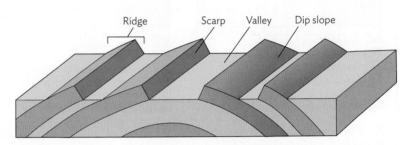

(b) A block diagram showing how resistant layers hold up ridges.

(c) A satellite photo of the Valley and Ridge Province of Pennsylvania (from Google Earth).

parallel to the bedding forms on one side, and a scarp cutting across the bedding forms on the other. In the Appalachian Mountains of Pennsylvania, ridges trace out the shape of plunging folds (**FIG. 6.13C**). A region in which the structure of bedrock strongly influences topography is called a **structurally controlled landscape**, and you will see how this is manifested in Exercise 6.11.

EXERCISE 6.11 | **Interpreting Structurally Controlled Landscapes**

Name: _____   Section: _____

Course: _____   Date: _____

The figure on the next page shows a region of central Pennsylvania that includes the boundary between two different structural provinces: the Valley and Ridge Province to the southeast and the Plateau Province to the northwest. Enlargements of the two provinces are shown below the main map. Based on the general shape of the land surface, as indicated by the maps, answer the following questions:

(a) In which of the two provinces is the landscape structurally controlled?

_____

(b) Compare the pattern of stream valleys in the Plateau Province with that in the Valley and Ridge Province. Approximately what is the dip of the beds beneath the Plateau Province?

_____

(c) In the Valley and Ridge Province, are the folds plunging or nonplunging? What is your evidence?

_____

_____

_____

_____

(d) What is the overall trend of fold hinges in the portion of the Valley and Ridge Province depicted in the enlargement?

_____

_____

(e) If fold hinges trend roughly perpendicular to the direction of compression, in what direction was the compressive "push" during the development of the folds in the Valley and Ridge Province? (Geologists have determined that these folds formed when Africa collided with eastern North America at the end of the Paleozoic Era.)

_____

_____

(continued)

EXERCISE 6.11 Interpreting Structurally Controlled Landscapes *(continued)*

Name: ____________________ Section: ____________
Course: ____________________ Date: ____________

Shaded relief of central Pennsylvania.

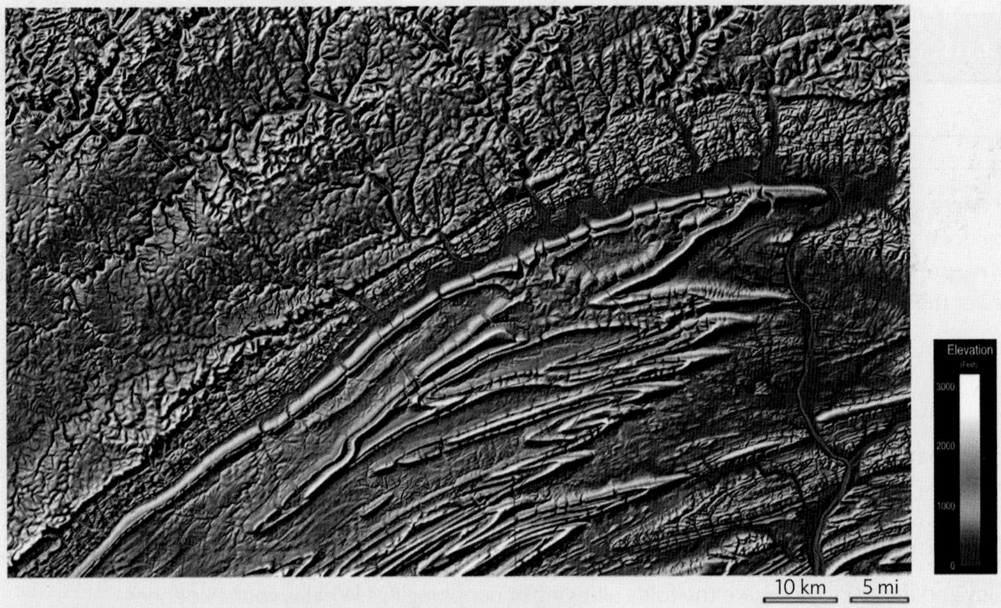

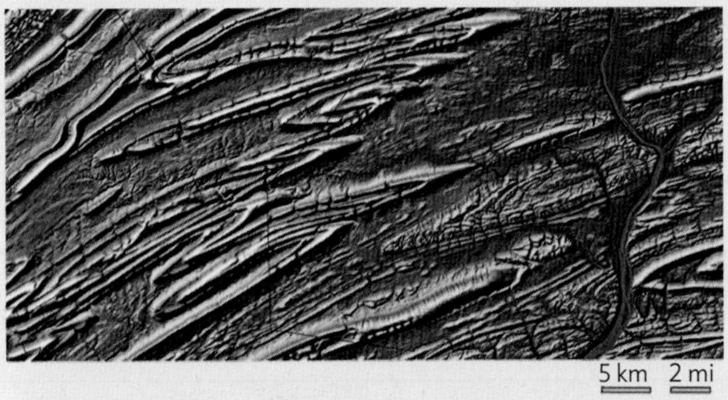

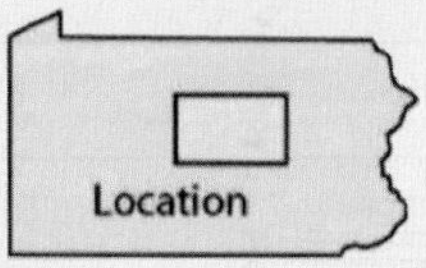

Pennsylvania Valley and Ridge Province, the ridges are underlain by resistant sandstone layers.

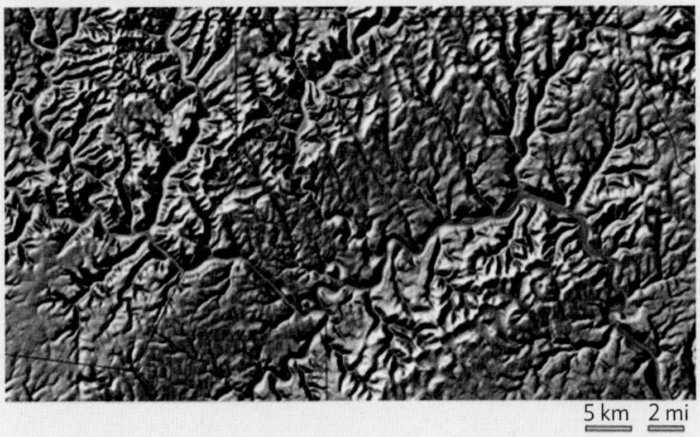

Pennsylvania Plateau Province, the pattern of the rivers and tributaries is called a dendritic drainage pattern.

Name: _____   Section: _____

Course: _____   Date: _____

? **What Do You Think** Knowledge of an area's
deformation history helps people avoid potentially
calamitous situations, such as building a school directly on an
active fault. Structural information can also pay off—big
time—in our search for energy and mineral resources. For
example, geologists have learned that oil and natural gas are
often trapped in the crests of anticlines. Because it costs
millions of dollars to drill an exploratory well, knowing where
the anticlines are (or aren't) can mean the difference between
a fortune and bankruptcy.

Imagine that a company has dug a shaft for an underground mine into a coal bed (gray layer in the figure below).
The coal bed and all beds above and below it dip 10° W. A
drill hole through the coal layer has provided the stratigraphic
sequence and the thicknesses of the layers, as represented
in the column on the left (drawn to scale). Mapping at the
ground surface has revealed two faults and the traces of two distinct beds. The miners have found that the coal layer
terminates at the faults, as shown in the front cross-section face. It is not economical to mine at a depth greater than the
red line. Based on the data available, would you recommend the mining company buy the mineral rights beneath Region A
(west of the western fault) or Region B (east of the eastern fault)? Complete the front cross-section face for help in visualizing the answer. Explain your decision on a separate piece of paper.

Practical application problem.

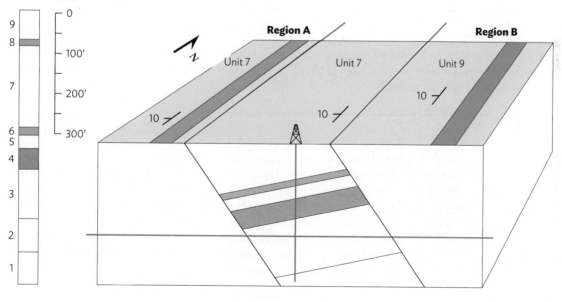

Name: _____   **Section:** _____
Course: _____   **Date:** _____

Exploring Earth Science Using Google Earth

1. Visit **digital.wwnorton.com/labmanualearthsci**
2. Go to the **Geotours** tile to download Google Earth Pro and the accompanying Geotours exercises file.

Expand the Geotour06 folder in Google Earth by clicking the triangle to the left of the folder icon. First, check and double-click on the USGS Map Credit placemark to fly to Split Mountain, east of Vernal, Utah. Then, take turns turning on the Split-Mtn_TopoMap and SplitMtn_GeoMap folders to answer questions about the geologic structure of this portion of Dinosaur National Monument (note that you can select a folder/overlay and use the transparency slider at the bottom of the Places panel to make items semi-transparent).

(a) Check the SplitMtn_GeoMap folder. Then check and double-click the North Fold placemark to fly to a nose-like protrusion of Split Mountain. Given that the blue/purple rocks are older than the green rocks, what type of fold is Split Mountain, and is it plunging or nonplunging? (If plunging, provide the general direction.)

_____ fold type

_____ plunging/nonplunging (and direction, if applicable)

(b) Check and double-click the South Fold placemark to fly to a nearby location south of Split Mountain. Given that the blue/purple rocks are older than the green rocks, what type of fold is this, and is it plunging or nonplunging? (If plunging, provide the general direction.)

_____ fold type

_____ plunging/nonplunging (and direction, if applicable)

(c) Check the SplitMtn_TopoMap folder. Then check and double-click the Flatiron placemark to fly to an example of a flatiron, a laundry iron–shaped landform commonly associated with dipping sedimentary bedding surfaces. Ignoring the valleys carved by the branching tributary that flows into the Green River (i.e., use the straight segment of the 5800 contour line pointed to by the placemark), and using the **Heading** information from the **Line** tab on the **Ruler** tool, draw lines to estimate both the strike direction (approximately parallel to the topographic contour) and the dip direction (approximately perpendicular to the topographic contour, away from the flatiron "point"). *Hint:* Check out Sections 07, 12, 13, or 18 to the SE to see topographic contours on a dip slope that is less eroded and/or turn on the SplitMtn_GeoMap and look for a strike/dip symbol with the amount of dip included.

_____ strike direction (0-360°)

_____ dip direction (0-360°)

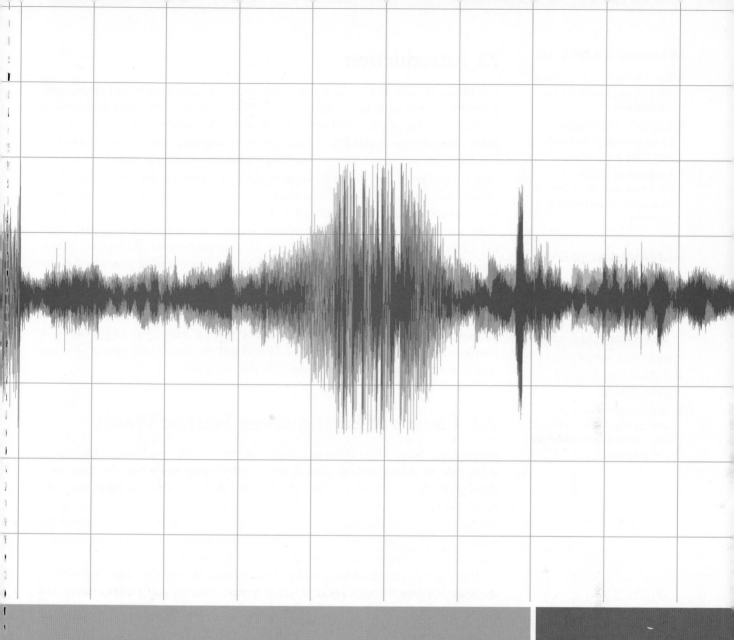

Earthquakes and Seismology

7

Earthquakes are produced by different kinds of seismic waves shown in this seismogram. Geologists analyze these waves to determine the earthquake's strength and location.

LEARNING OBJECTIVES

- Describe how faulting causes the ground to move during earthquakes
- Explain how earthquakes cause damage to buildings and other structures
- Locate an earthquake's epicenter and determine its magnitude and when it occurred

MATERIALS NEEDED

- Sharp pencil
- Clear plastic ruler with divisions in tenths of an inch and millimeters (included in the GeoTools section at the back of this manual)
- Architect's compass (or piece of string)
- Seismic overlay reading tool (in your toolkit in the GeoTools section at the back of this manual)

7.1 Introduction

Few things are as fearsome as a major earthquake. Unpredictable and enormously powerful, a great earthquake destroys more than buildings and other structures. It shakes our sense of safety and stability as it shakes the solid rock beneath our feet. But it is not just the shaking that is dangerous. Devastating tsunamis remind us that oceanic earthquakes can ravage coastlines thousands of miles from the earthquake origin. Landslides and mudslides triggered by earthquakes can engulf towns and villages, and the loss of water when rigid pipes break beneath city streets can cause health problems and make it difficult to fight fires.

In our attempt to understand earthquakes, we have developed tools that reveal the Earth's internal structure, define the boundaries between lithosphere plates, prove that the asthenosphere exists, and track the movement of plates as they are subducted into the mantle. In this chapter, we look at what an earthquake is and why it causes so much damage, and we learn how seismologists locate earthquakes and estimate the amount of energy they release. You will learn to read a seismogram and use it to locate an earthquake, determine when it happened, and measure its strength. First, let's review some basic facts about the causes and nature of earthquakes that are discussed in more detail in your textbook.

7.2 Causes of Earthquakes: Seismic Waves

Earthquakes occur when rocks in a fault zone break, releasing energy. The energy is brought to the surface by seismic waves called **body waves** because they travel through the body of the Earth. It is this energy that causes the ground to shake initially. The point beneath the surface where the energy is released is called the **focus** (or hypocenter) of the earthquake. The point on the surface directly above the focus is called the **epicenter** (**FIG. 7.1**). In most cases, the epicenter, being closest to the focus, is the site of the greatest ground motion and damage.

There are two kinds of body waves, distinguished by how they cause particles to move as they pass through rocks: **P-waves** (primary waves) and **S-waves** (secondary

FIGURE 7.1 The focus and epicenter of an earthquake.

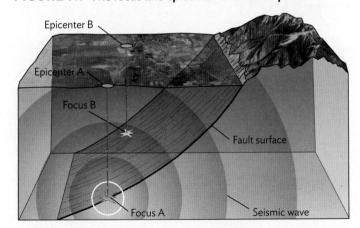

The focus is the point on the fault where slip begins. Seismic energy starts radiating from it. The epicenter is the point on the Earth's surface directly above the focus.

waves). In P-waves (**FIG. 7.2A**), particles in rock vibrate back and forth (red arrow) *in the overall direction that the wave is traveling* (green arrow). A wave that moves this way is called a *longitudinal wave*. You can demonstrate a P-wave by stretching a Slinky on a table and pushing and pulling on one end while keeping the other end in place. As the P-wave passes through the Slinky, the coils move as shown in Figure 7.2a: instead of remaining equal distances apart, they bunch together in some places and move farther apart in others.

FIGURE 7.2 Different types of earthquake waves.

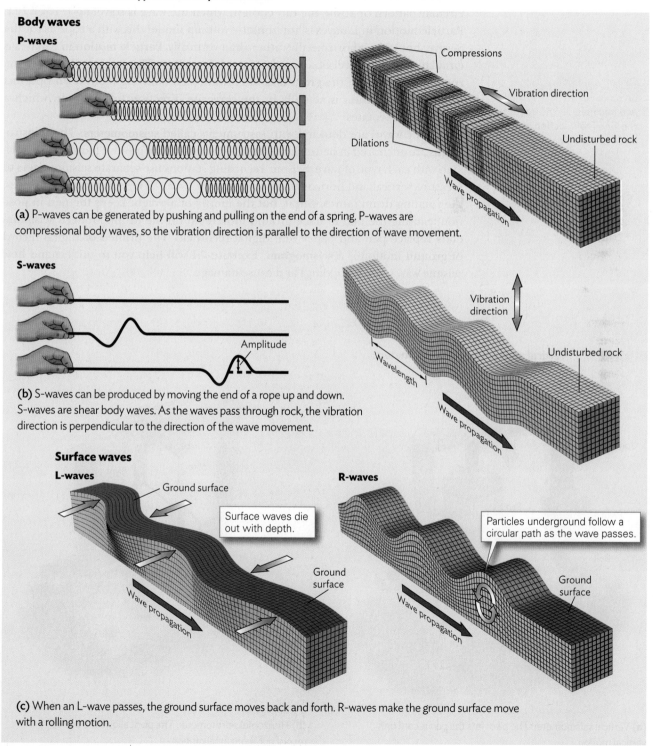

Body waves

P-waves

(a) P-waves can be generated by pushing and pulling on the end of a spring. P-waves are compressional body waves, so the vibration direction is parallel to the direction of wave movement.

S-waves

(b) S-waves can be produced by moving the end of a rope up and down. S-waves are shear body waves. As the waves pass through rock, the vibration direction is perpendicular to the direction of the wave movement.

Surface waves

L-waves   **R-waves**

(c) When an L-wave passes, the ground surface moves back and forth. R-waves make the ground surface move with a rolling motion.

S-waves (**FIG. 7.2B**) are seismic waves in which the particle movement *is perpendicular to the direction in which the wave is traveling.* A wave that moves this way is called a *transverse wave.* You can demonstrate an S-wave by having two people hold the ends of a rope and asking one to whip the rope in an up-and-down motion. As the S-wave passes, the rope wriggles like a snake.

P-waves travel faster through rock than S-waves and therefore reach the surface first. When P- or S-waves reach the surface, some of their energy is converted to two types of **surface waves**: the **L-wave** (Love wave), a shear wave similar to an S-wave, in which particles vibrate horizontally (parallel to the ground surface); and the **R-wave** (Rayleigh wave), a unique wave type in which particles move in a circular pattern opposite the direction in which the wave is traveling (**FIG. 7.2C**). Particle motion in L-waves is horizontal—you can model this with a rope as for the S-wave, but whip it horizontally rather than vertically. Particle motion in R-waves is circular, like the wheel of a bicycle as the bike moves. This analogy describes the rotational motion of the ground as the R-wave passes through, but it isn't perfect because the rotation is actually in the opposite direction from that in which a bicycle wheel rotates.

Seismic waves are detected with instruments called **seismometers**. These instruments are anchored in bedrock to measure the amount of ground movement associated with each type of wave. Seismic recording stations use separate seismometers to measure vertical and horizontal ground motion (**FIG. 7.3**). As the ground vibrates, the rotating drum moves with it, but the inertia of a weight keeps the pen in position, causing the pen to trace ground movement on the drum. Modern seismometers replace pen and paper with digital recorders. The printed or digital record of ground motion is a **seismogram**. Exercise 7.1 will help you to understand how seismic waves affect buildings and cause damage.

FIGURE 7.3 Vertical and horizontal seismometers.

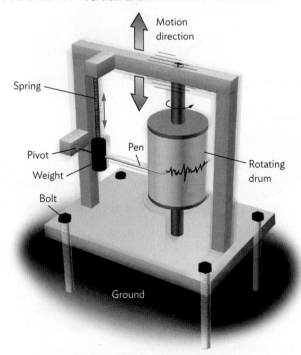

(a) Vertical seismometer: The pivot lets the pen record only vertical motion.

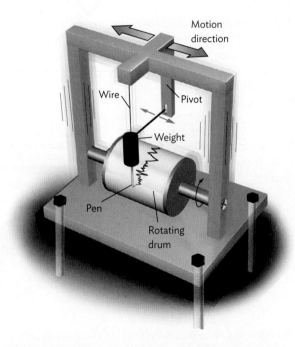

(b) Horizontal seismometer: The pivot allows the pen to record only horizontal motion.

Name: _____   Section: _____
Course: _____   Date: _____

When P- and S-waves reach the surface, the ground vibrates, and anything built on it is shaken in directions that depend on which wave type is involved and the distance from the epicenter. The figure below shows the arrival of P-, S-, L-, and R-waves at a skyscraper.

Ground shaking caused by seismic waves.

| P-wave at epicenter | P-wave far from epicenter | S-wave at epicenter | S-wave far from epicenter | L-wave | R-wave |

(a) Draw arrows to indicate every direction in which each of the buildings will move in response to the different types of waves.

(b) Why do the P- and S-waves at an earthquake epicenter make the ground shake differently from the ground in an area far from the epicenter?

_____

_____

_____

7.2.1 Investigating Earth's Interior with the Properties of Seismic Waves

Since the early 1900s, the measurement and study of seismic waves has led to a better understanding of the properties of the materials beneath the Earth's surface. One major discovery was made by Croatian seismologist Andrija Mohorovičić (pronounced *Moho-ro-vi-chich*), who recognized that separate sets of seismic waves arrived several seconds apart from each other at different recording stations. Further research led to the discovery that the waves also changed speeds at various depths within the Earth. In working to interpret these findings about seismic waves, scientists determined that:

- Seismic waves speed up when passing through solids, and slow down when encountering liquids.
- Velocities are slower in hot materials than cooler ones.
- Velocities generally increase with increasing pressure.
- S-waves cannot pass through liquids.

By studying these characteristics of seismic waves, scientists could then infer the nature of the materials that make up the Earth. Where the seismic waves change speed, "boundaries" are drawn to represent a change in rock properties. You will use the same seismic information in Exercise 7.2 to observe these changes and understand the basis for dividing the Earth into at least four distinct layers.

EXERCISE 7.2	What Do Seismic Waves Indicate About the Earth's Interior?

Name: _____   Section: _____
Course: _____   Date: _____

The diagram on the right is a graph of the speed of P- and S- waves with increasing depth in the Earth. Use this graph and what you have learned about the properties of P- and S-waves to answer the questions that follow.

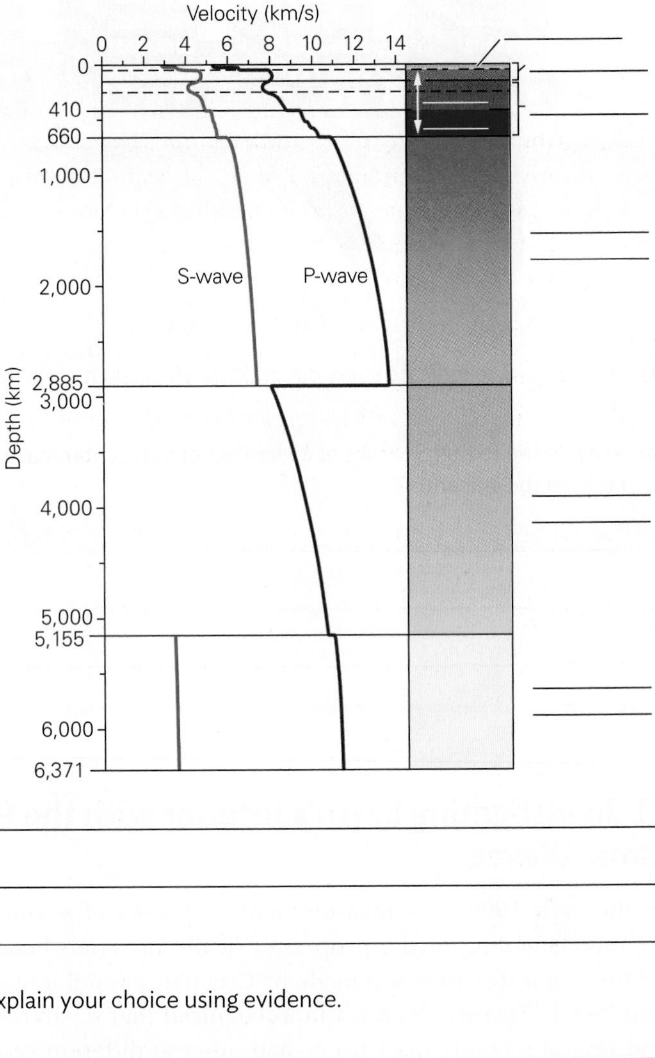

(a) In the spaces next to the diagram, label the lower mantle, asthenosphere, outer core, lithosphere, upper mantle, inner core, and crust.

(b) What happens to the speed of the seismic waves as they pass from the crust to the mantle? Why do you think this occurs?

_____

_____

_____

_____

_____

_____

_____

_____

_____

_____

(c) Where do P-waves decrease speed dramatically? Explain your choice using evidence.

_____

_____

(d) Which layer of the Earth is thought to be composed of material in liquid form? Use evidence to support your answer.

_____

_____

(continued)

Name: _____   Section: _____

Course: _____   Date: _____

(e) What can you tell about the composition, temperature, and pressure of the inner core based on the seismic wave velocities?

_____

_____

_____

7.3 Locating Earthquakes

Locating earthquakes helps us to understand what causes them and to predict whether an area will experience more in the future. Most earthquakes occur in linear belts caused by faulting at the three kinds of plate boundaries (**FIG. 7.4**) and are used to define those boundaries. But earthquakes also occur within plates, and their causes are less well understood. How can we locate earthquake epicenters, especially when they are in remote areas? Seismologists triangulate epicenter locations by using sophisticated mathematical analysis of the arrival times of the different seismic wave types at many seismic recording stations.

In this manual, we use a much simpler method to locate an epicenter, and determine precisely when the earthquake occurred, with data from only three seismic

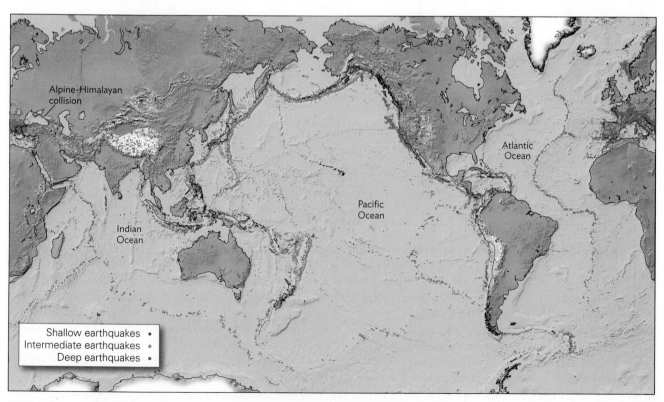

FIGURE 7.4 Worldwide distribution of earthquake epicenters. Most earthquakes occur in distinct belts along plate boundaries.

stations. The basis for this method is the fact that the four types of seismic waves travel at different velocities. Let's look at location first, then timing.

Exercise 7.3 illustrates the reasoning used to determine the distance from a seismometer (the streetlight in questions c and d) to an earthquake epicenter (the starting line of each car). But in this exercise, you have been given the velocity of the cars. How do seismologists determine these measurements?

EXERCISE 7.3 **The Logic of Locating Earthquake Epicenters**

Name: _____   Section: _____

Course: _____   Date: _____

This exercise leads you through the reasoning used to calculate the distance from a seismic station to an earthquake epicenter. But instead of two different seismic waves, let's see first how this works with two cars that start along a road at exactly the same time (see the figure below). Both cars use cruise control set at 1 mile per minute (60 miles per hour), but the controls are not exactly the same. Car 1 covers each mile in exactly 60 seconds and Car 2 in 61 seconds. Because Car 1 arrives a second before Car 2 for each mile they travel, the delay between the arrival times of Car 1 and Car 2 increases at each mile marker.

To see how this works, complete the illustration below:

(a) In each of the boxes to the right of Car 1 and Car 2, note the travel time (in seconds) needed to arrive at that milepost—the first one has been completed.

(b) In the boxes at the bottom of the figure, indicate the delay time between the arrival of Car 1 and Car 2 at each milepost.

The logic of locating earthquake epicenters based on the arrival times of different types of seismic waves.

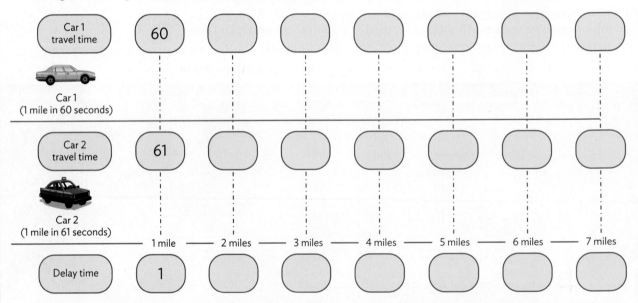

Of course, geologists don't know at first how far seismic waves have traveled or when they were generated by an earthquake. But as with the cars in this exercise, geologists do know (1) the velocities of the different types of seismic waves; (2) that the waves started at the same time (when the fault moved); and (3) the precise times when the different wave types were recorded by seismometers. To figure out how far away an epicenter was and when the earthquake happened, we work backward to derive the information.

You can see how this works by continuing to use the two imaginary cars instead of seismic waves and working backward like a seismologist. Imagine you are waiting for Car 1 and Car 2 to drive by a streetlight outside your window. You don't know where they came from or what time they left, but you do know (1) their velocities (you have been told that Car 1 will arrive traveling at 1 mile in 60 seconds and Car 2 at 1 mile in 61 seconds) and (2) that they started at exactly the same time. As they drive by, you (3) record the precise delay between the times they pass the streetlight outside your window.

(continued)

Name: _____   Section: _____
Course: _____   Date: _____

(c) Using that information, *how far* did the two cars travel if Car 2 passed the streetlight
    (i) 25 seconds after Car 1? _____ miles
    (ii) 45 seconds after Car 1? _____ miles

(d) Now determine the *precise time* that the cars started their trip (hour:minute:second) if Car 1 passed the streetlight at 12:25:00 and was
    (i) 25 seconds before Car 2: _____ : _____ : _____
    (ii) 45 seconds before Car 2: _____ : _____ : _____
    (*Hint:* Determine how far the cars traveled and then use the known velocity of Car 1 to estimate how much travel time it needed to go that distance.)
Now put the two pieces of reasoning together.

(e) If Car 1 arrives at precisely 1:45:22, followed by Car 2 at 1:46:37, how far did they travel? _____ miles

(f) At what precise time did they leave? _____ : _____ : _____

Seismologists have measured velocities of P-, S-, L-, and R-waves in different rock types and know that P-waves average approximately 6.3 kilometers per second at the surface and S-waves about 3.6 km/s. A graph of these velocities, called a travel-time diagram, is used to calculate the distance to an epicenter (**FIG. 7.5**). The black curves in Figure 7.5 show how much time it takes (vertical axis) the four types of seismic waves to travel the distances shown on the horizontal axis. To find the time it takes a P-wave to reach a point 4,000 km from an epicenter, find the intersection of the 4,000-km vertical line with the P-wave travel-time curve. Draw a horizontal line from the intersection to the time axis on the left and read the time required— 7 minutes in this example. Exercise 7.4 provides practice in reading travel-time curves, and Exercise 7.5 takes you through the steps of locating an earthquake's epicenter.

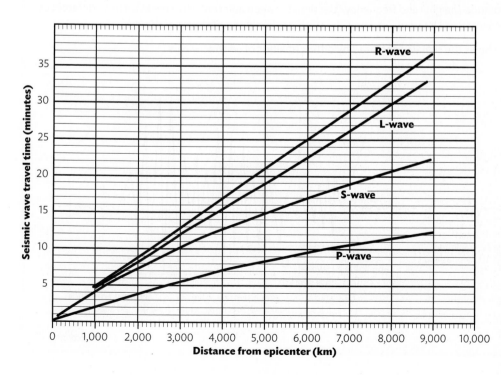

FIGURE 7.5 Time-travel curves show the relationship between distance traveled and relative seismic wave velocity. Because the P- and S-waves travel at a range of speeds, they plot as curves on the graph. R and L waves travel at relatively constant speeds, and so plot as straight lines on the graph.

EXERCISE 7.4   Reading a Travel-Time Diagram

Name: _____   Section: _____

Course: _____   Date: _____

Refer to the travel-time diagram in Figure 7.5 to answer the following questions:

(a) How long does it take a P-wave to travel 5,000 km? _____ minutes

(b) How long does it take an S-wave to travel 5,000 km? _____ minutes

(c) How long does it take an L-wave to travel 5,000 km? _____ minutes

(d) How long does it take an R-wave to travel 5,000 km? _____ minutes

---

EXERCISE 7.5   Locating an Earthquake's Epicenter and Determining When It Occurred

Name: _____   Section: _____

Course: _____   Date: _____

Park the cars, and let's tackle an earthquake. Use the following series of steps to get all the information you need to identify the location of the earthquake and pinpoint when it occurred:

1. Identify the four different seismic wave types on seismograms from three seismic recording stations.

2. Determine the arrival times and measure the delays between different wave types.

3. Use these data and the travel-time diagram to estimate each station's distance from the epicenter.

4. Use triangulation to locate the epicenter.

5. Determine the time of faulting with the travel-time diagram.

Step 1: Reading a Seismogram

You will need to know what P-, S-, L-, and R-waves look like on a seismogram in order to identify them correctly. Each seismic wave causes the ground to shake differently (see Fig. 7.2), and this produces a different appearance on a seismogram. The differences are in wave amplitude (height) and frequency (the time between adjacent wave peaks). These differences are summarized in the figure below and will help you interpret the seismograms in the next part of this exercise.

This close-up of a seismogram shows the signals generated by different kinds of seismic waves.

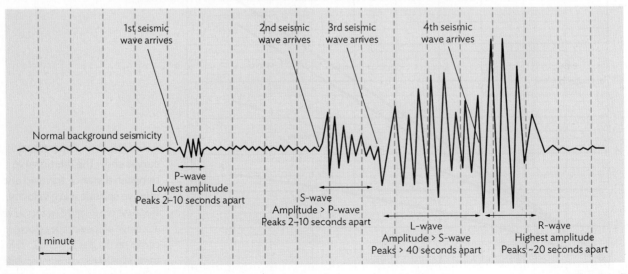

(continued)

Name: _____   Section: _____

Course: _____   Date: _____

Now that you know what different types of waves look like on a seismogram, let's see how to measure their arrival times. The figure below shows how waves look on a typical seismogram printout. The dashed vertical lines are time markers and are 1 minute apart. The waves reflect time moving forward from left to right on each row, and as a row ends, the time starts again on the left side of the next row below it (just like lines in a book). Seismic waves rarely arrive precisely on a minute marker, so you have to estimate the number of seconds before or after each minute. [*Note:* To avoid confusion involving time zones and changes to and from daylight saving time, seismogram times are recorded in Greenwich mean time (GMT).]

Recognizing seismic wave arrivals and measuring time on a seismogram.

Sudden change in amplitude and frequency from previous vibration indicates arrival of a new seismic wave type

Using the figure above, determine the arrival times of each of the four different types of waves. Place a P next to the place where you believe the P-wave arrives, an S where the S-wave arrives, an L where the L-wave arrives, and estimate and place an R next to where the R-wave appears to arrive. Then determine the correct arrival time for each wave. Use the inset circles to help you determine precise times that include fractions of minutes, as needed.

Arrival of P-wave _____ : _____ : _____

Arrival of S-wave _____ : _____ : _____

Arrival of L-wave _____ : _____ : _____

Estimated arrival of R-wave _____ : _____ : _____

Looking at the amplitudes of the waves recorded on the seismogram, which wave types do you think might cause the most damage in an earthquake? Explain.

_____

(continued)

EXERCISE 7.5 Locating an Earthquake's Epicenter and Determining When It Occurred *(continued)*

Name: ______________________ Section: ____________
Course: ______________________ Date: ____________

Step 2: Measuring the Delay between Arrival of Different Waves

Geologists use all four wave types to locate an earthquake, but to save time, your instructor may suggest using only the P- and S-wave data, so modify the following instructions as appropriate. The figure on the next page shows seismograms from three stations. Identify each of the wave types and record their arrival times in the table below. Use the seismic reading overlay tool found in the back of the manual to get precise arrival times for each wave. (Place the overlay over each seismogram, aligning the vertical dashed lines, to see the 15-second intervals.) Then calculate the times between arrivals by subtracting the P-wave arrival time from that of the S-wave, the S-wave from that of the L-wave, and the L-wave from that of the R-wave.

Arrival times and delays between seismic waves.			
Seismic wave arrival times	**Seattle**	**Boston**	**Los Angeles**
P-wave			
S-wave			
L-wave			
R-wave			
Delays between seismic waves			
S-P			
L-S			
R-L			

(continued)

Seismograms for Exercise 7.5

Seattle, Washington

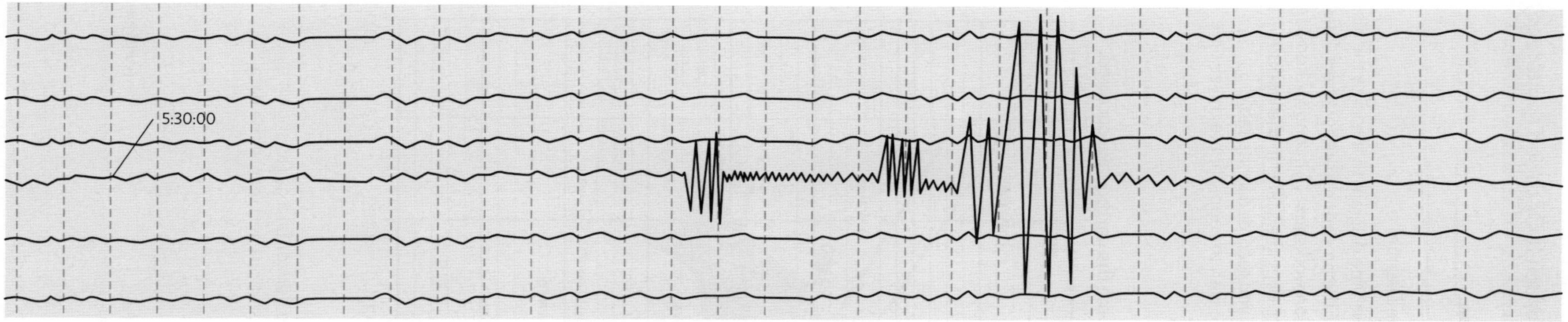

5:30:00

Boston, Massachusetts

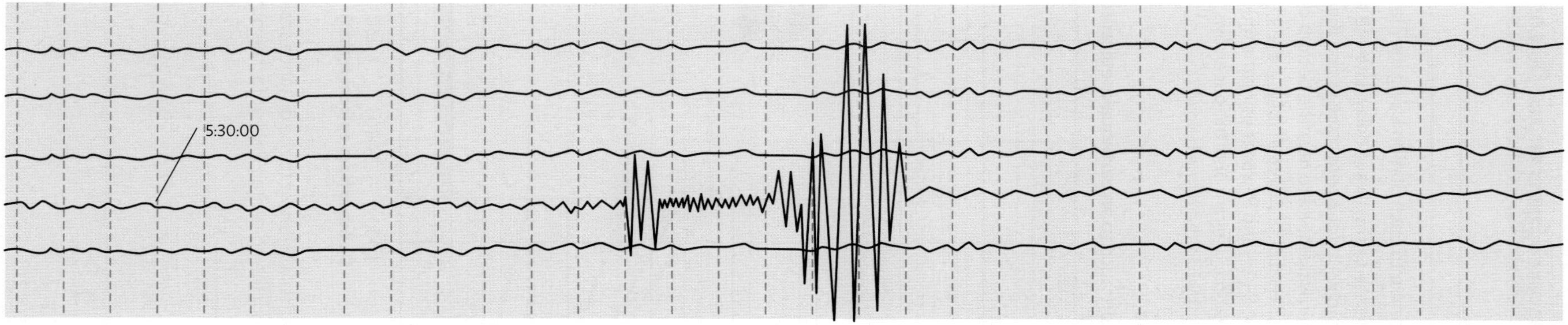

5:30:00

Los Angeles, California

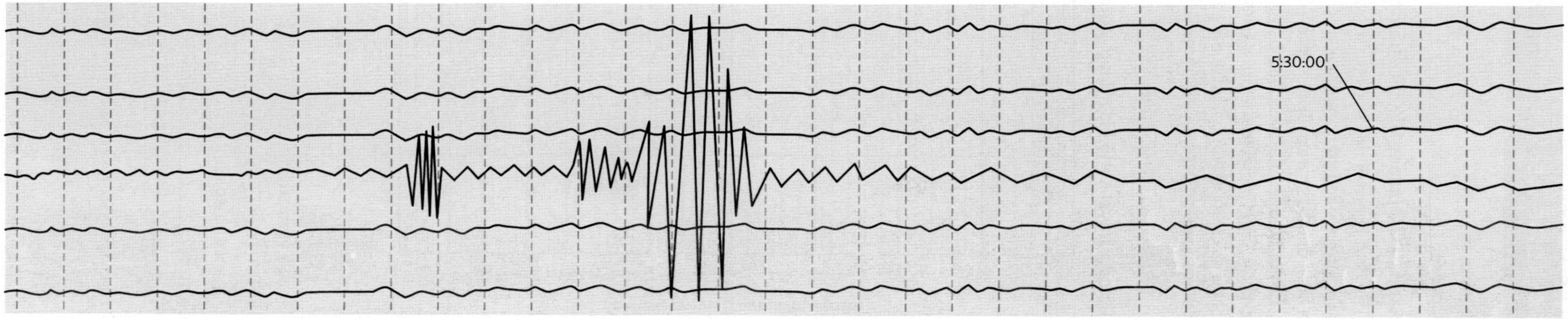

5:30:00

(continued)

Name: _____   Section: _____
Course: _____   Date: _____

Step 3: Estimating Distance from the Epicenter to Each Station Using the Travel-Time Diagram

The next figure in this exercise shows how to use seismic wave delay data and the travel-time diagram to measure the distance from each station to the epicenter. Start with one of the stations and then repeat the procedure for the others. Draw a horizontal arrow to represent the P-wave arrival anywhere near the bottom on a station worksheet (Appendix 7.1 on page 205) or create your own scale using a separate sheet of paper. In this case, make sure to copy the scales in the worksheets *exactly*. Then indicate the arrival times of the S-, L-, and R-waves using the time delays you recorded in the table. Slide the station worksheet across the travel-time diagram until all four arrows coincide with the appropriate travel-time curves (one possible worksheet position is shown in the figure below).

Using a travel-time diagram to determine distance to an earthquake epicenter and time the earthquake occurred.

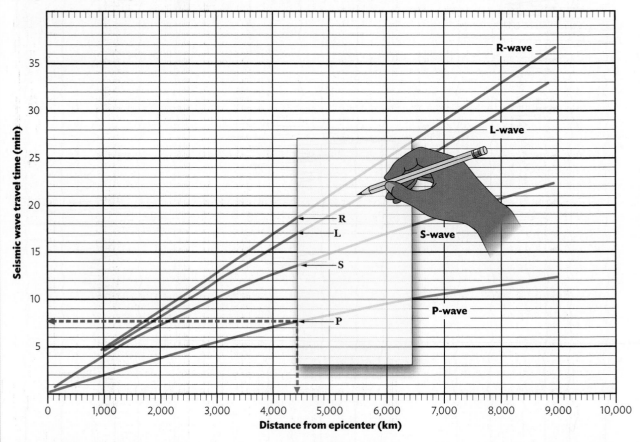

Draw a vertical line from the edge of the station worksheet to the distance scale on the horizontal axis (as shown by the dashed red arrow) and read the distance from the epicenter. Repeat for the other two stations and record the data.

Distance to epicenter from Seattle: _____ km

Boston: _____ km

Los Angeles: _____ km

(continued)

Name: _____   Section: _____

Course: _____   Date: _____

Step 4: Determining the Location of an Earthquake with Triangulation

It's one thing to know *how far* an epicenter is from a seismic recording station, but quite a different thing to know exactly *where* it is. Seismologists from Nova Scotia's Cape Breton Island calculated that an earthquake occurred 4,000 km from their station. Therefore, the epicenter must lie somewhere on a circle with a radius of 4,000 km centered on their station—but that could be in the Atlantic Ocean, Hudson Bay, Mexico's Yucatán Peninsula, or the Front Ranges of the Rocky Mountains (see the map to the right).

A second station in Caracas, Venezuela, calculated a distance of 6,000 km to the epicenter—somewhere on a circle with a radius of 6,000 km centered on Caracas. The two circles cross in two places, one in the Rockies, the other somewhere in the eastern Atlantic Ocean (outside the figure). The *epicenter* must be at one of these intersections, but both locations are equally possible. A third station is needed to settle the question, in this case pinpointing the epicenter in the Front Ranges of the Rocky Mountains. *Data from at least three seismic stations are needed to locate any epicenter, and the process is called* **triangulation**. The more stations used, the more accurate the location.

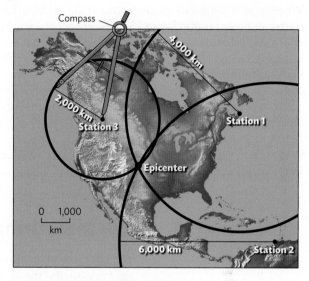

Locating an earthquake epicenter.

To locate the approximate epicenter (at last!) from your data, use an architect's compass or piece of string scaled to the appropriate distance for the map below. Draw an arc representing the distance from the first station and repeat the process for the other two stations. Review your work. Because of the tools you are using your three arcs might not intersect perfectly, but they should be near one another. Draw a small circle with a 300-km diameter showing the area where you would expect to find the epicenter.

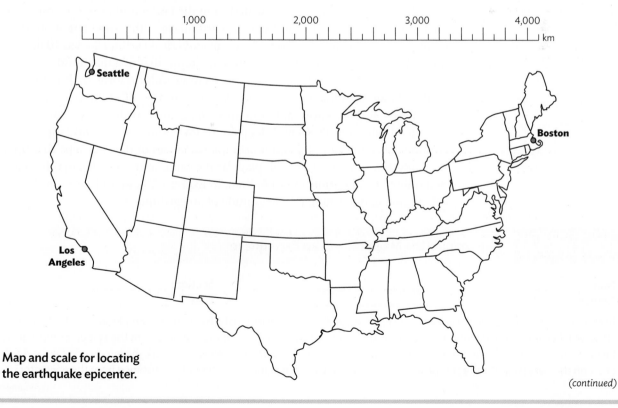

Map and scale for locating the earthquake epicenter.

(*continued*)

Name: _____   Section: _____
Course: _____   Date: _____

Step 5: Determining the Time When an Earthquake Occurred

When, exactly, did the earthquake occur at the epicenter? Align one of your station worksheets from Appendix 7.1 with the travel-time curves in Figure 7.5. Draw a horizontal line from the P-wave intercept on your worksheet to the time scale on the vertical axis, as shown by the green dashed arrow in the travel-time diagram on page 194. Read directly how many minutes the P-wave took to travel to the station, using a millimeter ruler to estimate the number of seconds—in this case, 9 minutes and 35 seconds.

   Subtract that amount of time from the P-wave arrival time in the data table of this exercise to get the time when the earthquake occurred. Repeat for the other two stations—the answer should be the same for each.

Time of the earthquake based on

Seattle _____ : _____ : _____
Boston _____ : _____ : _____
Los Angeles _____ : _____ : _____

7.4 Measuring the Strength of an Earthquake

The last piece of information we need is the strength of the earthquake. There are two ways to describe the strength of an earthquake. The **Mercalli intensity scale** is based on the amount of damage sustained by buildings. The larger the number (written as a Roman numeral), the greater the damage. But because damage to buildings can depend on factors unrelated to the energy released by an earthquake—such as the quality and nature of construction and the type of ground beneath the buildings—this method is not very useful in our study of the Earth. In 1935, Charles Richter designed the first widely accepted method for estimating the energy released in an earthquake: the **Richter magnitude scale**, which is based on the amount of energy released during faulting and calculated from the amount of actual bedrock motion. Each level of magnitude indicates an earthquake with ground motion 10 times greater than the next lower level. Thus, a magnitude 4 earthquake has 10 times more ground motion than a magnitude 3 and one-tenth that of a magnitude 5.

   We now know that Richter's method is accurate only for local, shallow-focus earthquakes. Modern estimates of earthquake strength use different methods involving body waves, surface waves, and a **moment magnitude scale** to calculate accurately the magnitude of shallow- and deep-focus, local and distant, and large and small earthquakes. They also depend on complex analyses of the rock type that broke in the fault, how much offset took place at the fault, and other factors that can't be determined from seismic records alone. In Exercise 7.6, we will use a simple graphical method to determine the magnitude of an earthquake.

EXERCISE 7.6   Determining the Magnitude of an Earthquake

Name: _____   Section: _____
Course: _____   Date: _____

In this simplified exercise, you will estimate m_b, the magnitude of an earthquake based on the amplitude of a body wave (P-wave). Because ground motion decreases with distance from a seismic station, distance from the epicenter must also be taken into account. To determine m_b of the earthquake you measured in Exercise 7.5, mark the left-hand scale of the chart on the next page at the appropriate S-P delay for one of your stations to account for distance from the epicenter.

(continued)

Name: _____   Section: _____
Course: _____   Date: _____

Measure the maximum P-wave amplitude and mark it on the right-hand scale. Now draw a line connecting these two points. The value for m_b is found where the line intersects the center magnitude scale.

(a) What is m_b for an earthquake with exactly the same **P-wave amplitude** as the example in the chart below (red line) but with an S-P delay of 30 seconds? _____   10 seconds? _____

(b) What is the relationship between wave amplitude, magnitude, and distance from the epicenter?

_____

_____

_____

(c) What is m_b for an earthquake with exactly the same **S-P delay time** as the example in the chart below, but with a P-wave amplitude of 2 mm? _____   100 mm? _____

(d) What is the magnitude of the earthquake you measured in Exercise 7.5, based on the following data from three stations close to the epicenter? Station A S-P delay = 20 s, P-wave amplitude = 100 mm; Station B: S-P delay = 40 s; P-wave amplitude =12 mm; Station C: S-P delay = 50 s, P-wave amplitude = 7 mm. _____

Note: The seismograms in Exercise 7.5 are artificial. If they were from a real earthquake, the values for m_b calculated from each should be nearly identical.

Determining the magnitude of an earthquake by body wave and amplitude.

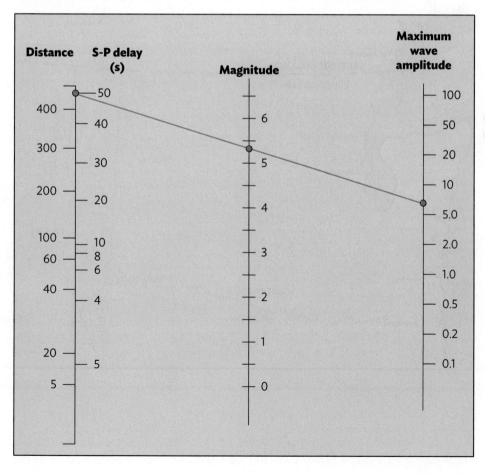

Name: _____   Section: _____

Course: _____   Date: _____

With the modern worldwide network of seismic stations, any earthquake epicenter can be located quickly and accurately. But what about earthquakes that took place *before* seismometers were invented? How can we define the Earth's zones of seismic activity if we can't include earthquakes that occurred as (geologically) recently as 100 or 200 years ago? If there are records of damage associated with those events, geologists can estimate epicenter locations by making *isoseismal maps* based on the modified Mercalli intensity scale that show the geographic distribution of damage. First, historical reports of damage are analyzed for each location and given an approximate intensity value. These data are plotted on a map, and the map is contoured to show the variation of the damage. Ideally, the epicenter is located within the area of greatest damage.

The figure below shows Mercalli intensity values for an 1872 earthquake that shook much of the Pacific Northwest. Draw contour lines on the map to show areas of equal damage (isoseismal areas) and indicate with an x the location for the epicenter of this earthquake.

Modified Mercalli intensity values for the December 1872 Pacific Northwest earthquake.

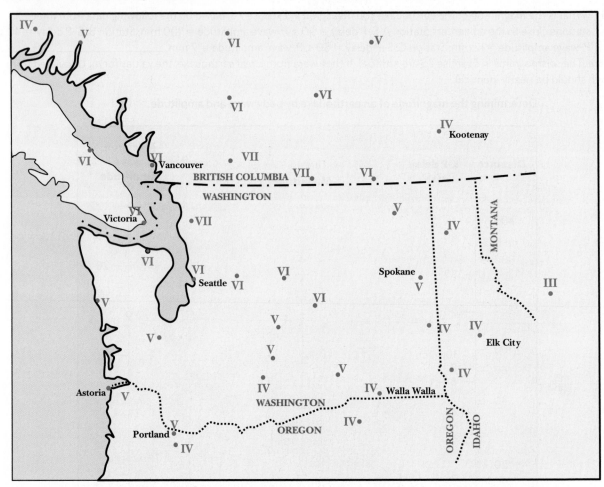

7.5 Predicting Earthquake Hazards: Liquefaction

Liquefaction is a process in which unconsolidated sediment loses its structural strength and can no longer support buildings. At a small scale, this can be annoying (**FIG. 7.6A**), but on a larger scale, it can be catastrophic (**FIG. 7.6B**). Liquefaction occurs during an earthquake when vibration causes water in pore spaces between loose sand grains to coat those grains completely, lessening the friction that held them together. The sand loses its ability to support heavy objects and behaves like quicksand. Liquefaction requires several key conditions: unconsolidated sediment, pores saturated or nearly saturated with water, and ground vibration. The sediment may be natural, such as rapidly deposited coastal sediments, or landfill used to create new land for shoreline development. Landfill is increasingly used in crowded cities for new housing and business districts, but these areas could be severely damaged by liquefaction during an earthquake. Emergency planners must prepare for such an event, and they begin by identifying areas in which conditions for liquefaction are likely. Exercise 7.8 examines one such case.

FIGURE 7.6 Results of liquefaction.

(a) Upwelling liquefied sand in San Francisco broke through the sidewalk and curbs, causing minor disruption.

(b) Apartment houses in Niigata, Japan, rotated and sank when an earthquake caused liquefaction of the ground beneath them.

Name: _____   Section: _____
Course: _____   Date: _____

Liquefaction has the greatest impact when all pore spaces between loose grains are filled with water. The *water table* separates zones below ground where all the pores are saturated with water from higher zones where some pores are dry. The water table and its importance in our ability to use underground water are discussed in Chapter 12. Here, we examine its possible role in predicting damage due to liquefaction.

The figure below shows data from a study of liquefaction potential for San Francisco County. It distinguishes areas where bedrock is exposed at the surface from those underlain by unconsolidated sediment, and it shows the depth to the water table in the sediment. Lower numbers mean the water table is closer to the ground surface; you'll work more closely with contour maps in Chapter 9. Examine the map carefully to locate places where liquefaction has occurred in the past and for clues to why it happened in those locations.

(a) Briefly describe how the water table relates to past episodes and locations of liquefaction in San Francisco. Why does it seem important to study the water table when considering the possibility of liquefaction?

_____

_____

_____

_____

_____

_____

_____

_____

_____

_____

_____

_____

_____

_____

_____

_____

_____

Liquefaction history in San Francisco related to bedrock and water table height.

Bedrock   Depth to water table (10-, 30-, 50-foot contours)   × Previous site of earthquake-generated liquefaction   1 mile

The map on the left (on the next page) shows locations susceptible to liquefaction as well as historical liquefaction events in the San Francisco area during the 1906 and 1989 earthquakes. Some of the 1989 events were in the same area as those of 1906, but several new areas were affected.

(b) Compare the locations of the 1989 liquefaction events with the water table elevations in the previous diagram. Does the relationship between liquefaction and water table elevation you discovered in question (a) also apply to all of these areas? If not, suggest possible explanations for the difference.

_____

_____

(continued)

Name: _____   Section: _____
Course: _____   Date: _____

(c) Now look at the locations of the 1989 liquefaction events that occurred in places not affected by the 1906 earthquake.

  (i) Are they randomly distributed throughout the region or restricted to specific locations? Explain. _____

_____

  (ii) Are they located in areas of varied susceptibility to liquefaction or in areas with the same level of susceptibility? Explain. _____

_____

  (iii) Compare the 1989 liquefaction sites with the map of shoreline changes on the right below. Why did the 1989 earthquake affect these areas, but not the 1906 earthquake? _____

_____

_____

Dots show liquefaction sites during the 1906 and 1989 earthquakes. Shading shows the degree of susceptibility to liquefaction.

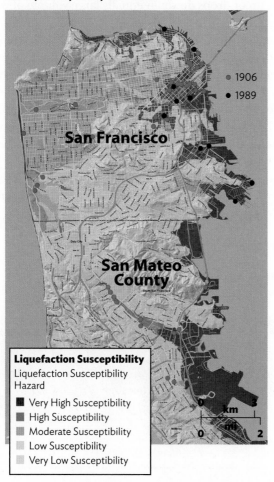

● 1906
● 1989

San Francisco

San Mateo County

Liquefaction Susceptibility
Liquefaction Susceptibility Hazard

■ Very High Susceptibility
■ High Susceptibility
■ Moderate Susceptibility
Low Susceptibility
Very Low Susceptibility

Shoreline changes in the San Francisco area. Pink shading shows landfill added since the 1906 earthquake.

(continued)

Name: _____ Section: _____

Course: _____ Date: _____

? **What Do You Think** If San Francisco is hit by another major earthquake, many roads will be blocked, and much of the relief effort will have to come by ship or plane. As adviser to the San Francisco disaster preparedness group, you have been asked to consider what might happen during the quake that could make supplies delivered by ship and plane difficult to receive in San Francisco. On a separate piece of paper, write a memo to the City Council describing the steps that you would propose the city take in order to deal with these risks.

7.6 Tsunami!

The Japanese word *tsunami* is now familiar to the entire world because of the devastation caused by these waves along Indian Ocean shorelines in 2004 and more recently in Japan in 2011. A tsunami occurs when the floor of the ocean is offset vertically—moving up or down dramatically—during an underwater earthquake. This displacement pushes up or drops down an equally large volume of water, causing a series of waves to travel across the ocean at about 500 miles per hour. However, if you were sitting in a small boat in the middle of the ocean, you wouldn't even know it passed beneath you. Unlike a typical wind-generated wave, a tsunami's wave height may be a foot or less; but its wavelength is thousands of feet. These waves pose no danger in mid-ocean, but can result in coastal disasters when, like ordinary waves, they begin to "break" as they near land (see Fig. 15.7). The word *tsunami* means harbor wave, and it is in harbors and estuaries that the damage is magnified (**FIG. 7.7**). Tsunamis striking a relatively straight coastline distribute their energy along the entire shore. When the wave enters an embayment such as a harbor or river mouth, however, the water is funneled into a narrow space, and the wave can build to heights greater than 10 m (more than 33 feet) and cause unimaginable damage.

FIGURE 7.7 Concentration of tsunami energy along coastlines. Tsunami waves are concentrated in embayments, resulting in walls of water much higher than along straight coastal segments.

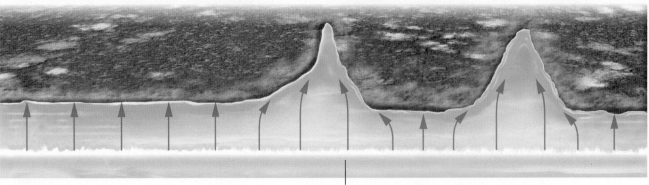

Tsunami wave crest

After watching dramatic images of tsunamis coming ashore near Sendai, Japan, some of our students asked how "just water" could cause so much damage. The amount of energy carried by a tsunami is truly unimaginable, but Exercise 7.9 offers a comparison that may help.

EXERCISE 7.9 **Why Is a Tsunami so Powerful?**

Name: _____   Section: _____
Course: _____   Date: _____

A cube of water 1 foot on a side (1 cubic foot, or 1 ft³) weighs approximately 62 pounds. Imagine being hit by a 62-pound weight—it would certainly hurt.

(a) Now imagine a low wall of water 10 feet wide, 1 foot deep, and 1 foot high; it would weigh _____ lb.

(b) A wall of water 30 feet high (like the Sendai tsunami), 1 foot deep, and 10 feet wide would weigh _____ lb.

(c) A wall of water 30 feet high, 5,280 feet wide (a mile), and 1 foot deep would weigh _____ lb.

(d) To approximate a tsunami better: a wall of water 30 feet high, 5,280 feet wide, and "only" 2,640 feet deep would weigh _____ lb.

(e) One of the most powerful man-made objects is a modern railroad locomotive. A large locomotive weighs about 200 **tons** = 400,000 pounds. The impact of the conceptual tsunami in question (d) is equivalent to being hit by _____ locomotives.

This simple calculation should help you understand why "just water" can cause so much damage. And remember that a tsunami moves much faster than a locomotive.

Name: _____   Section: _____

Course: _____   Date: _____

Exploring Earth Science Using Google Earth

1. Visit digital.wwnorton.com/labmanualearthsci
2. Go to the Geotours tile to download Google Earth Pro and the accompanying Geotours exercises file.

Expand the Geotour07 folder in Google Earth by clicking the triangle to the left of the folder icon. Check and double-click on the North Anatolian fault, Turkey, folder to fly to northern Turkey. The segments within the folder show the traces of a series of faults that compose the North Anatolian fault system. The segments are labeled and colored to show a relatively brief historical record of earthquake activity along the North Anatolian fault (based on approximate regions that were active during each major seismic event; gray faults have not been active during these major events).

(a) What is the dominant general pattern of major earthquakes along the main portion of the North Anatolian fault from 1939 to 1999? (*Hint:* Look for a temporal sequence of earthquakes related to direction.)

_____

_____

_____

(b) What might cause this pattern?

_____

_____

_____

(c) Assuming you are a geologist/geophysicist consulting with the Turkish government, where would you predict would be the likeliest area to next experience a major earthquake in association with this fault system (based on the data shown)? Check and double-click the c: placemarks to fly to each area location. _____ area

Seismic Analysis Worksheets

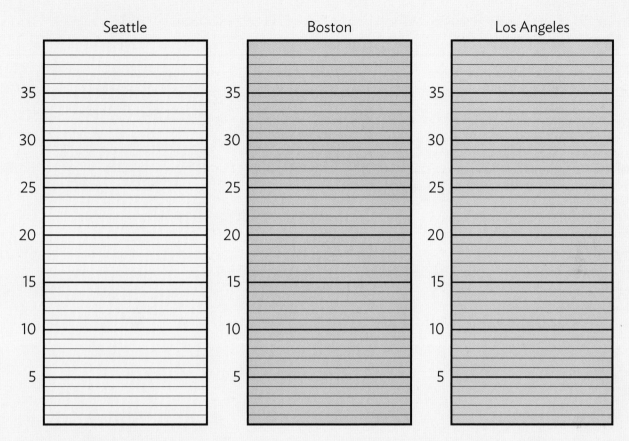

Cut out and use separately for the appropriate seismograph station on the travel-time diagram on p. 193.

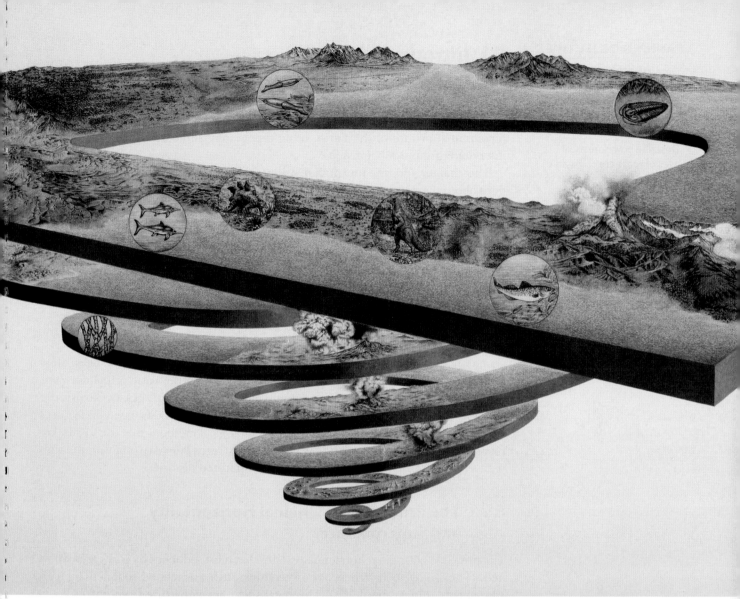

Interpreting Geologic History

The Earth has undergone many changes in its 4.56-billion-year history, as depicted in this spiral timeline.

■ Determine the relative ages of rocks and geologic processes and use these data to interpret complex geologic histories

■ Describe how numerical ages of rocks are calculated and apply these methods to dating geologic materials and events

■ Explain how geologists piece together Earth history from evidence found in widely separated locations

MATERIALS NEEDED

■ Pen, pencils, calculator

■ Metric ruler

8.1 Introduction

You've learned to identify minerals, use mineralogy and texture to interpret the origin of rocks, deduce what types of deformation have affected a given area, and recognize evidence of tectonic events. With these skills, you can construct a three-dimensional picture of the Earth, using topography and surface map patterns to infer underground relationships.

This chapter adds the fourth dimension: time. Geologists ask two different questions about the ages of rocks and geologic processes: (1) Is a rock or process older or younger than another? (its **relative age**) and (2) exactly how many years old is a rock or process? (its **numerical age**). In this chapter, we'll look first at how relative ages are determined, then at methods for calculating numerical age, and finally combine them to decipher geologic histories of varying complexity.

8.2 Physical Criteria for Determining Relative Age

Common sense is the most important resource for determining relative ages. Most of the reasoning used in relative dating is intuitive, and the basic principles were used for hundreds of years before we could measure numerical ages. Geologists use two types of methods to determine relative age: physical methods based on features in rocks and relationships among them, and biological methods that use fossils. We focus first on the physical methods and return to fossils later, starting with the simplest situations and building toward more complicated histories.

8.2.1 The Principles of Original Horizontality and Superposition

The principle of original horizontality states that most sedimentary rocks are deposited in horizontal beds (there are exceptions, such as inclined sedimentation in alluvial fans, dunes, and deltas). This means that any departure from horizontality in these layers is due to processes that acted upon the rocks after deposition. And sedimentary rocks are not the only ones that form in horizontal layers. Lava flows and volcanic ash falls also form layers that are typically horizontal (**FIG. 8.1**). Given the principle of original horizontality, it is easy to determine the relative ages in a sequence of rocks using common sense, *as long as the layers are still in their original horizontal position*. Try for yourself in Exercise 8.1.

FIGURE 8.1 Original horizontality in sedimentary and igneous rocks. The vertical grooves in some layers are a result of weathering and erosion.

(a) Horizontal sedimentary rocks in the Grand Canyon, Arizona.

(b) Horizontal lava flows, Columbia Plateau, Washington State.

(c) Horizontal ash fall deposits, Long Valley, California.

EXERCISE 8.1   Relative Ages of Horizontal Rock: The Principle of Superposition

Name: _____   Section: _____

Course: _____   Date: _____

Using common sense, label the *oldest and youngest* layers in each of the three photographs in Figure 8.1. Explain your reasoning.

_____

_____

_____

Congratulations! You have just repeated a discovery by Niels Stensen (also known as Nicolaus Steno) almost 400 years ago that is one of the foundations for relative dating. This **principle of superposition** states that in a sequence of strata that is still in its original horizontal position, the youngest beds will be at the _____ and the oldest will be at the _____ .

8.2.2 The Principle of Cross-Cutting Relationships

The principle of superposition cannot be used if layers of sedimentary or igneous rock have been tilted or folded from their original horizontal attitudes, nor can it be used for most metamorphic rocks and igneous rocks that weren't originally formed in horizontal layers. The principle of cross-cutting relationships helps to determine the relative ages of such rocks. Like the principle of superposition, it is based on common sense: if one rock or feature cuts across another rock, it must be younger than the rock that it cuts. Similarly, if a process affects a rock or group of rocks, it must have occurred after those rocks formed. You'll apply this principle in Exercise 8.2.

EXERCISE 8.2   Relative Ages in Original Horizontality and Cross-Cutting Situations

Name: _____   Section: _____

Course: _____   Date: _____

The following questions refer to the three photographs shown below.

Dikes intruding the granite of the Sierra Nevada batholith, Yosemite National Park, California.

Fault offsetting volcanic ash layers in Kingman, Arizona.

Folded sedimentary rocks on the island of Crete.

(a) Using the principle of cross-cutting relationships, label the granite and the two dikes in the left photo to indicate the sequence in which they intruded (e.g., 1, 2, 3).

(continued)

Name: _____   Section: _____

Course: _____   Date: _____

(b) The fault in the center photo must be _____ than the rocks it offsets. While you're at it, use the principle of superposition to label the oldest and youngest horizontal volcanic ash layers.

(c) The principle of original horizontality suggests that the process that folded the sedimentary rocks in the right photo must be _____ than the rocks. What can you say about the relative ages of the sedimentary rock layers? Explain.

_____

_____

_____

(d) When magma flows onto or into existing layers of rock, it "bakes" the rock it touches, causing *contact metamorphism.* Knowing this, how could you tell whether a basalt layer between two shale beds was a lava flow or a sill?

_____

_____

_____

8.2.3 The Principle of Inclusions

Pieces of one rock type are sometimes included in (enclosed in) another rock. Inclusion is most common in clastic sedimentary rocks, when fragments of older rocks are incorporated into conglomerates, breccias, and sandstones. It also happens where igneous rocks intrude an area and enclose pieces of the host rock. Inclusions of different rock types in an igneous rock are called **xenoliths** (from the Greek *xenos*, meaning stranger, and *lith*, meaning rock) (**FIG. 8.2**). Applying our common sense again, the **principle of inclusions** states that such inclusions must have existed before the intrusion or before the sedimentary rock formed, and therefore must be older than the intrusion or sedimentary rock that encloses them.

FIGURE 8.2 Inclusions in igneous and sedimentary rocks.

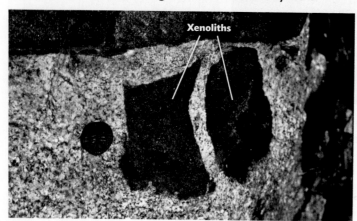

(a) Gabbro xenoliths in the Baring Granite in Maine.

(b) Fragments of rhyolite tuff in conglomerate from Maine.

EXERCISE 8.3 Applying the Principle of Inclusions

Name: ______________________ Section: ____________
Course: ______________________ Date: ____________

(a) Which is older, the granite in Figure 8.2a or the xenoliths? Explain.

____

____

____

(b) Clasts in sedimentary rock must be ______________ than the sedimentary rock in which they are included. Explain.

____

____

____

(c) Based on the block diagram below, state which rock is older, the sandstone or the rhyolite, and support your answer with evidence using the principle of inclusions.

____

____

____

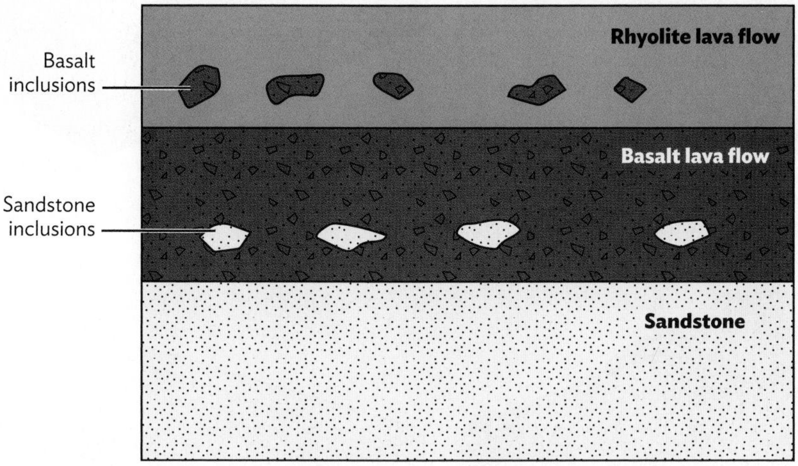

8.2.4 Sedimentary Structures

Some sedimentary structures indicate where the top or bottom of a bed was at the time it was deposited, so that it's possible to tell whether a bed is older or younger than another even if the beds are vertical or have been completely overturned. These top and bottom features include cross bedding, mud cracks, graded bedding, ripple marks, and impressions such as animal footprints. **FIGURES 8.3** and **8.4** show some of the features that help determine relative ages of sedimentary rocks, whether or not they are in their original horizontal positions.

 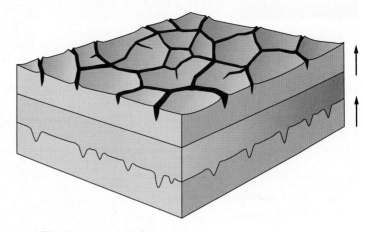

(a) Mud cracks: Mud cracks are widest at the top and narrow downward. The diagram at the right is therefore right-side up (as indicated by the arrows), so the bottommost bed is older than the bed in the middle.

 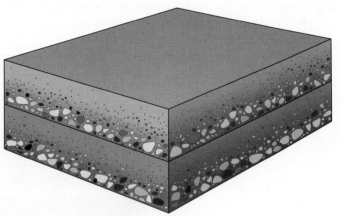

(b) Graded beds: The coarsest grains of sediment settle first and lie at the bottom of the bed. They are followed by progressively smaller grains, producing a size gradation.

 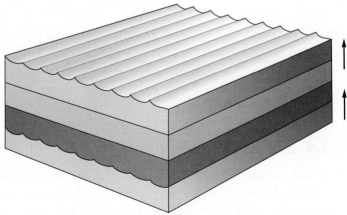

(c) Symmetric ripple marks: The sharp points of the ripple marks point toward the top of the bed.

FIGURE 8.4 Impressions can indicate the top or bottom of a sedimentary bed.

Features such as dinosaur footprints (left) and raindrop impressions (right) formed when something (here, a dinosaur and raindrops, respectively) sank into soft sediment exposed at the Earth's surface.

EXERCISE 8.4　　Using Sedimentary Structures to Unravel Geologic History

Name: _____　　Section: _____

Course: _____　　Date: _____

(a) Sketch diagrams showing how each of the three sedimentary structures (labeled below each sketch box) would appear in beds that had been turned upside down.

Mud cracks　　　　　　　　**Graded beds**　　　　　　　　**Symmetric ripple marks**

(b) Are the sedimentary features in Figure 8.4 right-side up or upside down as shown? Explain your reasoning.

_____

_____

The figure on the next page shows the value of sedimentary structures in relative dating. A cliff face exposing horizontal rock layers (left side) might be incorrectly interpreted as undeformed horizontal strata without the top and bottom features that prove that the layers have been folded, causing some to be overturned. The right side of the figure shows what the geologist might have seen had the adjacent rocks not been eroded away or covered by glacial deposits.

(continued)

Name: _____   Section: _____
Course: _____   Date: _____

(c) Number the layers, with 1 representing the oldest rock. Which is the youngest?

_____

_____

Reversals of top and bottom features reveal folding.

〜〜 Upright symmetrical ripple marks

°°° Upright graded bedding (smallest grains on top)

8.2.5 Unconformities: Evidence for a Gap in the Geologic Record

When rocks are deposited continuously in a basin without interruption by tectonic activity, uplift, or erosion, the result is a stack of parallel horizontal beds. Beds in a continuous sequence are said to be **conformable** because each conforms to, or has the same shape and orientation as, the others, as in the photograph of the Grand Canyon in Figure 8.1a.

Tilting, folding, and uplift leading to erosion interrupt this simple history and break the continuity of deposition. In these cases, erosion may remove large parts of the rock record, leaving a gap in an area's history. We may recognize that deposition was interrupted or that erosion took place, but we can't always tell how long the interruption lasted—it could have been a million, a hundred million, or a billion years—or what happened during that time. As we noted in Section 6.3.4, a contact indicating a gap in the geologic record is called an **unconformity** because the layer above it was not deposited conformably and continuously on those below.

There are three kinds of unconformities. An **angular unconformity** forms when rocks are tilted or folded, eroded, and covered by younger horizontal strata (**FIG. 8.5A**). The layers above the angular unconformity are at an angle to the older folded or tilted beds below. A **nonconformity** forms when a pluton intrudes host rock and crystallizes, and erosion then removes the host rock and part of the pluton (**FIG. 8.5B**). Later deposits of sedimentary rock sit directly on the eroded intrusive rock; the absence of contact metamorphism in this instance shows that the pluton had cooled before the sedimentary rock was deposited. A **disconformity** can be the subtlest type of unconformity because the strata above and below may be parallel. Disconformities form when a sequence of sedimentary rock is deposited and

FIGURE 8.5 How the three kinds of unconformities form.

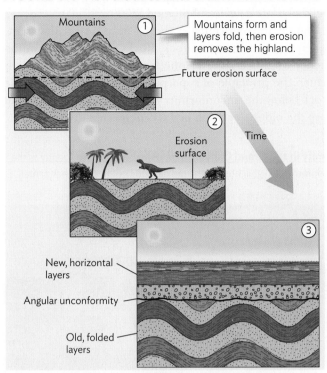

Mountains form and layers fold, then erosion removes the highland.

(a) An angular unconformity: (1) layers undergo folding; (2) erosion produces a flat surface; (3) sea level rises and new layers of sediment accumulate.

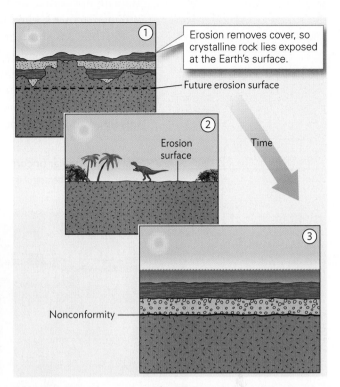

Erosion removes cover, so crystalline rock lies exposed at the Earth's surface.

(b) A nonconformity: (1) a pluton intrudes; (2) erosion cuts down into the crystalline rock; (3) new sedimentary layers accumulate above the erosion surface.

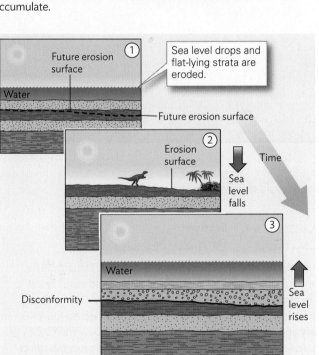

Sea level drops and flat-lying strata are eroded.

(c) A disconformity: (1) layers of sediment accumulate; (2) sea level drops and an erosion surface forms; (3) sea level rises and new sedimentary layers accumulate.

(d) This road cut in Utah shows a sand-filled channel cut down into floodplain mud. The mud was exposed between floods, and a soil formed on it. When later buried, all the sediment turned into rock. The channel floor is an unconformity. Note that the channel cuts across the paleosol (ancient soil). Thus, the paleosol is also an unconformity, as it represents a time during which deposition did not occur.

remains horizontal, undergoes erosion if sea level drops, and is later covered by new sedimentary rock when the area is once again submerged (**FIG. 8.5C, D**).

FIGURE 8.6 shows an angular unconformity in the Grand Canyon that separates two conformable sequences of sedimentary rock. Rocks above the unconformity are a horizontal sequence of sandstone, siltstone, and shale that is essentially undeformed and within which the principle of superposition can be applied. The sequence of sedimentary rock below the unconformity was originally horizontal but was tilted and eroded before the oldest overlying bed was deposited.

FIGURE 8.6 An angular unconformity in the Grand Canyon. The lower beds are tilted gently to the right and are separated by the angular unconformity (highlighted by the red dashed line) from the horizontal beds that lie above them.

EXERCISE 8.5	Deciphering Geologic History Using Physical Principle Relative Dating

Name: _____   Section: _____

Course: _____   Date: _____

In this exercise, you will combine the physical principles of relative dating with your knowledge of geologic structures to interpret the geologic histories of four areas with various degrees of complexity. This is the ultimate in "cold case" forensic science, as the events involved happened hundreds of millions of years ago.

Rock units in the four cross sections that follow are labeled with letters that have no meaning with respect to their relative ages (i.e., A can be older or younger than B). Sedimentary and metamorphic rocks are labeled with *UPPERCASE* letters, igneous rocks with *lowercase* letters. Note that the rocks and selected contacts are labeled, but the *processes affecting the rocks* are not.

In the timeline to the right of each cross section, list the rock units in order from oldest (at the bottom—remember superposition) to youngest (at the top). Where an event such as folding, tilting, erosion, or intrusion has occurred, draw an arrow to indicate its place in the sequence and label the event.

Briefly explain which principles you used to determine the relative ages of the rock units in each diagram. In many cases, you will have no doubt about the sequence of events, but in some cases, there may be more than one possibility. In the space provided after each diagram, briefly explain the reason for this uncertainty.

(continued)

Name: _____ Section: _____

Course: _____ Date: _____

Geologic cross section 1.

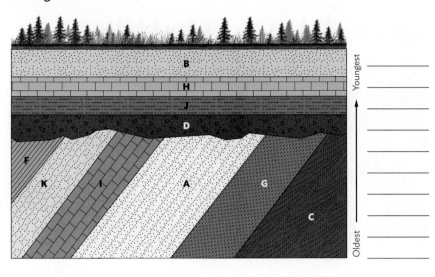

Geologic cross section 2.

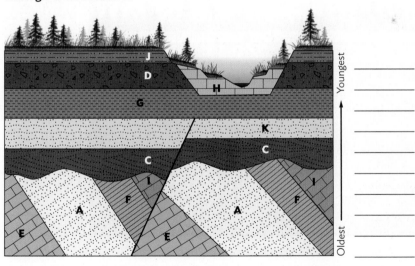

(continued)

Name: _____   Section: _____
Course: _____   Date: _____

Geologic cross section 3.

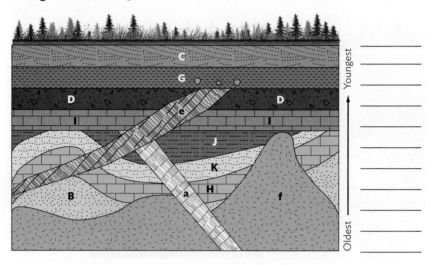

Geologic cross section 4.

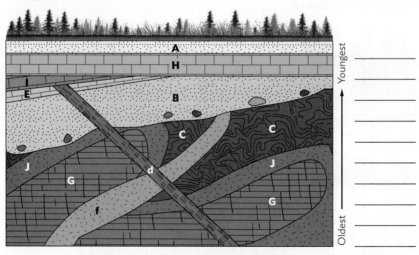

We will see later in this chapter how it is possible to date the rocks above and below an unconformity and therefore estimate the amount of geologic time missing between their deposition.

8.3 Biological Methods for Relative Age Dating

In 1793, British canal builder and amateur naturalist William Smith noticed different fossils in the layers of rock he was excavating. He discovered wherever he worked that particular fossils were always found in rocks that lay above others, and he suggested that fossils could be used to tell the relative ages of the rocks.

8.3.1 The Principle of Faunal and Floral Succession

Other geologists confirmed Smith's hypothesis and formulated the **principle of faunal and floral succession**, which states that fossil animals (fauna) and plants (flora) throughout the world record a specific, predictable order from older rocks to younger ones. Not all fossils can be used to date rocks, however, because some, such as blue-green algae, have existed over most of geologic time and are not specific to any particular time span.

Index fossils are remains of plants or animals that were distributed widely throughout the world but existed for only a short span of geologic time before becoming extinct. When we find an index fossil, we know that the rock in which it is found dates from that unique span of time. This is like knowing that a Ford Edsel could have been made only in the three years between 1957 and 1960, or the Model A between 1903 and 1931. *Tyrannosaurus rex*, for example, lived only in the span of geologic time known as the Cretaceous Period; it should not have been cast as a villain in a Jurassic theme park (representing an earlier time period).

8.3.2 The Geologic Time Scale

Determining the relative ages of rock units by physical methods and using index fossils to place those units in their correct time spans allowed geologists to construct *a geologic time scale* (**FIG. 8.7**). The scale divides geologic time into progressively smaller segments called eons, eras, periods, and epochs (not shown). The names of eras within the Phanerozoic Eon reveal the complexity of their life forms: Paleozoic, meaning *ancient* life; Mesozoic, meaning *middle* life; and Cenozoic, meaning *recent* life. The end of an era is marked by a major change in life forms, such as an extinction in which most life forms disappear and others fill their ecological niches. For example, nearly 90% of fossil genera became extinct at the end of the Paleozoic Era, making room for the dinosaurs, and the extinction of the dinosaurs at the end of the Mesozoic Era made room for us mammals. Several period names come from areas where rocks of that particular age were best documented: Devonian from Devonshire in England, Permian from the Perm Basin in Russia, and the Mississippian and Pennsylvanian from U.S. states.

The geologic time scale was originally based entirely on *relative* dating. We knew that Ordovician rocks and fossils were older than Silurian rocks and fossils, but had no way to tell how much older, or where exactly in geologic time they belonged. The numerical ages shown in Figure 8.7 provide that information *today*, but were calculated using methods discovered more than 100 years after the original scale was constructed.

FIGURE 8.7 The geologic time scale.

Eon	Era	Numerical age (Ma)	Period	Age ranges of selected fossil groups
Phanerozoic	Cenozoic	2.6	Quaternary	Flowering plants · Mammals · Humans
		66	Tertiary	
	Mesozoic	145	Cretaceous	Dinosaurs · Reptiles
		201	Jurassic	
		252	Triassic	
	Paleozoic	289.9	Permian	Amphibians
		323.2	Pennsylvanian	
		359	Mississippian	
		419	Devonian	Shelled animals · Fish
		443.4	Silurian	
		485	Ordovician	Trilobites
		541	Cambrian	
Proterozoic	Several eras	2,500 (2.5 Ga)		
Archean	Several eras	4,000 (4.0 Ga)		
Hadean		4.56 Ga		

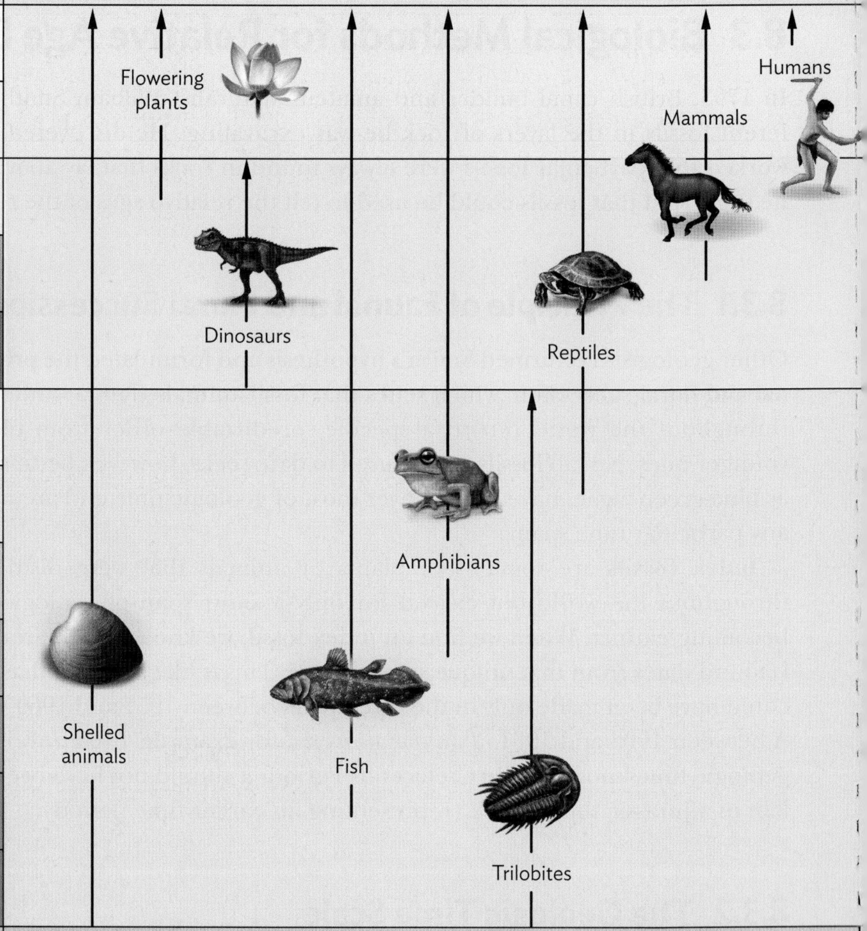

EXERCISE 8.6   Exploring Faunal Succession over Geologic Time

Name: _____   Section: _____

Course: _____   Date: _____

To better understand the immensity of time over which life has changed during the Earth's history, you will create a scale model of the geologic time scale to show the proportion of the Earth's history that each division represents. You will also add some of the changes seen in the fossil record that are the main basis for defining the divisions. The specific organisms listed for a geologic era or period in the table below are not the complete list of organisms that existed during the indicated times; rather, they are only a few examples of organisms that may be familiar to you—it would be impossible to label them all in a limited lab period. (*Note:* These ages are based on current fossil studies and are subject to slight change with ongoing research.)

Your instructor has provided you with paper (sheets or rolls) and a metric ruler. Follow the procedure outlined below and then answer analysis questions based on your observations. (*Note:* If this exercise is available and being conducted online, the setup, materials, and directions may differ. See the instructions in the online lab for how to conduct this exercise.)

Step 1: Lay your paper out on a long table or floor and tape the ends to the table or floor. The paper should be running in a line horizontally in front of you.

Step 2: Make a **scale** within the first 20 centimeters in the **top left corner**.

> 1 meter (m) = 1 billion years
>
> 1 centimeter (cm) = 10 million years
>
> 1 millimeter (mm) = 1 million years

Step 3: Starting on the **left side** of the paper, measure 20 cm to the right and make a vertical line down the paper. Label this line "Today," as shown in the example below:

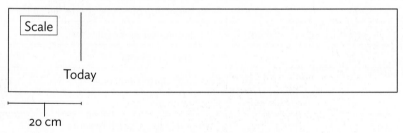

Step 4: Using the information from the table on the next page, measure, mark, and label the **major events** on your timeline. Use different colored pencils or pens as indicated:

Cenozoic Era: Orange
Mesozoic Era: Green
Paleozoic Era: Blue
Precambrian: Red

(a) According to your scale, did life on Earth emerge closer to the formation of the planet or closer to the emergence of modern humans? _____

(b) For what percentage of the Earth's history (round to the nearest whole number) has there been life on Earth? _____

(c) How much time passed between the first appearance of early life (single-cell) and multicellular organisms? _____

(continued)

Name: _____   Section: _____

Course: _____   Date: _____

(d) What marks the ends of the Paleozoic and Mesozoic Eras? _____

(e) When in geologic time did the first *Stegosaurus* appear on Earth? Provide both the age and time period. _____
_____

When did *Tyrannosaurus rex* first appear? _____

When did modern humans first appear in geological time? _____

Is *Tyrannosaurus rex* closer in time to *Stegosaurus* or to humans? _____

Did dinosaurs and humans exist at the same time? _____

Examples of major events in geologic time

Geologic time division	Time (million years, approximate)	Scale	Major event or appearance of fossil organisms
Cenozoic Era	0.2	0.2 mm	First *Homo sapiens* (modern humans)
	2.5	2.5 mm	First saber-toothed cats
	5	5 mm	First hippos
	40	4 cm	First giant whales
	55	5.5 cm	First armadillos
Mesozoic Era	66	6.6 cm	Mass extinction of ~ 75% of species on Earth
	68	6.8 cm	First *Tyrannosaurus rex* (dinosaur)
	130	13 cm	First flowering plants
	176	17.6 cm	First *Stegosaurus* (dinosaur)
	200	20 cm	First mammals
	240	24 cm	Start of the age of the dinosaurs
Paleozoic Era	252	25.2 cm	Mass extinction of up to ~ 90% of species on Earth (primarily marine)
	315	31.5 cm	First reptiles
	390	39 cm	First sharks
	412	41.2 cm	First insects
	470	47 cm	First land plants (mosses, etc.)
	500	50 cm	First jellyfish
	542	54.2 cm	First shelled marine organisms
Precambrian (Proterozoic)	1,600	1.60 m	First multicellular organisms
Precambrian (Archean)	3,800	3.8 m	First organisms (single-celled bacteria)
Precambrian (Hadean)	~ 4,550	4.6 m	Earth forms

8.3.3 Fossil Age Ranges

Some index fossil ages are very specific. The trilobite *Elrathia*, for example, lived only during the Middle Cambrian, whereas the trilobite *Redlichia* was restricted mostly to the Early Cambrian. Others lived over a longer span, such as the trilobite *Cybele* (Ordovician and Silurian) and the brachiopod *Leptaena* (Middle Ordovician to Mississippian). But even index fossils with broad age ranges can yield specific information if they occur with other index fossils whose overlap in time limits the possible age of the rock. This can be seen even in the broad fossil groups shown in Figure 8.7. Trilobite, fish, and reptile fossils each span several periods of geologic time, but if specimens of all three are found together, the rock that contains them could only have been formed during the Pennsylvanian or Permian periods. Likewise, knowing the ages of rock layers that surround a geologic event (based on the index fossils the rock layers contain) allows us to bracket the range of ages for when that event occurred. For example, if a fault cuts through rocks of Permian age but not the rocks above that were formed in the Jurassic Period, we could state that the fault occurred sometime between the Permian and Jurassic Periods.

EXERCISE 8.7 **Dating Rocks by Overlapping Fossil Ranges**

Name: _____ Section: _____
Course: _____ Date: _____

Brachiopods are marine bivalves (two-shelled invertebrates) that flourished during the Paleozoic Era but have been mostly replaced by modern clams. The figure on the next page shows selected Paleozoic brachiopod species used as index fossils and their age ranges.

(a) Based on the overlaps in brachiopod age ranges shown in the graph on the next page, what brachiopod assemblage in rock would indicate

• a Permian age?

_____

_____

_____

• a Silurian age?

_____

_____

_____

• an Ordovician age?

_____

_____

_____

(continued)

Name: _____   Section: _____
Course: _____   Date: _____

Selected Paleozoic brachiopod index fossils.

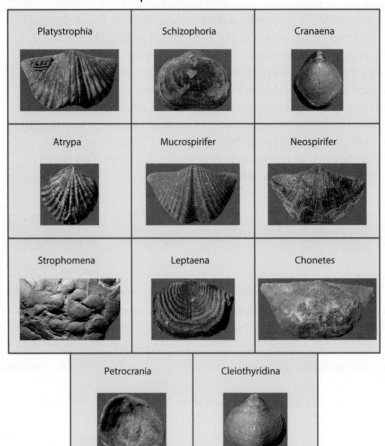

Age ranges for the brachiopod index fossils.

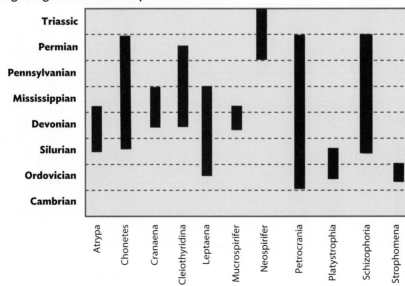

(continued)

Name: _____   Section: _____

Course: _____   Date: _____

(b) Now apply these overlaps to cross section 1 in Exercise 8.5.

• If *Neospirifer* is found in Unit D, *Platystrophia* in Unit F, and *Strophomena* in Unit A, suggest an age for Unit C. Explain your reasoning.

_____

_____

_____

_____

• What is the length of the gap in geologic time represented by the angular unconformity below Unit D?

_____

_____

_____

_____

(c) In cross section 2 of Exercise 8.5, *Strophomena* is found in Unit E, *Platystrophia* and *Petrocrania* in Unit F, and *Petrocrania*, *Chonetes*, and *Neospirifer* in Unit C.

• When were Units E, A, F, and I tilted? Explain your reasoning. (Remember that the answer may be a range of time as opposed to a specific time period.)

_____

_____

_____

_____

• What is the length of the gap in the geologic record represented by the contact between Unit C and the tilted rocks beneath it? Explain your reasoning.

_____

_____

_____

_____

• When in geologic time did the fault cutting Units K, C, I, F, A, and E occur? Explain your reasoning.

_____

_____

_____

_____

8.4 Determining Numerical Ages of Rocks

The geologic time scale (see Fig. 8.7) was used to describe the relative ages of rocks for about 100 years before numerical ages (in years) could be added. The principal method of determining numerical ages, called *radiometric dating*, is based on the fact that the nuclei of some atoms of elements found in minerals (**parent elements**) decay to form atoms of new elements (**daughter elements**) at a fixed rate *regardless of conditions*. This decay is called radioactivity, and the atoms that decay are radioactive isotopes (atoms of the same element but with different numbers of neutrons in their nuclei).

The radiometric "clock" begins when a mineral containing the parent element crystallizes during igneous or metamorphic processes. Over time, the amount of the parent element decreases and the amount of the daughter element increases (**FIG. 8.8**), in a pattern much like that of sand draining from the upper half of an hourglass into the lower half. An isotope's **half-life** is the amount of time it takes for half of the parent atoms in a mineral sample to decay to an equal number of daughter atoms (the center hourglass in Fig. 8.8).

How long the half-life of an isotope is depends on the rate at which that particular isotope decays. The faster the decay rate, the shorter the half-life; the slower the decay rate, the longer the half-life (**FIG. 8.9**). Generally, after only 4 half-lives have gone by, fewer than 10% of the parent atoms will be left. After 10 half-lives, there won't be enough parent atoms left to measure accurately, so a particular isotope is useful for determining dates only within time spans up to 10 times its half-life. Logically, isotopes with short half-lives are used to date relatively young rocks, whereas those with very long half-lives are used to date relatively old rocks.

FIGURE 8.8 Changing the parent:daughter ratios during radioactive decay.

Parent (%)	100	75	50	25	0
Daughter (%)	0	25	50	75	100
Parent:daughter ratio	—	3:1	1:1	1:3	—

FIGURE 8.9 Radioactive decay curves.

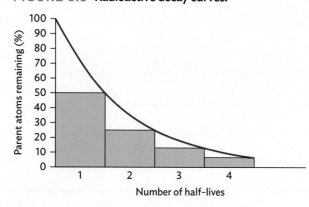

(a) The radioactive decay curve for an isotope.

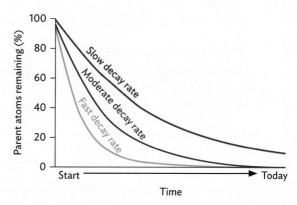

(b) Decay curves for isotopes with different decay rates.

TABLE 8.1 Geologically important radioactive isotopes used for radiometric dating

Parent isotope	Daughter decay product	Half-life (years)	Useful dating range (years)	Datable material
Samarium-147	Neodymium-143	106 billion	>10,000,000	Garnets, micas
Rubidium-87	Strontium-87	48.8 billion	>10,000,000	Potassium-bearing minerals (mica, feldspar, hornblende)
Uranium-238	Lead-207	4.5 billion	>10,000,000	Uranium-bearing minerals (zircon, apatite, uraninite)
Uranium-235	Lead-207	713 million	>10,000,000	Uranium-bearing minerals (zircon, apatite, uraninite)
Potassium-40	Argon-40	1.3 billion	>10,000	Potassium-bearing minerals (mica, feldspar, hornblende)
Carbon-14	Nitrogen-14	5,730	100–70,000	Organic materials

TABLE 8.1 lists the isotopes commonly used in radiometric dating, their half-lives, and the minerals in which they can be found. Many people who have heard of carbon-14 dating may be surprised to learn that this isotope has such a short half-life that it can be used to date only relatively recent materials. It is therefore a valuable tool for archeologists studying cultures tens of thousands of years old, but not for geologists studying rocks millions, hundreds of millions, and billions of years old.

Conversely, isotopes with very long half-lives, such as samarium-147, decay so slowly that they can be used to date only very old rocks. To calculate the numerical age of a rock, geologists first crush it and separate out the mineral containing the desired isotope. A mass spectrometer determines the parent:daughter ratio, and then a logarithmic equation is solved to calculate the rock's age. Your calculations will be easier: once you know the parent:daughter ratio, you can use **TABLE 8.2** to calculate age by multiplying the half-life by the number in the fourth column.

TABLE 8.2 Calculating the numerical age of a rock from the half-life of an isotope

Parent atoms remaining (%)	Parent: daughter ratio	Number of half-lives elapsed	Multiply half-life by _____ to determine age	Parent atoms remaining	Parent: daughter ratio	Number of half-lives elapsed	Multiply half-life by _____ to determine age
100	—	0	0	35.4	0.547	1½	1.500
98.9	89.90	1/64	0.016	25	0.333	2	2.000
97.9	46.62	1/32	0.031	12.5	0.143	3	3.000
95.8	22.81	1/16	0.062	6.2	0.066	4	4.000
91.7	11.05	1/8	0.125				
84.1	5.289	1/4	0.250				
70.7	2.413	1/2	0.500	0.05		11	Don't bother! There are too few parent atoms to measure accurately enough.
50	1.000	1	1.000	0.025		12	

Piecing together the 4.56-billion-year history of the Earth requires geologists to combine all the methods presented in this chapter: the physical principles of relative dating, fossil-based ages from sedimentary rocks, and numerical ages from igneous and metamorphic rocks. In Exercise 8.8, you will use these methods exactly the same way a professional geologist does.

EXERCISE 8.8   Putting It All Together to Decipher Earth History

Name: _____   Section: _____

Course: _____   Date: _____

(a) First, let's practice calculating ages using Tables 8.1 and 8.2.
We've done an example below for a rock containing a uranium-238:lead-206 ratio of 22.81.
To determine the rock's age, we need to know both the half-life of U-238 and the number of half-lives that have passed.

From Table 8.1, each half-life of U-238 = 4,500,000,000 years.
From Table 8.2, the U-238 to Pb-206 ratio of 22.81 represents 1/16th or 0.062 half-lives elapsed.
0.062 half-lives $\times$ (4,500,000,000 yrs / half-life) = 279,000,000 years
Therefore, the rock is 2.79 $\times$ 10^8 (279 million) years old.

Your turn! How old is a rock if it contains:
- a uranium-235:lead-207 ratio of 46.62? _____ years
- a rubidium-87:strontium-87 ratio of 89.9? _____ years
- 6.2% of its original carbon-14 atoms? _____ years
- 97.9% of its original potassium-40 atoms? _____ years

Now let's add some numerical ages to the cross sections in Exercise 8.5.

(b) In geologic cross section 3 of Exercise 8.5, *Dike e* contains zircon with a uranium-235:lead-207 ratio of 11.05; and Pluton f contains hornblende with 84.1% of its parent potassium-40. (Remember that when determining these ages, the answer for some may be a range of time as opposed to a specific date or time period.)

- How old is Dike e? _____ Pluton f? _____
- How old (in years and using period names) are Units B, H, K, and I?

_____

_____

- When were these layers folded?

_____

_____

- When did the unconformity separating Unit I from the underlying rocks form?

_____

_____

- How old are Units I and D?

_____

_____

8.5 Correlation: Fitting Pieces of the Puzzle Together

Imagine how difficult it would be for an alien geologist visiting the Earth 200 million years from now to reconstruct today's geography. Plate-tectonic processes could have moved some continents, split some apart and sutured others together, or opened new oceans and shrunk or closed others. In what is now North America, there would be evidence of mangrove swamps, forests, grassy plains, large lakes and rivers, shoreline features, alpine glaciers and deserts, and active volcanoes and other mountains. Numerical dating would show rocks that today range from more than 3 billion years old to as little as 1 year old.

This is the challenge facing geologists today as we try to read the record of Earth history. There is no single place where all 4.56 billion years of Earth history are revealed—not even in the Grand Canyon. Geologists working in California can work out the history of their field areas using all of the skills you've learned to interpret the formation and deformation of rocks, but how can they compare their results with those of someone working in Maine? In Japan? In Africa? Before we understood index fossils and learned how to date rocks numerically, the best we could do was to recognize that a sequence of rocks in Japan looked like a sequence of rocks in Britain—but were the rocks the same age, or did they just represent the same processes acting hundreds of millions of years apart?

With numerical dating tools, we can show that different kinds of rock from different areas around the globe are actually the same age—a process called **correlation** —and then compile paleogeographic maps showing where ancient shorelines and continents were located at that time. Exercises 8.9 and 8.10 examine a smaller-scale version of this process.

EXERCISE 8.9 | **Correlation**

Name: _____ Section: _____

Course: _____ Date: _____

Five years ago, geologists determined the relative ages of rocks in two areas of midcontinental North America, but the sequences (shown in the figure on the next page) were not identical. Unfortunately, similar rocks appear at several places in each sequence, making it difficult to know exactly which limestone unit, for example, in the western sequence correlates with a particular limestone unit in the eastern sequence. It is better to compare *sequences* of units, which reflect sequences of depositional environments, than to look only at similarities between individual rock units.

(a) Suggest correlations between rock units in the two sequences by drawing lines connecting the tops and bottoms of units in the western sequence to what you think are their matching units in the eastern sequence. (Look at the figure on the next page for an example of how to do this.) As in the real world, there may be more than one possibility, so explain your hypotheses and how you might test them.

_____

_____

_____

_____

(continued)

Name: _____ Section: _____

Course: _____ Date: _____

(b) Do you think the environments indicated by these sedimentary rocks were the *same* in both areas? Was deposition continuous (conformable) in both areas? Explain your reasoning.

_____

_____

_____

_____

Lithologic correlation of two sequences. Dashed lines show one possible correlation.

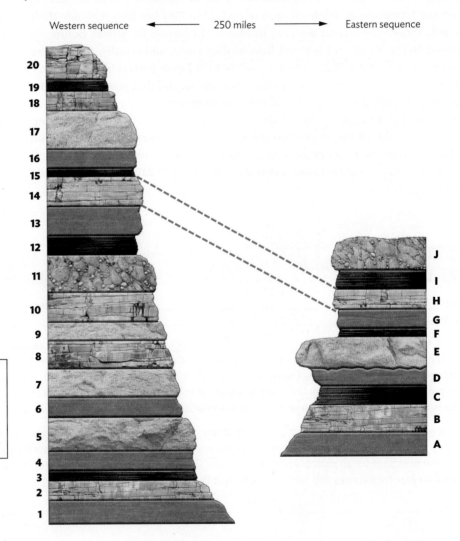

Western sequence   ◄——— 250 miles ———►   Eastern sequence

Legend:
- ▢ Limestone
- ▢ Siltstone
- ▢ Shale
- ▢ Sandstone
- ▢ Conglomerate

(continued)

Name: _____   Section: _____
Course: _____   Date: _____

During field mapping in 2019, trilobite fossils were found in limestone units in both areas (see the figure below). Paleontologists reported that these trilobites were identical index fossils from a narrow span of Middle Cambrian time.

(c) Using this additional information, draw lines indicating matching layers on the figure.

(d) Why does the presence of index fossils make correlation more accurate than correlation based on lithologic and sequence similarities alone?

_____

_____

_____

_____

Lithologic correlation of two sequences assisted by index fossils.

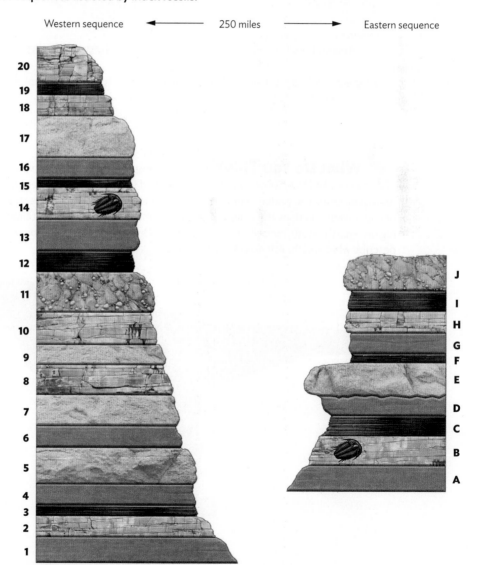

Name: _____   Section: _____

Course: _____   Date: _____

After more than 200 years of study, geologists have gathered an enormous amount of data about the geologic history of North America. This exercise presents some of those data and asks you to imagine what the geography of the continent looked like toward the end of the Cretaceous Period, just before an event that led to the extinction of most life forms—including, most famously, the dinosaurs.

The map on the next page summarizes data from the distribution of sedimentary rocks that form in specific environments (such as those you examined in Lab 4); results from index fossil and numerical dating of sedimentary and igneous rocks; and interpretations of the folding, faulting. and metamorphic history of Cretaceous and older rocks. We saw earlier that index fossils are valuable for pinpointing the relative ages of the rocks in which they are found. Here, we also use **facies fossils,** fossils of organisms that could live only in a narrowly restricted environment and therefore reveal the surface geography at the time they lived.

On the map, draw boundaries delineating the following North American geographic settings in the late Cretaceous Period:
- Continental interior
- Old, eroded mountain ranges
- Active tectonic zone
- Continental shoreline
- Shallow ocean (continental shelf and parts of the continent flooded by shallow seas)
- Deep ocean

Evidence for each setting is described just below the map. The facies fossils used as symbols on the map are, wherever possible, Cretaceous organisms.

? What Do You Think We all have impressions of what the Earth may have looked like during the Cretaceous Period from watching movies and television. Using the evidence from the Cretaceous map, describe on a separate sheet of paper what you think the United States and Canada looked like at that time, by region—north, south, central, and coastal areas. Then find your state or province and describe what you think it was like at that time.

(continued)

Name: _____   Section: _____

Course: _____   Date: _____

Map of North America today showing evidence for its Cretaceous paleogeography.

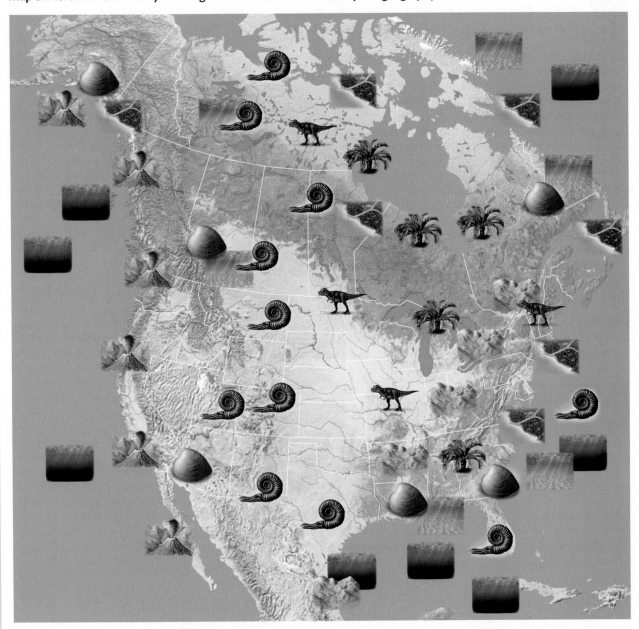

Continental Land Environments

Continental Interior	**Active Tectonic Zone**	**Old Mountains**	**Shorelines**
Terrestrial plants and animals	Volcanic arc rocks	Coarse clastic sedimentary rocks	Beach sandstones and coastal organisms

Marine Environments

Shallow Ocean	**Deep Ocean**
Shallow marine organisms and sedimentary rocks	Deep marine organisms and sedimentary rocks

Name: _____   Section: _____
Course: _____   Date: _____

Exploring Earth Science Using Google Earth

1. Visit **digital.wwnorton.com/labmanualearthsci**
2. Go to the **Geotours** tile to download Google Earth Pro and the accompanying Geotours exercises file.

Expand the Geotour08 folder in Google Earth by clicking the triangle to the left of the folder icon. The folder contains placemarks for features in the Grand Canyon of Arizona that provide natural examples to help understand some of the key concepts/principles used in relative dating.

(a) Check all of the **a**: placemarks and double-click one of the placemark icons to fly to the eastern part of the Grand Canyon. According to the principle of original continuity, which bed corresponds to the bed labeled "Source"?

_____

(b) Check all of the **b**: placemarks and double-click one of the placemark icons to fly to an area with three beds highlighted. According to the principle of superposition, order the beds from oldest to youngest.

_____

_____

(c) Check and double-click the Great Unconformity I placemark to fly to an exposure of the Great Unconformity (a surface that separates older layered Precambrian units from the younger sedimentary Cambrian unit with up to a 1-billion-year gap in the rock record due to erosion). What type of unconformity is it at this location? (Trace the pink layer to the left and see what happens to it.)

_____

(d) Check and double-click the Great Unconformity II placemark to fly to another exposure of the Great Unconformity. Here crystalline metamorphic rock and a white igneous dike are truncated and overlain by the same tan sedimentary Cambrian unit as above. Using these cross-cutting relationships, which is older: the dike or the tan Cambrian rock layer? Which likely contains inclusions of the other?

_____ is older

_____ contains inclusions of other unit

(e) What type of unconformity is the Great Unconformity at this location?

_____

Canyonlands National Park

COLORADO RIVER

×1593

×1473

×1525

Elephant Canyon

×1513

×1550

×1556

×1528

×1509

Spring Canyon

The Slide

RIVER

COLORADO

×1525

1543

| 0 | | 1 | | 2 miles |
| 0 | 5 | | 2 km | |

Studying Earth's Landforms: Working with Topographic Maps and Other Tools

9

A topographic map of Canyonlands National Park in southeastern Utah.

9.1 Introduction

Rapidly changing technology makes it easier to study the Earth's surface today than at any time in history. We can now make detailed models of the surface from satellite elevation survey data and download images of any area on the planet with the click of a mouse. Earth scientists were quick to understand the scientific value of satellite imaging technology and adopted new methods as fast as they were developed. Some of the images in this manual were not even available to researchers a decade ago. Thus the study of the Earth's surface today is nearly as dynamic as the surface itself.

This chapter is an introduction to traditional maps and aerial photographs, as well as to some of the newer methods that geologists and other Earth scientists use to view the Earth's surface and understand how its landscapes form. Much of the new technology is free for you to use on your computer. Google Earth and NASA WorldWind provide free satellite images of the entire globe and can generate three-dimensional views of landforms. Digital versions of the newest topographic maps and several generations of historical maps are available for free for most U.S. states from the U.S. Geological Survey (USGS) at https://store.usgs.gov. And with Google Maps, you can zoom to maps showing any location on Earth.

9.2 Ways to Portray the Earth's Surface

The ideal way to study landforms would be to first fly over them for a bird's-eye view and then walk or drive over them to see them from a human perspective. Since we can't do that in this class, we will have to bring the landforms to you instead, using topographic maps, aerial photographs, satellite images, and digital elevation models.

FIGURE 9.1 portrays an area of eastern Maine using these four different methods. **Topographic maps** (Fig. 9.1a) use contour lines (described in Section 9.6) to show landforms. They used to be drawn by surveyors who measured distances, directions, and elevations in the field, but they are now made by computers from aerial photographs and radar satellite elevation data. **Aerial photographs** (Fig. 9.1b) are photographs taken from a plane and pieced together to form a mosaic of an area. **Landsat images** (Fig. 9.1c) are made by a satellite that takes digital images of the Earth's surface using visible light and other wavelengths of the electromagnetic spectrum. Scientists can adjust the wavelengths to color the image artificially and to emphasize specific features; for example, some infrared wavelengths help reveal the amount and type of vegetation present. **Digital elevation models** (**DEMs**) (Fig. 9.1d) are computer-generated, three-dimensional views of landforms made from radar satellite elevation readings spaced at 10-m or 30-m intervals on the Earth's surface. A new generation of DEMs using 1-m intervals is now being released that provides a more accurate model of the surface than anything available to the public 5 years ago. Each kind of image has strengths and weaknesses that make it useful for different purposes, as you will see in Exercise 9.1.

FIGURE 9.1 An area in eastern Maine shown by different imaging methods. The scale
bar and north arrow in part (a) apply to all four images.

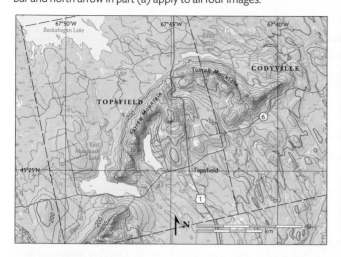

(a) Topographic map.

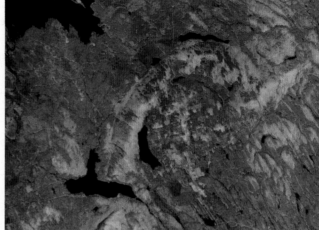

(b) Aerial photograph.

(c) Landsat image (artificial color).

(d) Digital elevation model.

EXERCISE 9.1   **Which Image Works Best?**

Name: _____   Section: _____
Course: _____   Date: _____

(a) Examine the images in Figure 9.1 and use the table below to rank them on a scale of 1 to 4 by how well they show the
map elements indicated (1 is most effective and 4 is least effective). Images may be assigned the same rank.

Map element	Topographic map	Aerial photograph	Landsat image	DEM
Location				
Distance				
Direction				
Elevation				
Changes in slope				
Names of features				

(continued)

Name: _____   Section: _____

Course: _____   Date: _____

(b) Which of the images enables you to recognize the topography most easily? Why?

_____

_____

(c) Which is *least* helpful in trying to visualize hills, valleys, and lakes? Why?

_____

_____

(d) Erosional processes often produce aligned ridges, valleys, and elongated hills called the *topographic grain*. Which images show the topographic grain in this area most clearly? Once you've seen it on those images, can you recognize it on the others? Use a marker or colored pencil to draw the alignment direction on each image.

_____

_____

(e) Which images show highways most clearly?

_____

_____

(f) Which images show unpaved lumber roads most clearly?

_____

_____

(g) Study the details of what is shown in each image. Which image do you think is the oldest? The most recent? Explain your reasoning.

_____

_____

_____

9.2.1 Map Projections

The images in Figure 9.1 are flat, two-dimensional pictures, but the Earth is a nearly spherical three-dimensional body. Only a three-dimensional representation—a globe—can accurately show the *areas* and *shapes* of the Earth's features and the *directions* and *distances* between them. The process by which the three-dimensional Earth is converted to a two-dimensional map is called **projection**. There are many different projections, each of which distorts one or more map elements, and each of which is useful for some purposes but unusable for others. Three common map projections are shown in **FIGURE 9.2**, and **TABLE 9.1** indicates what they distort and for what purposes they are best used.

FIGURE 9.2 **Three common map projections and their different views of the world.**
The equator is indicated in each projection by a red line.

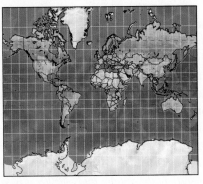

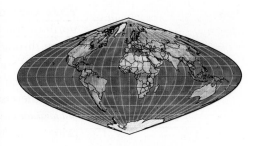

(a) Orthographic.

(b) Mercator.

(c) Uninterrupted sinusoidal.

Table 9.1 **The strengths and weaknesses of three common map projections**

	Orthographic	Mercator	Uninterrupted sinusoidal
Strengths	Directions between points are preserved. East-west distances are accurate.	Accurate near the equator. Directions preserved in most areas.	Areas of continents are represented accurately.
Weaknesses	Shapes and areas are distorted, especially near the edge of the projection. Distances other than in east-west direction are distorted.	Severe shape and area distortions away from the equator. Useless for northern and southern polar areas.	Scale is constant only along a central north-south line and the equator, and changes elsewhere. Shapes of features distant from these reference lines are distorted.
Comments	Perspective view is similar to view of a globe from a great distance. Note the accurate representation of the size of Greenland compared with the Mercator projection.	Note vastly distorted polar areas of Greenland (shown much more accurately on the other projections) and Antarctica.	Note that the area of Greenland (much smaller than Africa or South America) is portrayed accurately as compared with the Mercator projection.
Uses	Often used to provide context for images of the Earth taken from space.	Nautical navigation charts, because point-to-point directions are accurate.	Commonly used for features elongated north-south (e.g., maps of Africa and South America).

9.3 Map Elements

An accurate portrayal of the Earth's surface must contain three basic elements: (1) location, a way to show precisely where on the planet a feature is; (2) the distance between features; and (3) an accurate depiction of the directions between features. In many instances, it is also important to know the elevations of hilltops and other features and the steepness of the slopes in mountains, hills, and valleys.

9.3.1 Map Element 1: Location

Road maps and atlases use a simple grid system to locate cities and towns; for example, Chicago is in grid square A8. Many other places may be in the same grid square, but this location is good enough for someone driving to Chicago from Los Angeles. More sophisticated grid systems are used for more precise locations. Maps published by the USGS use three different grid systems: latitude/longitude, the Universal Transverse Mercator (UTM) grid, and for most states, the Public Land Survey (or Township and Range) System. The UTM grid is least familiar to Americans but is used extensively in the rest of the world.

9.3.1a Latitude and Longitude The latitude/longitude grid is based on location north or south of the **equator** and east or west of the **prime meridian**, a line that passes through the Royal Observatory in Greenwich, England, as well as both the North and South poles (**FIG. 9.3**). A *parallel of latitude* connects all points that are the same angular distance north or south of the equator. The maximum value for latitude is therefore 90° N or 90° S (the North and South Poles, respectively). A *meridian of longitude* connects all points that are the same angular distance east or west of the prime meridian. The maximum value for longitude is 180° E or 180° W (the international date line). *Remember: When stating a location, you must always indicate whether a point is north or south of the equator **and** whether it is east or west of the prime meridian.*

FIGURE 9.3 The latitude/longitude grid.

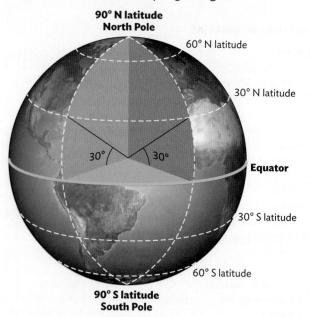

(a) Latitude is measured in degrees north or south of the equator.

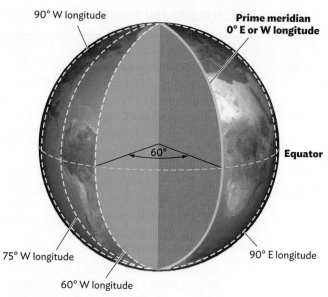

(b) Longitude is measured in degrees east or west of the prime meridian (Greenwich, England).

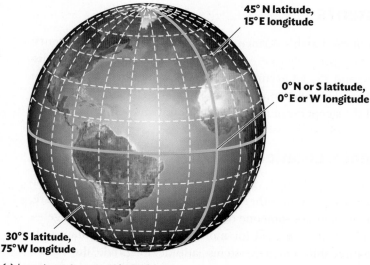

(c) Locating points using the grid.

Latitude and longitude readings are typically reported in **degrees** (°), **minutes** ('), and **seconds** (") (for example, 40°37'44" N, 73°45'09" W). There are 60 minutes in a degree and 60 seconds in a minute. For reference, 1 degree of latitude is equivalent to approximately 69 miles (111 km), 1 minute of latitude is about 1.1 mile (1.85 km), and 1 second of latitude is about 100 feet (31 m). The same kind of comparison can be made for longitude *only at the equator* because the meridians converge at the poles, so the distance between degrees of longitude decreases gradually toward the poles. While this is the standard way of recording latitude and longitude, many modern navigation applications on computers and portable electronic devices use decimals instead of minutes and seconds for partial degrees, and substitute (+) for N latitudes and E longitudes, and (−) to denote S latitudes and W longitudes.

EXERCISE 9.2   **The Latitude/Longitude Grid**

Name: _____   Section: _____
Course: _____   Date: _____

(a) Describe the locations of the points indicated by the stars in the figure below, on which parallels of latitude and meridians of longitude are spaced 15° apart. Remember to specify N or S for latitude, W or E for longitude.

(b) With the aid of a globe or map, determine the geographic features that are located at

  • 45°00'00" N, 90°00'00" W _____

  • 15°00'00" N, 30°00'00" E _____

  • 30°00'00" S, 90°00'00" W _____

(c) With the aid of a globe or map, determine the latitude and longitude of your geology laboratory as accurately as you can. How might you locate the laboratory more accurately?

_____

_____

_____

(d) If you have access to a GPS receiver, locate the corners of your laboratory building. Draw a map (on a separate piece of paper) showing the building's location, orientation, and distances between the corners.

(continued)

Name: _____   Section: _____
Course: _____   Date: _____

(e) Use the latitude/longitude grid to locate the following U.S. and Canadian cities as accurately as possible. For reference, use a globe, map, Google Earth, or Google Maps.

City	Location	City	Location
Nome, AK		Seattle, WA	
Chicago, IL		Los Angeles, CA	
St. Louis, MO		Houston, TX	
New York, NY		Miami, FL	
St. John's, Newfoundland		Ottawa, Ontario	
Calgary, Alberta		Victoria, British Columbia	

(f) Which of the cities in question (e) do you think is closest in latitude to each of the following world cities? Predict first without looking at a map, globe, or Google Earth, and then check. Were you surprised by any?

City	Predicted best match	Latitude and longitude	Actual best match
Oslo, Norway			
Baghdad, Iraq			
London, England			
Paris, France			
Rome, Italy			
Beijing, China			
Tokyo, Japan			
Quito, Ecuador			
Cairo, Egypt			
Cape Town, South Africa			

9.3.1b Public Land Survey (Township and Range) System   The Public Land Survey System was created in 1785 to provide accurate maps as the United States expanded westward from its 13 original states. Much of the country is covered by this system; the exceptions are the original 13 states, Kentucky, Maine, Tennessee, West Virginia, Alaska, Hawaii, Texas, and parts of the southwestern states surveyed by Spanish colonists before they joined the Union. Points can be located rapidly in this system to within an eighth of a mile (**FIG. 9.4**).

The grid is based on accurately surveyed north–south lines, called **principal meridians**, and east–west **base lines** for each survey region. Lines drawn parallel to these at 6-mile intervals create a grid of squares 6 miles on a side, forming east–west rows called **townships** and north–south columns called **ranges**. Each 6-mile square is

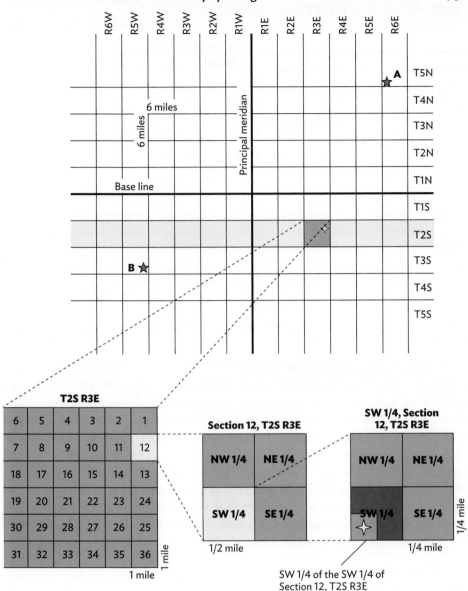

FIGURE 9.4 The Public Land Survey System grid. Points A and B are for use in Exercise 9.3.

SW 1/4 of the SW 1/4 of
Section 12, T2S R3E

divided into 36 **sections**, 1 mile on a side, numbered as shown in Figure 9.4. Sections are divided further into **quarter sections** ½ mile on a side, and these in turn are each divided into four squares ¼ mile on a side.

Township rows are numbered north or south of the baseline (e.g., T2S), and ranges are numbered east or west of the principal meridian (e.g., R3E). The location of the yellow starburst in Figure 9.4 is described in the series of blowups:

- **T2S R3E** locates it somewhere within an area of 36 square miles (inside a 6-mi × 6-mi square).
- **Section 12, T2S R3E** locates it somewhere within an area of 1 square mile.
- **SW ¼ of Section 12, T2S R3E** locates it somewhere within an area of ¼ square mile.
- **SW ¼ of the SW ¼ of Section 12, T2S R3E** locates it within an area of ¹⁄₁₆ of a square mile.

In Exercise 9.3, you will practice using the Public Land Survey System.

EXERCISE 9.3 Locating Points with the Public Land Survey System

Name: _____   Section: _____

Course: _____   Date: _____

(a) Determine the locations of points A and B in Figure 9.4.

A: _____   B: _____

(b) Locate and label the following points in Figure 9.4.
- NE ¼ of Section 36, T3N R4W
- SE ¼ of Section 18, T1N, R1 E
- NW ¼ of Section 3, T45 R5E

(c) Now you're ready for the real thing. Determine the locations of the points indicated by your instructor on the topographic maps provided. (*Note:* If this exercise is available and being conducted online, the setup, materials, and directions may differ. See the instructions in the online lab for how to conduct this exercise.)

_____

_____

9.3.1c Universal Transverse Mercator Grid   The UTM grid divides the Earth into 1,200 segments, each containing 6° of longitude and 8° of latitude (**FIG. 9.5**). East–west rows of segments are assigned letters (C–X), and north–south columns of segments, called **UTM zones**, are numbered 1–60 eastward from the international Date Line (180° W longitude). Thus, UTM zone 1 extends from 180° to 174° W longitude, zone 2 from 174° to 168° W longitude, and so on. The 48 contiguous United States lie within UTM zones 10 through 19, at roughly 125° to 67° W longitude (**FIG. 9.6**). The UTM grid is based on a Mercator projection, in which the north and south polar regions are extremely distorted, making Greenland and Antarctica look much larger than they actually are. Because of this distortion, the UTM grid does not extend beyond 84° N and 80° S latitudes.

FIGURE 9.5 The UTM grid.

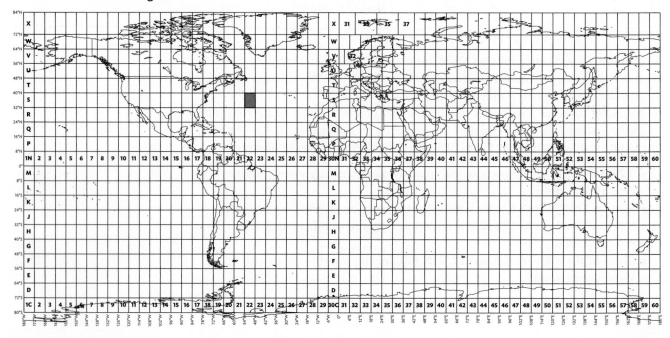

FIGURE 9.6 UTM zones for the 48 contiguous United States.

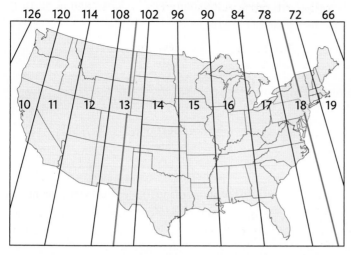

FIGURE 9.6 UTM zones for the 48 contiguous United States.

The red line is the central meridian (105° W) for UTM zone 13; the blue line is the central meridian for zone 18.

To locate a feature, begin with the grid segment in which the feature is located. For example, the red box in Figure 9.5 is grid segment S22. UTM grid readings tell us *in meters* how far north of the equator (*northings*) and east of a central meridian for each UTM zone (*eastings*) a point lies. The **central meridian** for each UTM zone is a line of longitude that runs through the center of the zone, as shown in Figure 9.6, and is arbitrarily assigned an easting of 500,000 m so that no point has a negative easting. Points east of a central meridian thus have eastings greater than 500,000 m, and those west of a central meridian have eastings less than 500,000 m. The red line in Figure 9.6 is the central meridian for UTM zone 13, the blue line is the central meridian for zone 18.

FIGURE 9.7 shows how the UTM grid appears on the most recent USGS topographic maps. Each grid square is exactly 1,000 m (1 km) on a side. Labels along the top or (as here) bottom of the map are **eastings (E)**—distances in meters from the central meridian for the UTM zone (in this case, UTM zone 19, as shown in Fig. 9.6). Labels along the west or (as here) east side of the map are **northings (N)**—distances in meters north or south of the equator. These values are written out fully near the corner of the map and elsewhere in a shorthand form. For example, the easting near the bottom right-hand corner of the map, $^{5}48^{000\text{m}}$E, is the full value: 548,000 m east of the central median for UTM zone 19. The eastings to the west are the shorthand version: $^{5}47$ and $^{5}45$ are, respectively, 547,000 and 545,000 m east of the central meridian. Note that the grid label for 546 is missing; this is because it should have appeared in the place near where a longitude value is shown (25′, representing longitude 68°25′00″ W). Similarly, the full northing value just above the 45°00′ is $^{49}83^{000\text{m}}$N (4,983,000 m north of the equator), and the values above it are the shorthand version (e.g., $^{49}84$).

It is possible to measure locations on a topographic map to within 100 m or better using a simple UTM tool calibrated for the appropriate map scale. A UTM tool for the map scale used in Figure 9.7 can be found in your geologic toolkit at the end of this manual. To determine the location of the blue dot in Figure 9.7, place the lower left corner of the UTM tool on the map so that it covers the grid square in which the point lies. As we've seen, the side of each grid square is 1,000 m, and the UTM tool divides it into smaller boxes 100 m on a side. The tool shows that the easting for the blue point is between 545,000 and 546,000, and that the northing is between 4,985,000 and 4,986,000. The blue point's easting is approximately

550 m (5.5 boxes) east of the 545,000 mark = **545,550** m E, and its northing is 120 m (1.2 boxes) south of the 4,986,000 mark = **4,985,880** m N. In Exercise 9.4, you will practice using the UTM grid.

EXERCISE 9.4   **Locating Points with the UTM Grid**

Name: _____   Section: _____

Course: _____   Date: _____

Refer to Figure 9.7 to complete this exercise.

(a) Using the 1:24,000 UTM tool (from the geologic toolkit in the back of this manual), give the location in northings and eastings of the red star on Figure 9.7.

_____ N   _____ E

(b) What is the location of the point where the stream leaves Olamon Pond and flows to the southwest?

_____ N   _____ E

(c) What feature is located at UTM 547,575 m E and 4,983,450 m N?

_____

FIGURE 9.7 Southeast corner of the Greenfield quadrangle, Maine, showing UTM grid squares and marginal UTM grid values. Map scale = 1:24,000. In the upper left corner, a UTM tool is shown in position to locate the blue point.

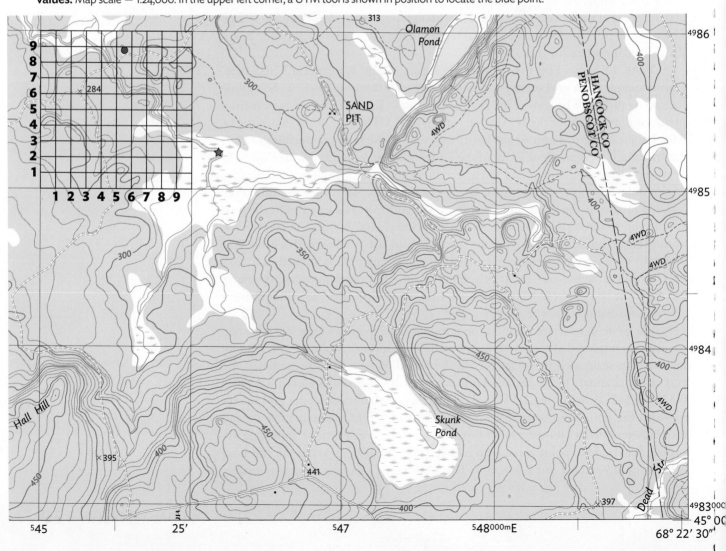

9.3.2 Map Element 2: Direction

Geologists use the **azimuth method**, which is based on the dial of a compass, to indicate direction (**FIG. 9.8**). The red-tipped compass needle in Figure 9.8a is pointing northeast—that is, somewhere between north and east—but how much closer to north than to east? On an azimuth compass (Fig. 9.8b), 0° or 360° is due north, 090° due east, 180° due south, and 270° due west. The direction 045° is exactly halfway between north and east. The direction of the needle in Figure 9.8b can be read as 032°, meaning that the direction from the white end of the needle to the red end is 32° east of north.

FIGURE 9.8 Using the azimuth method to describe the direction between two points.

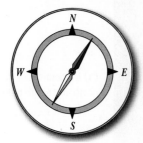

(a) A simple compass.

(b) A compass with azimuth markings.

(c) A circular protractor can be used to determine the direction between points.

Name: _____   **Section:** _____

Course: _____   **Date:** _____

Use the circular (azimuth) protractor from your geologic toolkit at the back of this manual to determine the directions between the following pairs of points.

(a) In the figure below, from:

A to C _____°   C to A _____°   B to C _____°   C to B _____°

A

N

B

C

(b) In Figure 9.4, from A to B _____°

(c) In Figure 9.7, what is the direction from the red star to the blue dot? _____°

(c) In Figure 9.7, what is the direction from the peak of Hall Hill near the southwest corner of the map to the northernmost point of Skunk Pond? _____°

But which north is north? No, this isn't a silly question. A compass needle points to the *magnetic* north pole, the point where the lines of force from the Earth's magnetic field enter the planet (**FIG. 9.9A**). The **true** or **geographic north pole** is the north end of the Earth's rotational axis. The two "north poles" are not located in the same place, as shown in Figures 9.9a and 9.9b, so which do we use in discussing direction? As its name suggests, directions are given relative to **true north**.

In some places, as **FIGURE 9.9B** shows, compass needles (red arrows) point to a direction *west* of a line pointing to true north (black arrows), but in other places,

FIGURE 9.9 Magnetic north, true north, and magnetic declination.

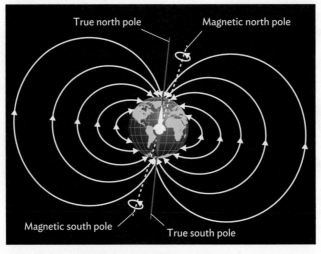

(a) Magnetic and true north and south poles.

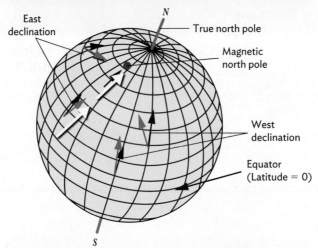

(b) Magnetic declination is the angle between true north and magnetic north.

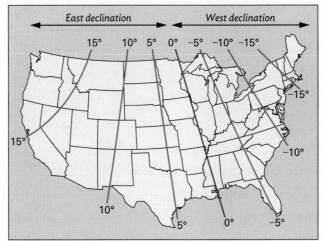

(c) Magnetic declination map for the contiguous United States.

compass needles point in a direction east of true north. The angular difference between true north and magnetic north is called **magnetic declination**. There are still other places where a compass needle pointing toward magnetic north also points toward true north (white arrows)—in other words, where magnetic declination is zero. **FIGURE 9.9C** shows how magnetic declination varies in the contiguous United States. Everywhere along the heavy blue line, declination is zero. Areas west of that line have east declination, and areas east of that line have west declination, with the declination increasing with distance from the zero line.

How does this variation affect our use of a compass? To determine the true direction (or bearing) from one place to another, you have to adjust from magnetic north to true north by adding the declination to or subtracting it from the azimuth reading. Geologists use a compass that can be adjusted to account for declination.

9.3.3 Map Element 3: Distance and Scale

A map of your campus, or of the entire world, can fit onto one sheet of paper—if we scale the Earth down so it fits. A **map scale** indicates how much an area has been scaled down so that we can relate inches on the map to real distances on the ground. Map scale may be expressed verbally, proportionally, or graphically, depending on the purpose of the map and the accuracy desired.

9.3.3a Verbal Scales A **verbal scale**, used on many road maps, uses words, such as "1 inch equals approximately 6.7 miles," to describe the scaling of map and real distances. This kind of scale is good enough for the purpose of allowing drivers to estimate distances between cities.

9.3.3b Proportional Scales The most accurate way to describe scaling is with a **proportional scale** (**FIG. 9.10**), one that tells exactly how much the ground has been scaled down to fit onto the paper. For example, a proportional scale of 1:100,000 ("one to one hundred thousand") means that distances on the ground are 100,000 times greater than the distance measured on the map. This ratio is the same for all units of measurement, so that 1 inch on such a map corresponds to 100,000 inches on the ground (1.58 miles) *and* 1 cm on the map corresponds to 100,000 cm on the ground (1 km). The scales in Figures 9.10b and 9.10c, 1:62,500 and 1:24,000, respectively, indicate that those maps have been scaled down less than the map in Figure 9.10a. The larger the number in a proportional scale, the more the map has been scaled down, and the less space is needed to portray an area on the map.

Note that the more a map is scaled down, the less detail it can show. Thus, Figure 9.10a is accurate enough to show the general locations and outlines of the lakes, but not accurate enough for a canoer looking for the exact point closest to a road. Figure 9.10c, which shows the irregularities of the shorelines much more accurately, is far better for that purpose.

The metric system is ideally suited for scales such as 1:100,000,000 or 1:100,000 because it is based on multiples of 10. In the United States, distance is measured in miles, but map distance is measured in inches. Unfortunately, relationships among inches, feet, and miles are not as simple as those in the metric system. There are 63,360 inches in a mile (12 inches per foot x 5,280 feet per mile), so the proportional scale 1:63,360 means that 1 inch on a map represents exactly 1 mile on the ground. Old USGS topographic maps use a scale of 1:62,500. For most purposes, we can interpret this scale to be approximately 1 inch = 1 mile, even though an inch on such a map would be about 70 feet short of a mile. Other common map scales are 1:24,000 (1 in. = 2,000 ft), 1:100,000 (1 in. = 1.58 mi), 1:250,000 (1 in. = 3.95 mi), and 1:1,000,000 (1 in. = 15.8 mi).

FIGURE 9.10 An area in eastern Maine shown at three common proportional map scales. The red bar on each map represents 1 km.

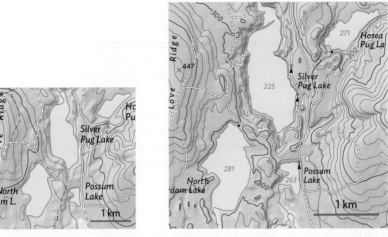

(a) Scale = 1:100,000.

(b) Scale = 1:62,500.

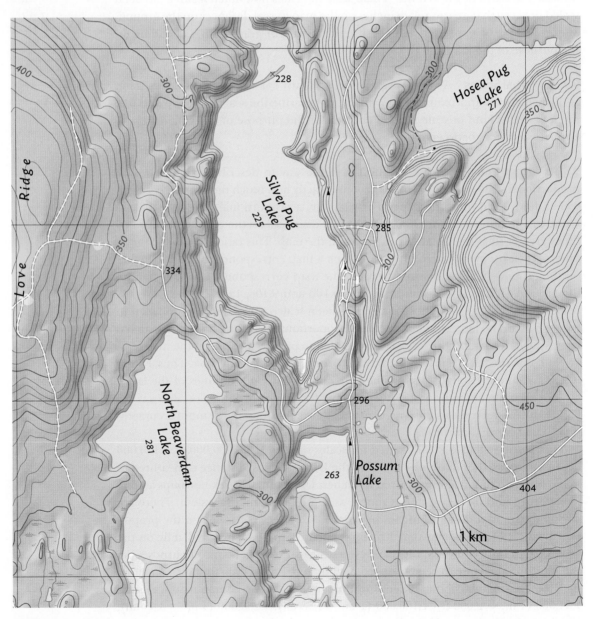

(c) Scale = 1:24,000.

FIGURE 9.11 Three common graphical scales. Nautical miles as well as statute (standard) miles are shown on maps of coastal areas.

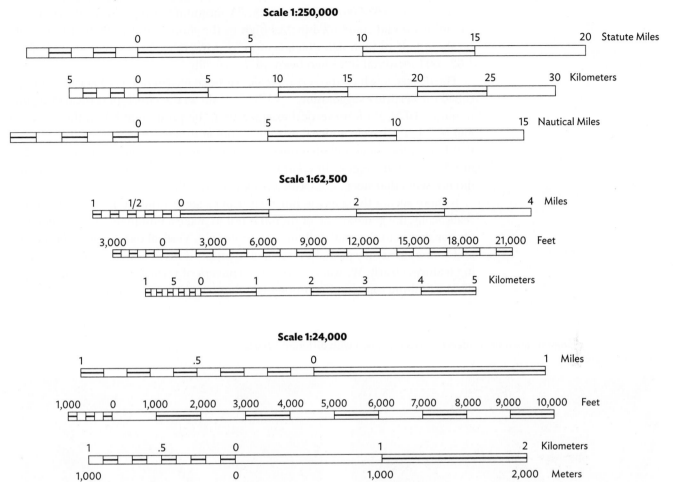

9.3.3c Graphical Scales Map scale can also be shown graphically, using a **bar scale** (FIG. 9.11). A bar scale can be more accurate than a verbal scale, depending on the map scale and how carefully you measure, but is not as accurate as a proportional scale. Graphical scales on USGS topographic maps typically show distances in miles and kilometers, but sometimes show distances in feet for maps covering small areas (that is, maps that are scaled down less).

9.4 Vertical Exaggeration: A Matter of Perspective

Digital elevation models show the land surface in three dimensions and must therefore use an appropriate *vertical* scale to indicate how much higher one feature is than another. It would seem logical to use the same scale for vertical and horizontal distances, but we don't usually do so because then mountains wouldn't look much like mountains and hills would barely be visible. Landforms are typically much wider than they are high, standing only a few hundred or thousand feet above or below their surroundings. At a scale of 1:62,500, 1 inch represents about a mile. If we used the same scale to make a three-dimensional model, a hilltop 400 feet above its surroundings would be less than one-tenth of an inch high, and a mountain rising a mile above its base would be only 1 inch high. We therefore exaggerate the vertical scale compared with the horizontal to show features from a human perspective. For a three-dimensional model of a 1:62,500 map, a vertical scale of 1:10,000 would exaggerate apparent elevations by a little more than six times (62,500/10,000 = 6.25). A mountain rising a mile above its surroundings would stand 6.25 inches high in the model, and a 400-foot hill would be about half an inch tall, which is more realistic than the tenth of an inch if the 1:62,500 horizontal scale had been used vertically.

The degree to which the vertical scale has been exaggerated is, logically enough, called the **vertical exaggeration**. FIGURE 9.12 shows the effects of vertical exaggeration on a DEM. With no vertical exaggeration, the prominent hill in the center of Figure 9.12a is barely noticeable. One of the authors of this manual has climbed that hill several times and guarantees that climbing it is far more difficult than Figure 9.12a would suggest. In contrast, Figure 9.12d exaggerates too much; the hill did not seem that steep, even with a pack loaded with rocks.

Is there such a thing as too much vertical exaggeration? To portray a landform realistically, the general rule of thumb is not to make a mountain out of a molehill. Or, for that matter, a molehill out of a mountain. Vertical exaggeration between 2 and 5 times generally preserves reasonable proportions of landforms while presenting features clearly. We will return to the concept of vertical exaggeration when we discuss drawing topographic profiles from topographic maps in Section 9.8.

FIGURE 9.12 Digital elevation models for an area drawn with different vertical exaggerations (VE).

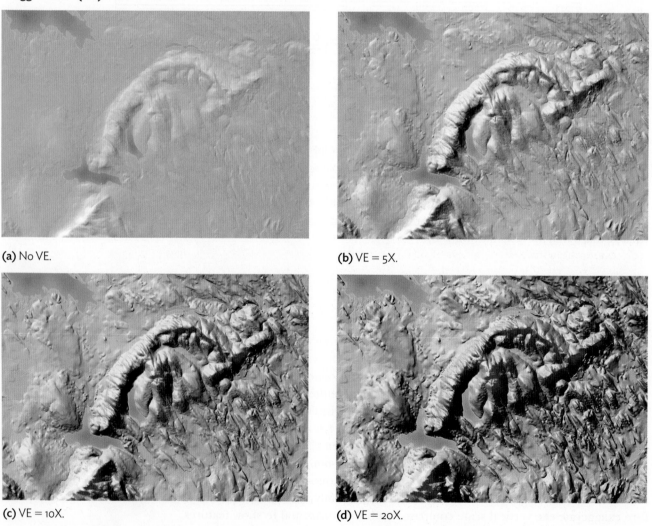

(a) No VE.

(b) VE = 5X.

(c) VE = 10X.

(d) VE = 20X.

9.5 Introduction to Topographic Maps

Thus far in this chapter, we have become very familiar with digital tools that help picture the world around us. With the click of a mouse, Google Earth and NASA WorldWind provide satellite images of any point on the planet; digital elevation models (DEMs) are available to help us visualize topography; Global Positioning System (GPS) satellites circling the Earth can help you locate exactly where you are standing; and sophisticated geographic information system (GIS) software can locate points, give elevations, measure lengths of meandering streams, and construct topographic profiles of an area. But what would you do if you were in the field and did not have access to any sophisticated equipment or could not get a reliable signal? A geologist would use a **topographic map**—a special type of map that uses contour lines to show landforms. Topographic maps cost almost nothing, weigh much less than a digital device, and withstand rain, swarming insects, and being dropped better than computers do. A recent topographic map gives the names and elevations of lakes, streams, mountains, and roads, and it outlines fields and distinguishes swamps and forests as well as a satellite image does. With a little practice, you can learn more about landforms from topographic maps than from a satellite image or DEM. The following sections of this chapter explain how topographic maps work and help you develop map-reading skills for identifying landforms, planning hikes, solving environmental problems, and possibly even saving your life.

9.6 Contour Lines

Like aerial photographs and satellite images, topographic maps show location, distance, and direction very accurately. Topographic maps also show the shapes of landforms, elevations, and the steepness of slopes with a special kind of line called a **contour line**. A contour line is a line on a map that connects points of the same value for whatever has been measured: elevations on topographic maps, temperatures on weather maps (**FIG. 9.13**), even population densities or average incomes. Thus, the temperature at every point on the 60° isotherm (temperature contour line) in Figure 9.13 is 60°F, every point on the 40° isotherm is 40°F, and so forth. Each isotherm separates areas where the temperature is higher or lower than that along the line. Thus, all points between the 50° and 60° isotherms have temperatures between 50°F and 60°F. The map in Figure 9.13 has a **contour interval** of 10°, meaning that contour lines represent temperatures at 10° increments. You will work with such maps in Chapter 18.

FIGURE 9.13 Contour lines on a weather map show temperature distributions. These contour lines are called *isotherms* and represent lines of constant (or equal) temperature.

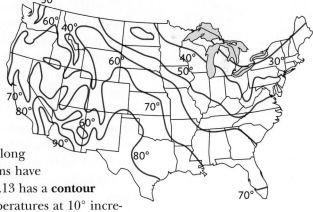

9.6.1 Contour Lines on Topographic Maps

Now that you understand the basic concept of contour lines, let's look at how they work on topographic maps. A contour line on a topographic map connects points that have the same *elevation above sea level* (sea level is the reference for all elevations). Topographic maps have contour intervals chosen to most clearly illustrate the land surface they portray: small contour intervals are used where there isn't much change in elevation, and large contour intervals are used where variation in elevation is high. Typical contour intervals used on USGS topographic maps are 10, 20, 50, and 100 feet.

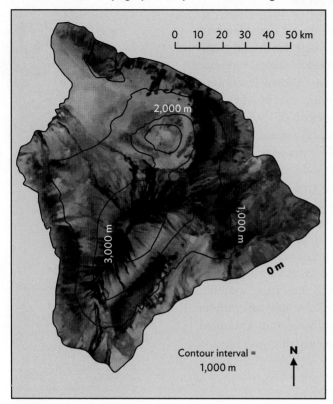

(a) 1,000-m contour interval.

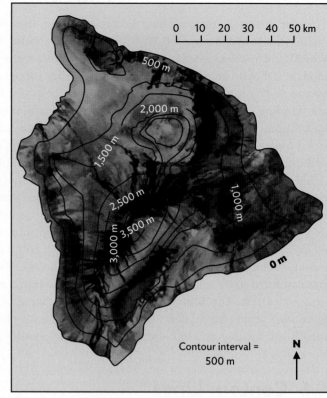

(b) 500-m contour interval.

FIGURE 9.14 shows the island of Hawaii mapped with two different contour intervals, superimposed on a DEM so that you can clearly see the topography that the lines portray. The lowest contour line, 0 m above sea level, is the same in both maps because sea level is the reference for all elevations and because the shoreline defines the outer shape of the island. A large contour interval, 1,000 m, is used in Figure 9.14a because Hawaii is mountainous, with thousands of feet of relief. If the contour interval had been 10 m, the map would have been unreadable, with 100 contour lines for each contour line you see. The 1,000-m contour interval can give only a rough idea of the topography, however. The smaller contour interval (500 m) in Figure 9.14b shows more details of the land surface.

Geologists use the term **relief** to describe the range in elevation between the highest and lowest points in an area. **High relief** characterizes mountainous areas with great elevation differences between mountaintops and valleys. **Low relief** characterizes flat areas such as plains and plateaus where there is very little elevation change over broad expanses.

9.7 Reading Topographic Maps

Topographic maps come in many scales, cover areas of different sizes, and use different contour intervals, so the first thing you should do when looking at a map is to examine its borders for useful information such as the contour interval, scale, and location. FIGURE 9.15 provides a guide for finding this information on the most recent USGS topographic maps (along with an example), and Appendix 9.1 shows standard map colors and symbols used to represent natural and artificial

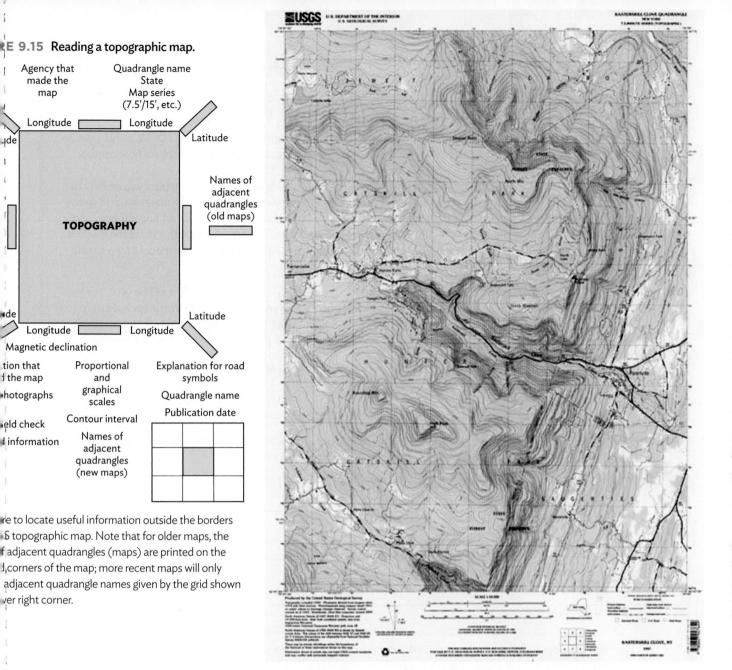

E 9.15 **Reading a topographic map.**

Agency that made the map

Quadrangle name
State
Map series (7.5'/15', etc.)

Longitude — Longitude

Latitude

Names of adjacent quadrangles (old maps)

TOPOGRAPHY

Latitude

Longitude — Longitude

Magnetic declination

tion that
the map

Proportional and graphical scales

Explanation for road symbols

photographs

Quadrangle name

Contour interval

Publication date

eld check

d information

Names of adjacent quadrangles (new maps)

e to locate useful information outside the borders
topographic map. Note that for older maps, the
adjacent quadrangles (maps) are printed on the
corners of the map; more recent maps will only
adjacent quadrangle names given by the grid shown
er right corner.

(b) Topographic map of Kaaterskill Quadrangle in New York.

features. Digital versions of the newest USGS topographic maps are available (free) at https://store.usgs.gov and contain recent aerial photographs as well as the traditional map.

9.7.1 Contour Lines and Slope

A DEM uses shading and perspective to show the configuration of the land surface, whereas a topographic map uses contour lines. One way to understand how contour lines portray hills, valleys, rivers, and other landforms is to compare a topographic map with a DEM on which the same landscape is clearly visible, as shown previously in Figure 9.14. There are rules for reading contour lines that will allow you to see how they can provide accurate images of complex three-dimensional landforms. The next four exercises will help you learn those rules, starting with how contour lines portray the slopes of hills and valleys in Exercise 9.6.

Name: _____   Section: _____

Course: _____   Date: _____

Now that you have seen the way in which a topographic map represents landforms represented on a DEM, in this exercise you will make your own topographic map from a scaled-down 3-D landform model in order to investigate how contour lines represent landform shapes.

Your instructor will provide you with the landform model secured to the inside of a box, or will instruct you on how to construct your own model. (*Note:* If this exercise is available and being conducted online, the setup, materials, and directions may differ. See the instructions in the online lab for how to conduct this exercise.) You will be adding water at increments to act as the horizontal surface representing heights (elevations) relative to sea level. While this is not the method used to determine elevation points in the real world, it is a handy tool for working on a scaled-down model of the Earth's surface in a laboratory class.

Step 1: Carefully pour water into the box until the water level reaches the first (lowest) marking. This will be your "sea level" elevation. Place the overhead transparency (taped to a plastic lid or whatever you or your instructor have used) on top of the box. Look straight down into the box and, using your marker, trace the contact where the water touches the land surface. You have just made your first contour line! Label this line as instructed for future reference.

Step 2: Remove the lid and fill the box with water to the next marking. Repeat Step 1.

Step 3: Repeat Steps 1 and 2 until you have completed tracing the water–land contact at the top-most marking. Congratulations! You have made a topographic map of your 3-D landform model.

(a) **Slope:** Find a flat place on the landform model and a place where the slope is steep. Now locate these places on the topographic map you created. Compare the contour lines in the flat and steep places.

  i. How does the spacing of the contour lines show the difference between gentle and steep slopes?

  _____

  _____

  ii. Describe the slopes on all/both sides of your landform. Are the slopes equally steep on all sides or is one side steeper than the other? Explain your reasoning.

  _____

  _____

  iii. Describe the slopes of your landform in words, and draw a sketch (using the graph paper provided at the end of this chapter) showing what it would look like to climb over the landform from northwest to southeast, or the direction given by your instructor.

  _____

  _____

(b) **Nested contour lines:** Find a place on your topographic map where there are a series of concentric (nested) contour lines.

  i. What type of feature do nested contour lines indicate?

  _____

  ii. Do all of your nested contour lines show elevation changing constantly either up or down, or is it inconsistent? Explain your reasoning.

  _____

  _____

9.7.2 Contour Lines and Elevation

Topographic maps provide several key features to help determine elevations. Two of the easiest to use are benchmarks and index contour lines. A benchmark is an accurately surveyed point marked on the ground by a brass plaque cemented in place. It records the latitude, longitude, and elevation of that spot. On topographic maps, elevations of selected points (such as hilltops, lake surfaces, and highway intersections) are indicated with a symbol and a number, such as $X_{1438"}$ or Δ_{561}. The latter symbol represents the location of a benchmark.

To find the elevations of other points on the map, however, we must use the contour lines. Take a look at **FIGURE 9.16**, which is a larger version of the map in Figure 9.1a; but first, remember to check for the contour interval at the bottom of the map for the difference in elevation between adjacent contour lines. On Figure 9.16, as on most topographic maps, every fifth contour line is darker than those around it and has its elevation labeled. These lines are *index contours*, which represent elevation intervals that are five times the contour interval (multiples of 5 × 20 in Fig. 9.16— 300 feet, 400 feet, and so on). To determine the elevation of an unlabeled contour line, determine its position relative to an index contour. For example, a contour line immediately adjacent to the 500-foot index contour must be 480 or 520 feet—20 feet lower or higher than the index contour on a map with a contour interval of 20 feet.

FIGURE 9.16 Topographic map of the Topsfield area in Maine.

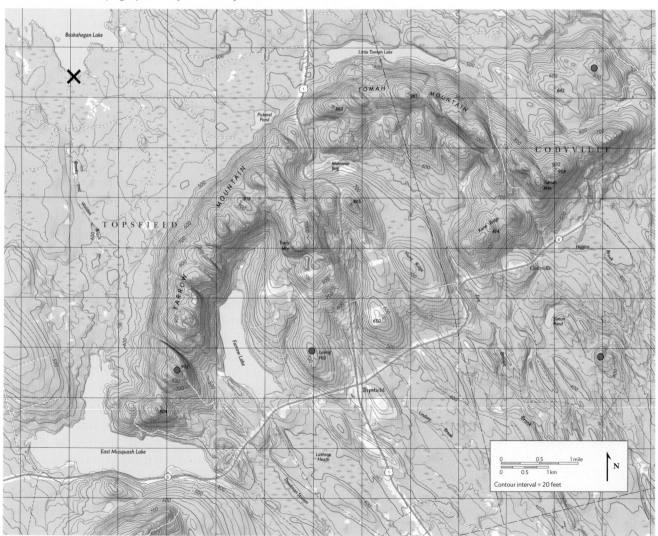

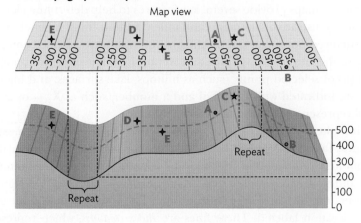

FIGURE 9.17 Reading elevations of hills and valleys from topographic maps.

What about the elevation of points between two contour lines? All points on one side of a contour line are at higher elevations than the line itself, and those on the opposite side are at lower elevations—but how much higher or lower? The contour interval in **FIGURE 9.17** is 50 feet, so a point between two adjacent contour lines must be less than 50 feet higher or lower than those lines. Estimate the elevation by how close a point is to either contour line: a point midway between the 500- and 550-foot contour lines would be estimated as 525 feet above sea level; a point ¾ of the way to the 550-foot contour would be about 538 feet above sea level, and so forth.

Point C on Figure 9.17 lies between two 500-foot contour lines, and the adjacent contours indicate that C is near the crest of a hill. What can we say about the elevation of the highest point on the hill? First, that it is over 500 feet, and second, that it must be less than 550 feet, because there isn't a 550-foot contour line. There is no further information that would allow us to be more precise.

The elevation of Point A is between 400 and 450 feet. Because the slope of the hill is constant between those contour lines, and A is halfway between them, the estimated elevation of A is 425 feet above sea level.

FIGURE 9.18 Hachured contour lines indicate depressions.

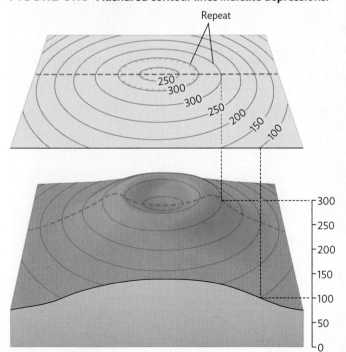

Point B lies between 350 and 400 feet. Here, however, the spacing of the contour lines indicates that the slope is gentler near the 350-foot line than near the 400-foot line. The estimated elevation of B is 355 feet above sea level.

Concentric contour lines (as in Exercise 9.6) may indicate either a hill or a depression. To avoid confusion, it must be clear that concentric contour lines outlining a depression indicate progressively *lower* elevations toward the center rather than the *higher* elevations that would indicate a hill. To show this, small marks called *hachures* are added to the line, pointing toward the lower elevation (see the innermost two lines in **FIG. 9.18**).

The feature in Figure 9.18 could be a volcano with a crater at its summit. Note that the nested contour lines on the flanks of the feature indicate increasing elevation, just like those of Figure 9.16, but that the 300-foot contour line is repeated and the inner one is hachured. This shows that the rim of the crater is higher than 300 feet but lower than 350 feet. With this background, you are now ready to estimate elevations on topographic maps in Exercise 9.7.

EXERCISE 9.7   Determining Elevations from Topographic Maps

Name: _____   **Section:** _____

Course: _____   **Date:** _____

To determine the elevation of a point on a topographic map:

1. Determine the contour interval for the map.

2. Find the known elevation closest to the point. This may be a benchmark or a contour line. Remember that every fifth contour line is a heavier index contour, but you may have to follow contour lines some distance before finding an elevation label.

3. Determine whether the point is higher or lower than the known elevation by examining the sequence of adjacent contour line elevations and by taking note of nearby feature names, such as "fire tower," "valley," and so forth.

4. Estimate the distance between the point and the contour lines that bracket it to get the elevation.

Using this procedure and what you have learned in the text, answer the following questions:

(a) In Figure 9.18, how high is the highest point on the rim of the crater? Explain how you made this estimation.

_____

_____

_____

(b) In Figure 9.18, what is the elevation of the lowest point of the crater? Explain how you made this estimation.

_____

_____

_____

(c) A black "X" in the figure at right (near the top left of the figure) marks a spot on the shore of Baskahegan Lake, but the printed DEM doesn't say anything about its elevation. Use the topographic map (Fig. 9.16) to estimate the elevation of this point as accurately as possible. _____ feet

(d) In Figure 9.16, what are the elevations of
- the highest point on Hunt Ridge? _____ feet
- the highest point on Farrow Mountain? _____ feet
- the intersection of U.S. Route 1 and Maine Route 6 in Topsfield? _____ feet
- Malcome Bog?_____ feet

(e) What is the relief between Little Tomah Lake and the top of Tomah Mountain?_____ feet

(f) What is the relief between the surface of East Musquash Lake and the intersection between U.S. Route 1 and Maine Route 6 in Topsfield?_____ feet

9.7.3 Contour Lines and Streams: Which Way Is the Water Flowing?

Geologists study the flow of streams and drainage basins to prevent the downstream spread of pollutants, and to determine where to collect water samples to find traces of valuable minerals. Streams flow downhill from high elevations to low elevations. You could determine the flow direction by looking for benchmarks along the stream or by comparing elevations where different contour lines cross the stream, but there is also an easier way, as you will see in Exercise 9.8.

EXERCISE 9.8	Understanding Stream Behavior from Topographic Maps

Name: _____ Section: _____
Course: _____ Date: _____

(a) Look again at Figure 9.16 and find the unnamed stream at the north end of Farrow Lake. Based on the elevations of the contour lines that cross the stream and the general nature of the topography, state the compass direction in which you think the stream flows. _____

(b) Do the same for the stream at the east end of Malcome Bog. _____

(c) Now look at the contour lines as they cross these two streams. Their distinctive V shape tells us which way the streams are flowing. Suggest a "rule of V" that describes how those Vs reveal the direction of streamflow.

_____

(d) Based on your rule of V, does the stream at the west side of Pickerel Pond flow into or away from the pond? _____

(e) In what direction does Jim Brown Brook (in the southeast corner of the map) flow? _____

9.7.4 Rules and Applications of Contour Lines on Topographic Maps

In the last few sections, you learned how contour lines show topography. The basic "rules" for reading contour lines are mostly common sense, and you figured them out for yourself in Exercises 9.6, 9.7, and 9.8. In Exercise 9.9, you will summarize what you have learned. In later chapters, you will use these rules to study landforms produced by streams, glaciers, groundwater, wind, and shoreline currents, and to assess how those agents of erosion create the Earth's varied landscapes.

EXERCISE 9.9	Rules for Reading Contour Lines on Topographic Maps

Name: _____ Section: _____
Course: _____ Date: _____

Complete the following sentences to summarize what you've learned about contour lines.

(a) Two different contour lines cannot cross because _____.

(b) The spacing between contour lines on a map reveals the _____ of the ground surface. Closely spaced contour lines indicate _____ and widely spaced contour lines indicate _____.

(c) Concentric contour lines indicate a _____.

(d) Concentric *hachured* contour lines indicate a _____.

Topographic maps help geologists solve a wide variety of problems. For example, they can trace the spread of contaminated groundwater, and they can locate the epicenters of old earthquakes from the amount of damage they caused. Some exercises in later chapters will require topographic map *making* skills, in addition to your ability to *read* topographic maps, in order to understand the shapes and origins of landforms.

In Exercise 9.10 you will make a topographic map using elevation data provided by surveyors. Exercise 9.11 shows that reading a topographic map can be a lifesaving skill.

EXERCISE 9.10   **Making a Topographic Map**

Name: _____   Section: _____
Course: _____   Date: _____

Now that you understand the basic rules of contour lines, you can make your own topographic map from elevation data. The map below shows surveyed elevations for a coastal area. Using these elevation data, construct a topographic map on the figure that shows the topography of the area, using a contour interval of 20 feet. One line has been done to get you started. Remember that contour lines must follow the rules you have just learned, including the guideline that contour lines don't go straight across streams—the lines must form a V *oriented correctly* to indicate the direction of streamflow.

Points at elevations of 20, 40, 60, and 80 feet will lie on contour lines and help you locate the lines. You will have to estimate where those contour lines pass between the other elevation points. (A contour line will be nearer to a point whose value is close to that of the contour line, and farther away from a point whose value is farther from that of the contour line.)

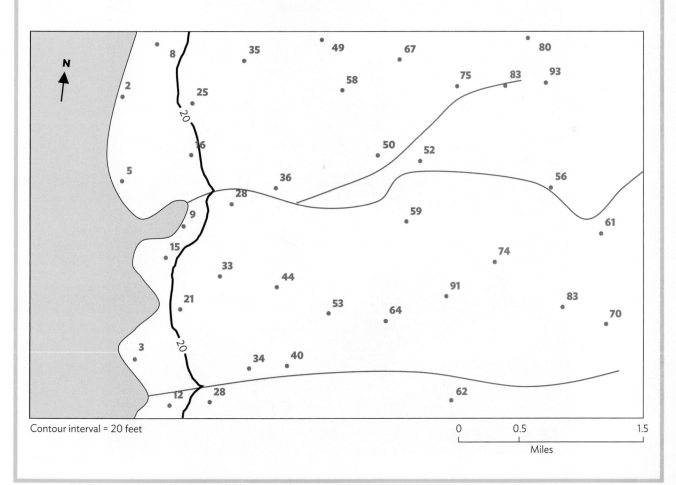

Contour interval = 20 feet

0   0.5   1.5
Miles

Name: _____ Section: _____
Course: _____ Date: _____

A small plane carrying you and a friend crashes near the northeast corner of an island at the spot indicated on the following map. The nearest humans are a lighthouse keeper and his family living on the opposite side of the island. Unfortunately, the plane's radio was destroyed during the crash, and your cell phones don't work. No one knows where you are, and the only way to save yourselves is to walk to the lighthouse.

   You and your friend were injured in the crash and can't climb hills higher than 75 feet or swim across rivers. You can only walk about 12 miles a day and carry enough water to last 3 days. It gets worse: the rivers have crocodiles, there's a large area filled with deadly quicksand, and the jungle is filled with—yes—lions, tigers, and bears. The good news is that you have a compass and a topographic map showing these hazards. You have figured out exactly where the plane crashed, and you know where you have to go. The map also shows a well where you can refill your canteens—if you can get there in 3 days.

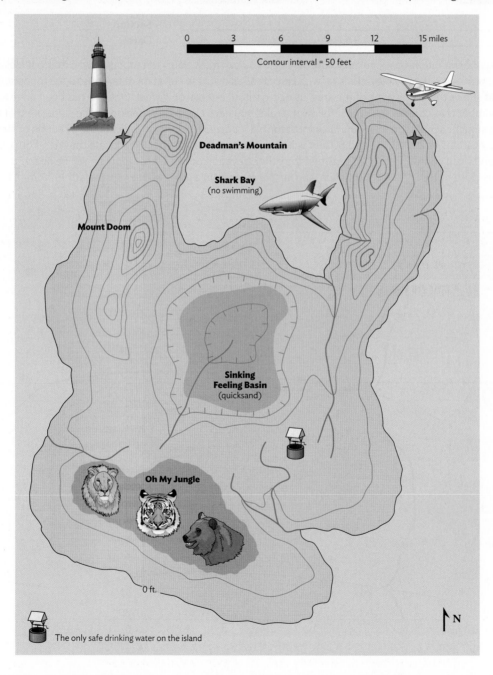

0   3   6   9   12   15 miles

Contour interval = 50 feet

Deadman's Mountain

Shark Bay
(no swimming)

Mount Doom

Sinking
Feeling Basin
(quicksand)

Oh My Jungle

0 ft.

N

The only safe drinking water on the island

(continued)

Name: _____   Section: _____
Course: _____   Date: _____

(a) With a pencil and the azimuth protractor and ruler in the geologic toolkit at the end of this manual, plan the shortest route to the lighthouse, avoiding steep hills, rivers, quicksand, and hungry jungle carnivores. Record the direction and distance of each leg in the table below and answer the following questions:

Your route to safety on Survivor Island.					
Leg #	Direction of leg (in degrees)	Length of leg (in miles)	Leg #	Direction of leg (in degrees)	Length of leg (in miles)
1			11		
2			12		
3			13		
4			14		
5			15		
6			16		
7			17		
8			18		
9			19		
10			20		
Total distance					

(b) How many days will the trip take? _____

(c) Will you have to stop for water? _____ If so, on what day do you get to the well? _____

(d) If you do need to stop for water, how many days will it take to get from the well to the lighthouse? _____

(e) Where is the highest point on the island? Give its elevation as precisely as you can. _____

(f) What is the elevation of the lowest point in Sinking Feeling Basin? _____

(g) Which is the steepest side of Deadman's Mountain? _____

9.8 Topographic Profiles

Topographic maps provide what is called a **map view** of an area—what the topography looks like from above. It is often useful to visualize what that area would look like in the third dimension as well. This section shows how to create a simple profile using an appropriate vertical scale.

9.8.1 Drawing a Topographic Profile

One benefit of using a topographic map is that it is easy to construct an accurate **topographic profile**, a cross-sectional view of the topography along a particular line. GIS software, such as MICRODEM or Google Earth, can draw the profile for you from digital data, but to learn what a profile can and cannot do, it is best to draw profiles by hand. Following the steps outlined below will create the profile that you

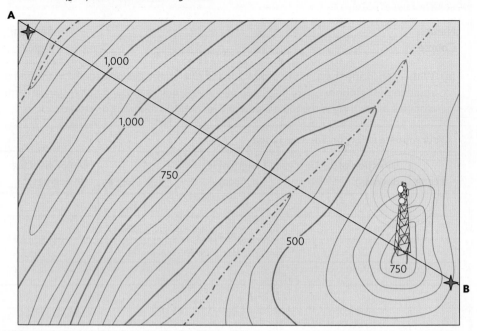

FIGURE 9.19 **Map of the microwave tower project area.**
Scale = 1:62,500; contour interval = 50 feet.

will use in Exercise 9.12 to analyze a real-life project involving cell-phone tower placement. **FIGURE 9.19** shows the area for which a profile is needed.

STEP 1: Place a strip of paper along the line of profile (A–B) shown in Figure 9.19 and label the starting and ending points.

STEP 2: Make a mark on the strip of paper where it crosses contour lines, streams, roads, and so on, and label each mark with its elevation (**FIG. 9.20**). For clarity, index contours are labeled in Figure 9.19.

FIGURE 9.20 **Collecting data for the profile.**

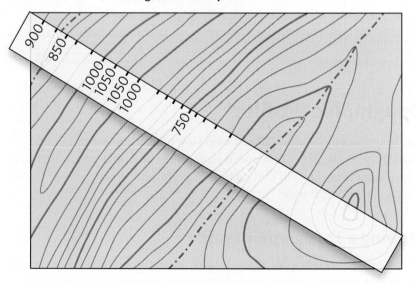

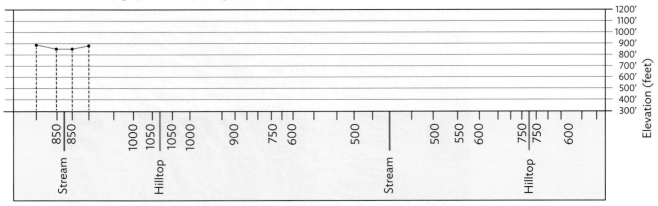

FIGURE 9.21 Setting up and completing the profile.

STEP 3: Select the vertical scale (see Section 9.8.2 below) and prepare one of the graph sheets at the end of the chapter for the profile by labeling the appropriate horizontal graph lines with elevations corresponding to those of the contour lines in Figure 9.19. **FIGURE 9.21** illustrates this step.

STEP 4: Place your strip of paper with the labeled marks (from Step 2) at the bottom of the vertical scale and use a ruler to transfer the elevations to their correct positions on the graph sheet. Place a dot where each contour line mark on the strip intersects the corresponding elevation line on the graph sheet (see Fig. 9.21).

STEP 5: Connect the dots, using what you know about contour lines to estimate the elevations at the tops of hills and bottoms of stream valleys along the line of profile (see Fig. 9.21).

9.8.2 Choosing the Vertical Scale

The horizontal scale of a profile must be the same as that of the topographic map it is made from, but care must be taken in choosing the vertical scale so that the profile will be a reasonable representation of the area. Recall from earlier in this chapter that choosing an inappropriate vertical scale can make a mountain look like a molehill or a molehill look like Mt. Everest.

We therefore use vertical scales that exaggerate the vertical dimension with respect to the horizontal, in order for hills to look like hills and mountains to look like mountains. **FIGURE 9.22** shows three profiles drawn from the same DEM, but using different vertical scales, to illustrate how much difference vertical exaggeration can make.

To find out how much a profile exaggerates the topography, we need to know the vertical scale used in profiling. To determine the vertical scale of the sample profile in Figure 9.21, we first measure the vertical dimension with a ruler; this shows that 1 *vertical* inch on the profile represents nearly 900 feet of elevation, a proportional scale of 1:10,800 (900 ft × 12 in. per foot = 10,800 in.). To calculate vertical exaggeration, we divide the map proportional scale by the profile proportional scale:

$$\text{Vertical exaggeration} = \frac{\text{map scale}}{\text{profile scale}} = \frac{62,500}{10,800} = 5.8\times$$

FIGURE 9.22 The effect of vertical exaggeration on topographic profiles.

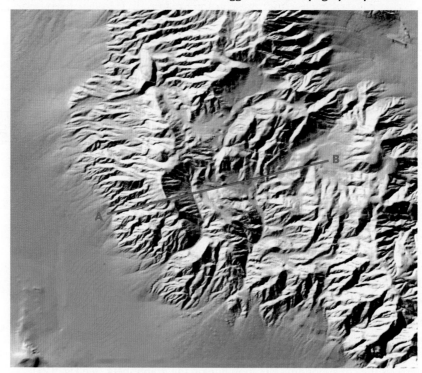

(a) DEM of Hanging Rock Canyon quadrangle, California, showing a line of profile (A–B).

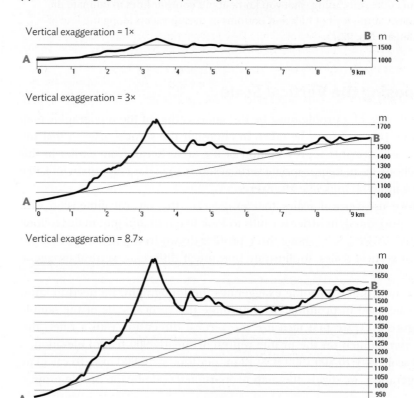

(b) Topographic profiles along the line of profile drawn with different vertical scales. Note that the horizontal scale is the same in all three profiles. Only the vertical exaggeration changes.

EXERCISE 9.12 Your First Job as a Geologist: Avoiding Cell-Phone Dead Zones

Name: _____   Section: _____
Course: _____   Date: _____

A telecommunications company wants to place microwave relay towers in a new region to improve cell-phone reception and plans to put one on the hilltop in the southeast corner of the area mapped in Figure 9.19 (where the bottom red starburst is). The project managers are concerned that a prominent ridge might block the signal to areas northwest of the tower (where the top red starburst is). It is not immediately obvious from the map whether there will be a "dead zone"— a place where the signal from the tower will be blocked by that ridge.

   You have been hired as a consultant to answer this question. The best way to do so is to construct a topographic profile northwestward from the tower across the hills and valleys.

(a) Draw a topographic profile along the line A–B in Figure 9.19 using the graph paper at the end of the chapter. The horizontal scale is set on that map, but you must choose a vertical scale for the profile that will not over exaggerate relief. (Recall that a vertical exaggeration between 2 and 5 times generally preserves reasonable proportions of landforms while presenting features clearly.)

(b) Calculate the vertical exaggeration of your profile. _____ x

(c) Now draw another profile along the same line with twice the vertical exaggeration.

? **What Do You Think** Now that you've completed the profile, what will you report to the telecommunications company? Is there a direct line of sight into the valley at the northwest corner of the map that will permit customers there to have cell-phone coverage, or does the ridge east of the valley block the signal? If the ridge blocks the signal, how high would the tower have to be to guarantee service to that valley? On a separate sheet of paper, write a brief report containing your recommendations (and submit your invoice to the company for services rendered).

EXERCISE 9.13 Working with a Real Topographic Map

Name: _____   Section: _____
Course: _____   Date: _____

Now it's time to put what you've learned into practice on a local topographic map provided by your instructor. (*Note:* If this exercise is available and being conducted online, the setup, materials, and directions may differ. See the instructions in the online lab for how to conduct this exercise.)

(a) With your map in hand, begin by gathering basic information:

Map or quadrangle name	State	Map date	Latitude of NE corner	Longitude of NE corner	Minutes of latitude/ longitude covered
Scale	Contour interval	Magnetic declination	Quadrangle adjacent to east	Agency that made the map	Last update (if applicable)*

*Particularly in urban areas, many maps show changes in pink and indicate when the original map was updated.

(b) Follow your instructor's directions and answer the questions your instructor poses for this map.

Name: _____　Section: _____

Course: _____　Date: _____

Exploring Earth Science Using Google Earth

1. Visit **digital.wwnorton.com/labmanualearthsci**
2. Go to the **Geotours** tile to download Google Earth Pro and the accompanying Geotours exercises file.

Expand the Geotour09 folder in Google Earth by clicking the triangle to the left of the folder icon. First, check and double-click on the USGS Map Credit and Instructions placemark to fly to Denver, CO. Then check the USGSTopo folder, which overlays seamless topographic maps over the viewing area.

(a) Check and double-click the Dinosaur Ridge placemark to fly to where the ridge is cut by Interstate 70. Using the map, what is the contour interval of the map at this zoom level? _____ ft

(b) Using the map, what is the relief (difference in elevation) from the placemark location to the bottom of the gap where the interstate passes through the ridge? _____ ft

(c) Check and double-click the Depression placemark to fly to (you guessed it . . .) a depression. If you hike into the depression (actually a mine) from the northeast, to the west of the stream that flows through Jackson Gulch, you hike upslope over a rim and down into the depression, repeating contours when you change gradient direction. If you hike from the southwest into the depression, you do not see a repeat in contours. Why? *Hint:* Think about contour rules for depressions.

_____

_____

_____

(d) Check and double-click the Horsetooth Reservoir path to fly to an area just west of Fort Collins, CO. Toggle the USGSTopo folder on/off and change your view to look at the sedimentary rock layers here. Right-click on the Horsetooth Reservoir path in the Places panel and select **Show Elevation Profile** (this draws a topographic profile). As you move your cursor over the profile, a corresponding arrow tracks across the terrain in the Google Earth viewer window. On the profile, move your cursor over the three ridges with parallel east-facing slopes and observe the nature (i.e., orientation and spacing) of the contour lines on the topographic map. What does this suggest about the relative amount and direction of tilt/dip of these three sedimentary layers?

_____

_____

_____

Topographic Map Symbols

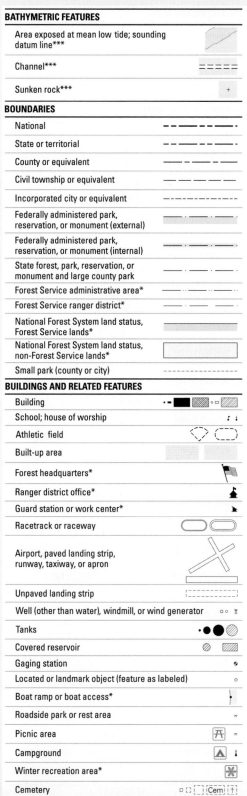

BATHYMETRIC FEATURES

Area exposed at mean low tide; sounding datum line***	
Channel***	
Sunken rock***	

BOUNDARIES

National	
State or territorial	
County or equivalent	
Civil township or equivalent	
Incorporated city or equivalent	
Federally administered park, reservation, or monument (external)	
Federally administered park, reservation, or monument (internal)	
State forest, park, reservation, or monument and large county park	
Forest Service administrative area*	
Forest Service ranger district*	
National Forest System land status, Forest Service lands*	
National Forest System land status, non-Forest Service lands*	
Small park (county or city)	

BUILDINGS AND RELATED FEATURES

Building	
School; house of worship	
Athletic field	
Built-up area	
Forest headquarters*	
Ranger district office*	
Guard station or work center*	
Racetrack or raceway	
Airport, paved landing strip, runway, taxiway, or apron	
Unpaved landing strip	
Well (other than water), windmill, or wind generator	
Tanks	
Covered reservoir	
Gaging station	
Located or landmark object (feature as labeled)	
Boat ramp or boat access*	
Roadside park or rest area	
Picnic area	
Campground	
Winter recreation area*	
Cemetery	

COASTAL FEATURES

Foreshore flat	
Coral or rock reef	
Rock, bare or awash; dangerous to navigation	
Group of rocks, bare or awash	
Exposed wreck	
Depth curve; sounding	
Breakwater, pier, jetty, or wharf	
Seawall	
Oil or gas well; platform	

CONTOURS

Topographic

Index	
Approximate or indefinite	
Intermediate	
Approximate or indefinite	
Supplementary	
Depression	
Cut	
Fill	
Continental divide	

Bathymetric

Index***	
Intermediate***	
Index primary***	
Primary***	
Supplementary***	

CONTROL DATA AND MONUMENTS

Principal point**	
U.S. mineral or location monument	
River mileage marker	

Boundary monument

Third-order or better elevation, with tablet	
Third-order or better elevation, recoverable mark, no tablet	
With number and elevation	

Horizontal control

Third-order or better, permanent mark	
With third-order or better elevation	
With checked spot elevation	
Coincident with found section corner	
Unmonumented**	

Topographic Map Symbols

CONTROL DATA AND MONUMENTS – *continued*

Vertical control

Third-order or better elevation, with tablet	BM $\times$ 5280
Third-order or better elevation, recoverable mark, no tablet	$\times$ 528
Bench mark coincident with found section corner	BM + 5280
Spot elevation	$\times$ 7523

GLACIERS AND PERMANENT SNOWFIELDS

Contours and limits	
Formlines	
Glacial advance	
Glacial retreat	

LAND SURVEYS

Public land survey system

Range or Township line	
Location approximate	
Location doubtful	
Protracted	
Protracted (AK 1:63,360-scale)	
Range or Township labels	R1E T2N R3W T4S
Section line	
Location approximate	
Location doubtful	
Protracted	
Protracted (AK 1:63,360-scale)	
Section numbers	1 - 36 1 - 36
Found section corner	
Found closing corner	
Witness corner	WC
Meander corner	MC
Weak corner*	

Other land surveys

Range or Township line	
Section line	
Land grant, mining claim, donation land claim, or tract	
Land grant, homestead, mineral, or other special survey monument	
Fence or field lines	

MARINE SHORELINES

Shoreline	
Apparent (edge of vegetation)***	
Indefinite or unsurveyed	

MINES AND CAVES

Quarry or open-pit mine	
Gravel, sand, clay, or borrow pit	
Mine tunnel or cave entrance	
Mine shaft	
Prospect	X
Tailings	Tailings
Mine dump	
Former disposal site or mine	

PROJECTION AND GRIDS

Neatline	39°15' 90°37'30"
Graticule tick	55'
Graticule intersection	
Datum shift tick	

State plane coordinate systems

Primary zone tick	640 000 FEET
Secondary zone tick	247 500 METERS
Tertiary zone tick	260 000 FEET
Quaternary zone tick	98 500 METERS
Quintary zone tick	320 000 FEET

Universal transverse mercator grid

UTM grid (full grid)	273
UTM grid ticks*	269

RAILROADS AND RELATED FEATURES

Standard gauge railroad, single track	
Standard gauge railroad, multiple track	
Narrow gauge railroad, single track	
Narrow gauge railroad, multiple track	
Railroad siding	
Railroad in highway / Railroad in road / Railroad in light-duty road*	
Railroad underpass; overpass	
Railroad bridge; drawbridge	
Railroad tunnel	
Railroad yard	
Railroad turntable; roundhouse	

RIVERS, LAKES, AND CANALS

Perennial stream	
Perennial river	
Intermittent stream	
Intermittent river	
Disappearing stream	
Falls, small	
Falls, large	
Rapids, small	
Rapids, large	
Masonry dam	
Dam with lock	
Dam carrying road	

Topographic Map Symbols

RIVERS, LAKES, AND CANALS – *continued*

Perennial lake/pond	
Intermittent lake/pond	
Dry lake/pond	
Narrow wash	
Wide wash	Wash
Canal, flume, or aqueduct with lock	
Elevated aqueduct, flume, or conduit	
Aqueduct tunnel	
Water well, geyser, fumarole, or mud pot	
Spring or seep	

ROADS AND RELATED FEATURES

Please note: Roads on Provisional-edition maps are not classified as primary, secondary, or light duty. These roads are all classified as improved roads and are symbolized the same as light-duty roads.

Primary highway	
Secondary highway	
Light-duty road	
Light-duty road, paved*	
Light-duty road, gravel*	
Light-duty road, dirt*	
Light-duty road, unspecified*	
Unimproved road	
Unimproved road*	
4WD road	
4WD road*	
Trail	
Highway or road with median strip	
Highway or road under construction	Under Const
Highway or road underpass; overpass	
Highway or road bridge; drawbridge	
Highway or road tunnel	
Road block, berm, or barrier*	
Gate on road*	
Trailhead*	

SUBMERGED AREAS AND BOGS

Marsh or swamp	
Submerged marsh or swamp	
Wooded marsh or swamp	
Submerged wooded marsh or swamp	
Land subject to inundation	Max Pool 431

SURFACE FEATURES

Levee	Levee
Sand or mud	Sand
Disturbed surface	
Gravel beach or glacial moraine	Gravel
Tailings pond	Tailings Pond

TRANSMISSION LINES AND PIPELINES

Power transmission line; pole; tower	
Telephone line	Telephone
Aboveground pipeline	
Underground pipeline	Pipeline

VEGETATION

Woodland	
Shrubland	
Orchard	
Vineyard	
Mangrove	Mangrove

* USGS–USDA Forest Service Single-Edition Quadrangle maps only.
In August 1993, the U.S. Geological Survey and the U.S. Department of Agriculture's Forest Service signed an Interagency Agreement to begin a single-edition joint mapping program. This agreement established the coordination for producing and maintaining single-edition primary series topographic maps for quadrangles containing National Forest System lands. The joint mapping program eliminates duplication of effort by the agencies and results in a more frequent revision cycle for quadrangles containing National Forests. Maps are revised on the basis of jointly developed standards and contain normal features mapped by the USGS, as well as additional features required for efficient management of National Forest System lands. Single-edition maps look slightly different but meet the content, accuracy, and quality criteria of other USGS products.

** Provisional-Edition maps only.
Provisional-edition maps were established to expedite completion of the remaining large-scale topographic quadrangles of the conterminous United States. They contain essentially the same level of information as the standard series maps. This series can be easily recognized by the title "Provisional Edition" in the lower right-hand corner.

*** Topographic Bathymetric maps only.

Topographic Map Information

For more information about topographic maps produced by the USGS, please call 1-888-ASK-USGS or visit us at http://ask.usgs.gov/.

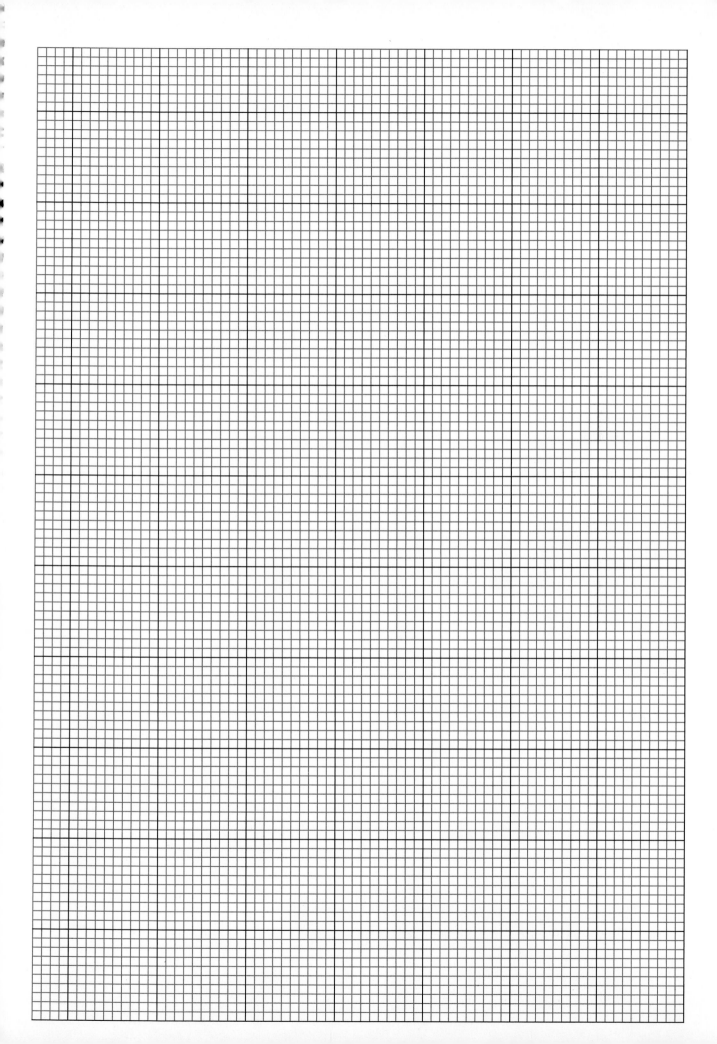

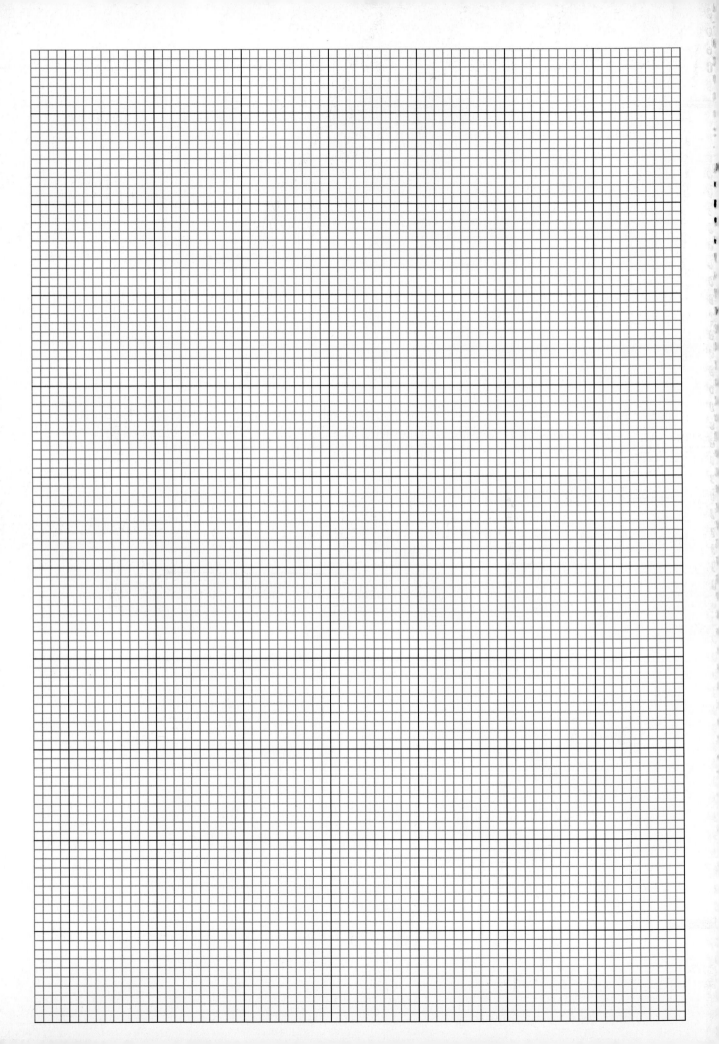

Mass Wasting and Landscape Evolution

10

A 2005 landslide in Bluebird Canyon, in Laguna Beach, California, devastated homes.

MATERIALS NEEDED

- Metric ruler
- Sample of sandstone and of shale
- Water
- Angled board
- Wet sediment
- Magnifying glass for viewing images

10.1 Introduction

The downslope movement of Earth material—rocks, rock debris, ice and snow—mainly by the force of gravity is called **mass wasting**. These events can take place very slowly over long time spans, or occur suddenly with catastrophic consequences. People and news media often use the general term landslide to describe all such events, though as we'll see later in this chapter, geologists distinguish between several types of mass wasting. The rapid movement of rock debris during a landslide can be triggered by natural phenomena such as heavy rainfall or severe weather events, earthquakes, or volcanic activity. The threat of mass wasting is increased in areas with steep topography, such as mountainous regions or areas surrounding volcanoes (**FIG. 10.1**). It is also affected by human activities, such as the modification of the landscape through construction of roads and other land development. When the land slides, it takes with it not only rocks and debris; trees and anything else on the surface, including homes and bridges, are also moved. When this debris comes to rest, it can destroy bridges, dam rivers, and cause flooding, or bury a neighborhood.

Mass wasting processes can be extremely costly in terms of lives and money. In 2017 alone, more than 400 landslides took place worldwide, resulting in over 2,000 deaths. The higher fatality figures tend to occur in less economically developed countries where the population density is greater (**TABLE 10.1** and **FIG. 10.2**). In the United States, landslides cause 25 to 50 deaths per year, and very few American insurance companies cover mass wasting or any other type of ground motion. Yet the economic damage from these events is immense; in 2017, landslides within the US alone accounted for more than $3.5 billion in damage. As the world population rises and more people settle in less suitable (or less stable) areas, and as the global climate continues to change, major disasters like these increase. Therefore, it is more critical than ever to understand the factors that contribute to mass wasting processes, so that we can create better plans to reduce our vulnerability to the hazards they present.

In this chapter, you will investigate the different types of mass wasting, identify how they are recognized by scientists, and then use that information to estimate the dangers to the general population in various locations around the world.

FIGURE 10.1 The landslide (technically a debris flow, as we'll see later) in the town of Yungay, Peru, in 1970.

(a) Before the landslide, the town of Yungay perched on a hill near the ice-covered mountain Nevado Huascarán.

(b) The landslide completely buried the town beneath debris. A landslide scar is visible on the mountain in the distance.

Table 10.1. Recent major catastrophic landslides worldwide

Date	Location	Reported casualties
February 2006	Southern Leyte, Philippines	1,126
August 2009	Siaolin, Taiwan	439–600
August 2010	Gansu, China	1,287
May 2014	Badakustan, Afghanistan	350–500
July 2014	Malin, India	136
August 2014	Sunkoshi, Nepal	156+
October 2015	El Cambray Dos, Guatemala	220
April 2017	Mocoa, Colombia	314
June 2017	Rangamati, Chittagong, and Bandarban, Bangladesh	152
September 2018	Palu, Sulawesi, Indonesia	1,944+

Data from the World Atlas: https://www.worldatlas.com/articles/the-deadliest-landslides-of-the-21st-century.html.

FIGURE 10.2 Recent catastrophic landslides worldwide.

(a) A world map showing the locations of the landslides listed in Table 10.1.

(b) An image of the April 2017 landslide (technically a mudflow, as we'll see later) in Mocoa, Columbia.

(c) The 2017 Mocoa landslide brought devastation to 17 neighborhoods, killing 314 people.

10.2 Controls on Landscape Evolution

Changes in the landscape are generated by processes that uplift and expose rocks at the Earth's surface, and then remove and redeposit that debris. As you investigated in Chapter 2, the most extensive source of the processes that uplift rocks is plate tectonics. Once the rocks are exposed at the Earth's surface, they then undergo *weathering*, both physically (being broken down into smaller pieces) and chemically (reacting with water, the atmosphere, and organisms). The dominant agents that move and redeposit this weathered material include water, wind, and ice. However, for mass wasting, the dominant agent is the force of gravity, which pulls Earth material downhill—slower on gentle slopes, and faster on steeper slopes. In this section, you'll examine how gravity operates on weathered material, and the role water plays in contributing to mass wasting events.

10.2.1 The Influence of Gravity on Mass Wasting

The force of gravity acts everywhere on the Earth's surface, pulling everything on it toward the center of the Earth. When the Earth's surface is flat, gravity pulls material on it downward (**FIG. 10.3A**). So, if the surface remains flat, the surface material is held in place. On the other hand, if the surface is sloped, then the force of gravity pulls on the surface material both downward (perpendicular to the slope) and at an oblique angle (parallel to the slope) as shown in **FIGURE 10.3B**.

So, how can we determine whether particular slopes are more or less likely to be affected by mass wasting events? We can make some simple measurements and plug them into an equation, as shown below:

Force of friction = coefficient of friction × normal force

$$F = f \times N$$

Said another way, the force of friction (F) is the product of the **coefficient of friction** (f) times the **normal force** (N). The coefficient of friction is the resisting force: a number between 0 and 1 that is the force needed for an object to move or slide against another, and which depends on the material involved (sand, silt, rock, etc.). The greater the coefficient of friction, the more resistant to moving or sliding a material is. The normal force is the arrow in Figure 10.3b that is perpendicular to the slope. Mass wasting occurs when the force moving parallel to the slope (the **driving force, *D***) is greater than the force of friction (F), or simply when $D > F$. You will practice applying these formulas for yourself in Exercise 10.1.

FIGURE 10.3 The role of gravity in mass movement.

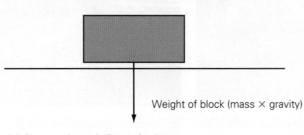

(a) Slope angle = 0° (flat surface).

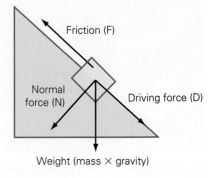

(b) Slope angle > 0°

Name: _____   Section: _____
Course: _____   Date: _____

Study the two figures below. Intuitively, you can probably guess which of the slopes displays stable conditions and which shows unstable conditions. But let's test that gut feeling with some simple measurements and calculations. To help you decide, you will need to determine the force of friction in each case. The coefficient of friction is already given in the first table.

(a) In the two figures to the right, label all forces (shown by arrows) in each diagram.

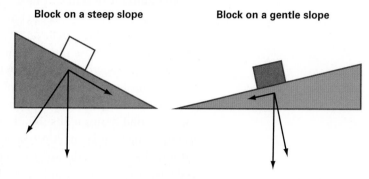

Block on a steep slope   **Block on a gentle slope**

(b) Measure the length (in millimeters) of the normal force in each diagram. Enter these values in the first table below. (*Note:* In reality, forces are measured in units of force, not length. We're using length here to allow you to measure and compare these values.)

(c) Use the equation for the force of friction (*F*) to calculate this force for each diagram. Enter the products in the first table below. Then, using a ruler, draw an arrow in each diagram with the correct length of *F*.

Figure	Coefficient of friction (*f*)	Length of normal force N (mm)	Force of friction F (mm)	Which is greater: F or D?	Stable or unstable condition?
Block on a steep slope	0.25				
Block on a gentle slope	0.50				

(d) Look at the arrows in each diagram and determine which arrow is longer, the driving force or the friction force. Use this answer to decide whether each diagram is illustrating a stable or unstable condition; in other words, which case will result in mass wasting. Write your answer in the last column of the table above.

(e) From what you have observed here, summarize how the slope angle influences slope stability:

_____

_____

_____

(f) Now suppose we changed the materials of the slopes in the examples above, such that the coefficient for friction in the left figure is 0.75, and the coefficient for the figure on the right is 0.25. Record your new calculations in the table below and determine whether each diagram is depicting a stable or unstable condition.

Figure	Coefficient of friction (*f*)	Length of normal force N (mm)	Force of friction F (mm)	Which is greater: F or D?	Stable or unstable condition?
Block on a steep slope	0.75				
Block on a gentle slope	0.25				

(g) From what you have observed here, summarize how the underlying material influences slope stability (*Hint:* Recall what a greater coefficient of friction corresponds to):

_____

_____

_____

10.2.2 The Role of Water on Mass Wasting

Although gravity is the ultimate force driving mass wasting, water also plays a role. Water that fills the spaces between rock particles (*pores*) or fractures in crystalline rocks adds weight to the surface material (**FIG. 10.4A**). Water can also add weight when it clings to the surface of rock particles or is absorbed into minerals (**FIG. 10.4B**). Adding weight to the surface material increases the role of gravity—heavier material moves faster downslope.

The amount of water in loose Earth materials will also have an impact on the steepness of their slopes. If you have ever tried to build a sandcastle at the beach, you will have observed this behavior. Trying to build a block of sand using a pail and dry sand results in a loose pile or cone of sand particles that slide or roll downhill and are not stuck together (**FIG. 10.5A**). However, add just a little water and the slightly wet unconsolidated materials exhibit a very high *angle of repose* (the maximum slope angle that loose particles can maintain), because surface tension between the water and the solid grains tends to hold the grains in place (**FIG. 10.5B**). Add too much water, though, and your construction may be in danger. When the material becomes saturated with water, the angle of repose is reduced to very small values and the material tends to flow like a fluid (**FIG. 10.5C**). This is because the water gets between the grains and eliminates grain-to-grain frictional contact. This process is also observed during earthquakes, where ground that is saturated with water behaves like a liquid while it is shaken—a process referred to as *liquefaction* (see Chapter 7).

In addition to adding weight to Earth materials and changing the amount of grain-to-grain frictional contact, water weakens materials through chemical change. For this reason, water is often referred to as the "universal solvent," because, given enough time, the attractive electrical forces on water molecules are strong enough to break the bonds between the other atoms making up solid substances.

Finally, we can consider fluid pressure. Water filling pore spaces exerts a force of pressure on the surrounding particles. In the ground, as the level of water rises and falls—from natural fluctuations in the water cycle or through withdrawal by humans—this changes the supporting forces on the rock material. Lower water levels can ultimately result in the compaction and lowering of the ground surface, thereby producing steeper slopes (**FIG. 10.6**). In Exercise 10.2, you will take a closer look at the role of water in mass wasting events.

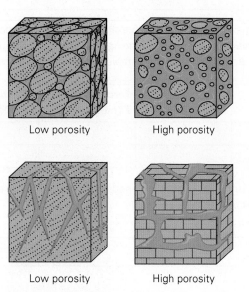

Low porosity   High porosity

Low porosity   High porosity

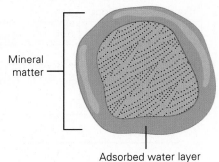

Mineral matter

Adsorbed water layer

FIGURE 10.4 The role of water on mass wasting.

(a) Water fills the pore spaces between sediment particles (top) or fractures rock (bottom).

(b) Water clings to the surfaces of particles or is absorbed into minerals.

FIGURE 10.5 Sand and water mixtures affect slope steepness and stability.

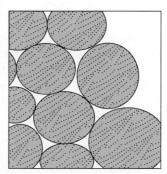

(a) Dry sand in a mound. The inset shows frictional contact between the grains.

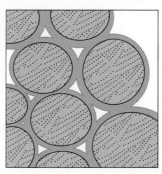

(b) Damp sand (perfect for building a sandcastle). The inset shows that there is surface tension between the thin layers of water surrounding the grains, holding them together.

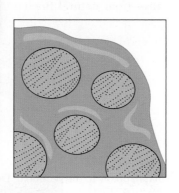

(c) Saturated sand (collapsed sandcastle). The inset shows water completely surrounding the grains, eliminating any contact between them.

FIGURE 10.6 Fluid pressure and compaction.

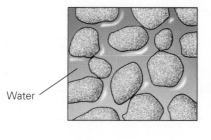

Water

(a) Fluid pressure supports grains.

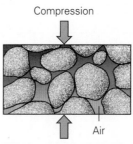

Compression

Air

(b) Lowering fluid pressure results in grain compaction.

Name: _____   Section: _____
Course: _____   Date: _____

Your instructor has provided you with the materials to set up a simple experiment. (*Note:* If this exercise is available and being conducted online, the setup, materials, and directions may differ. See the instructions in the online lab for how to conduct this exercise.) You should have a board (or other surface) that is propped up at an angle. On top of this board is a layer of wet sediment. The sediment should be staying in place before you begin.

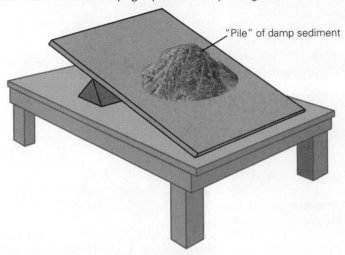

"Pile" of damp sediment

(a) Near the bottom of the slope, cut and scoop away a section of sediment so that there is now a "cliff" at the bottom.

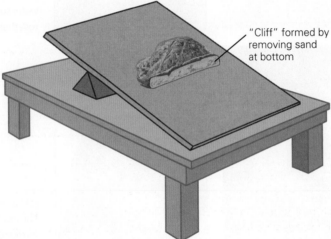

"Cliff" formed by removing sand at bottom

(b) Wait a few moments and observe what happens to the sediment. Describe what you observed:

_____

_____

_____

(c) Why do you think this mass wasting occurred?

_____

_____

_____

(continued)

Name: _____   Section: _____
Course: _____   Date: _____

(d) Now you will simulate a flash flood. Make it "rain" by carefully pouring water at the top of the slope, or where directed by your instructor, and watch what happens. Briefly summarize your observations:

_____

_____

_____

(e) Why do you think this mass wasting occurred?

_____

_____

_____

For the next experiment, your instructor has provided you with a piece of sandstone, a piece of shale, and some water. (*Note:* If this exercise is available and being conducted online, the setup, materials, and directions may differ. See the instructions in the online lab for how to conduct this exercise.)

(f) Drip a small amount of water on the sandstone and on the shale. What happens when water is placed on the sandstone?

_____

_____

_____

What happens when water is added to the shale?

_____

_____

_____

Now submerge the sandstone and shale in the container of water and leave them for 5 to 10 minutes. Then remove the rocks from the container. Slide your fingers across the sandstone and across the shale and describe how they feel:

Wet sandstone: _____

Wet shale: _____

(g) In the "real world," it is common to find layers of sandstone stacked on top of layers of shale. Considering what you have observed here, how do you think the properties of wet sandstone and shale could influence mass wasting? What if the rocks were tilted or on a steep slope?

_____

_____

_____

_____

10.3 Types of Mass Wasting

Factors such as the style of movement and type of material determine the type of mass wasting and its effects. The character of movement can be as a cloud of particles, as a cohesive or interconnected mass, or as a soupy mixture called a *slurry*. The type of material involved can be either snow, ice, rock, or rock debris. These in turn determine the speed of movement (categorized as fast or slow). Below, we first examine the faster-moving types of mass wasting, before turning to the slower-moving types. The fast-moving types include:

- **Rock falls**—when rock material from a steep cliff or overhang moves through the air by gravity, it is referred to as a rock fall (**FIG. 10.7A**). Areas with many fractures (such as cracks and joints) are particularly vulnerable to rock falls.
- **Flows**—when debris (rock particles and soil) or liquid mud move downhill like a fluid, it is referred to as a **debris flow** or **mudflow**, respectively (**FIG. 10.7B**; and see Fig. 10.2B). Material within the flow typically moves at different speeds from other parts of the flow.

FIGURE 10.7 **Examples of mass wasting categories.**

(a) A rock fall consisting of sandstone blocks that have tumbled down a slope in Utah.

(b) The Illgraben debris flow in a catchment basin, Swiss Alps.

(c) A snow avalanche in Alaska tumbling down as a turblulent cloud.

(d) A rock glacier flowing down a mountainside in Alaska.

(e) The Oso, Washington, landslide of 2014.

- **Snow avalanches**—the downslope movement of a mass of snow, either as a turbulent cloud of powder or as a slurry of snow and water, is an avalanche (**FIG. 10.7C**).
- **Rock glaciers**—in cold climates where water is frozen as ice for long periods of time, rock debris can be cemented together by ice. This mass moves slowly downhill as a rock glacier (**FIG. 10.7D**).
- **Slides**—often the term *landslide* is used to describe any rapid mass wasting. But **rockslides**, **debris slides**, and **mudslides** are specifically defined as a cohesive mass of material that moves by sliding parallel to the surface it is moving across (**FIG. 10.7E**). (While often used interchangeably in news reports, mudslides differ from mudflows by being less liquidy and more cohesive.)

Recognizing the signs of dangerous mass wasting events, such as slides, can prepare you for their potential impact. Landslide-prone areas include those on existing old landslide deposits, at the bottom or top of a steep-cut slope, and at the base or in the hollow of minor drainage areas, such as streams or creeks (whether filled with water or dry). In Exercises 10.3 and 10.4, you will practice identifying the factors that make areas susceptible to mass wasting and the types of mass wasting that occur.

EXERCISE 10.3	Recognizing Different Types of Mass Wasting

Name: _____   Section: _____
Course: _____   Date: _____

For each of the two locations shown in the photos below and on the next page, you will determine the factors contributing to the mass wasting and the type of mass wasting, and then give evidence to support your answers.

Location 1: Yosemite National Park, California, on highway 140, just east of the Arch Rock Entrance. This event occurred in January 2016.

An area along a section of Highway 140 near Yosemite National Park both before (left) and after (right) a mass wasting event.

(a) In the photo on the left above, what geologic features can you identify that contribute to the high likelihood mass wasting will occur? Explain your answer.

_____

_____

_____

(continued)

Name: _____   Section: _____

Course: _____   Date: _____

(b) What type of mass wasting occurred at Yosemite? List supporting evidence.

_____

_____

_____

(c) What forces do you think influenced this mass movement most? Why?

_____

_____

_____

(d) What type of damage is visible in the right-hand image on the previous page?

_____

_____

(e) Do you think this mass wasting was predictable to occur? Why or why not?

_____

_____

_____

Location 2: An area in North Carolina, adjacent to the Appalachian Mountains. Following the passage of Hurricane Ivan in September 2004, the Peek's Creek mass wasting event occurred.

Origin of the mass wasting event. The inset shows a close-up view.

A section of Peek's Creek after the mass wasting event. Bare areas appear brown; vegetated areas are green. The dashed blue line marks the extent of the rock material from the event. The red arrow marks the original position of the house in the inset photo and the direction it traveled. The white arrow points to the final position of the house.

(continued)

Name: _____   Section: _____

Course: _____   Date: _____

(f) What type of mass wasting event took place in the photos below? Use evidence to support your answer.

_____

_____

_____

_____

Based on the damage shown in the images, why do you think some of the houses remained untouched by the event while others were severely damaged or destroyed?

_____

_____

_____

(g) What forces do you think most likely influenced this mass wasting event? Explain your answer using evidence.

_____

_____

_____

_____

EXERCISE 10.4   **The Oso Landslide on March 22, 2014**

Name: _____   Section: _____

Course: _____   Date: _____

The photo below shows an aerial view of the 2014 landslide along the North Fork Stillaguamish River in northwest Washington State. The two photos on the next page are satellite images taken before and after the landslide occurred. According to the USGS, the landslide's average speed was estimated at 40 miles per hour, and it covered one half square mile of land area, moving enough material to cover about 600 football fields 10 feet deep. Nearly 40 homes and other structures were buried in debris, including approximately 1 mile of State Route 530. In the nearby community, 43 fatalities were reported.

Aerial photograph of the Oso landslide.

(a) From these three landslide images, briefly describe what happened during this landslide and use evidence to support your answer.

_____

_____

_____

_____

_____

_____

(continued)

Name: _____   Section: _____

Course: _____   Date: _____

A satellite image of the region before the Oso landslide.

A satellite image of the region after the Oso landslide.

(b) The area had experienced precipitation 1.5 to 2 times greater than the long-term average for this time of year (February to March). Explain why this is thought to be the trigger for the Oso landslide.

_____

_____

_____

_____

_____

_____

_____

(c) What negative impact(s) beyond the landslide itself could result from another large mass wasting event in this region? (*Hint:* What does the vulnerable land lie alongside?)

_____

_____

_____

_____

_____

_____

_____

Less obvious examples of flows and slides are those movements that occur slowly and across less-steep terrain, a type of mass wasting called **creep** (**FIG. 10.8A**). If a mass of material detaches from its substrate along a spoon-shaped, sliding surface and slips downward as a semi-coherent block, it's called a **slump** (**FIG. 10.8B**).

Soil creep occurs on most slopes that have a greater than 5° angle. The topsoil or loose rock (and anything on the surface, such as trees, poles, fences, etc.) moves downslope under the force of gravity while the deeper soil and rock remain relatively stable. The rate of soil creep is imperceptibly slow and steady. When measured over several years, the average maximum movement is roughly 1 meter (0.3 feet) per decade. The rate of downslope movement depends on many factors, but the most common cause of creep is the repeated expansion and contraction of the soil from freeze-thaw cycles or alternating cycles of hydration and dehydration (**FIG. 10.8C**). You'll examine the impacts of soil creep more closely in Exercise 10.5.

FIGURE 10.8 Creep and slump.

(a) Trees bend as soil slowly creeps downhill in Mt. Rainer National Park, Washington. Because creep occurs slowly enough, the trees can continue to grow straight even as the movement of the soil pulls their bases downslope.

(b) Slumping has dumped sediment into this river in Costa Rica.

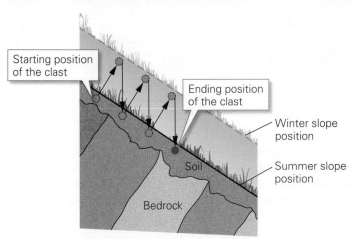

(c) Creep due to seasonal freezing and thawing: A clast rises perpendicular to the ground during freezing, but sinks vertically during thawing. After three years, it migrates to the position shown.

Investigating the Impact of Soil Creep

Name: _____ Section: _____

Course: _____ Date: _____

Consider the following scenario:

In order to build a road for a growing mountain community, a local river was diverted and the former riverbed was reclaimed as the road's foundation. After decades of repeated cold winters with heavy snowfall followed by warm, wet summers, the landscape changed to that shown in the figure below. As a consultant, you measure the rate at which the topsoil is creeping and find that it is moving an average of 2 cm/yr (0.8 in/yr).

(a) Using the rate of creep and the information in the figure, calculate how long it will be before all the remaining vegetated topsoil (that is, to the top of the figure) reaches the river channel. Remember to include units in your answer.

_____

(b) What is the ultimate fate of the telephone poles, asphalt road, and fence shown in the figure?

_____

_____

_____

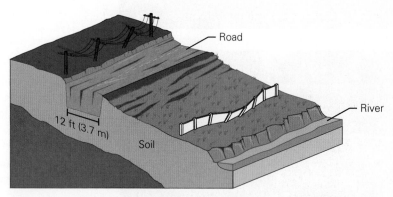

Soil creep occurring on reclaimed land. Note the positions of the telephone poles, road, topsoil (vegetation), and fence.

Mass Movement and Its Human Impact

Name: _____ Section: _____

Course: _____ Date: _____

Congratulations! Your company just gave you a promotion and, as part of the deal, it wants to relocate you to its main office on the island of Oahu, Hawaii. You need to work fast to find somewhere to live. While the company will assist you with the move, you consult a real estate agent to locate some potential properties. You have instructed the agent that you want a short commute to work. While you are aware that volcanic eruptions, heavy rainfall, hurricanes, and mass wasting are hazards in the Hawaiian Islands, you think you have a pretty good idea where a relatively safe location would be to live.

(continued)

Name: _____   Section: _____
Course: _____   Date: _____

Your agent offers you three locations all within the same commuting distance to your new office:

Location 1 is at the bottom of a valley nestled between rocky cliffs.

Location 2 is at the top of a ravine with spectacular views of the Pacific Ocean.

Location 3 is right off the coast with a view of the beach.

? What Do You Think What are the various mass wasting vulnerabilities associated with each location? (Use evidence in your answer.) Choose the location you would opt to live in, and explain what could be done to help mitigate the hazard(s) it faces. Write your answers on a separate sheet of paper.

Name: _____   Section: _____

Course: _____   Date: _____

Exploring Earth Science Using Google Earth

1. Visit **digital.wwnorton.com/labmanualearthsci**
2. Go to the Geotours tile to download Google Earth Pro and the accompanying Geotours exercises file.

Expand the Geotour10 folder in Google Earth by clicking the triangle to the left of the folder icon. The folder contains placemarks keyed to questions that explore different types of mass movement.

(a) Check and double-click the Bingham Copper Mine – Present placemark to fly to the largest open-pit copper mine in the world. The placemark highlights a disturbed area on the flank of the pit. Now check and double-click the Bingham Copper Mine – 2013 placemark to see historical imagery of the mine. What type of mass wasting occurred here?

_____

(b) Suggest a reason why the feature highlighted by the "2013" placemark occurred.

_____

_____

_____

_____

(c) Click the "X" on the historic imagery slider or the clock icon in the menu bar to return to present-day imagery, and make sure that Layers > Borders and Labels are checked. Check and double-click the Vajont Dam (also known as the Vaiont Dam) placemark to fly to Monte Toc, Italy, site of a catastrophic landslide in 1963. Check and double-click the Monte Toc placemark. In what general direction do the rock layers on Monte Toc tilt? *(N, S, E, or W; fly around the area to look at the mountain and use the compass in the viewer window; N is toward the valley containing the Vajont Dam.)*

_____

_____

_____

(d) Prior to the slope failure, the Vajont Dam retained water high up on the valley walls. Given this and the general inclination of the rock layers, explain why the north face of Monte Toc failed and moved downslope.

_____

_____

_____

Streams and Landscapes

11

Streams come in different sizes and shapes. They create different landscapes through erosion and by moving and depositing material.

LEARNING OBJECTIVES

- Learn how streams erode and deposit material
- Become familiar with landforms created by stream erosion and deposition
- Interpret active and ancient stream processes from landscape features

MATERIALS NEEDED

- Thin string
- Clear plastic ruler with divisions in tenths of an inch and millimeters (included in the GeoTools section at the back of this manual)
- Graph paper (provided at the end of this chapter) for constructing topographic profiles
- Colored pencils
- Magnifying glass or hand lens to read closely spaced contour lines
- Plastic box, stream sand, and water

11.1 Introduction

Water flowing in a channel is called a **stream**, whether it is as large as the Amazon River or as small as the smallest creek, run, rill, or brook. Streams are highly effective agents of erosion and may move more material after one storm in an arid region than wind does all year. This chapter explores how streams carry out geologic work, why not all streams behave the same way, and how different streams may produce very different landscapes.

11.2 How Do Streams Work?

All streams operate according to a few simple principles regardless of their size:

- Water in streams flows downhill because of gravity.
- Streams normally flow in a well-defined channel, except during floods, when the water overflows the channel and spills out across the surrounding land.
- The motion of water gives a stream kinetic energy, enabling it to do the geologic work of erosion and deposition. The amount of energy depends on the amount (mass) of water and its velocity (kinetic energy = $\frac{1}{2}mv^2$), so big, fast-flowing streams erode more material than small, slow-flowing streams.
- Streams transport sediment particles, from tiny clay-sized grains to small boulders. These particles slide or roll on the bed of the stream, bounce along, or are carried in suspension within the water.
- The flow of water erodes unconsolidated sediment from the walls and bed of the channel. That sediment, carried by the flowing water, abrades solid rock.
- Streams deposit sediment when their flow slows down (or when the stream evaporates). The heaviest particles are deposited first, then the smaller grains, as the stream's energy decreases.

Now for a few geologic terms: A stream **channel** is the area within which the water is actually flowing. A stream **valley** is the region within which the stream has eroded the land. In some cases, the channel completely fills the bottom of the valley; in others, it is much narrower than the broad valley floor. Some valley walls are steep, others gentle.

FIGURE 11.1 shows that streams may follow the same basic rules but may nevertheless look very different. Exercise 11.1 asks you to observe and describe the ways in which streams differ.

FIGURE 11.1 A tale of two streams.

(a) Yellowstone River, Wyoming.

(b) River Cuckmere, England.

Name: _____   Section: _____
Course: _____   Date: _____

The basic principles listed above control how all streams work, but those principles may be displayed in different ways by different streams. The result is that streams like those in Figures 11.1a and 11.1b may appear to be acting quite differently from one another. Look carefully at the streams in Figure 11.1 and compare the Yellowstone River and the River Cuckmere based on the questions below.

(a) Which stream has the wider channel?

_____

(b) Which stream has the broader valley?

_____

(c) Which stream has the more clearly developed valley walls?

_____

(d) Describe the relationship between valley width and channel width for both streams.

_____

_____

(e) Which stream has the straighter channel?

_____

Which has a more sinuous (meandering) channel?

_____

(f) Which stream appears to be flowing faster?

_____

What evidence did you use to determine this?

_____

_____

(g) Which stream appears to be flowing more steeply downhill?

_____

A brief lesson in stream anatomy will help to explain stream erosion and deposition. A stream begins at its **headwaters** (or head), and the point at which it ends—by flowing into another stream, the ocean, or a topographic low point—is called its **mouth**. The headwaters of the Mississippi River, for example, are in Lake Itasca in Minnesota, and its mouth is in the Gulf of Mexico in Louisiana. The *longitudinal profile* of a stream from headwaters to mouth is generally a smooth, concave-up curve (**FIG. 11.2**). The **gradient** (steepness) of a stream may vary from a few inches to hundreds of feet of **vertical drop** per mile and is typically steeper at the head than at the mouth.

FIGURE 11.2 Longitudinal stream profile.

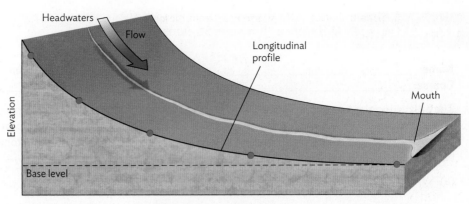

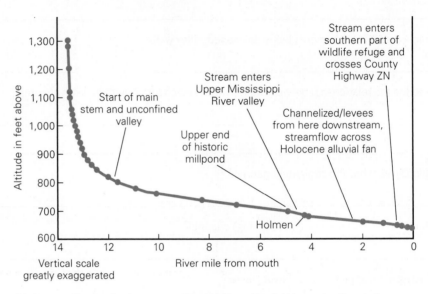

(a) An idealized depiction of a stream profile.

(b) An actual longitudinal stream profile of Halfway Creek, Wisconsin. (Data is from USGS. Vertical coordinate information is referenced to the National Geodetic Vertical Datum of 1929. So, "Altitude in feet above" the NGVD-29 datum.)

A stream can erode its channel only as low as the elevation at its mouth, because if it cut any deeper, it would have to flow uphill to get to the mouth. The elevation at the mouth, called the **base level**, thus controls erosion along the entire stream.

Sea level is the ultimate base level for streams that flow into the ocean; base level for a *tributary* that flows into another stream is the elevation where the tributary joins the larger stream.

One difference between the Yellowstone River and the River Cuckmere (see Fig. 11.1) is the straightness of their channels: the Yellowstone has a relatively straight channel, whereas the Cuckmere channel meanders across a wide valley floor. The **sinuosity** of a stream measures how much it meanders, as shown in the following formula:

$$\text{Sinuosity} = \frac{\text{Length of stream channel (meanders and all) between two points}}{\text{Straight-line distance between the same two points}}$$

Because sinuosity is a ratio of the two lengths, it has no units. An absolutely straight stream would have a sinuosity of 1.00 (if such a stream existed), whereas streams with many meanders have high values for sinuosity (**FIG. 11.3**). Exercise 11.2 examines what causes some streams to meander whereas others have straight channels.

Low sinuosity (straight).

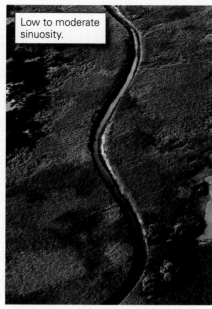

Low to moderate sinuosity.

High sinuosity.

EXERCISE 11.2 **Why Some Streams Meander but Others Are Straight**

Name: _____ Section: _____

Course: _____ Date: _____

In this exercise, you will use the maps in **FIGURES 11.4, 11.5,** and **11.6.** Approximate mile measurements to the nearest tenth of a mile.

(a) Compare the course of the Bighorn River between points A and B with that of its tributary between points C and D (Fig. 11.4). Fill in the table below.

Note: To measure the true total channel length, especially when it meanders, you can use a ruler or string. With a ruler, measure each straight segment separately, recording each length, and then add them up at the end and convert to miles using the scale on the map. With string, trace the river and tape down the string at each place where it bends. At the end, remove the tape, straighten the string, and measure the length of string you used.

	Bighorn River	Unnamed tributary
Channel length (miles)		
Straight-line length (miles)		
Sinuosity (no units) Channel length divided by straight-line length		
Highest elevation* (feet)		
Lowest elevation* (feet)		
Vertical drop Highest elevation minus lowest elevation (feet)		
Gradient* (feet per mile) Vertical drop divided by channel length		

*Streams aren't considerate; they don't begin and end on contour lines. Scan the entire stream looking for and estimating the highest and then the lowest point on each stream. Then calculate the vertical drop and gradient.

(continued)

FIGURE 11.5 The Genesee River south of Rochester, New York.

0   0.5   1 mile

0   0.5   1 km

Contour interval = 10 feet

N

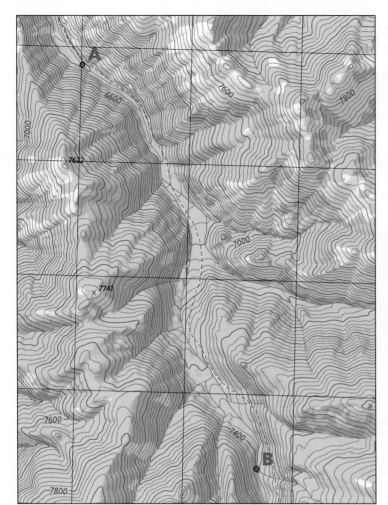

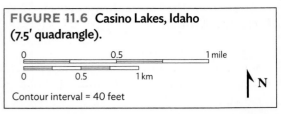

FIGURE 11.6 Casino Lakes, Idaho (7.5′ quadrangle).

Contour interval = 40 feet

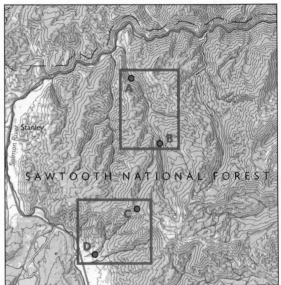

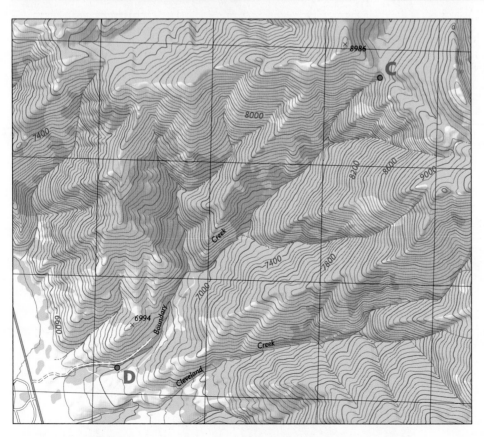

Name: _____   Section: _____
Course: _____   Date: _____

(b) What is the apparent relationship between a stream's gradient and its sinuosity?

_____

_____

(c) Test this hypothesis on the Genesee River of New York (Fig. 11.5) and the Casino Lakes area of Idaho (Fig. 11.6). Fill in the table below and describe how the Genesee River differs from the Idaho streams.

	Genesee River (A–B)	Casino Lakes area	
		Stream A–B	Stream C–D
Valley shape (V-shaped or broad with flat bottom)			
Gradient (feet/mile)			
Valley width (miles)		Channel essentially fills valley floor	
Channel width (feet)			
Valley width/channel width		~1.0	~1.0
Sinuosity Channel length divided by straight-line length			

(d) Did the streams in Figures 11.5 and 11.6 support your hypothesis about the relationship between sinuosity and gradient? Explain.

_____

_____

(e) What is the apparent relationship between sinuosity and the valley width/channel width ratio?

_____

_____

(f) What is the apparent relationship between stream gradient and the shape of a stream valley?

_____

_____

Now apply what you've learned to the photographs of the streams in Figures 11.1 and 11.3.

(g) Which probably has the steeper gradient, the River Cuckmere or the Yellowstone River? Explain your reasoning.

_____

_____

(h) Which of the streams in Figure 11.3 probably has the steepest gradient? The gentlest gradient? Explain your reasoning.

_____

_____

11.3 Stream Valley Types and Features

The Yellowstone River and the River Cuckmere (see Fig. 11.1) illustrate the two most common types of stream valleys: steep-walled, V-shaped valleys whose bottoms are occupied fully by the channel, and broad, flat-bottomed valleys much wider than the channel, within which the stream meanders widely between the valley walls.

FIGURE 11.7 shows how these valleys form. When water is added to a stream in a V-shaped valley, the channel expands and fills more of the valley. When water is added to a stream with a broad, flat valley and a relatively small channel, it spills out

FIGURE 11.7 Evolution of stream valleys.

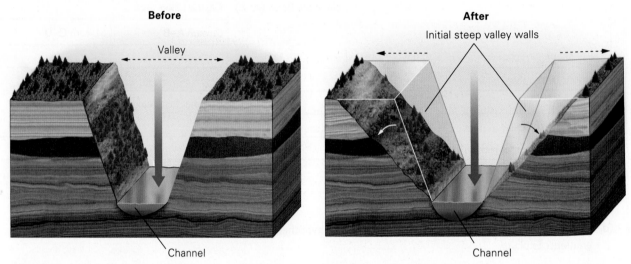

(a) Steep, V-shaped valley. Vertical erosion carves the channel and valley downward vertically (large blue arrow), producing steep valley walls. Mass wasting (slump, creep, landslides, and rockfalls) reduces slope steepness to the angle of repose (curved arrows) and widens the top of the valley (dashed arrows).

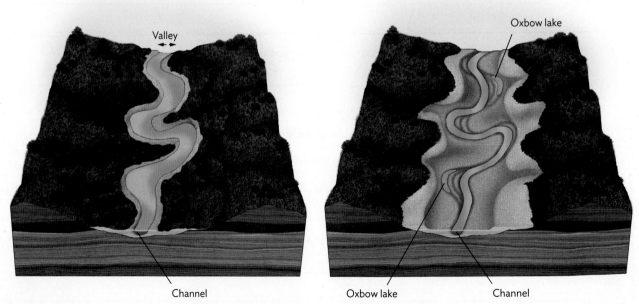

(b) Broad, flat-bottomed valley. As the stream meanders, lateral erosion widens the valley (arrows). Mass wasting reduces the slope of the valley walls, as in part (a). Oxbow lakes mark the position of former meanders.

of the channel onto the broad valley floor in a **flood**. Sediment carried by the flood-water is deposited on the **floodplain** (an area of flat land on either side of a stream that is covered with water during a flood—when the stream overflows its banks, the sediment it is carrying is deposited within the floodplain, enriching its fertility).

Many depositional and erosional features of floodplains can be easily recognized on topographic maps or in photographs (**FIG. 11.8**). **Natural levees** are ridges of sediment that outline the stream channel; they form when a stream overflows its banks and deposits its coarsest sediment next to the channel. Several generations of natural levees are visible in Figure 11.8, showing how the meanders changed position with time. On the inside of a meander loop, more water molecules rub against the river bed, causing the water to slow down. This slower velocity results in sediment deposition, forming **point bars**. At the same time, erosion occurs on the outside of the meander loop, forming a **cut bank**. The water on the outside of the meander is deeper and must travel a greater distance in the same amount of time as the water on the inside of the curve, so the water moves faster. If you have ever ridden on a carousel, you may have noticed the difference in speed of the riders: those on the inside travel slower, while those on the outside travel faster. As a result of erosion and deposition occurring along the same point of a river, on opposite banks, meanders migrate over time, moving outward and downstream. Sometimes a stream cuts off a meander loop and straightens itself. The natural levees that formerly flanked the meanders help outline the former position of the river, leaving **meander scars**. Meander scars that have filled with water are called **oxbow lakes**.

FIGURE 11.8 A floodplain and its associated features.

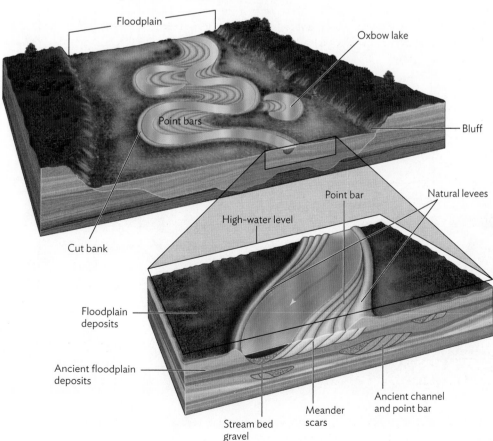

Identifying Stream Features

Name: _____ Section: _____

Course: _____ Date: _____

FIGURE 11.9 depicts a topographic map of the St. Francis River in Arkansas and Mississippi, as well as a portion of the Mississippi River. Label one example of each of the following stream erosional and depositional features on the map: valley, channel, meander, point bar, oxbow lake, and meander scar.

FIGURE 11.9 The St. Francis River in Arkansas and Mississippi.

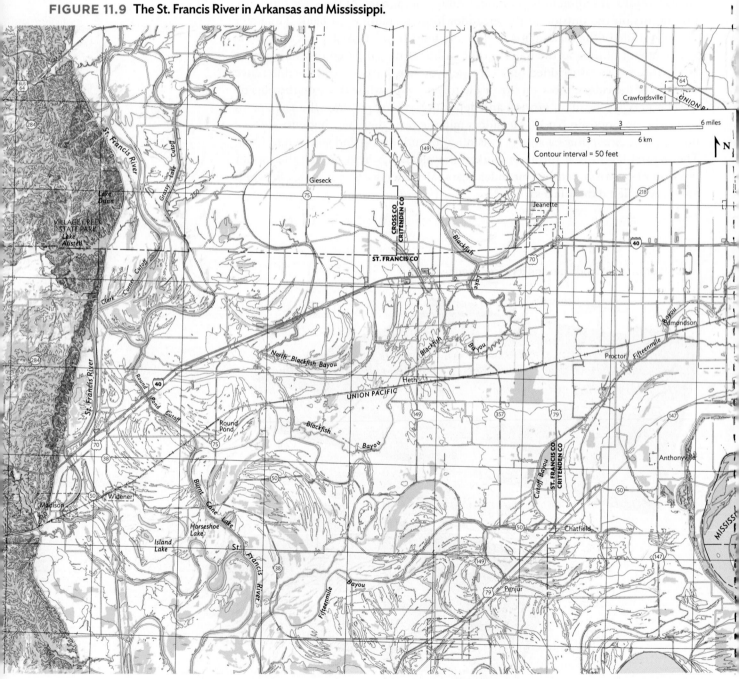

11.4 Changes in Streams over Time

Streams erode *vertically* over time by leveling the longitudinal profile to the elevation of the mouth. Streams also erode *laterally*, broadening their valleys by meandering (see Fig. 11.7). Most streams erode both laterally and vertically at the same time, but the balance between vertical and lateral erosion commonly changes as the stream evolves.

The headwaters of a high-gradient stream are much higher than its mouth, and the stream's energy is largely expended on vertical erosion, lowering channel elevation all along its profile. Over time, the gradient lessens as erosion lowers the headwaters area to elevations closer to that of the mouth. The stream still has enough energy to alter the landscape, however, and that energy is expended on lateral erosion. The valley then widens by a combination of mass wasting and meandering. Even when headwaters are lowered to nearly the same elevation as the mouth, a stream still has energy for geologic work, but because it cannot cut vertically below its base level, most of its energy at this stage is expended on lateral erosion.

Initially, a stream meanders within narrow valley walls, but over time, it erodes those walls farther and farther, eventually carving a very wide valley. As its gradient decreases, a stream redistributes sediment that it has deposited, moving it back and forth across the floodplain.

Although most streams follow this sequence of progressively lowering their gradients and therefore changing their erosional behavior, not all do so. Some streams have very gentle gradients and meander widely from the moment they begin to flow. Streams on the Atlantic and Gulf coastal plains are good examples of this kind of behavior.

As a stream cuts downward, it sometimes encounters rock that is much more resistant to erosion than the rocks downstream. The stream will eventually cut through the resistant rock, but until it does, there will be an irregularity (a **nick point**) in its concave-up longitudinal profile. In small streams, nick points may appear as sets of rapids, but in large rivers they can be huge waterfalls, such as Niagara, Victoria, and Iguaçu (**FIG. 11.10**). Over time, the falls retreat as the stream carves through them.

Some streams have varying amounts of **discharge** (the volume of water flowing through a river channel at a

FIGURE 11.10 The development of waterfalls.

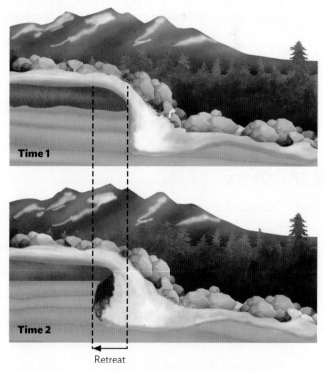

Retreat

(a) Iguaçu Falls on the Iguaçu River, the boundary between Argentina and Brazil, is caused by a thick, highly resistant layer of basalt.

(b) The waterfall will retreat over time as the river cuts through the basalt.

given point) over a season and so, rather than meandering, will form many interwoven channels—these streams are said to be **braided**. The braiding occurs when the stream has too much sediment—usually a large amount of gravel—to be carried during low-water stages, and so it deposits the sediment, forming gravel bars. These bars block the flow of water in the stream channels, and therefore the water must go around, forming continuously braided channels.

Whenever the flow of a stream slows—as with braiding, the inside of meander bends, a decrease in stream gradient, or discharge—the amount of energy the stream has decreases, resulting in the deposition of sediment. The varying sizes and densities of the different sediments the stream carries determine when each type is deposited. Larger, heavier sediments such as gravel or sand will be deposited first, followed by smaller and smaller grains until only the finest-grained sediment can be carried by the stream. Given time, even these fine particles—like mud—may settle out of the water. These deposits may be temporary, only to be later carried away by increasing discharge from precipitation or the melting of snow and ice. Exercise 11.4 investigates the behavior of a stream that you will create.

EXERCISE 11.4	Interpreting Stream Behavior

Name: _____   Section: _____
Course: _____   Date: _____

Let's look at how a stream behaves—eroding or depositing sediment—as the amount of water added to a stream increases over time. Your instructor will provide materials to create your own stream model with sediment and a source of water. (*Note:* If this exercise is available and being conducted online, the setup, materials, and directions may differ. See the instructions in the online lab for how to conduct this exercise.) Follow the setup instructions below, or as given by your instructor.

Step 1: Pack the plastic box with damp sand, to make the riverbed.

Step 2: Elevate one end of the box by placing it over a stack of books or other stable object.

(a) Before adding water to your stream model, develop a hypothesis about what you think will happen to the shape of the stream and stream valley as water begins and continues to flow in the stream.

_____

_____

_____

In a few sentences, explain the reasoning for your prediction.

_____

_____

_____

(b) Gently add water to the top of the stream bed, either by pouring from a glass or watering can or using a dropper. Observe the characteristics of the river channel and valley that form.

Sketch a diagram of your stream in the boxed space on the next page. Label stream features that you can identify.

Where does the greatest amount of erosion take place? _____

Why do you think this is so? _____

Where does the greatest amount of deposition occur? _____

Why do you think this is so? _____

(continued)

Name: _____   Section: _____
Course: _____   Date: _____

┌───┐
│   │
│   │
│   │
│   │
│   │
│   │
│   │
│   │
│   │
│   │
│   │
└───┘

Is there anywhere that both erosion and deposition are taking place? Explain your choice.

_____

_____

_____

(c) Describe whether your stream is meandering, straight, or braided and cite examples to support your choice.

_____

_____

_____

(d) In a few sentences, summarize how your stream changed with time (as water continued to flow); consider the characteristics of streams and stream valleys and compare them with your model.

_____

_____

_____

(e) Why might understanding the behavior of a stream be important to society and the environment?

_____

_____

_____

_____

Name: _____   Section: _____
Course: _____   Date: _____

FIGURES 11.11 and 11.12 on the following pages show two meandering streams, each of which balances energy use differently between vertical and lateral erosion.

(a) Draw a profile for each stream along line A-A'. Use only index contour lines. *Note:* Section 9.8.1 in Chapter 9 contains detailed instructions for creating topographic profiles.

Meadow River, West Virginia

A   A'

feet
- 1,800
- 1,750
- 1,700
- 1,650
- 1,600
- 1,550
- 1,500
- 1,450
- 1,400

Arkansas River near Tulsa, Oklahoma

A   A'

feet
- 1,300
- 1,250
- 1,200
- 1,150
- 1,100
- 1,050
- 1,000
- 950
- 900

(b) Which stream do you think is doing the most lateral erosion? Explain your answer.

_____

_____

(c) Which stream do you think is doing the most vertical erosion? Explain your answer.

_____

_____

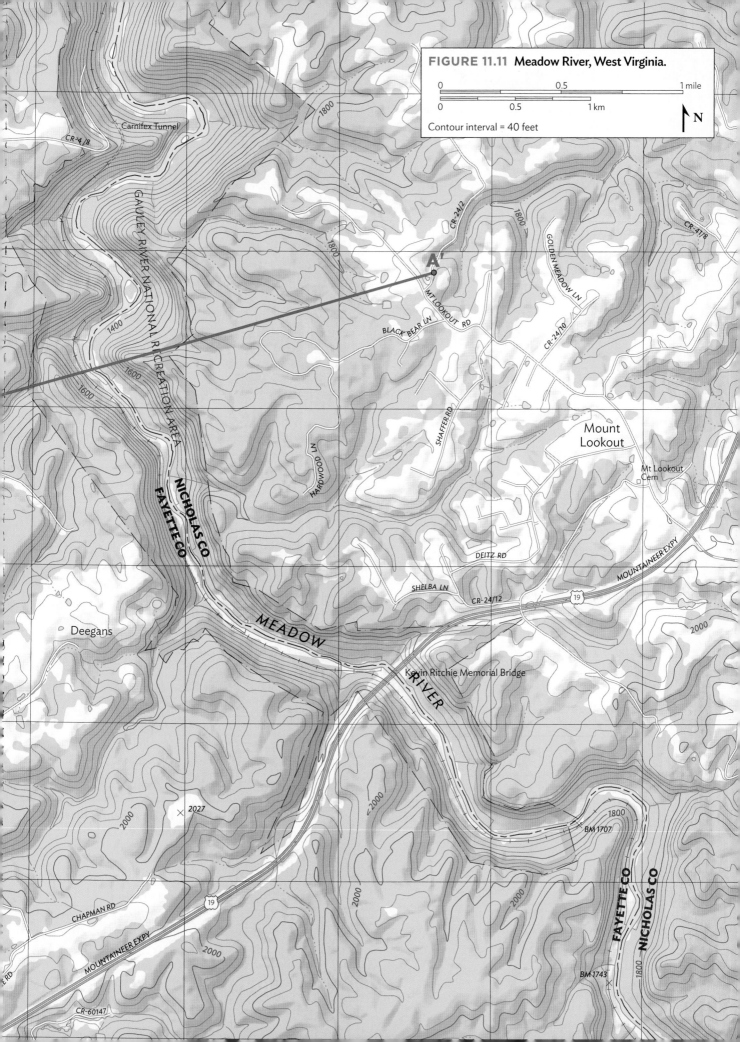

FIGURE 11.11 Meadow River, West Virginia.

Contour interval = 40 feet

FIGURE 11.12 Arkansas River near Tulsa, Oklahoma.

Contour interval = 10 feet

N

11.5 Stream Networks

Streams are particularly effective agents of erosion because they form networks that cover much of the Earth's surface. Rain falling on an area runs off into tiny channels that carry water into bigger streams and eventually into large rivers. Each stream—from tiniest to largest—expands headward over time as water washes into its channel, increasing the amount of land affected by stream erosion. Understanding the geometric patterns of stream networks and the way they affect the areas they drain is important in understanding how to prevent or remedy stream pollution, soil erosion, and flood damage.

11.5.1 Drainage Basins

The area drained by a stream is its **drainage basin**, which is separated from adjacent drainage basins by highlands called **drainage divides**. The drainage basin of a small tributary may cover a few square miles, but that of the master (main) stream in the region may be hundreds of thousands of square miles. **FIGURE 11.13** shows the six major drainage basins of North America. Five deliver water to an ocean directly or indirectly (Pacific, Arctic, Atlantic, Hudson Bay, and Gulf of Mexico), but the Great Basin is bounded by mountains and there is no exit for the water.

The yellow arrows show the dominant flow direction—note that some of these drain *northward*. The Mississippi River (one of many rivers in the Gulf of Mexico drainage basin) has the largest drainage basin of any river on the continent and drains much of the interior of the United States. The *continental divide* separates streams that flow into the Atlantic Ocean from those that flow to the Pacific. The Appalachian Mountains are the divide separating the Gulf of Mexico and direct Atlantic Ocean basins; the Rocky Mountains separate the Gulf of Mexico and Pacific basins. A favorite tourist stop in Alberta, Canada, is a *triple* divide that separates waters flowing north to the Arctic Ocean, west to the Pacific, and northeast to Hudson Bay.

11.5.2 Drainage Patterns

Master and tributary streams in a network typically form recognizable patterns (**FIG. 11.14**). **Dendritic** patterns (from the Greek *dendron*, for tree, as they resemble veins in a leaf) develop where the materials at the surface are equally resistant to erosion. This may mean horizontal sedimentary or volcanic rocks; loose, unconsolidated sediment; or igneous and metamorphic areas where most rocks erode at the same rate. **Trellis** patterns form where ridges of resistant rock alternate with valleys underlain by weaker material. **Rectangular** patterns indicate zones of weakness (faults, fractures) perpendicular to one another. In **radial** patterns, streams flow either outward (**centrifugal**) from a high point (e.g., a volcano) or inward (**centripetal**) toward the center of a large basin. **Annular** drainage patterns occur where there are concentric rings of alternating resistant and weak rocks—typically found in structures called domes and basins. (You can distinguish an annular drainage pattern from a centripetal radial one by the circular flow pattern at the inner part of the crater.) **Parallel** drainage patterns occur on a uniform slope when several streams with parallel courses develop simultaneously.

FIGURE 11.13 Major drainage basins of North America.

- - - - Mississippi River basin limit
—— Drainage divide

Arctic Ocean

Arctic drainage basin

Hudson Bay

Pacific Ocean

Hudson Bay drainage basin

Atlantic drainage basin

Pacific drainage basin

Atlantic Ocean

Gulf of Mexico drainage basin

Great Basin drainage basin

Mississippi River

Continental Divide

Gulf of Mexico

Boundary of Mississippi River drainage basin

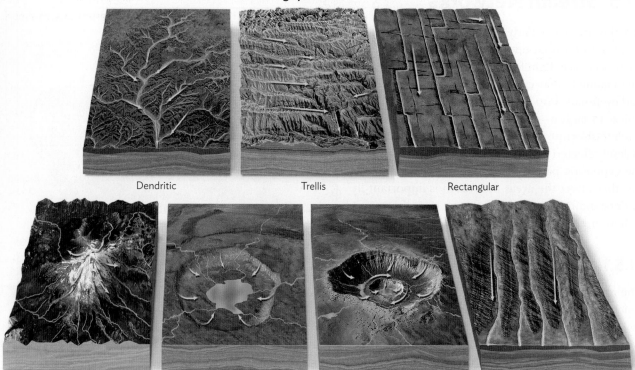

FIGURE 11.14 Common drainage patterns.

Dendritic   Trellis   Rectangular

Radial: centrifugal   Radial: centripetal   Annular   Parallel

<table>
<tr><td>EXERCISE 11.6</td><td>Drainage Basins and Drainage Divides</td></tr>
</table>

Name: _____   Section: _____
Course: _____   Date: _____

FIGURE 11.15 is a map showing several tributaries on the north and south sides of the Missouri River near Jefferson City, Missouri. One large tributary, the Osage River, joins the Missouri from the south, near the eastern margin of the map, but most of the tributaries on the north side are much smaller.

(a) With a colored pencil, trace one of the tributary streams feeding *directly* into the Missouri River from the north. With the same pencil, trace the tributaries that flow directly into that stream, and then the tributaries of those smaller streams. Remember the "rule of V" (see Chapter 9) as you trace the smaller streams to their headwaters. As an example, a segment of one small stream has been traced in red on the map.

(b) With a different colored pencil, trace an adjacent tributary of the Missouri, and trace its tributaries in turn. Repeat for more streams and their tributaries on the north side of the Missouri River, using a different color for each stream.

(c) You have just outlined most of the drainage on the north side of the Missouri. Now, again with a different colored pencil, trace the divides that separate the individual drainage basins for each master stream. This should be easy because you've already identified streams in each drainage basin with a different color. An example of this step has been done in blue for the stream segment traced in red on the map.

Note that some drainage divides are defined sharply by narrow ridges, but others that lie within broad upland areas, where most of the headwaters are located, are more difficult to trace.

(d) What drainage pattern is associated with the Mississippi River drainage basin in Figure 11.13? What does that tell you about the materials that underlie the central part of the United States?

_____

_____

FIGURE 11.15 Divides in the drainage of the Missouri River near Jefferson City, Missouri.

Contour interval = 65 feet

11.6 When There's Too Much Water: Floods

A flood occurs when more water enters a stream than its channel can hold. Many floods are seasonal, caused by heavy spring rains or melting of thick winter snow piles. Others, called **flash floods**, follow storms that can deliver a foot or more of rain in a few hours. **FIGURE 11.16** is a map compiled by the Federal Emergency Management Agency (FEMA) showing the estimated flood potential in the United States, based on the number of square miles that would be inundated.

It might appear that states lightly shaded in Figure 11.16 would have little flood damage, but that would be an incorrect reading of the map because two of the worst river floods in U.S. history occurred in Rapid City, South Dakota, and Johnstown, Pennsylvania. Indeed, these two cities have been flooded many times. The *area* of potential flooding may not be as large as in some other states, but the *conditions* for flooding may occur frequently. South Dakota's Rapid Creek has flooded more than 30 times since the late 1800s, including a disastrous flash flood on June 9, 1972, triggered by 15 inches of rain in 6 hours. The creek overflowed or destroyed several dams, ruined more than 1,300 homes and 5,000 cars, and killed more than 200 people in Rapid City (**FIG. 11.17**).

FIGURE 11.16 Flood potential in the United States.

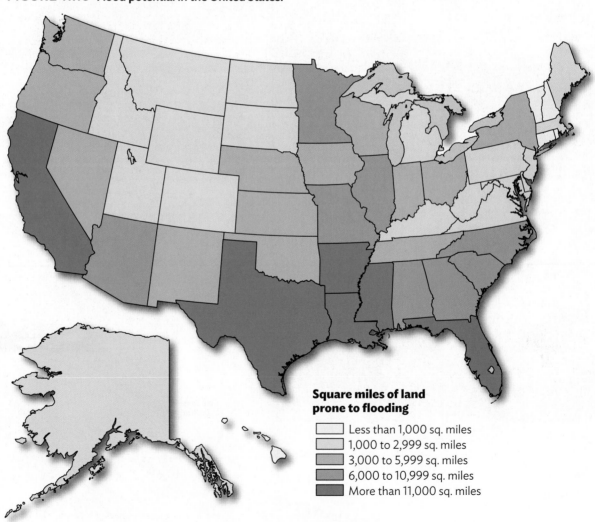

Square miles of land prone to flooding

- Less than 1,000 sq. miles
- 1,000 to 2,999 sq. miles
- 3,000 to 5,999 sq. miles
- 6,000 to 10,999 sq. miles
- More than 11,000 sq. miles

FIGURE 11.17 Effects of the June 9, 1972, flood in Rapid City, South Dakota.

(a) Rapid Creek has a narrow floodplain where it flows through Dark Canyon, 1 mile upstream of Rapid City. Floodwater filled the entire floodplain. Concrete slabs (arrows) are all that is left of the homes built in the floodplain.

(b) The spillway of the Canyon Creek Dam (arrow) became clogged with debris carried by the floodwater, causing water to flow over the dam, which then failed completely.

(c) Not the usual lineup of cars for gasoline.

(d) This railroad trestle, along with highway bridges, was washed away, slowing relief efforts.

Name: _____   Section: _____

Course: _____   Date: _____

The streams shown on maps throughout this chapter are all subject to flooding, but the potential problems faced by cities along the banks of different streams are not the same. In this exercise, you will use the maps from this chapter to answer the following questions about potential problems caused by floods.

(a) Refer to the maps in Figures 11.9, 11.11, and 11.12. For which of these streams would a flood 20 feet higher than the current stream channel cause damage to human-built structures and roads? In each case, describe what would happen.

_____

_____

_____

_____

(b) Describe the potential effect of a flash flood that raised the level of the Bighorn River (see Fig. 11.4) 20 feet above its banks. Estimate and describe the areas that would be affected.

_____

_____

_____

_____

(c) Would Geneseo Airport on the east bank of the Genesee River (see Fig. 11.5) have to be evacuated if water rose 20 feet? Explain.

_____

_____

_____

_____

(d) Compare the topographic map of the Missouri River with the recent satellite image of the same area in **FIGURE 11.18**. What kind of information relevant to flooding potential does the map provide? What information does the satellite image add?

_____

_____

_____

_____

(e) What structures have been built in the Missouri River floodplain that could be destroyed or made unusable in a flood in which the river rose 20 feet? How would the loss of these structures affect relief efforts?

_____

_____

_____

_____

FIGURE 11.18 Topographic map (top) and satellite image (bottom) of the Missouri River near Jefferson City, Missouri.

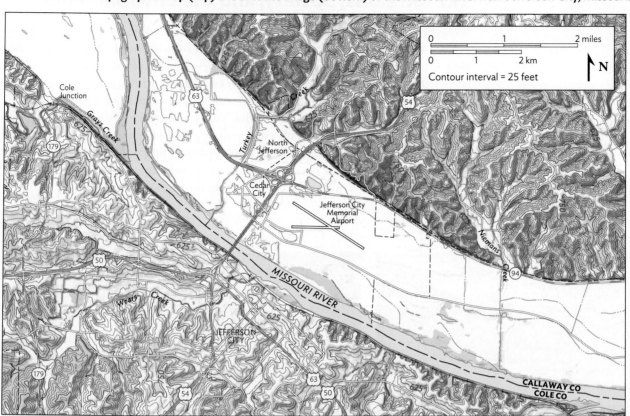

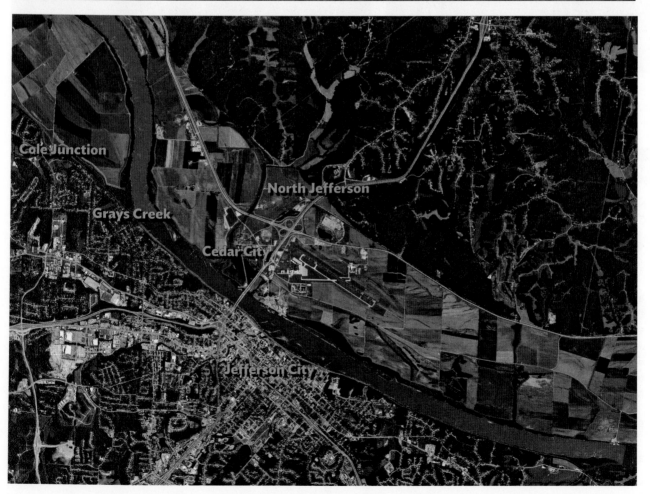

11.7 Streams, Society, and the Environment

Streams have been a vital natural resource since the beginnings of human society, serving as a source of water and food, as avenues for exploration and transportation, and as political boundaries. From medieval times, streams drove the water wheels that provided the energy for mills to grind grain into flour and cut trees into lumber, and eventually, the more sophisticated versions that powered the Industrial Revolution in the 18th and 19th centuries. Today, streams contribute to power generation directly, by turning turbines that generate hydroelectricity, and indirectly, by cooling nuclear reactors. But building dams and diverting water for industrial cooling interfere with natural stream processes, and pollutants have all too frequently had disastrous effects on fish and other stream dwellers (**FIG. 11.19**).

FIGURE 11.19 When streams are polluted.

(a) Before (left) and after (right) the 2015 Gold King Mine wastewater spill in Colorado.

(b) The heavily polluted Calumet River near Chicago.

EXERCISE 11.8 Unforeseen Consequences of Human Interaction: Aswan Dam

Name: _____   Section: _____
Course: _____   Date: _____

? What Do You Think Unintended or unforeseen consequences of human interactions with streams can cause serious problems for human and wildlife communities. Consider the Aswan High Dam, which was built across the Nile River in the 1960s to generate electricity for Egypt, a nation lacking oil and gas resources. The dam created an enormous lake (Lake Nasser) upstream. As is the case with every dam, the lake acts as a local base level for the Nile. The dam also diverts about 55 trillion liters of freshwater for irrigation and hydroelectric power generation every year, preventing that water from reaching the Mediterranean Sea, into which the Nile drains.

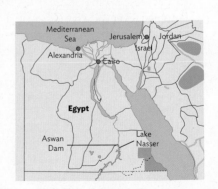

For thousands of years, annual spring floods replenished fertile soil in the floodplain of the Nile. The Aswan High Dam prevents these floods.

On a separate sheet of paper, describe the positive and negative effects you think the dam might have had on Egyptian agriculture. In addition, describe the effects the dam might have had on aquatic organisms in the Mediterranean Sea.

Name: _____   Section: _____
Course: _____   Date: _____

Exploring Earth Science Using Google Earth

1. Visit **digital.wwnorton.com/labmanualearthsci**
2. Go to the **Geotours** tile to download Google Earth Pro and the accompanying Geotours exercises file.

Expand the Geotour11 folder in Google Earth by clicking the triangle to the left of the folder icon. The folder contains placemarks from three different locations that highlight spectacular landforms associated with streams.

(a) Check and double-click the Rio Ucayali River, Peru, placemark to fly to a river in South America. Study the time-lapse animation in the placemark balloon to observe the rapid changes in meanders that occurred between 1984–2012. In addition to migrating laterally and cutting off some meander necks, do the meanders migrate downstream or upstream?

Google Earth

_____

(b) Check and double-click the GoosenecksSP_TopoMap overlay to fly to a section of the San Juan River near Mexican Hat, UT. Given the extensive meanders and the closely spaced contours for the stream valley walls, explain how these features likely evolved. [*Just for fun:* Check and double-click the Lake Powell placemark to fly downstream (west) to where the San Juan River flows into the backwaters of Lake Powell (caused by the Glen Canyon Dam on the Colorado River) to see what human-induced changes to base level have caused.]

_____

_____

_____

_____

(c) Check and double-click the Cumberland_TopoMap overlay to fly to the Cumberland, MD, area. Braddock Run used to follow approximately along the Interstate 68 path across Haystack Mountain to flow into the North Branch of the Potomac River (NBPR). (Check the Interstate 68 former Braddock Run and Haystack Mountain placemarks.) Now that path is a wind gap through the mountain, and Braddock Run flows to the river at Eckhart Junction (check the Eckhart Junction placemark). Given that Braddock Run originally had to erode down through the resistant ridge of Haystack Mountain while another tributary of the NBPR (*check to turn on Tributary path*) could headwardly erode along the less resistant shale valley, propose an explanation for what created the wind gap and altered Braddock Run's direction of flow.

_____

_____

_____

_____

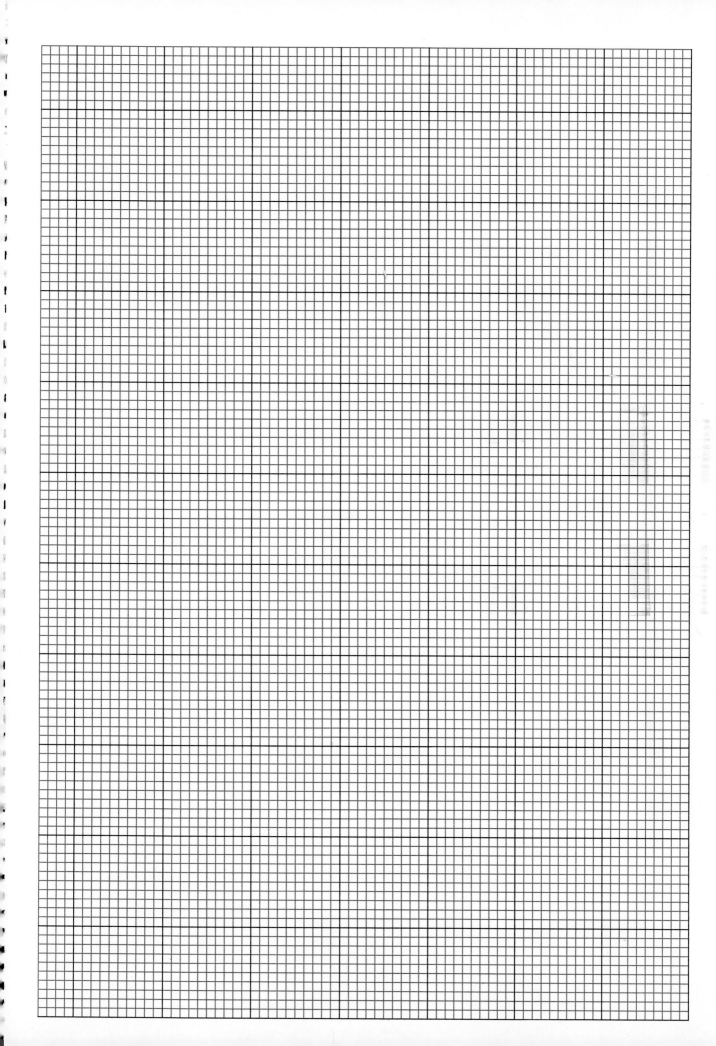

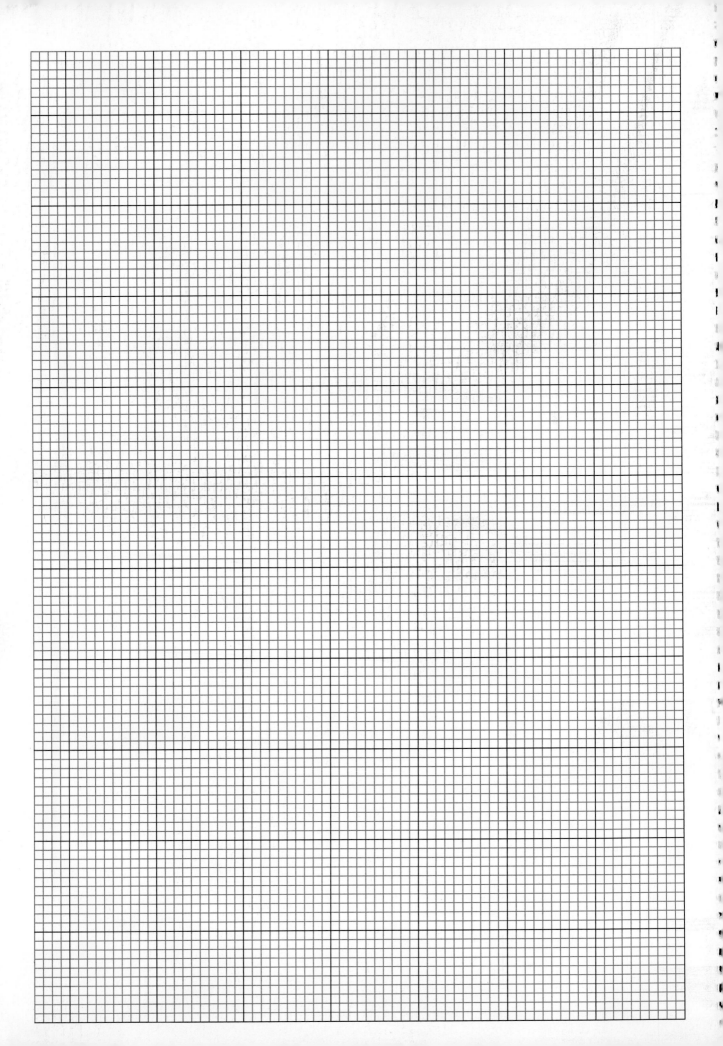

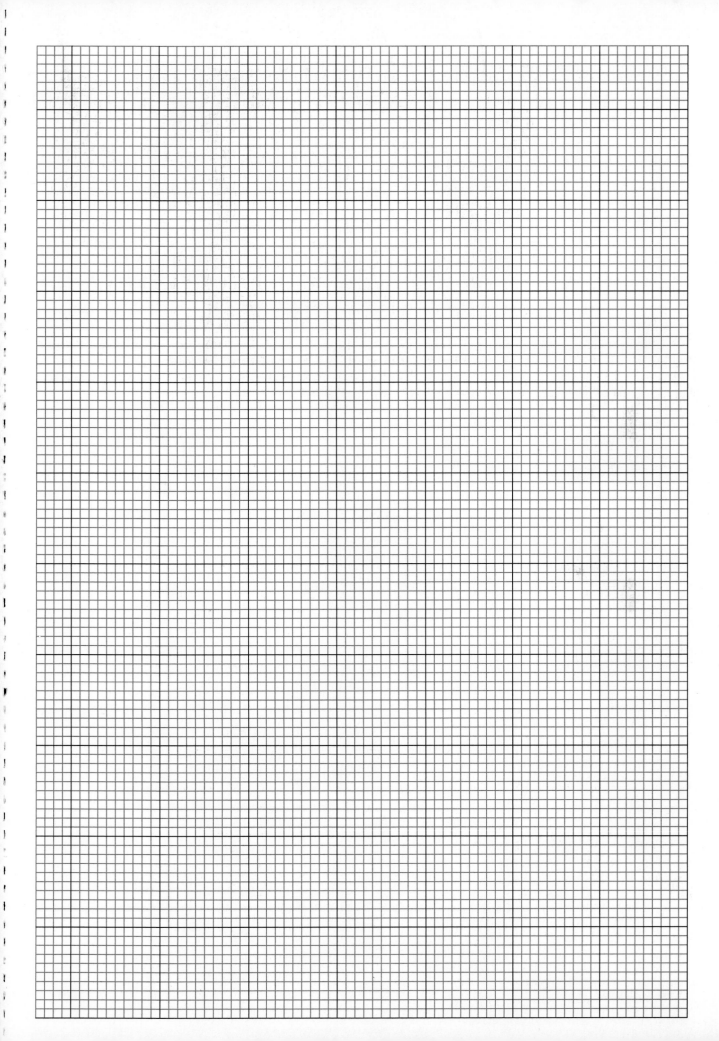

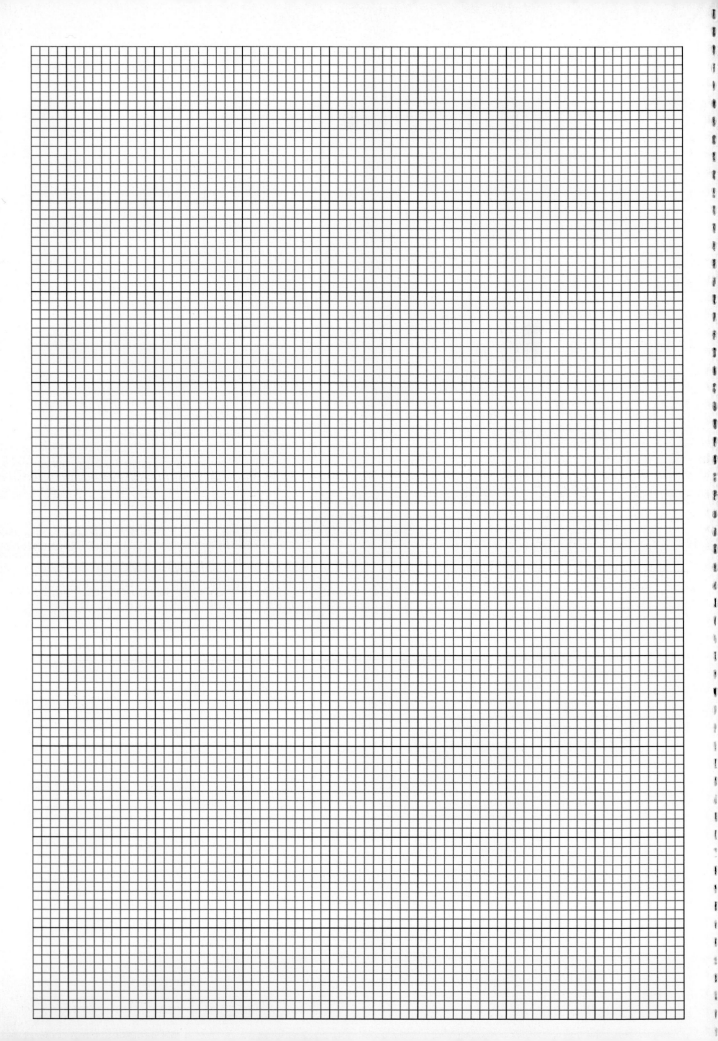

Groundwater: A Resource and Landscape Former

12

Stalactites, stalagmites, and columns in Luray Caverns, Virginia.

- Describe how groundwater infiltrates and flows through Earth materials
- Explain why groundwater erodes and deposits materials differently from streams and glaciers
- Recognize landscapes formed by groundwater and interpret groundwater flow direction from topographic features
- Describe how geologists carry out groundwater resource and pollution studies

MATERIALS NEEDED

- Colored pencils
- Specimens of four different Earth materials, containers, and graduated cylinders for the infiltration exercise

12.1 Introduction

When rain and snow fall on the land, some runs off into streams, some evaporates into the air, and some is absorbed by plants. The remainder sinks into the ground and is called **groundwater**. Groundwater is a vital resource, used throughout the world for drinking and washing, and for irrigating crops. Strict rules govern its use in many places because it flows so slowly that renewal of the groundwater supply takes a long time. In addition, its slow flow and underground location make it difficult to purify once it has been polluted.

Water moves much more slowly underground than in streams because it must drip from one pore space to another rather than flow in a channel. Groundwater therefore has much less kinetic energy than stream water and cannot carry the particles with which streams scrape and wear away at bedrock. Instead, groundwater erodes *chemically*, by dissolving soluble rocks such as limestone, dolostone, and marble, which are largely made of carbonate minerals like calcite and dolomite.

If groundwater erodes underground, how can it form landscapes at the surface? Unlike all other agents of erosion, it does so from below. As groundwater dissolves rocks underground, it undermines their support of the surface above, and the land may collapse to produce very distinctive landscapes, often pockmarked with cavities called **sinkholes** (**FIG. 12.1A**). Landscapes in areas of extreme groundwater erosion are among the most striking in the world, with narrow, steep-sided towers unlike anything produced by streams or glaciers (**FIG. 12.1B**).

But before exploring how we make use of this vital resource or how it creates new landscapes, we first need to understand how water sinks into and flows through the ground. Two concepts are particularly relevant here: porosity and permeability. **Porosity** refers to the volume of empty space, or pore space, a material has. **Permeability** refers to the degree to which those pores are connected. Exercise 12.1 explores how these and other factors control the movement of water underground from the surface and from point to point below ground.

FIGURE 12.1 Landscapes carved by groundwater.

(a) Karst topography, characterized by sinkholes and underground caves, in southern Indiana.

(b) Karst towers in the Guilin area of southern China.

Name: _____ Section: _____
Course: _____ Date: _____

Let's begin with a look at factors that control the flow of water underground. Your instructor will provide four containers with screened compartments filled with Earth materials that have different grain shapes and sizes, as in the figure below. (*Note:* If this exercise is available and being conducted online, the setup, materials, and directions may differ. See the instructions in the online lab for how to conduct this exercise.)

(a) If equal amounts of water were poured into each of the four containers, which material do you think would permit the greatest amount of water to pass through the screened compartment to the graduated cylinder? Rank your predictions here:

Greatest amount of water   *Smallest amount of water*

_____   _____   _____   _____

In a few sentences, explain your reasons for these choices.

_____

_____

_____

(b) Which material would transmit water fastest? Slowest? Explain.

Fastest water transmission   *Slowest water transmission*

_____   _____   _____   _____

In a few sentences, explain your reasons for these choices.

_____

_____

_____

(continued)

EXERCISE 12.1 **Factors Affecting Infiltration and Groundwater Flow *(continued)***

Name: ______________________ **Section:** __________

Course: ______________________ **Date:** __________

In the top section of the table below, describe the indicated properties of the contents of each container. Now check the predictions you have just made: Pour equal amounts of water into the four containers and measure the amount that passes through each container every 30 seconds. Then calculate the rates of flow. Record your observations in the table.

	Observation	Container A	Container B	Container C	Container D
Properties of material that might affect water flow	Grain size				
	Sorting				
	Grain shape				
	Porosity				
	Permeability				
Volume of water transmitted (ml)	30 seconds				
	60 seconds				
	90 seconds				
	120 seconds				
	150 second				
	180 seconds				
	210 seconds				
	240 seconds				
Other observations	Average rate of flow (total volume transmitted ÷ total time for transmission)				
	Water color				
	Amount of water retained				

(continued)

Name: _____   Section: _____

Course: _____   Date: _____

(c) Which materials transmitted the most water? Which retained the most water?

_____

_____

_____

_____

(d) What properties of the materials are correlated with good water transmission? Which are correlated with water retention?

_____

_____

_____

_____

12.2 Aquifers and Aquitards

Materials that transmit water well are called **aquifers,** from the Latin *aqua* (water) and *fero* (to carry). Being porous is not necessarily enough for a material to be a good aquifer. Scoria, for example, is very porous, but its pores are not connected—meaning it's not very permeable. Pore spaces must be connected for water to move from one to another. Materials that do not transmit water are called **aquitards** because they retard water, or **aquicludes** because they exclude water from passing through (**FIG. 12.2**). An *unconfined aquifer* starts at the ground surface and extends downward, whereas a *confined aquifer* is separated from the ground surface by an aquitard.

FIGURE 12.2 Aquifers and aquitards.

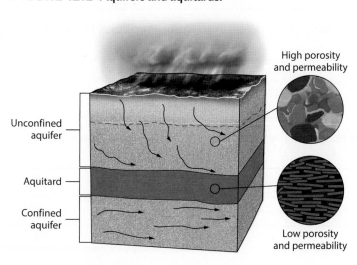

The materials in Exercise 12.1 that transmitted water easily had to be both porous (there was room for the water between grains) *and* permeable (the water could move through the material). *An aquifer must be both porous and permeable.* Exercise 12.2 explores these concepts.

EXERCISE 12.2	The Difference between Porosity and Permeability

Name: _____   Section: _____
Course: _____   Date: _____

Are all porous rocks aquifers? Hold pieces of highly porous pumice and sandstone above two beakers, or rest them on the rims as shown in the following figure. Using a water dropper, slowly add water to the top of each rock and carefully observe how much water passes into the beaker. (*Note:* If this exercise is available and being conducted online, the setup, materials, and directions may differ. See the instructions in the online lab for how to conduct this exercise.)

Porosity and permeability in pumice and sandstone.

(a) Are pumice and sandstone equally permeable? Explain.

_____

_____

_____

(b) Is a material that is similar to sandstone more likely to function as an aquifer or an aquitard? Explain your answer.

_____

_____

_____

(c) Is a material that is similar to pumice more likely to function as an aquifer or an aquitard? Explain your answer.

_____

_____

_____

Groundwater's slow pore-to-pore movement helps chemical erosion because the longer water sits in contact with minerals, the more it can dissolve them. Similarly, even a few drops of water placed on a sugar cube will eventually dissolve their way through the cube. Unfortunately, chemical erosion can cause dangerous health problems if humans or animals drink water in which poisonous material is dissolved. For example, when groundwater wells were drilled in Bangladesh to provide safer drinking water than that from streams contaminated with bacteria, the groundwater turned out to contain high concentrations of arsenic. An international effort is under way to identify the source of the arsenic and devise methods for removing it from the groundwater.

12.3 The Water Table and Land Subsidence

Gravity pulls groundwater downward until it reaches an aquitard that it cannot penetrate, and the water begins to fill pores in the aquifer just above the aquitard. More water percolating downward finds pores at the bottom of the aquifer already filled, and so it saturates those even higher. The **water table** is the boundary between the **zone of saturation** (or the **saturated zone**), where all pore spaces are completely filled with water, and the **zone of aeration** (or the **unsaturated zone**), where some pores are partly filled with air (**FIG. 12.3**). You can model the water table by adding water to a beaker filled with sand and watching the level of saturation change.

The water table generally mimics the surface topography, being closest to the surface under hills and lower under valleys (**FIG. 12.4A**). Its position changes through natural processes and human activity. The water table drops naturally during times of drought because there isn't enough rain to recharge the groundwater supply. During wet periods, additional water causes the water table to rise (**FIG. 12.4B, C**). Both changes can cause problems for those living in an area where the water table is normally close to the surface (see Exercise 12.3).

FIGURE 12.3 **The water table separates the zone of saturation and the zone of aeration.**

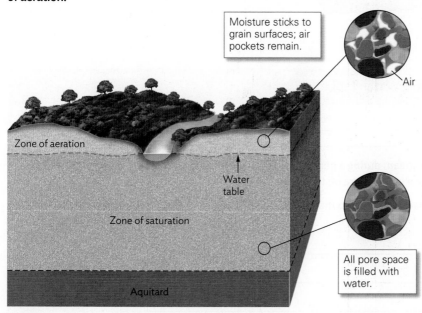

Moisture sticks to grain surfaces; air pockets remain.

Air

Zone of aeration

Water table

Zone of saturation

All pore space is filled with water.

Aquitard

A stream or lake forms where the water table intersects the land surface.

FIGURE 12.4 The position of the water table.

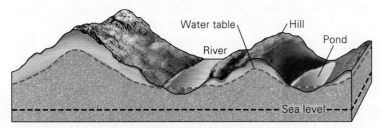

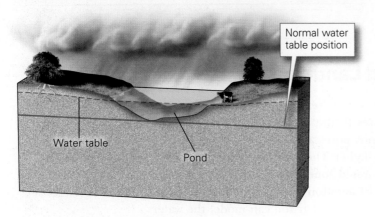

(a) The shape of a water table beneath hilly topography.

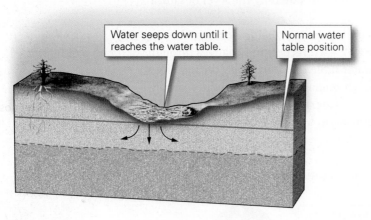

(b) Where the water table intersects the ground surface, ponds remain filled—the surface of the pond represents the water table.

(c) In dry regions and during dry seasons, the water table drops beneath the surface. So, water that had formed ponds sinks into the expanded zone of aeration and the ponds dry up.

EXERCISE 12.3	Effects of a Changing Water Table

Name: _____   **Section:** _____

Course: _____   **Date:** _____

A family builds a home on a small rise near a stream during a period of normal rainfall and gets its water from a well drilled into the zone of saturation (like the one shown in the figure on the next page).

(a) During an extended drought, the water table will drop to the level shown by the blue line in the figure. What effect will this have on the local pond and stream?

_____

_____

_____

(continued)

EXERCISE 12.3 Effects of a Changing Water Table *(continued)*

Name: ______________________________ Section: ____________
Course: ______________________________ Date: ____________

(b) How will the change in water table elevation affect the family?

____

____

____

(c) What do you suggest they could do to remedy this problem?

____

____

____

Historical records show that during abnormally wet years, the water table has risen as high as the elevation shown by the yellow line in the figure below.

Relationship between the water table and topography.

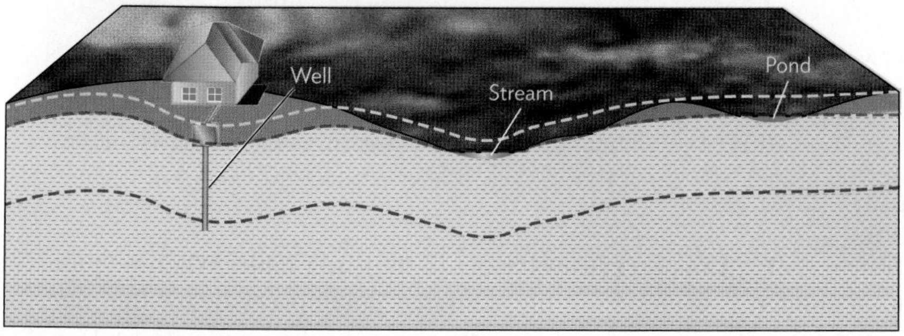

The dashed lines show positions of the water table under different circumstances: red = normal water table position; blue = position during drought; yellow = position during rainy periods.

(d) What effect will this have on the stream and pond?

____

____

____

(e) What effect will this have on the family's home?

____

____

____

(f) What do you suggest they could do to address this problem?

____

____

____

FIGURE 12.5 The impact from a cone of depression.

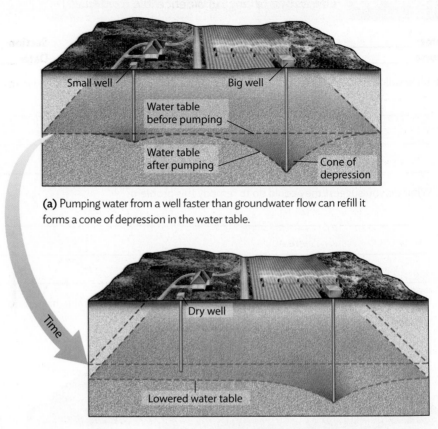

(a) Pumping water from a well faster than groundwater flow can refill it forms a cone of depression in the water table.

(b) Pumping by the big well may be enough to make the small well run dry.

Human activity can also cause changes in the water table. When we pump water from an aquifer, it comes from the saturated pore spaces below the water table, and water must flow downward from other parts of the aquifer to replace it. This lowers the water table, especially in the area immediately around the well, creating a **cone of depression** in the water table (**FIG. 12.5A**). You create a cone of depression every time you drink a thick milkshake from your local fast-food restaurant.

FIGURE 12.6 Land subsidence in Baytown, Texas.

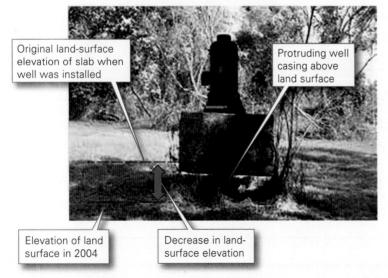

Next time, take the top off the shake and insert the straw just below the surface of the shake. Sip slowly and look at what happens to the surface. Because the shake flows sluggishly (not unlike groundwater), the area below the straw is drawn down more than the area near the edges of the cup. You've just made a small cone of depression. A cone of depression can form when water is pumped out of a well too fast; when this is done in a large well, it can cause shallower wells nearby to run dry (**FIG. 12.5B**).

Removing water from the ground also has the consequence of taking away the water pressure that supports the aquifer sediments. Without the supporting pressure of the groundwater, the sediment particles collapse together, causing the land surface to lower; this is referred to as **land subsidence** (**FIG. 12.6**).

Name: _____   Section: _____

Course: _____   Date: _____

Starting around the 1880s, areas of southeastern Texas, such as those shown in the figure at right, began extracting groundwater for public use and irrigation. As the population of the Houston–Galveston area increased, and with it the need for irrigation and the development of industry, groundwater remained the area's main water source. However, as the demand on the underlying aquifer systems increased, the sediments in the aquifers collapsed, resulting in the lowering of the land in these areas. By 1979, the land elevation in the area had subsided by as much as 3 meters (~10 ft)! Recognizing the effects that the increasing rates of groundwater pumping had on the land surface, the Texas legislature established the Houston–Galveston Subsidence District (HGSD), whose task was to regulate and reduce the withdrawal of groundwater. The dark lines on the map are the county lines. The light red lines on the map show five of the designated subsidence districts. In this exercise, you will study the use of groundwater in the Houston area and its consequences for the people who live there.

Map showing southwestern Texas and Harris County (Houston area).

Study the bar graph below that shows groundwater use in the Harris-Galveston Counties Regulatory Area 1 (near Houston, TX) for the years 1976 to 2016.

Groundwater withdrawals for the Harris-Galveston Subsidence District (Regulatory) Area 1 (shaded yellow below), grouped by use. Years of record are along the horizontal axis, groundwater withdrawal rates on the vertical axis.

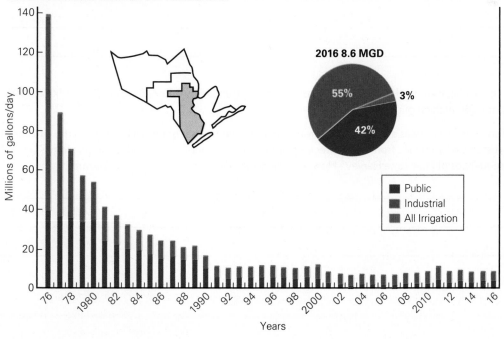

2016 8.6 MGD

55%   3%   42%

Public
Industrial
All Irrigation

Millions of gallons/day

Years

(continued)

Name: _____   Section: _____

Course: _____   Date: _____

(a) For what purpose was most of the groundwater used in the Harris-Galveston Regulatory Area 1 in the last decade?

_____

_____

(b) Since 1976, which uses of groundwater have declined the most?

_____

_____

Which use has decreased the least, or stayed about the same? What trend has this use exhibited over the last few years? Why might this be?

_____

_____

_____

_____

(c) What is the overall decrease in groundwater removal between 1976 and 2016, in millions of gallons per day?

_____

(d) Based on your answers above, do you think that ground subsidence has completely stopped in this area? Why or why not?

_____

_____

_____

_____

_____

In 2017, Hurricane Harvey moved over Texas and Louisiana and, trapped by other atmospheric conditions, stalled, dumping over 240 mm (60 inches) of rain in some areas.

(e) While Hurricane Harvey was a particularly extreme weather event, what potential hazards do the citizens of Harris and Galveston Counties face from heavy precipitation events in general, if the land is still subsiding? (*Hint:* Look up the consequences of Hurricane Harvey.) How does land subsidence contribute to this hazard?

_____

_____

_____

_____

12.4 Groundwater Flow

The lowering of the water table, either through natural processes of the hydrologic cycle or through human activities, can cause a change in the direction of groundwater flow to occur. Pollutants that are in the ground may initially not enter a well if the well is not in the direction of flow from the pollutants. But if the flow direction is changed, such as may result from the formation of a cone of depression, then the pollutants might now enter the groundwater, flow with the groundwater, and contaminate a well. Whether it is for the purpose of determining the source of groundwater pollution, or to be able to drill a sufficiently deep well, we need to know the depth of the water table and groundwater flow direction in an area.

FIGURE 12.7 shows groundwater flow paths. Groundwater flows from recharge areas to discharge areas, typically following curving concave-up paths. The weight of the water in an open (unconfined) aquifer exerts pressure. This pressure provides potential energy to drive the flow of a given amount of groundwater; this energy is known as the *hydraulic head*. To measure hydraulic head at a specific point, scientists drill a vertical hole down to the point and insert a pipe. They then measure the height to which water rises in the pipe relative to a reference elevation (such as sea level)—this represents the hydraulic head. In confined aquifers, the water is under higher pressure than in an unconfined aquifer and so there is more potential energy to push the water toward the land surface and up through an *artesian well* (FIG. 12.8). An artesian well is one that doesn't require a pump to bring water to the surface, relying instead on the pressure in the aquifer to bring the water up the well to the surface. Groundwater tends to flow from higher to lower pressure areas, and in the most direct path.

FIGURE 12.7 **The flow of groundwater.**

FIGURE 12.8 **The configuration of a regional artesian system.** Water rises above the confined aquifer because the water is under pressure.

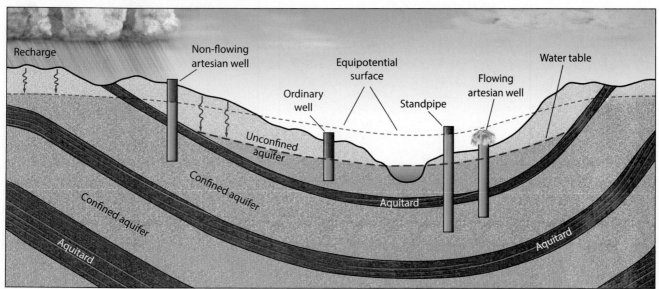

We can also map the water table surface in an area, as viewed from above, which allows us to determine groundwater flow direction. We can do so by taking measurements of the depth from the ground surface to the water table (or the hydraulic head), then plot these values on a map of the area and create contour maps of the water table—just like a topographic map. Just like water on the land surface, groundwater also flows "downhill." But remember that underground, these "hills" are the surface of the water table, and groundwater flows, or flow lines, are perpendicular to the water table contours (FIG. 12.9). In Exercises 12.5 and 12.6, we will investigate environmental issues with groundwater as a resource by using contour maps of the water table and determining groundwater flow direction.

FIGURE 12.9 **Example of water flow perpendicular to contours.**

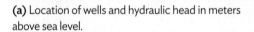

(a) Location of wells and hydraulic head in meters above sea level.

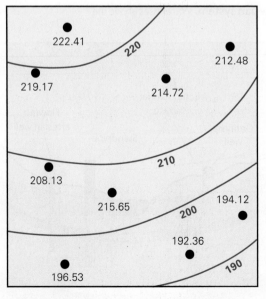

(b) Location of wells in (a) with water contours set at 10 meter intervals.

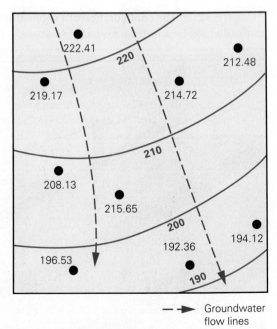

- - - → Groundwater flow lines

(c) Groundwater flow lines shown perpendicularly crossing water contours.

Name: _____   Section: _____
Course: _____   Date: _____

Imagine that there are three houses in a neighborhood, A, B, and C, as shown in the figure below. Each house has its own underground septic tank and a well. House A is newly built and the owners have decided to place their well as shown in the figure.

The location of three houses, their wells, and septic tanks in relation to water table contours (in meters above sea level).

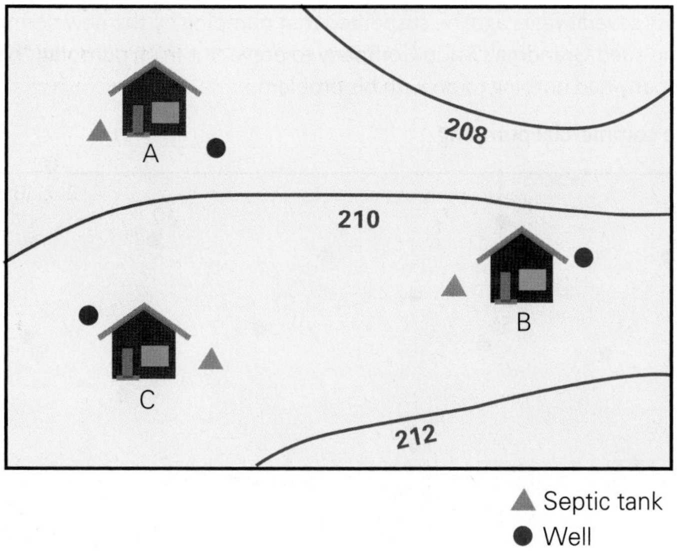

▲ Septic tank
● Well

(a) The owners of House A are concerned that their well could be contaminated with dissolved matter and bacteria if any of the neighborhood septic tanks leaked. Which house's septic tank is most likely to contaminate A's water well? Support your answer with evidence and explain.

_____

_____

_____

_____

(b) Assume that the aquifer is contained in gravel beds, through which groundwater has an average flow rate of 25 cm (10 in) per day. If a leaky septic tank is located 0.25 km (0.16 mi) up the groundwater flow path from A's well, how long would it take for dissolved matter to reach A's water well? Show your work.

_____

_____

(c) Suggest an alternate location for A's well and support your choice with evidence from the figure.

_____

_____

_____

_____

Name: _____ Section: _____
Course: _____ Date: _____

This exercise presents a real-life problem, although details have been changed to protect the privacy of the parties involved.

The following maps show an area in which farmers have relied on groundwater for generations for both home use and to irrigate their crops. In an attempt to diversify the local economy, the town council offered tax incentives to bring two new industries into the area: Grandma's Soup Company and the Queens Sand and Gravel Quarry. Trouble arose when Farmer Jones (see location on map) found that his well had run dry even though it was the rainy season. Having taken Geology 101 at the state university several years ago, he suspected that pumping by the new companies had lowered the water table. He hired a lawyer and sued Grandma's Soup Company to prevent it from pumping "his" water. Grandma's lawyers responded that the company had nothing to do with his problem.

Depth to the water table before commercial pumping.

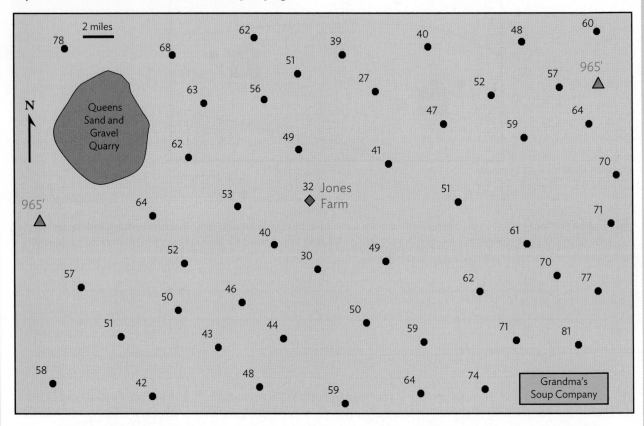

The judge, who has never taken a geology class, has appointed *you* to resolve the problem. Your tasks are (1) to find out why Farmer Jones has lost his water, and (2) to figure out how to keep everyone happy. After all, Jones and his friends have been benefiting from the fact that Grandma's Soup Company has been paying an enormous amount of taxes.

Your assistants have already gathered a lot of useful data. First, a survey shows that the land is as flat as a pancake, with an elevation of 965 feet. Second, there are records that show the depth to the water table in wells throughout the area *before* the companies opened (above map) and *after* the two companies had been pumping for a year (map on the next page).

(continued)

Name: _____   Section: _____

Course: _____   Date: _____

Depth to the water table after commercial pumping.

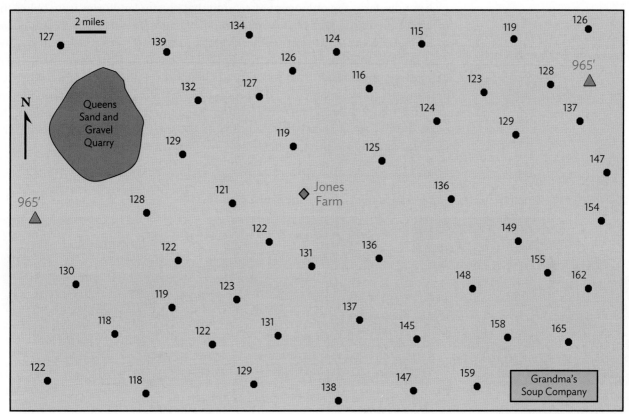

? **What Do You Think** Use these data to prepare a report for the judge that answers the following questions:

- Why has Farmer Jones's well run dry?
- What steps do you recommend to keep the Jones Farm *and* the two new industries in business?

Hint 1: Use your contouring skills to make a topographic map of the water table before and after the quarry and soup company began pumping.

Hint 2: Consider what the quarry and the soup company do with the water they pump.

12.5 Landscapes Produced by Groundwater

Groundwater-eroded landscapes are called **karst topography**, after the area in Slovenia and northeastern Italy where geologists first studied such landscapes in detail, but spectacular examples are also found in Indiana, Kentucky, Florida, southern China, Puerto Rico, and Jamaica.

Groundwater is the only agent of erosion that creates landforms from below the surface of the Earth, and all karst topographic features are *destructional* (produced by erosion). Caves and caverns, the largest groundwater erosional features, are underground, unseen, and often unsuspected at the surface. In some areas, extensive cave networks result from widespread chemical erosion. Sinkholes form when these cavities grow so large that the ground above them collapses because there is not enough rock to support it (**FIG. 12.10A**). **Karst towers** are what remain when the rock around them has been dissolved (**FIG. 12.10B**). **Karst valleys** form when several sinkholes develop along an elongate fracture (**FIG. 12.10C**).

FIGURE 12.10 Groundwater landscape erosion from beneath the Earth's surface.

Groundwater dissolves limestone, forming caves.

Caves grow, decreasing support of ground surface.

Ground collapses, forming steep-sided sinkholes.

(a) Origin of sinkholes by dissolution and collapse.

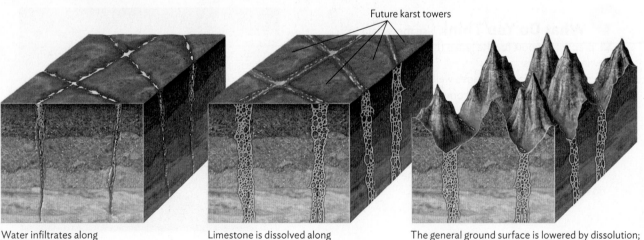

Future karst towers

Water infiltrates along intersecting fractures.

Limestone is dissolved along fractures.

The general ground surface is lowered by dissolution; limestone remaining between fractures stands above surface as karst towers.

(b) Origin of karst towers as erosional remnants.

FIGURE 12.10 Groundwater landscape erosion from beneath the Earth's surface. *(continued)*

(c) Origin of karst valleys along fractures

Unlike streams and glaciers, groundwater does not produce landforms by deposition. Groundwater deposition occurs in caverns when drops of water evaporate, leaving behind a tiny residue of calcite—sometimes on the roof of the cavern, sometimes on the floor. These deposits build up slowly over time to produce stalactites and stalagmites, respectively (**FIG. 12.11**). In some instances, a stalactite and stalagmite may grow together to form a column.

Groundwater resource quantity (whether there is enough groundwater to meet our needs) and quality (whether the groundwater is pure enough for us to drink or whether it has become contaminated through natural or human causes) are not the only groundwater issues that humans face. In Exercise 12.7 we'll learn about problems caused by groundwater erosion in populated karst areas.

FIGURE 12.11 Groundwater deposition in caverns.

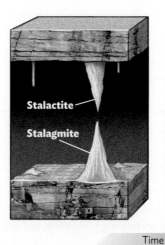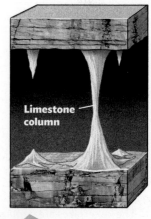

Stalactites grow *downward* from the roof; stalagmites grow *upward* from the floor. Columns form when stalactites and stalagmites merge.

Name: _____ Section: _____
Course: _____ Date: _____

Examine **FIGURE 12.12**, a topographic map of the Interlachen, FL, area. This increasingly populous region is underlain by limestone and exhibits classic karst features.

(a) What karst features are outlined by the concentric hachured contour lines?

_____

_____

(b) Explain why some of these features are dry whereas others are the sites of lakes.

_____

_____

_____

_____

(c) Interlachen's population has grown rapidly in the past 20 years, and a part of its suburban street grid can be seen in Figure 12.12. What special problems face someone planning to build in this karst region? Would you buy a house here?

_____

_____

_____

_____

_____

_____

(d) As the population has grown, some of the lakes have shrunk dramatically, and some have dried up entirely. Suggest possible reasons for these changes.

_____

_____

_____

_____

The larger lakes are popular for boating and fishing, but residents fear what might happen if one lake became polluted. Like streams, groundwater flows downhill, so pollutants would migrate from one lake to another. For example, a gasoline spill on any of the lakes along line A–B in Figure 12.12 could affect the other lakes. The topographic map doesn't provide enough detail to tell us the direction in which the pollutants would flow, so we'll use more modern methods. Using Google Earth, search for Lake Grandin, FL, and view the chain of lakes along line A–B. Move the cursor to each lake and record the elevation of its surface.

(e) Draw arrows on Figure 12.12 using different colored pencils to show the direction in which pollutants would flow from each of the lakes.

FIGURE 12.12 Karst topography in the Interlachen area of Florida.

0   0.25   0.5 mile

0   0.25   0.5 km

Contour interval = 10 feet

N

Boyds Lake

Lake Grandin

Long Pond

100

150

100

100

150

100

CO ROAD 315

B

Clearwater Lake

100

Silver Lake

100

150

Flamingo Lake

100

Hart Lake

Junior Lake

150

100

Perch Lake

Trout Lake

150

100

Lovers Lake

Hubbard Pond

Mariner Lake

100

Sand Hill Pond

100

A

150

Grassy Lake

Name: _____   **Section:** _____
Course: _____   **Date:** _____

Exploring Earth Science Using Google Earth

1. Visit **digital.wwnorton.com/labmanualearthsci**
2. Go to the **Geotours** tile to download Google Earth Pro and the accompanying Geotours exercises file.

Expand the Geotour12 folder in Google Earth by clicking the triangle to the left of the folder icon.

The folder highlights features common in karst landscapes where landforms develop due to dissolution of limestone by groundwater.

(a) Check and double-click the Bosnia and Herzegovina placemark to fly to a region in Europe where karst landforms were first described. What type of karst landforms dot this plain?

_____

(b) Check the OrleansIN_TopoMap overlay and then check and double-click the Feature placemark to fly to a similar karst terrain near Orleans, IN. The Feature placemark highlights the same kind of landform as shown in (a). Using the topographic map overlay, describe the topographic characteristics of this landform.

_____

_____

(c) Check and double-click the semi-transparent Feature Area polygon and note that many more of these same features are concentrated throughout this part of the map area. Turn the Feature Area polygon off, and check and double-click the Depth placemark to fly to an example near the Lost River. Approximately how deep (ft) is this feature? (*Note:* Leave the topographic map on.) *Hint:* Subtract the elevation of the center of the feature from the elevation of the edge.

_____

(d) Check and double-click the Spring placemark. From near the Depth placemark to the Spring placemark, the Lost River changes from a perennial stream to an intermittent stream that only has water in it during wet periods. West of the Spring placemark, the Lost River resumes flowing as a perennial stream at the surface. What happens in this interval that causes the Lost River to be "lost"? Toggle the topographic map on and off to view the area.

_____

_____

(e) Check and double-click the Dry Valley placemark to fly to a dry valley where the intermittent stream flow is diverted underground. What kind of stream is this? (*Note:* Leave the topographic map on.)

_____

Arid and Glacial Landscapes

13

Earth's extreme environments produce a variety of unique landforms, such as the glacier and sand dunes pictured here.

13.1 Introduction

In this chapter, we explore two of the most extreme environments on Earth: those that are very dry and those that are very cold. Sometimes these conditions exist in the same place! Arid regions are defined as those that receive less than 25 cm (10 in) of precipitation (rain, snow, etc.) per year, which provides very little water for chemical and physical weathering or for stream and groundwater erosion. This is especially true in very arid regions called **deserts**, which typically contain no permanent streams. Weathering, mass wasting, and wind and stream erosion are the major factors in developing arid landscapes, but because these processes are operating in very different proportions than in humid regions, their results are strikingly different from those in temperate and tropical landscapes (**FIG. 13.1**).

Movies and television portray arid regions with camels resting at oases amid mountainous sand dunes. Some arid regions, such as parts of the Sahara and Mojave

FIGURE 13.1 Arid landforms from the southwestern United States.

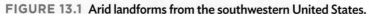

(a) Erosional remnants in Monument Valley, Arizona.

(b) Hoodoo panorama in Bryce Canyon, Utah.

(c) Delicate Arch in Arches National Park, Utah.

deserts, do look like that (minus camels in the Mojave). But some arid regions actually have more rock than sand, whereas others, surprisingly, lie next to shorelines, where there is no shortage of water. Even the North and South Poles rank among the Earth's most arid regions, as the air is so cold that it holds almost no moisture (**FIG. 13.2**). Although the polar regions are arid, they are also so cold that whatever snow falls there does not melt, allowing it to accumulate over thousands of years to form enormous continental glaciers. We will explore all glaciers in more detail in the second half of the chapter.

We will start by examining arid landscapes. Keeping the recently introduced characteristics of arid landscapes in mind, you will compare arid and humid landscapes in Exercise 13.1.

FIGURE 13.2 Types of arid regions.

(a) Landscape in Death Valley, in the Mojave Desert, California.

(b) Polar desert, Norway.

(c) Rocky desert, Utah.

Name: _____   Section: _____
Course: _____   Date: _____

The four photographs below show two typical arid landscapes (top) and two typical humid landscapes (bottom). Examine the photos and answer the following questions:

(a) In which areas are the landforms more angular? In which are they more rounded? Suggest an explanation for the difference.

_____

_____

(b) In which areas do you think landscape changes are more rapid? Are slower? Explain your answer.

_____

_____

(c) What other differences do you see between the arid and humid regions?

_____

_____

(d) What type of erosion appears to be dominant in the arid regions? In the humid regions? How is this possible?

_____

_____

Canyonlands National Park, Utah.

Grand Canyon, Arizona.

Delaware Water Gap, New Jersey–Pennsylvania.

Humid regions in Utah.

13.2 Processes and Landscapes in Arid Regions

The scarcity of water in arid regions causes them to have landscapes very different from those in humid regions. First, features tend to survive longer in arid areas than in humid ones because weathering, erosion, and deposition are slowed by the scarcity of water, which is needed to abrade, dissolve, and carry debris away. In the absence of water, soluble rocks such as limestone, dolostone, and marble—easily weathered and eroded in humid regions to form valleys—instead form ridges.

Second, while both physical and chemical weathering occur in arid regions, the dominant processes are usually different from those in humid areas. These unique processes cause the evolution of arid landscapes to differ from that of landscapes in humid regions. FIGURE 13.3 shows the stages in the development of an arid landscape, starting with a block made of resistant and nonresistant rocks that has been uplifted by faulting (Fig. 13.3a). Physical weathering takes advantage of fractures, causing cliffs to retreat and forming piles of rubble, called **talus**, on the plain below them (Fig. 13.3b). Streams in arid regions flow only after the few storms that occur during the rainy season, but they are still very effective agents of erosion. They erode materials from the highlands and redeposit them as **alluvial fans**, which begin to fill in the lowlands (Fig. 13.3c). With further deposition, alluvial fans coalesce to form a broad sand and gravel deposit (a **bajada**) flanking the uplifted block.

Where several fault blocks and basins are present, bajada sediment from neighboring blocks eventually fills in the intervening basins, burying all but a few remnants of the original blocks (Fig. 13.3d). These remnants stand out as isolated peaks, called **inselbergs**, above the sediment fill. Streams flowing into the basin from surrounding blocks have no way to leave and so form **playa lakes**, which may evaporate to form **salt flats**. Monument Valley (Fig. 13.3e) is a classic example of remnants of resistant rock isolated from their source layers by erosion.

Arid landscapes are typically more angular than humid landscapes. Compare, for example, the photos of the Grand Canyon and the Delaware Water Gap in Exercise 13.1. In the Grand Canyon, there is little soil. Rockfall is the dominant process of mass wasting, and sharp, angular slope breaks indicate contacts between resistant and nonresistant rocks. Nearly vertical slopes characterize the resistant sandstones, and more gently sloping surfaces characterize the less resistant shales. In contrast, a well-developed soil has formed above the bedrock in the Delaware Water Gap, partially masking erosional differences in the underlying rocks. Soil creep is the dominant mass wasting process, and it smooths and rounds the topography.

EXERCISE 13.2	Interpreting Arid Landscapes

Name: _____ Section: _____

Course: _____ Date: _____

In this exercise, you will practice interpreting aspects of several different arid landscapes.

Grand Canyon, Arizona

(a) Why hasn't soil creep smoothed the slopes of the Grand Canyon as pictured in this photo?

_____

_____

_____

_____

(continued)

FIGURE 13.3 Stages in the evolution of an arid landscape.

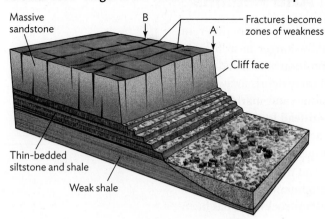

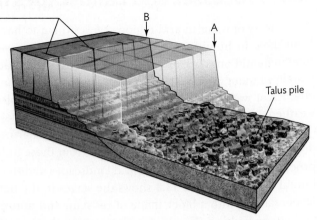

(a) An uplifted block contains resistant sandstone and more easily eroded shales.

(b) The cliff retreats from A to B as underlying rocks weather physically. Rockfall produces talus piles at the base of the cliff.

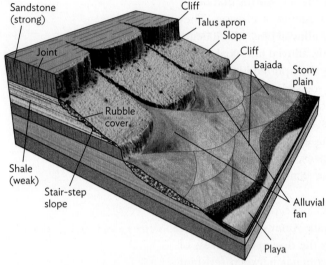

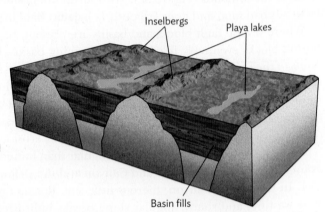

(c) An apron of debris covers the lowlands and protects the bedrock slope from further erosion.

(d) A late stage of landscape development involving several uplifted blocks.

(e) Buttes in Monument Valley, Arizona. Buttes are small to medium-sized erosional remnants with flat tops. Similar but larger features are called mesas, from the Spanish word for table.

Name: _____   Section: _____
Course: _____   Date: _____

Death Valley, California–Nevada

FIGURE 13.4 is a topographic overview of the Death Valley area. Death Valley is one of the driest places in North America and, with an elevation of more than 250 feet below sea level, one of the lowest spots on the continent. It also contains classic examples of arid landscapes. Examine the map and answer the following questions:

(b) Compare this region with Figure 13.3. Is there a single fault block, as in Figure 13.3a, or several blocks, as in Figure 13.3d? Explain your reasoning.

_____

_____

_____

_____

(c) Sketch in the border fault(s) with a colored pencil on both the map view and the profile in Figure 13.4.

(d) Why is Death Valley lower in elevation than the other major valleys shown in the topographic profile? Consider factors that played roles in forming the valley, such as stream erosion and tectonic activity.

_____

_____

_____

Now examine Death Valley more closely, starting with the Stovepipe Wells area (**FIG. 13.5**).

(e) What are the broad, rounded landforms that project into Death Valley from the mountains to the west and southeast? How did they form?

_____

_____

_____

(f) Several streams flow into Death Valley from these mountains. Where do they go? What is their base level?

_____

_____

(g) A stippled pattern identifies areas of rapidly moving sand dunes. Where did this sand come from?

_____

_____

(h) A more detailed view of the sand areas would show individual dunes, but these would not be accurate representations of the current land surface. Why not?

_____

_____

_____

(continued)

FIGURE 13.4 Topographic overview and profile of Death Valley National Park.

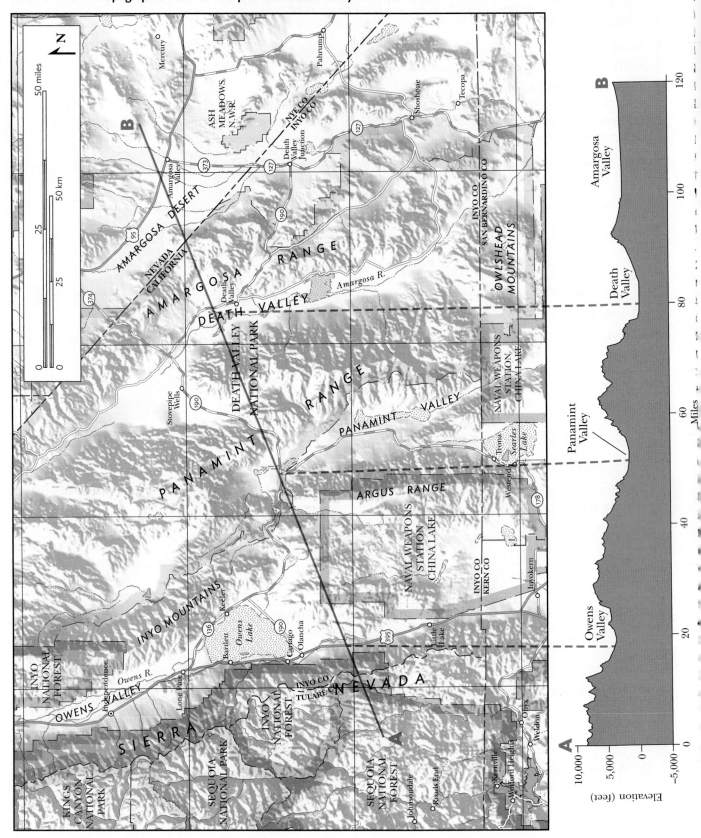

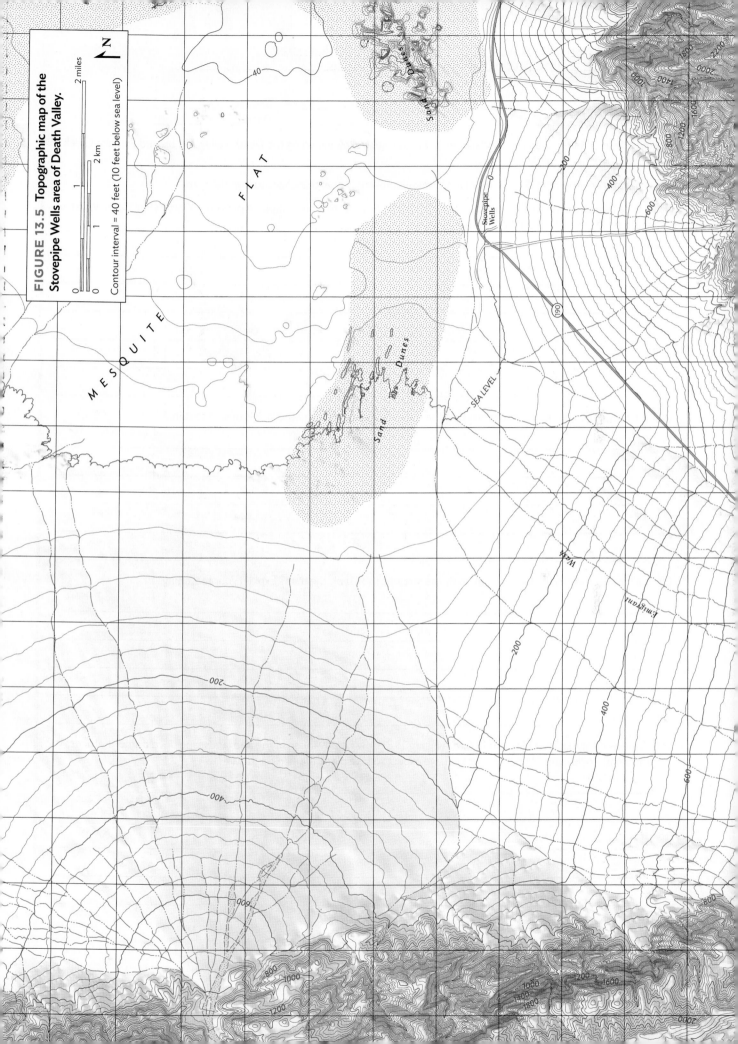

FIGURE 13.5 Topographic map of the Stovepipe Wells area of Death Valley.

Contour interval = 40 feet (10 feet below sea level)

2 miles

2 km

MESQUITE

FLAT

Sand Dunes

Sand Dunes

Stovepipe Wells

190

SEA LEVEL

West

Emigrant

40

0

200

400

600

200

400

600

800

800

1000

1200

1000

1200

1400

1600

1800

2000

200

400

600

800

1000

1200

1400

1600

1800

2000

2200

Name: _____   Section: _____
Course: _____   Date: _____

Examine the map of the area surrounding the town of Death Valley, including the Death Valley Airport and Furnace Creek Inn (**FIG. 13.6**).

(i) What arid environment depositional landforms are present? Label these features on the map.

(j) A large lake that is not present throughout the year is shown with a blue-stippled pattern in Cotton Ball Basin. What is this type of lake called, and how does it form?

_____

_____

_____

_____

_____

_____

(k) What evidence on the map suggests that the water in the streams and the lake is unsafe to drink?

_____

_____

_____

_____

_____

(l) Is all of the water in the lake brought in by the streams shown on the map? Explain your reasoning.

_____

_____

_____

_____

_____

(m) Compare Figures 13.4, 13.5, and 13.6 with Figure 13.3. How far advanced in the arid landscape development process is the Death Valley area? What changes can we expect to take place there in the future?

_____

_____

_____

_____

_____

(continued)

FIGURE 13.6 Topography near the town of Death Valley.

Contour interval = 100 feet

Name: _____   Section: _____

Course: _____   Date: _____

Buckeye Hills, Arizona

The Buckeye Hills, southwest of Phoenix, lie in an arid area similar to the Death Valley region, but the two areas differ in some ways. **FIGURE 13.7** is a topographic overview of the Buckeye Hills area.

(n) Describe the landscape shown in this overview.

_____

_____

_____

(o) How are the Buckeye Hills similar to the Death Valley region shown in Figure 13.4?

_____

_____

_____

(p) How are they different?

_____

_____

_____

Now look at the detailed map of the central part of the Buckeye Hills (**FIG. 13.8**).

(q) Identify three erosional and depositional features of arid landscapes. List them here and label them on the map.

_____

_____

(r) What is the probable origin of the small hills just east of Highway 85 that are pictured on this map?

_____

_____

(s) What evidence is there of current stream activity in the Buckeye Hills?

_____

_____

_____

(t) Refer to the diagram showing the evolution of arid landscapes (Fig. 13.3). Which area, Death Valley or the Buckeye Hills, is in a more advanced state of development? Explain your reasoning.

_____

_____

_____

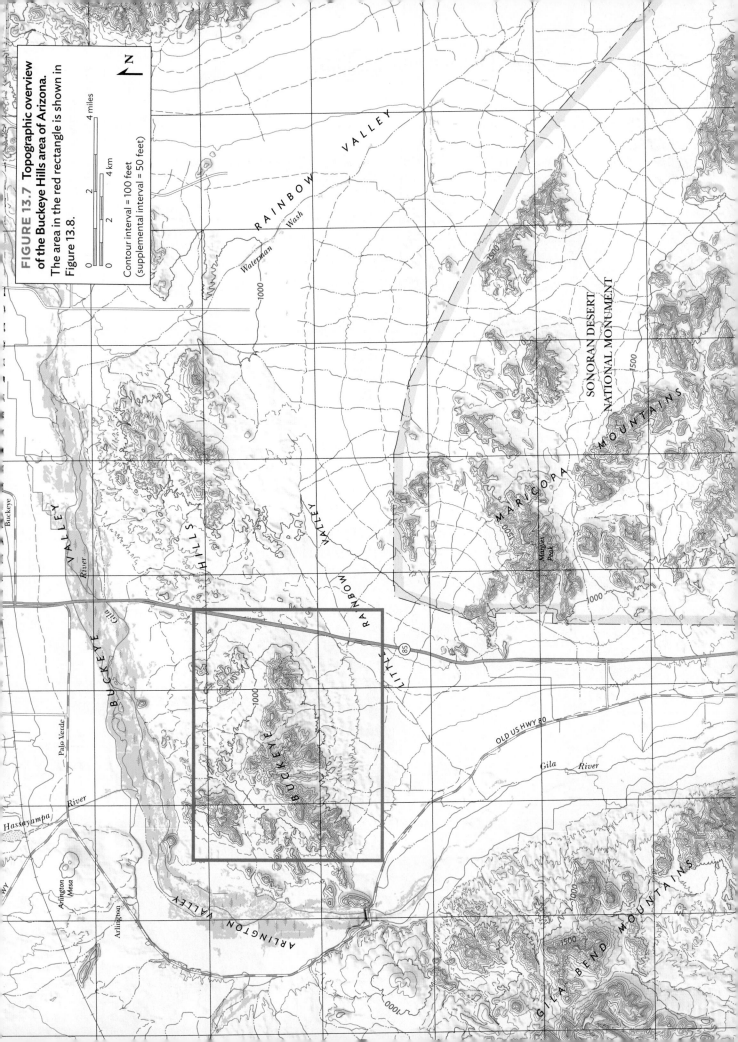

FIGURE 13.7 Topographic overview of the Buckeye Hills area of Arizona. The area in the red rectangle is shown in Figure 13.8.

Contour interval = 100 feet (supplemental interval = 50 feet)

N

4 miles
2
0

4 km
2
0

RAINBOW VALLEY

Waterman Wash

1000

SONORAN DESERT NATIONAL MONUMENT

1500

1500

MARICOPA MOUNTAINS

1500

Margies Peak

1000

BUCKEYE HILLS

RAINBOW VALLEY

LITTLE RAINBOW VALLEY

85

Buckeye

VALLEY

Gila River

BUCKEYE VALLEY

Palo Verde

River

Hassayampa

Arlington Mesa

Arlington

ARLINGTON VALLEY

BUCKEYE VALLEY

1000

OLD US HWY 80

Gila River

GILA BEND MOUNTAINS

1000

1500

1000

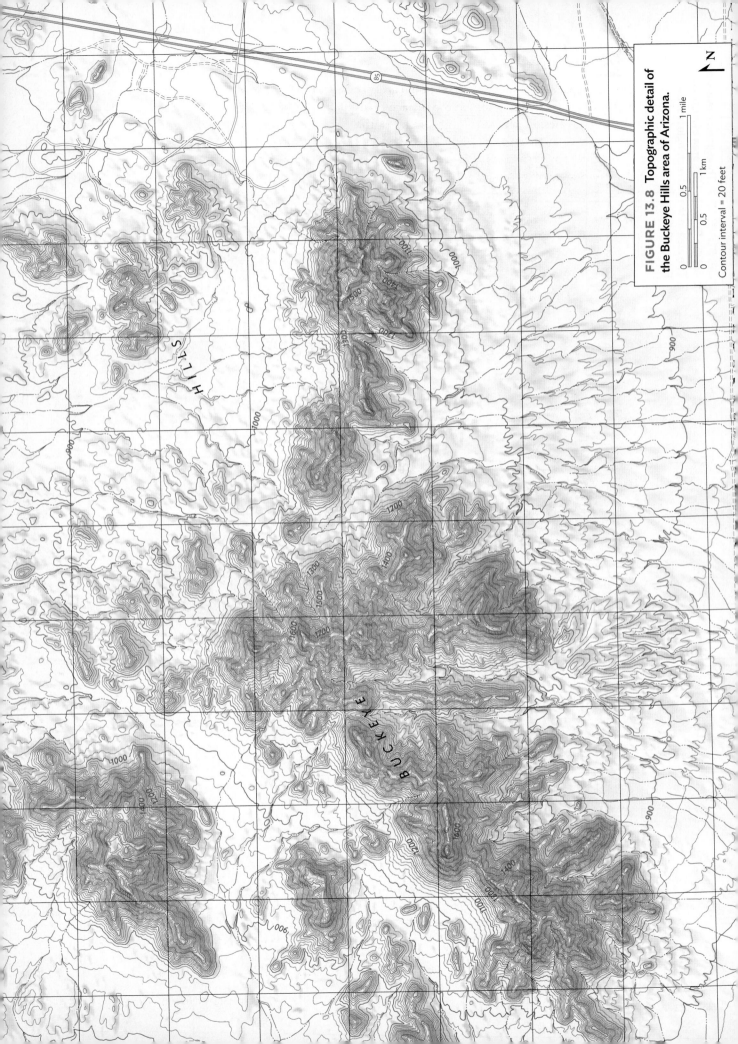

FIGURE 13.8 Topographic detail of the Buckeye Hills area of Arizona.

Contour interval = 20 feet

13.3 Wind and Sand Dunes

Wind has a greater impact in arid areas than in humid areas because, whereas a small amount of moisture between sand grains at the surface holds them together weakly in humid areas, that cohesion is absent in arid regions where such moisture isn't present. There are also fewer trees or other obstacles to block the full force of the wind. Wind erodes the way streams do, using its kinetic energy to pick up loose sand grains and using this sediment to abrade solid rock—literally sandblasting cliffs away.

Dunes are the major depositional landform associated with wind activity in deserts, but they also form along shorelines wherever there is an abundant source of

FIGURE 13.9 **Types of sand dunes. The red arrows indicate wind direction.**

(a) Barchan (constant wind, modest sand supply).

(b) Transverse (constant moderate wind, large sand supply).

(c) Star (multiple wind directions, modest sand supply).

(d) Longitudinal (constant strong wind, large sand supply).

sand. There are several kinds of dunes, depending on the amount of sand available, the strength of the wind, and how constant the wind direction is (**FIG. 13.9**). **FIGURE 13.10A** shows the ripple-marked surface of a large transverse dune from the stippled area on the map of Stovepipe Wells in Death Valley (see Fig. 13.5), and **FIGURE 13.10B** shows how such dunes develop.

FIGURE 13.10 Characteristics and evolution of a transverse dune.

(a) Transverse dunes at Stovepipe Wells in Death Valley. Small ripples may form on the dune surface.

(b) Movement of sand grains in a transverse dune. Sand is moved from the windward dune face to the slip face, causing the dune to migrate downwind.

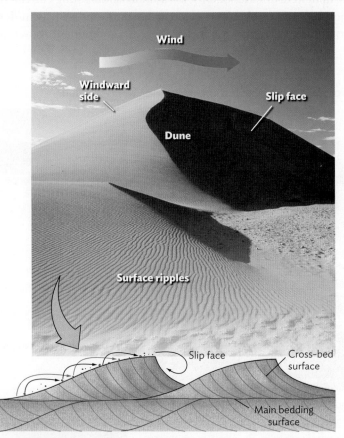

| EXERCISE 13.3 | Interpreting the History of the Nebraska Sand Hills |

Name: _____   Section: _____
Course: _____   Date: _____

The Nebraska Sand Hills are the remnant of a vast mid-continent sea of sand that existed approximately 30,000 years ago. More humid conditions today permit grasses, trees, and shrubs to stabilize the dunes so that they no longer move across the countryside like those in Death Valley or the Sahara, but the hills preserve evidence of the more arid conditions under which they formed. Examine the topographic overview of the Nebraska Sand Hills (**FIG. 13.11**) and answer the following questions:

(a) What evidence is there that this region is still relatively arid?

_____

_____

_____

(continued)

FIGURE 13.11 Topographic overview of part of the Nebraska Sand Hills. The area in the red rectangle is shown in Figure 13.12.

Contour interval = 100 feet

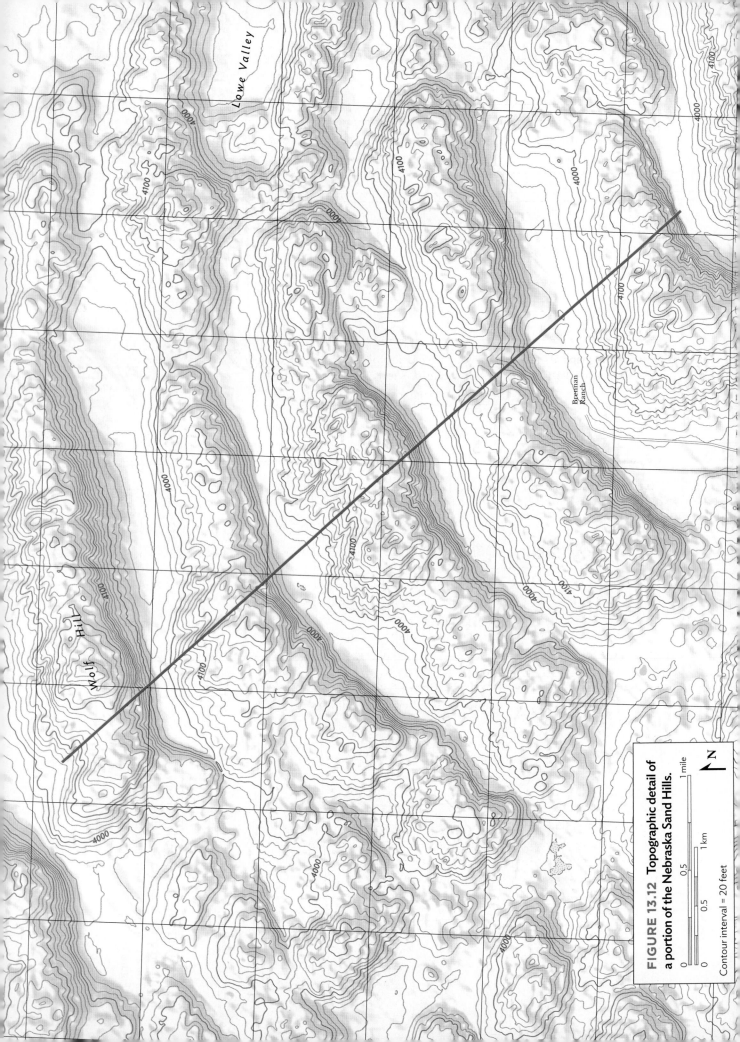

FIGURE 13.12 Topographic detail of a portion of the Nebraska Sand Hills.

Contour interval = 20 feet

Name: _____   Section: _____

Course: _____   Date: _____

(b) Describe the general topographic grain of the area, including the size and shape of the ridges and valleys present.

_____

_____

(c) The ridges are the remnants of the Pleistocene sand sea. What features do they represent?

Examine the detailed map (**FIG. 13.12**) of the area outlined in red in Figure 13.11.

_____

_____

_____

(d) Describe the shape of the ridges. Draw a topographic profile, using the graph paper provided at the end of this chapter, along the line indicated to help answer the next question.

_____

(e) What does the shape of the dunes tell you about the wind during the Pleistocene?

_____

_____

13.4 Desertification

Desertification is a process in which landforms, soils, vegetation, and animal life in an already dry region, change to those characteristic of arid regions. It can be the result of natural processes, as global climate changes and arid zones shift geographically; such processes can cause lakes to disappear and streams to run dry as the water table drops. But human activity has also become a major cause of desertification. For example, unsustainable agricultural practices can deplete soil nutrients to the point that the land is no longer able to support plant life, and irrigation systems in arid areas cause salt concentration in the soil, leading to the same result. Destruction of forests for wood-based resources or to make room for crops directly destroys habitats, but it also promotes soil erosion because replacing forest vegetation with fields that are tilled repeatedly for planting increases rates of soil erosion, eventually leading to less plant cover overall.

Anthropogenic (human-caused) desertification is a global problem. In some developing nations, the short-term need for wood for heating and cooking, coupled with an increasing population, drives the process. Archeologists have shown that desertification is not just a modern phenomenon—Easter Islanders denuded their island hundreds of years ago to the point that there are few trees on the island today. But North America, despite its wealth of energy resources, is also vulnerable to the combination of natural and human-driven desertification (**FIG. 13.13**). Note that some of the most vulnerable areas, such as California and the northern Great Plains, today produce much of the fruit, vegetables, and grains consumed by North Americans.

FIGURE 13.13 Vulnerability to desertification in North America.

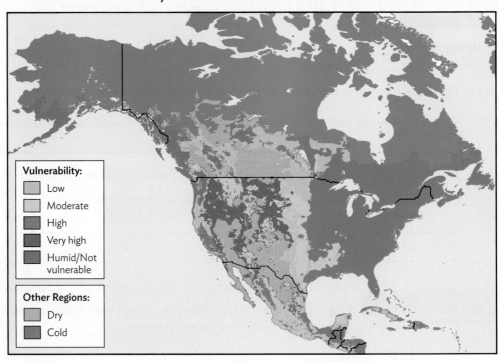

EXERCISE 13.4 Preventing Desertification

Name: _____ **Section:** _____

Course: _____ **Date:** _____

Examine Figure 13.13 and answer the following questions:

(a) Describe the geographic distribution of the areas of North America most vulnerable to desertification.

_____

_____

(b) Explain why half of the United States is vulnerable to desertification, but half is not.

_____

_____

_____

? What Do You Think As the director of your state's Department of Agriculture, you are alarmed by the high risk of desertification for most of your state. On a separate sheet of paper, describe the impacts that desertification would be likely to have on your state's agriculture and food supply—and on those of the United States in general. What strategies would you implement to at least slow the rate of desertification?

FIGURE 13.14 The Grand Teton Mountains in Wyoming display the jagged topography typical of areas affected by mountain glaciers.

FIGURE 13.14 The Grand Teton Mountains in Wyoming display the jagged topography typical of areas affected by mountain glaciers.

13.5 Introduction to Glaciers

Many of the arid landscapes we've looked at so far have been in very hot regions. Let's change gears and now look at some of the coldest places on the planet. Glaciers are broad sheets or narrow "rivers" of ice that last year-round and flow slowly across the land. They hold more than 21,000 times the amount of water in all unfrozen streams and account for about 75% of the Earth's freshwater. Today, huge continental glaciers cover most of one continent (Antarctica) and nearly all of Greenland. Smaller glaciers carve valleys on the slopes of high mountains, causing a characteristically sharp, jagged topography (FIG. 13.14). Distinctive landforms show that continental glaciers were even more widespread during the Pleistocene Epoch—the so-called ice ages from 700,000 to 12,000 years ago—when they covered much of northern Europe and North America. The Pleistocene Epoch is not the only ice age in Earth's history; glaciations happened in colder intervals of earlier eras, too.

Today, glaciers are receding at rates unprecedented in human history. As they shrink, the balance of the hydrologic cycle among glaciers, oceans, streams, and groundwater changes, altering our water supply and modifying ecological systems worldwide. Glaciated landscapes contain clues to past climate changes and therefore help us to understand what is happening now and to plan for the future. We know, for example, that continental glaciers advanced and retreated several times during the Pleistocene, and we know how fast those changes were. This knowledge enables us to measure the human impact on the rates of these processes.

13.5.1 The Formation and Evolution of Glaciers

When you think of ice forming in nature, rivers and lakes freezing during the winter may come to mind. But the ice that forms a glacier is not formed in this way. When abundant snow accumulates year after year in an area, such as at high altitude or in the polar regions, the individual crystals of ice that make up each snowflake are squeezed together by the weight of the overlying layers of snow so

FIGURE 13.15 The formation of glacial ice.
Snow compacts and melts to form firn, which recrystallizes into glacial ice. Crystal size increases with depth.

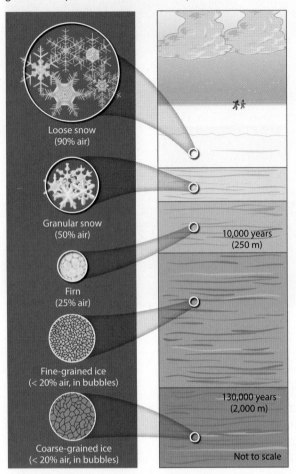

Loose snow
(90% air)

Granular snow
(50% air)

Firn
(25% air)

Fine-grained ice
(< 20% air, in bubbles)

Coarse-grained ice
(< 20% air, in bubbles)

10,000 years
(250 m)

130,000 years
(2,000 m)

Not to scale

that the individual flakes recrystallize into a granular snow called *névé* (**FIG. 13.15**). Névé that survives the melting season is compacted further into ice that is at least two-thirds as dense as liquid water. This intermediate stage between snow and glacial ice is called **firn**. Further compaction of firn causes the ice crystals to increase in size, squeezing most of the remaining air into very tiny bubbles trapped between ice crystals, forming solid glacial ice. Once this thickened mass of ice is heavy enough, it moves downhill under the force of gravity, forming a glacier.

The ability of a glacier to exist depends upon the *ice budget*—the rates of addition and loss of glacial ice. Specifically, it refers to the amount of snow *accumulation* compared with *ablation*, or wastage—processes that remove snow and ice, such as melting, sublimation, and calving (the breaking off of ice at the toe, or terminus, of a glacier). Glaciers **advance** when the rate of accumulation exceeds the rate of ablation, and they **retreat** when ablation is greater than accumulation. The location on the glacier where these two processes are balanced is the *equilibrium line*. **FIGURE 13.16** illustrates the fundamentals of glacial advance and retreat. These two terms specifically refer to the position of the **toe** (or **terminus**) of the glacier over time. As glaciers advance, there is plentiful ice to supply the glacier and the toe moves farther out or downhill. When the ice at the end of the glacier melts faster than it is being replaced at the top, the toe shifts up-valley or uphill. Note that glacial retreat doesn't mean that the ice flows uphill against the force of gravity—ice always flows from the snowfield toward the toe. When accumulation and ablation are equal, a glacier is *stagnating*, and the position of the toe doesn't change.

FIGURE 13.16 Glacial advance and retreat.

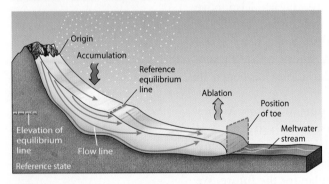

(a) The position of the toe represents a balance between accumulation and ablation.

(b) If accumulation exceeds ablation, the glacier advances, the toe moves farther from the origin, and the ice thickens.

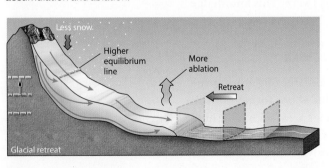

(c) If ablation exceeds accumulation, the glacier retreats and thins. The position of the toe moves back, even though ice continues to flow toward the toe.

FIGURE 13.17 Mountain and continental glaciers.

(a) The Crowfoot Glacier, a mountain glacier in Banff National Park, Canada.

(b) Continental glacier, Antarctica.

13.5.2 Types of Glaciers

Some small glaciers form on mountains and flow downhill, carrying out their erosional and depositional work in the valleys they carve. They are called, appropriately, **mountain**, **valley**, or **alpine glaciers** (**FIG. 13.17A**). Although they flow in valleys like streams, mountain glaciers work differently and form distinctly different landscapes. Conversely, the Antarctic and Greenland **continental glaciers** described earlier are not confined to valleys. They flow across the countryside as sheets of ice thousands of feet thick, dwarfing hills and burying all but the tallest peaks (**FIG. 13.17B**). The North American and European continental glaciers did the same during the Pleistocene.

13.5.3 How Glaciers Create Landforms

Glaciers erode destructional landforms and deposit constructional landforms. A glacier can act as a bulldozer, scraping regolith (soil and unconsolidated sediment) from an area and plowing it ahead. When it encounters bedrock, a glacier uses the sediment it carries like grit in sandpaper to abrade the rock and carve unique landforms. Glacial erosion occurs wherever ice is flowing, whether the glacier front is advancing, or retreating, or stagnating.

Depositional glacial landforms are equally distinctive: some are ridges tens or hundreds of miles long; others are broad blankets of sediment. All form when the ice in which the sediment is carried melts, generally when the glacier front is retreating. Some sediment, called **till**, is deposited directly by the melting ice, but some is carried away by meltwater; this sediment is called **outwash** (or *glaciofluviatile* [glacial + stream] *sediment*). Till and outwash particles differ in size, sorting, and shape (**FIG. 13.18**) because of the different properties of the ice and meltwater that deposited them. Exercise 13.5 compares the ways in which glaciers and water do their geologic work.

FIGURE 13.18 Till and outwash.

(a) Till is characterized by poor sorting (with particle sizes from small boulders through sand, silt, and mud) and by angular rock fragments.

(b) Layers of outwash. Each layer is very well sorted; most contain well-rounded, sand-sized grains, and some show cross bedding. Note the coarser pebbly layer (arrow) resulting from an increase in meltwater discharge.

EXERCISE 13.5	Comparison of Glaciers and Streams

Name: _____   Section: _____
Course: _____   Date: _____

Glaciers behave differently from stream water—even glaciers that occupy valleys. Based on what you know about streams and have learned about glaciers, complete the following table.

		Streams	Glaciers
State of matter		Liquid	
Composition		H_2O	
Areal distribution		In channels with tributaries flowing into larger streams	
Rate of flow		Fast or slow, depending on gradient	
Erodes by . . .		1. Abrasion using sediment load 2. Dissolving soluble minerals, rocks	
Maximum depth to which erosion can occur		Base level: sea level for streams that flow into the ocean; master stream for tributaries	
Deposits when . . .		Stream loses kinetic energy either by slowing down or evaporating	
Type(s) of sediment	**Maximum clast size range**	Small boulders to mud	
	Sorting	Generally well sorted	
	Clast shape	Moderately to well rounded, depending on distance transported	

13.6 Landscapes Produced by Continental Glaciation

Both continental and mountain glaciers produce unique landforms not found in areas affected by stream deposition and erosion. Landscapes formed by continental glaciers differ from those formed by streams in every imaginable way, including the shapes of hills, valleys, and stream divides, as well as the degree to which post-glacial streams are integrated into stream networks like those discussed in Chapter 11. **FIGURE 13.19** illustrates the different shapes of valleys carved by continental glaciers and by streams, and Exercise 13.6 explores their differences more fully.

Continental glaciers scour the ground unevenly as they flow, producing an irregular topography filled with isolated basins, which often become lakes when the ice melts. Many lakes inherited from the most recent Pleistocene glaciation are still not incorporated into normal networks of streams and tributaries, even though the ice melted thousands of years ago. Glaciers also deposit material unevenly, clogging old streams and choking new ones with more sediment than they can carry. *Poorly integrated drainage, with many swampy areas, is thus characteristic of recently glaciated areas.* In addition, hills scoured by continental glaciers commonly exhibit asymmetric, streamlined shapes called **drumlins** that reveal the direction in which the ice flowed. As the ice reaches these hills, it rises up, over, and down them, resulting in the steep (or blunt) side of the hill forming on the "upflow" side, and the tapered, gentler slopes on the "downflow" side, as illustrated in **FIGURE 13.20**. The blunt side of these hills points to the direction from which the ice originated.

FIGURE 13.19 Comparison of valleys carved by continental glaciers and by streams.

(a) Glacial landscape: an unnamed valley in Glacier National Park, Montana.

(b) Fluvial landscape: valley of the Yellowstone River, Wyoming.

FIGURE 13.20 Development of drumlins.

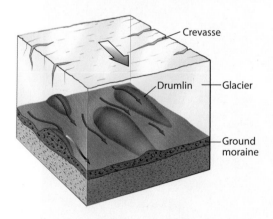

(a) Drumlins are sculpted beneath the ice of a glacier and are aligned with its flow direction.

(b) A drumlin formed by a glacier in Sodus, New York.

Name: _____   Section: _____
Course: _____   Date: _____

(a) What difference do you notice in the shapes of the valleys pictured in Figure 13.19? In particular, compare the overall shapes of the valleys pictured and the characteristics of each valley's base.

_____

_____

_____

_____

(b) Now compare the aerial photograph of an extensively glaciated area (below) with the two photographs of fluvial landscapes on the next page. Describe the differences among the photos without worrying about technical terms for the glacial features. *Consider:* Is there a well-developed stream network?

_____

_____

_____

_____

Glaciated landscape, Northwest Territories, Canada.

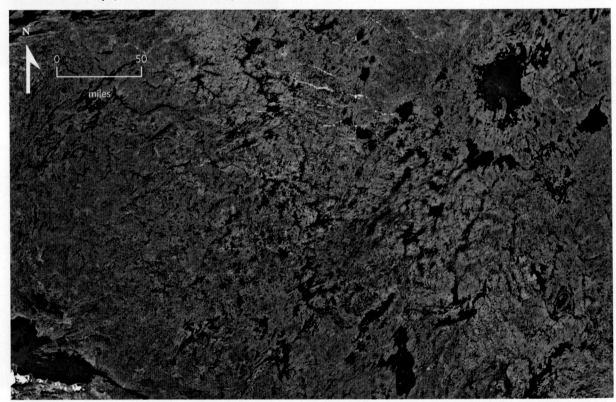

(continued)

Name: _____   Section: _____

Course: _____   Date: _____

Fluvial landscape, West Virginia.

Fluvial landscape, Colorado Plateau, Utah–Arizona.

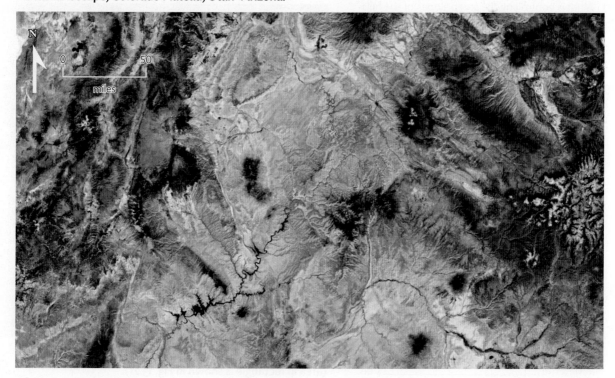

13.6.1 Erosional Landscapes

When a continental glacier passes through an area, it carves the bedrock into characteristic shapes, collectively referred to as *sculptured bedrock*. FIGURE 13.21 and the digital elevation model (DEM) in Exercise 13.7 show the topography of an area of sculptured bedrock eroded by continental glaciers in southern New York State. The hills are underlain by gneisses that were resistant to glacial erosion and the valleys by marbles and schists that were more susceptible to it.

EXERCISE 13.7   **Erosional Features of Continental Glaciation**

Name: _____   Section: _____
Course: _____   Date: _____

(a) Describe the topography in the DEM below in your own words, paying particular attention to the shapes of the bedrock hills. When features in a glacial landscape display a strong alignment, geologists say that the landscape has a "topographic grain" related to the direction in which ice moved. Describe the topographic grain in Figure 13.21 and the direction in which the ice probably moved.

_____

_____

(b) Construct a topographic profile, using the graph paper provided at the end of this chapter, from St. John's Church to Lake Rippowam (line A–B in Fig. 13.21), using 10x vertical exaggeration. Are the hills symmetric or asymmetric?

_____

(c) Are the steep and gentle slopes of hills distributed randomly or systematically? Describe their distribution.

_____

(d) What is controlling the steep slopes of the hills?

_____

_____

(e) Based on your description and observations, draw arrows on the DEM to the right and on Figure 13.21 showing the direction in which the continental glacier moved across the map area. Explain your reasoning.

_____

_____

_____

_____

(f) Fill in the blanks to complete the following sentence that describes how sculptured bedrock records glacial movement: The direction of ice flow indicated by sculptured bedrock is *from* the _____ side of the hill *toward* the _____ side.

DEM of the Peach Lake quadrangle in New York (area in rectangle is shown in Fig. 13.21).

FIGURE 13.21 Portion of the Peach Lake, New York, 7.5′ quadrangle.

0   0.25   0.5 mile

0   0.25   0.5 km

Contour interval = 20 feet

N

A

B

North Salem

Titicus River

116

121

KEELER LANE

LANE

HUNT LANE

Scott Ridge

584

608

952

900

Titicus Mountain

742

HAWLEY

ROAD

976

NEW YORK
CONNECTICUT

Lake Rippoowam

Twin Lakes Village

Lake Waccabuc

Lake Oscaleta

Round Pond

Pumping Station Swamp

121

116

13.6.2 Depositional Landscapes

Landscapes formed by continental glaciers contain several types of depositional landforms. We will look first at features made up of till deposited directly from a melting glacier, then at landforms deposited by meltwater.

13.6.2a Landforms Made of Till **Moraines** form when melting ice drops the boulders, cobbles, sand, silt, and mud that it has been carrying. Some moraines are irregular ridges; others are broad carpets pockmarked by numerous pits separated by isolated hills. The type of moraine—ridge or carpet—depends on whether the glacier is retreating or stagnating.

Till deposited by a retreating glacier forms an irregular carpet of debris called a **ground moraine**, which may be hundreds of feet thick. Large blocks of ice are isolated as a glacier retreats, and some may be buried by outwash or till. When a block melts, the sediment that covered it subsides and forms a depression in the ground moraine called a **kettle hole**. For this reason, the irregular surface of a ground moraine is often referred to as **knob-and-kettle** topography.

The terminus of a stagnating glacier may remain in the same place for hundreds or thousands of years, and the till piles up in a ridge that outlines the terminus. This ridge, called a **terminal moraine**, shows the maximum extent of the glacier. **FIGURE 13.22** shows some of the depositional features created at the termini of continental glaciers. Use these diagrams to identify landforms in the following exercises. A glossary of glacial depositional and erosional features can be found in Appendix 13.1 at the end of this chapter.

As noted earlier, drumlins are streamlined, asymmetric hills made of till that commonly occur in large groups. Four of the largest swarms of drumlins are in Nova Scotia; in southern Ontario and northern New York State; in southeastern New England (most famously Bunker Hill); and in Wisconsin, Iowa, and Minnesota.

13.6.2b Landforms Made of Outwash A melting continental glacier generates an enormous amount of water, which flows away from the terminus carrying all the clasts the water is powerful enough to carry—mostly sand and gravel, with perhaps a few small boulders. Meltwater streams round and sort this sediment and eventually deposit it beyond the terminal moraine as a generally flat **outwash plain** (see Fig. 13.22). The flat outwash plains visible in the southernmost part of **FIGURE 13.23** differ dramatically from the adjacent terminal moraines and from the knob-and-kettle topography of a ground moraine.

FIGURE 13.22 Formation of depositional landforms by continental glaciers.

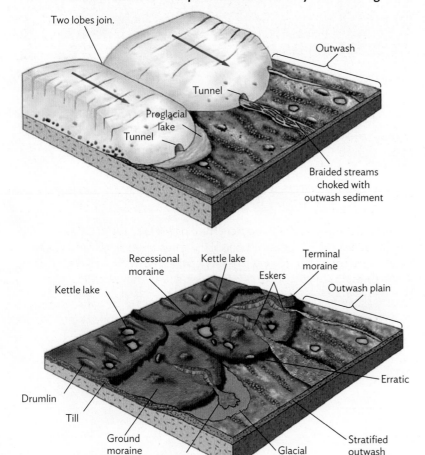

FIGURE 13.23 DEM of Long Island, New York, showing two terminal moraines and outwash plains.

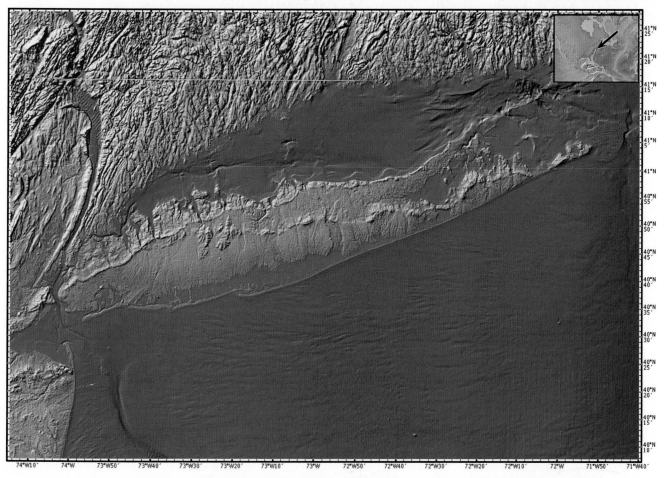

Landforms at the Terminus of a Continental Glacier

Name: _____ Section: _____

Course: _____ Date: _____

FIGURE 13.23 is a DEM of southeastern New York and southwestern Connecticut showing the irregular glacial topography at the southernmost extent of continental glaciation in North America. Long Island is composed largely of till and outwash deposited during the advance and retreat of *two* continental glaciers. This makes the topography of the area more complex than that shown in Figure 13.21, but the features that formed at the two glacial termini are recognizable.

(a) Outline or otherwise identify the two terminal moraines, the southernmost outwash plain, and the outwash and glacial deltas deposited between the terminal moraines during a time of glacial retreat. Use different colors to separate features deposited during the two different glacial advances.

(b) Which of the two moraines is younger? Explain your reasoning. (*Hint:* Recall the relative age dating technique of cross-cutting relationships.)

_____

_____

(continued)

Name: _____ Section: _____

Course: _____ Date: _____

FIGURE 13.24A is a map of the Arnott moraine in Wisconsin, formed approximately 14,000 years ago. It shows the classic irregular knob-and-kettle topography of terminal and ground moraines. Compare this morainal area with the fluvial landscape shown in **FIGURE 13.24B.**

(c) Are the stream networks and divides as well developed and clearly defined in the morainal area? _____

(d) Suggest an explanation for the absence of a well-integrated drainage system in the morainal area.

_____

_____

_____

_____

(e) Suggest an explanation for the numerous swampy areas in the morainal landscape.

_____

_____

_____

_____

(f) Compare the sizes and shapes of the hills (knobs) in the morainal landscape with those in the fluvial landscape.

_____

_____

_____

(g) Suggest an origin for the several small lakes in the morainal area. What will be their eventual fate?

_____

_____

_____

(h) There are several gravel pits in the morainal area. Why might gravel be more easily and profitably mined from these specific areas than from others?

_____

_____

_____

(i) What difficulties do you think might be involved in farming on a ground moraine?

_____

_____

_____

FIGURE 13.24 Comparing glacial and fluvial landscapes.
(a) Glaciated terrain in Wisconsin (Amherst and Blaine 7.5' quadrangles).

Contour interval = 20 feet

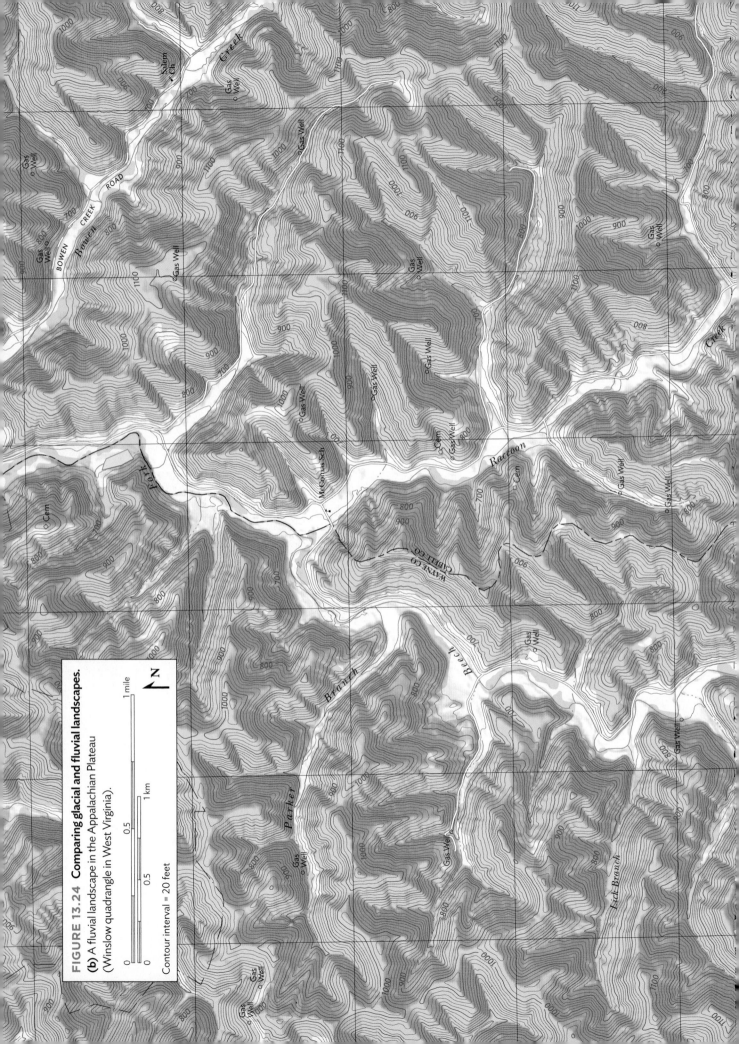

FIGURE 13.24 Comparing glacial and fluvial landscapes.

(b) A fluvial landscape in the Appalachian Plateau (Winslow quadrangle in West Virginia).

Contour interval = 20 feet

13.7 Landscapes Produced by Mountain Glaciation

Alpine, Himalayan, Rocky Mountain, and Sierra Nevadan vistas, with their jagged peaks, steep-walled valleys separated by knife-sharp divides, and spectacular water-falls, are the result of mountain glacier erosion. Small glaciers form in depressions high up in the mountains and flow downhill, generally following existing stream valleys. The depressions are deepened and expanded headward by glacial *plucking*, which eats into the mountainside, and the ice flows downhill, filling the former stream valleys and modifying them by abrasion (**FIG. 13.25**). When the ice melts, the expanded depressions become large, bowl-shaped valleys (**cirques**); formerly rounded stream divides are replaced by knife-sharp ridges (**arêtes**); and the head-ward convergence of several cirques leaves a sharp, pyramidal peak (a **horn peak**, or **horn**, named after the Matterhorn in the Alps).

Mountain glaciers transform pre-existing stream valleys in unmistakable ways, including their cross-sectional and longitudinal profiles and the way that their streams flow into one another. Figure 13.25c shows the characteristic U-shaped cross-sectional profile of a valley carved by a mountain glacier. Why are glaciated valleys U-shaped whereas stream valleys are V-shaped? Let's see if your ideas in Exercise 13.6 were right.

FIGURE 13.25 Erosional features formed by mountain glaciers.

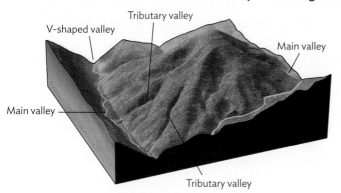

(a) Before mountain glaciation, the area has a normal fluvial topography.

(b) The extent of the glaciers during mountain glaciation.

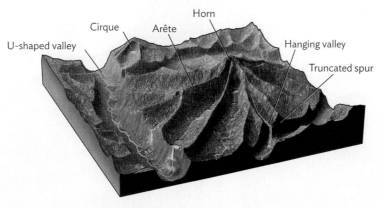

(c) The area's erosional features after mountain glaciation.

FIGURE 13.26 Glacial modification of a fluvial valley.

(a) Ice fills the entire valley, eroding everywhere, not just at the bottom like a stream. As it erodes, the valley is deepened and widened everywhere, resulting in the U shape on the right.

(b) One of the many spectacular fjords of Norway. The water is an arm of the sea that fills a U-shaped valley. Tourists are standing on Pulpit Rock (Preikestolen).

The answer lies in how the two agents of erosion operate (**FIG. 13.26A**). A stream actively erodes only a small part of its valley at any given time, cutting its channel downward or laterally (on the left in Fig. 13.26a). Mass wasting gentles the valley walls, creating the V shape. In contrast, a mountain glacier fills the entire valley, abrading the walls not only at the bottom, but everywhere it is in contact with the bedrock (center). When the ice melts, it leaves a U-shaped valley (right). When these glaciers retreat in coastal areas, the sea fills the former U-shaped valley, forming a **fjord** (**FIG. 13.26B**).

Most streams, as we saw in Chapter 11, have smoothly concave longitudinal profiles. Mountain glaciers, being solid, can overcome gravity locally and flow uphill for short distances. Their longitudinal profiles are therefore more irregular than a stream's, commonly containing a series of shallow scooped-out basins. Water can collect in some of these basins, producing a chain of lakes called **paternoster** (or **cat-step**) **lakes**—from the Latin *Our Father* because the lakes and the streams connecting them resemble a set of rosary beads (**FIG. 13.27**).

FIGURE 13.27 Longitudinal profiles of streams and mountain glaciers.

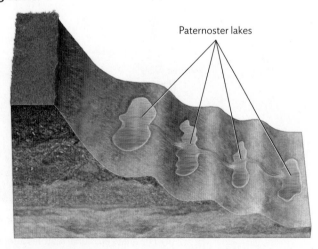

Paternoster lakes

(a) Smooth, concave stream profile.

(b) Irregular, scalloped profile of a mountain glacier, with paternoster lakes filling the basins.

The base level for a tributary stream is the elevation of the stream, lake, or ocean into which it flows, and most tributary streams flow smoothly into the water body at their base level. In contrast, glaciers have no base level, so large glaciers can carve downward more rapidly than smaller ones. Previously existing main stream valleys, with large glaciers, can therefore be carved more deeply than their tributary valleys.

When the ice melts, the result is a **hanging valley**, where the tributary stream hangs over the deeper main valley (**FIG. 13.28**). Today, Bridal Veil Creek falls almost 200 m to the floor of Yosemite Valley. The mouth of the creek and the floor of the main stream valley were once at the same elevation, but the small glacier that filled Bridal Veil Creek could not keep pace with the main glacier, which eroded downward more rapidly.

FIGURE 13.28
Bridal Veil Falls flows from a hanging valley in Yosemite National Park, California.

EXERCISE 13.9   **Recognizing Features of Mountain Glaciation**

Name: _____   Section: _____

Course: _____   Date: _____

(a) The following DEM shows part of Glacier National Park and several erosional features formed by mountain glaciers. Label examples of horn peaks, arêtes, cirques, and U-shaped valleys (with an H, A, C, and U, respectively), and look also for flat places (uniform color) that could be filled with lakes (label with an L).

Lowest elevation   Highest elevation

DEM (false color) of a portion of Glacier National Park showing classic features of mountain glaciation.

(b) Look at the chapter opening photograph and identify as many features of mountain glaciation as you can.

13.8 Glaciers and Climate Change

Intense weather events such as tornadoes, hurricanes, and typhoons; a rising sea level; and persistent drought have brought the issue of global climate change from the field sites and laboratories of geologists and meteorologists to media headlines and the world political arena. Glacial landforms and sophisticated studies of polar ice provide the best record of atmospheric changes over the past several hundred thousand years, enabling us to base hypotheses about future changes on solid fact rather than guesswork.

Glacial deposits provide evidence that the ice ages during what geologists call the Pleistocene Epoch involved four *glaciations*, in which continental glaciers advanced across the northern hemisphere, followed by *interglacials*, in which the ice retreated or melted completely. Carbon dating of wood from glacial and interglacial sediments shows that the first glaciation began about 700,000 years ago, and the last continental glaciers in Europe and North America disappeared about 12,000 years ago. The extent of the Pleistocene ice sheets is illustrated in **FIGURE 13.29**, along with the extent of sea ice coverage in the Northern Hemisphere.

These glaciations and interglacial periods don't occur at random, and changing climate conditions is one factor that controls their occurrences. In 1920, Serbian astronomer and geophysicist Milutin Milanković calculated the rates of change between glacial advance and retreat that occur within individual ice ages, such as the one observed during the Pleistocene Epoch. Based on these calculations, he

FIGURE 13.29 Pleistocene ice sheets of the northern hemisphere.

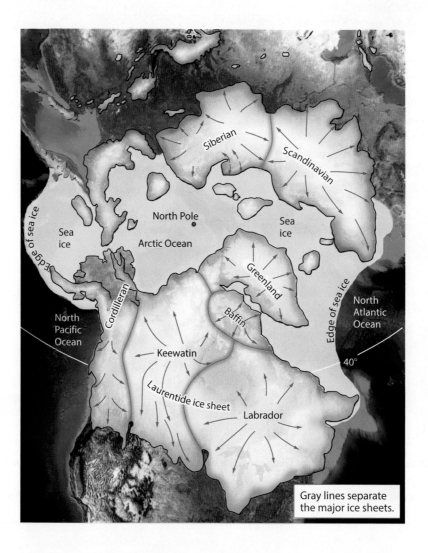

Gray lines separate the major ice sheets.

FIGURE 13.30 Milankovitch cycles cause the amount of insolation received at high latitudes to vary over time.

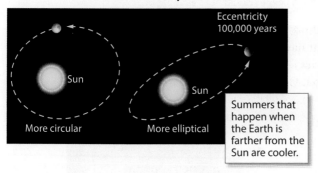

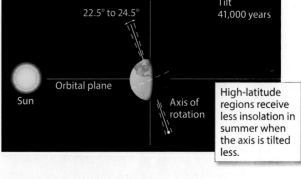

(a) Variations caused by changes in orbital shape. The shape of the orbit is exaggerated.

(b) Variations caused by changes in the tilt of the Earth's axis.

reasoned that the glacial-interglacial changes occurred in cycles taking place over three different time scales, each as a result of changes in the amount of energy the top of Earth's atmosphere receives from the Sun (called *insolation*). These cycles are now referred to as Milankovitch (to use the English spelling) cycles (**FIG. 13.30**):

- *100,000-year cycle* as a result of the changing shape of Earth's orbit around the Sun.
- *41,000-year cycle* due to the increasing and decreasing tilt of the Earth on its axis of rotation (or spin axis).
- *23,000-year cycle* when the Earth wobbles as it rotates, like a spinning top.

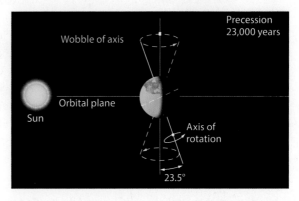

(c) Variations caused by the precession (wobbling) of the Earth's axis.

Furthermore, although the Pleistocene glaciers have disappeared from North America and Europe, the continental glaciers that remain in Antarctica and Greenland record atmospheric temperatures and concentrations of greenhouse gases (CO_2, methane) over the hundreds of thousands of years that they have existed. Geologists have drilled into these glaciers and collected ice cores (**FIG. 13.31**) for study in refrigerated laboratories. When the ice froze, it incorporated small bubbles of the air that existed at the time. The ice can be dated and the amounts of greenhouse gases incorporated in it can be measured, providing a timeline against which modern conditions can be compared.

Today, scientists are in a race against time to sample and analyze cores from the few remaining and accessible glaciers around the world, before they melt completely. These continuing studies should provide clues to what we may expect in the future as our climate changes. We will look at climate change in more detail in later chapters.

We don't need sophisticated instruments to understand how glaciers respond to climate change, however. All we have to do is look. The most dramatic example of glacial change was witnessed by people all over the globe, who watched on television as a huge piece of the Antarctic glacier—larger than the state of Rhode Island—broke off in the summer of 2017. Another notable example involves the Athabasca Glacier in Jasper National Park, Alberta, Canada, one of the most popular tourist destinations in the province and *the* most visited glacier in North America. The glacier has been retreating since 1843, and the Alberta tourist bureau is concerned that if the retreat continues, it won't be long before there won't be a glacier left to visit. **FIGURE 13.32** dramatically documents the glacier's retreat.

FIGURE 13.31 Using ice cores to assess changes in greenhouse gas concentrations over time.
A geologist removes an ice core that shows seasonal layering and which has dark layers of volcanic ash.

FIGURE 13.32 Retreat of the Athabasca Glacier, Alberta, Canada.

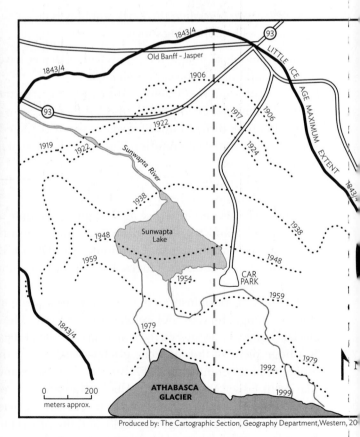

Produced by: The Cartographic Section, Geography Department, Western, 20

(a) The black-and-white photograph from 1906 (top) and the color photograph from 2006 (bottom) show approximately the same location. Notice that the glacier is visible only in the far left-hand portion of the 2006 image.

(b) Changes in the position of the glacier's terminus between 1843 and 1999.

Name: _____   Section: _____
Course: _____   Date: _____

You have been hired as a consultant to advise the Alberta tourist bureau and Parks Canada on how much longer the Athabasca Glacier will be viable as a tourist attraction. Note on the map in Figure 13.32b that the glacier's terminus does not follow a straight line, but flows around the contours of the landscape. So, to estimate the past rate of retreat, you need to follow a straight line (the red dashed line on the figure) from the current terminus (the farthest-extending point of the glacier) to the location where the 1843–1844 terminus intersects the Old Banff-Jasper Highway.

(a) Measure the amount of retreat from 1843 to 1999 along the line you drew: _____ m

(b) Determine the average rate of retreat during that period: Average rate of retreat _____ m ÷156 years = _____ m/yr

You note that the spacing of the dated terminus positions in Figure 13.32b suggests that the rate of retreat has not always been consistent: at some times, the glacier appears to have retreated more rapidly than at others.

(c) Compute the rates of retreat during different shorter time periods:
  (i) from 1843 to 1906: _____ m/yr
  (ii) from 1906 to 1999: _____ m/yr

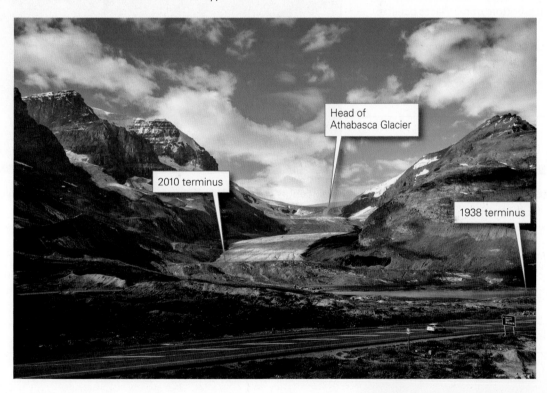

Head of
Athabasca Glacier

2010 terminus

1938 terminus

? What Do You Think The photo above is a recent view of the Athabasca Glacier. It is now approximately 6.2 km from the glacier's terminus to its source. You have measured rates of past retreat over three different time periods and can now use those rates to estimate how much longer the glacier will exist. The number of years it would take for the glacier to melt completely based on your estimate above from:
  • question (b) is _____ years.
  • question (c.i) is _____ years.
  • question (c.ii) is _____ years.

Name: _____   **Section:** _____
Course: _____   **Date:** _____

Exploring Earth Science Using Google Earth

1. Visit **digital.wwnorton.com/labmanualearthsci**
2. Go to the **Geotours** tile to download Google Earth Pro and the accompanying Geotours exercises file.

Expand the Geotour13 folder in Google Earth by clicking the triangle to the left of the folder icon. Check and double-click the Mono-Craters TopoMap overlay to fly to the eastern side of the Sierra Nevada range in California, a region that has experienced considerable alpine/valley glaciation (note that you can select the overlay and use the transparency slider at the bottom of the Places panel to make the overlay semi-transparent).

(a) Check and double-click the Feature placemark. Identify the glacial landform represented by this linear ridge.

_____

_____

_____

(b) Check and double-click the Ridges I placemarks. Identify the glacial landforms represented by these linear ridges.

_____

_____

(c) Check and double-click the Ridges II placemarks (leave the Ridges I placemarks checked). Which set of landforms (Ridges I or Ridges II) originated first? How do you know? *Hint:* You may want to toggle the topographic map on and off.

_____

_____

_____

(d) Check and double-click the Peak placemark. Identify the type of glacial landform represented by this triangular-shaped peak.

_____

(e) Check and double-click the Basin placemark. This bowl-shaped basin forms what type of glacial landform? What is its valley called (relative to the Walker Creek valley that is labeled on the topographic map)?

_____ basin

_____ valley

(f) Check and double-click the Curved Ridge placemark. What type of glacial landform is this curved ridge (which is made up of well-sorted sediment)? _____

Glossary of Glacial Landforms

Erosional landforms		
Landform	**Type of glacier**	**Description**
Arête	Mountain	Sharp ridge separating cirques or U-shaped valleys
Cirque	Mountain	Bowl-shaped depression on mountainside; site of snowfield that was source of mountain glacier
Hanging valley	Mountain	Valley with mouth high above main stream valley; formed by tributary glacier that could not erode down as deeply as larger main glacier
Horn peak	Mountain	Steep, pyramid-shaped peak; erosional remnant formed by several cirques
Sculptured bedrock	Continental	Asymmetric bedrock with steep side facing the direction toward which the ice flowed; also called roche moutonnée
U-shaped valley	Mountain	Steep-sided valley eroded by a glacier that once filled the valley
Depositional landforms		
Landform	**Type of glacier**	**Description**
Drumlin	Continental	Streamlined (elliptical), asymmetric hill composed of till; the long axis parallels glacial direction, and the gentle side faces the direction toward which the ice flowed
Esker	Continental	Narrow, sinuous ridge made of outwash deposited by a subglacial stream in a tunnel at the base of a glacier
Ground moraine	Mountain and continental	Sheet of till deposited by a retreating glacier
Kettle hole	Mountain and continental	Depression left when a block of ice that had been isolated from a melting glacier is covered with sediment and melts
Lateral moraine	Mountain	Till deposited along the walls of a valley when a mountain glacier retreats
Outwash plain	Mountain and continental	Outwash deposited by meltwater beyond the terminus of a glacier
Proglacial delta	Continental	Outwash deposited into a proglacial lake
Terminal moraine	Mountain and continental	Wall of till deposited when the terminus of a glacier remains in one place for a long time
Lakes associated with glacial landscapes		
Landform	**Type of glacier**	**Description**
Ice-marginal lake	Mountain and continental	Lake formed by disruption of local drainage by a glacier
Kettle lake	Mountain and continental	Lake filling a kettle hole
Paternoster lake	Mountain	One of a series of lakes filling basins in a glaciated valley
Tarn	Mountain	Lake occupying part of a cirque

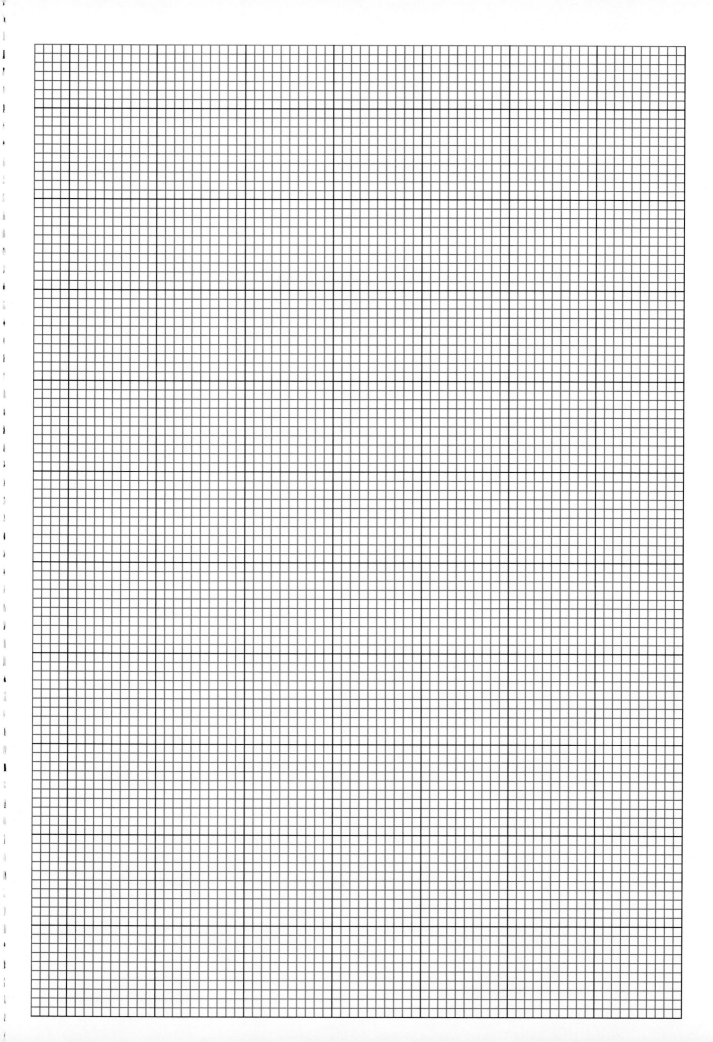

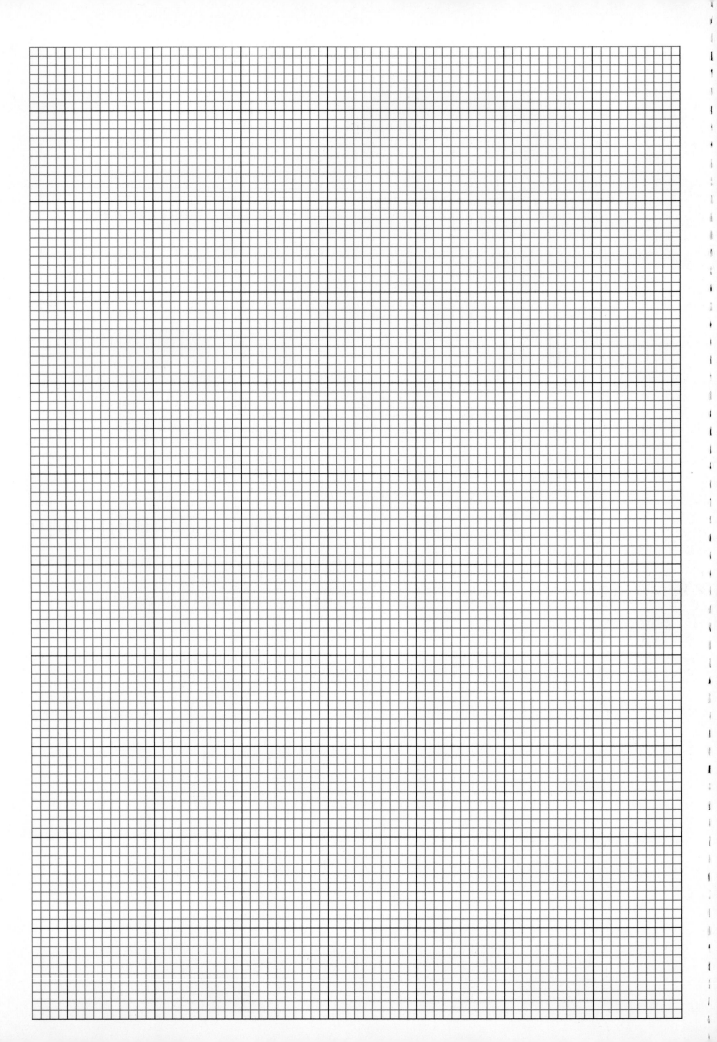

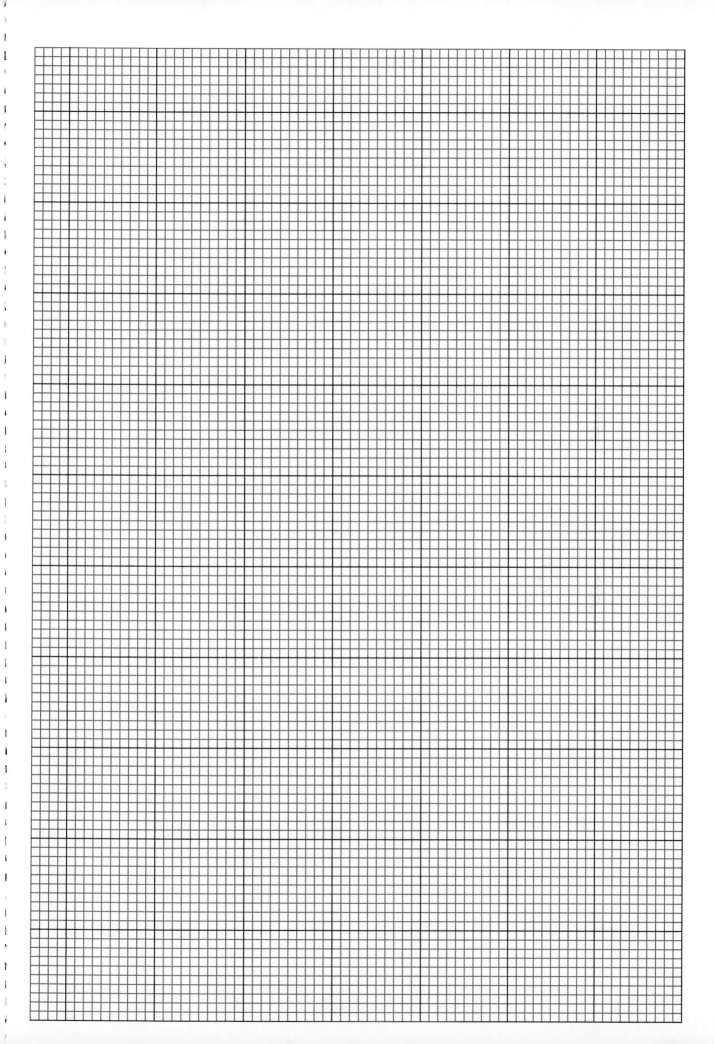

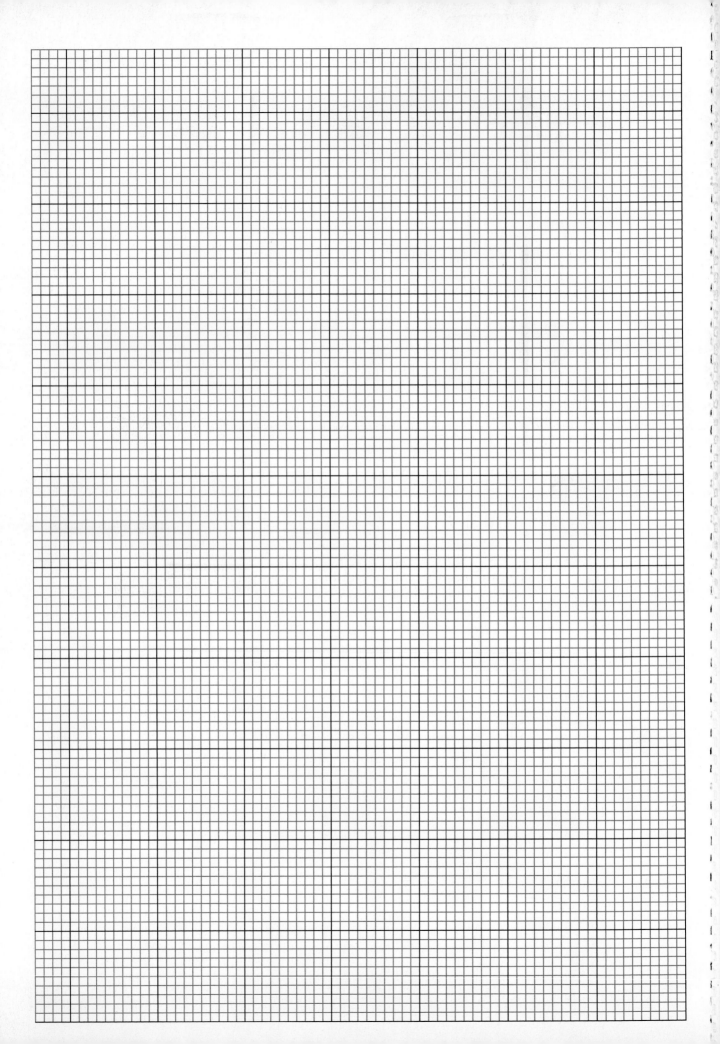

Oceanography: The Seafloor, Seawater Chemistry, and Ocean Circulation

14

Interactions between the ocean surface and the deep waters below drive the movement of the oceans.

- Describe the scale of the ocean and features on the seafloor
- Identify and describe the features of passive and active continental margins
- Describe the layering of ocean water
- Explain how the atmosphere drives surface ocean circulation and the formation of oceanic gyres
- Explain the relationship between the temperature and density of seawater
- Explain the relationship between the salinity and density of seawater

MATERIALS NEEDED

- Tennis ball and marker
- 100 ml graduated cylinders
- Food coloring
- Small test tubes
- 250 ml beakers
- Salt
- Ice
- Hot water or hot plate
- Colored pencils
- Metric ruler
- Mass balance
- Stopwatch

14.1 Introduction

As you investigated in the plate tectonics chapter, new seafloor is created as continents break apart along divergent plate boundaries. This new crust is composed of heavier chemical elements than continental crust, making it denser. As the plates continue to move, oceans continue to form and close, which can have profound effects on not only the features that form on the seafloor and surrounding landmasses, but also on the continual evolution of continental coastlines (which you will explore in the next chapter) and on the climate of the entire planet. This last effect will be explained through experiments and observations in Chapters 16–19 on the Earth's atmosphere.

In this chapter, we will investigate the physical features of the seafloor, the chemical nature of seawater, and the ways in which the oceans and atmosphere interact to move the sea surface.

14.2 Breadth of the Oceans

When we look at a map of the Earth today, it is obvious that most of the surface of the planet is covered in water, and that this water is primarily ocean (**FIG. 14.1**). In fact, the modern ocean covers approximately 71% of the Earth's surface. The Earth really has just one ocean, even though as humans we have divided it into individual parts in terms of geography. Of these parts, the largest bodies of water are referred to as oceans, while smaller areas of these oceans that are partially enclosed by land are often designated as seas. For example, the Irminger Sea is part of the Atlantic Ocean. As you will observe throughout the following exercises, these oceans and seas are connected not only in terms of water, but also energy. Let's first investigate the geographic locations of the major bodies of seawater on the planet, and then compare their distribution across the globe.

FIGURE 14.1 A satellite view of the Earth.

Name: _____   Section: _____

Course: _____   Date: _____

(a) On the map below, fill in the blanks with the geographic names of the oceans and seas, as provided in the list. (Write your answers with something that can be erased, as you may need to revise them!)

Oceans

1. North Pacific Ocean
2. South Pacific Ocean
3. North Atlantic Ocean
4. South Atlantic Ocean
5. Indian Ocean
6. Arctic Ocean
7. Southern Ocean

Seas and major water bodies

8. Hudson Bay
9. Gulf of Mexico
10. Baffin Bay
11. Bering Sea
12. Persian Gulf
13. North Sea
14. Sea of Japan
15. South China Sea
16. Coral Sea

17. Tasman Sea
18. Black Sea
19. Baltic Sea
20. Norwegian Sea
21. Caribbean Sea
22. Sea of Okhotsk
23. Philippine Sea
24. Mediterranean Sea
25. Arabian Sea
26. Red Sea

Equator

(b) Compare your map with a classmate's map. Discuss your reasoning for your choices. Check your answers against a globe, map, or atlas, and make changes as necessary.

(c) Which ocean appears to be the largest? _____

(d) Are the oceans connected? Give evidence to support your answer.

_____

_____

_____

(e) What boundaries have been used to divide the global ocean into regions?

_____

_____

_____

Name: _____   Section: _____
Course: _____   Date: _____

Look at the map below:

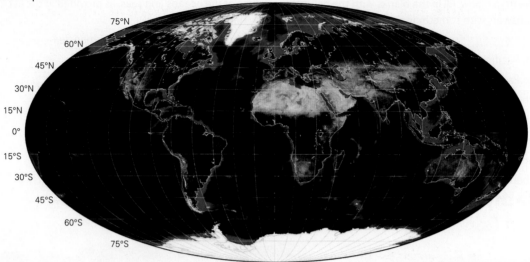

(a) For each 15-degree band of latitude, estimate the percentage of the surface that is water and the percentage that is land. Keep in mind that these two percentages for each band of latitude must add up to 100%. Write your answers in the table. (*Hint:* Determine how many coordinate boxes for each latitude band are composed of water and how many are composed of land, and divide each by the total number of boxes in the row. Round up or down to the nearest whole number percentage as necessary.)

Global Distribution of Land and Water by Latitude

Latitudes	Water (%)	Land (%)	Latitudes	Water (%)	Land (%)
0° to 15° N	_____	_____	0° to 15° S	_____	_____
15° to 30° N	_____	_____	15° to 30° S	_____	_____
30° to 45° N	_____	_____	30° to 45° S	_____	_____
45° to 60° N	_____	_____	45° to 60° S	_____	_____
60° to 75° N	_____	_____	60° to 75° S	_____	_____
75° to 90° N	_____	_____	75° to 90° S	_____	_____

(b) Based on your answers in the table, which hemisphere (northern or southern) has a greater percentage of land? _____ Which has more water? _____

(c) Let's compare just the polar regions (latitudes 75°–90°) in both hemispheres. Briefly explain the difference between the distribution of land and water in the two regions.

_____

_____

_____

(continued)

Name: _____   Section: _____

Course: _____   Date: _____

(d) Imagine you wanted to sail around the globe without the continents getting in your way. Using the map above, along which range of latitudes could you sail, going east or west, to accomplish your goal? Explain your choice(s).

_____

_____

_____

14.3 Bathymetry: Ocean Depths and Seafloor Features

As you just observed, the surface of the Earth is primarily ocean water. But to really grasp the immensity of the amount of seawater on the planet, we have to also consider the depth of the oceans, which on average reaches 3,622 meters (approximately 12,100 feet). As we move beneath the surface, there are places where the seafloor is as much as 11,035 meters (approximately 36,857 feet) deep, such as the Challenger Deep in the Mariana Trench (**FIG. 14.2**). In terms of the

FIGURE 14.2 **The depth of the Mariana Trench.**

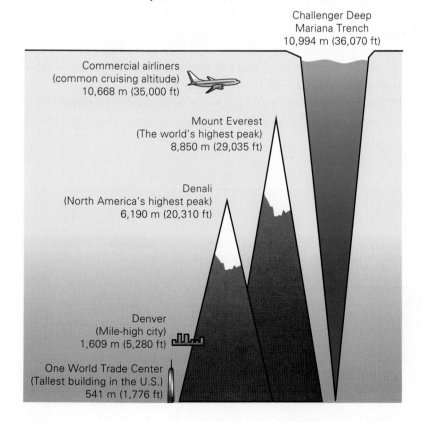

Challenger Deep
Mariana Trench
10,994 m (36,070 ft)

Commercial airliners
(common cruising altitude)
10,668 m (35,000 ft)

Mount Everest
(The world's highest peak)
8,850 m (29,035 ft)

Denali
(North America's highest peak)
6,190 m (20,310 ft)

Denver
(Mile-high city)
1,609 m (5,280 ft)

One World Trade Center
(Tallest building in the U.S.)
541 m (1,776 ft)

FIGURE 14.3 Satellite image of the Earth illustrating the amount of water on the planet. The blue sphere represents all water on the planet, which would be the equivalent of 1.386 billion cubic kilometers (332.5 million cubic miles).

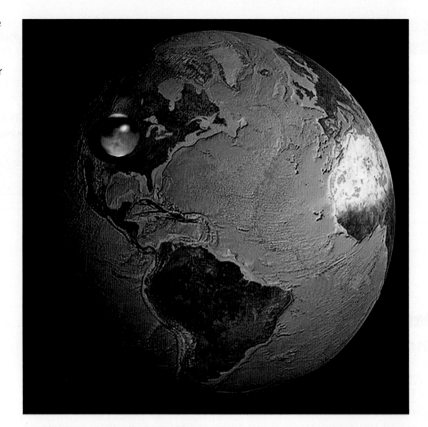

total volume, there is something like 1.3 billion cubic kilometers (300 million cubic miles) of water in the entire ocean! **FIGURE 14.3** illustrates that relative to the Earth as a whole, all the water on Earth would be the size of the blue sphere shown.

After the discovery and mapping of the Mid-Atlantic Ridge from sounding data collected during and after World War II, the US Naval Research Laboratory, along with other international agencies, conducted survey studies in an attempt to discover the **bathymetry**—the shape, depth, and features—of the seafloor across the globe. This was accomplished by using sonar technology, including echo sounders like the one shown in **FIGURE 14.4**. You will use data obtained from an echo sounder survey to visualize seafloor features in Exercise 14.3.

Instruments such as echo sounders and other sonar tools can give us direct information about the features of the seafloor. But what about in areas where we cannot

FIGURE 14.4 How does sonar work?

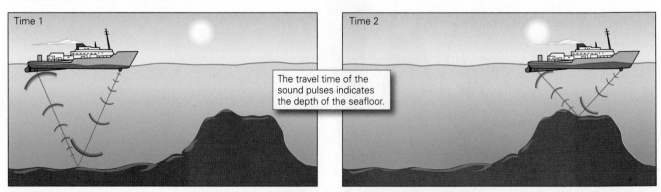

(a) As the ship moves, it constantly sends out sound pulses. These pulses bounce off the seafloor and return to the ship as echoes.

(b) If the ship goes over a shallow area, the travel time of the pulses decreases.

Name: _____   Section: _____

Course: _____   Date: _____

The data in the table below were collected from an echo sounding survey in the northern Pacific Ocean. The data shown are depths measured from the sea surface (sea level) down to the seafloor and are given in meters; so, the greater the negative number, the deeper the ocean depth.

(a) Use the graph below to plot data points for each of the lettered locations and their corresponding water depth. Depth is plotted on the vertical axis with sea level (0 meters) at the top and location along the horizontal axis. Connect the points with a smooth line. (Label the vertical axis with a scale of depth measurements, with 0 at the top.)

Echo sounder data in meters below sea level.

(Locations of measurements across the seafloor are given by the capital letters.)

Measurement location	Depth below sea surface, in meters (m)
A	−4,613
B	−4,571
C	−3,995
D	−3,314
E	−2,807
F	−2,755
G	−2,634
H	−2,587
I	−2,738
J	−2,815
K	−3,293
L	−3,763

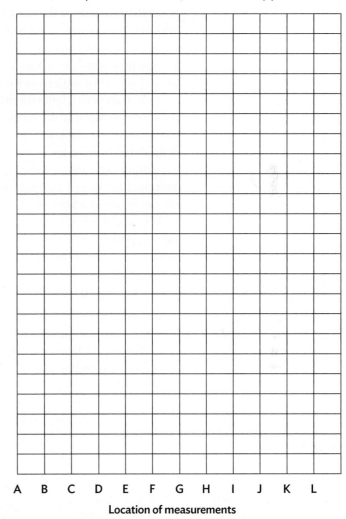

Depth below sea level (meters)

A B C D E F G H I J K L

Location of measurements

(b) At what data location is the seafloor the most shallow? _____ The most deep? _____

bring ships? In those cases, scientists utilize a system of Earth-orbiting satellites that measure the height of the sea surface and changes in sea-surface heights across the globe, as well as other measurements such as sea-surface temperature and the movements of ocean currents. This method works because sea-surface topography generally reflects seafloor topography; for example, a tall underwater mountain can push up the sea surface above it.

The satellites used to measure sea-surface height utilize an instrument called a *radar altimeter*. This device works in a similar way to sonar. The radar altimeter sends out a beam of energy (radio waves) down to the sea surface, which then bounces back to the satellite. The effects of tides and atmospheric conditions are accounted for to yield an apparent height of the sea surface at any point below the satellite. Scientists can then use this information to infer what the seafloor topography is like, so long as the sea surface truly mimics seafloor features.

14.4 Configuration of the Ocean Basins

Before we can observe the movement of ocean water, we need to identify the features that are part of the ocean basins—from the greatest depths to their junctions with the continents. As we will investigate in this chapter and the next, how ocean water moves—as currents, waves, and tides—is not only controlled by the sources of energy that cause the sea to move, but by the physical features that are present and act as barriers or deflectors for the flow of ocean water.

Where an ocean and continent meet is referred to as a *continental margin*. These locations are separated into two types determined by whether or not that margin lies along a tectonically active area—in other words, is that margin also a tectonic plate boundary? If the edge of the continent coincides with a convergent plate boundary (or, in rarer cases, a divergent plate boundary), then it is an **active continental margin**, like the one that surrounds the Pacific Ocean. **Passive continental margins** do not lie along a plate boundary, as is the case for both sides of the Atlantic Ocean.

You might ask why we divide up the continental margins into these types. The answer is that the features that form along the edges of continents will vary between active and passive margins—their geometric dimensions may be different, and some features are absent in one margin while present in the other, as illustrated in **FIGURE 14.5** and described below:

- **Continental shelf:** This is the area that you explore when you go to the beach for a swim or paddle. While still underlain by continental crust, it is relatively shallow, gradually sloping to a maximum angle of 0.3° and with water depths rarely exceeding 200 m (650 ft).

FIGURE 14.5 Features of passive and active continental margins.

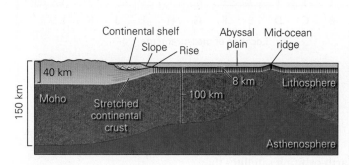

(a) Passive continental margin.

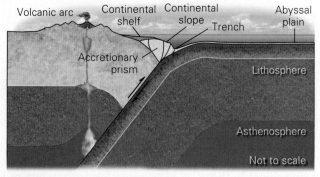

(b) Active continental margin.

- **Continental slope:** The seafloor drops more steeply here, marking the edge of the continental shelf. As the slope descends, water level increases from 200 m to nearly 4 km (2.5 miles).
- **Continental rise:** Where the continental slope ends, the seafloor angle decreases and water depths can reach up to 4.5 km (2.8 miles), marking the region called the continental rise.
- **Abyssal plain:** Here we find the flattest part of the seafloor, a vast, nearly horizontal region where water depths are around 4.5 km. The abyssal plain contains extinct volcanoes called **seamounts,** flat-topped seamounts known as **guyots,** and the mid-ocean ridges.
- **Deep-sea trench:** Along active margins, the continent drops steeply along the continental slope into a deep-sea trench marking the subduction zone. In these margins, the continental shelf is very narrow or completely absent, and there is no continental rise. From the trench, the seafloor rises up to the abyssal plain.

EXERCISE 14.4   Recognizing the Differences between Passive and Active Margins

Name: _____   Section: _____

Course: _____   Date: _____

Different sides of a continent can have different types of continental margins. For example, the west coast of the Americas is primarily an active margin, while the east coast is mainly passive.

(a) In the two images (below and on the next page), label the features you can identify for each continental margin, using the descriptions given above and Figure 14.5 to assist you. (If necessary, you can write your labels in the margins outside the map, with pointers to the feature.)

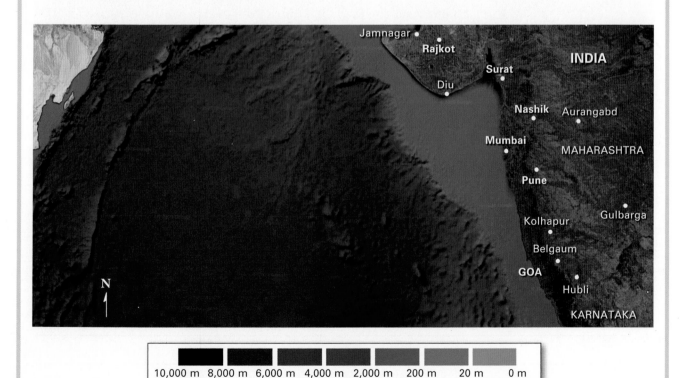

10,000 m 8,000 m 6,000 m 4,000 m 2,000 m 200 m 20 m 0 m

(continued)

Name: _____   Section: _____
Course: _____   Date: _____

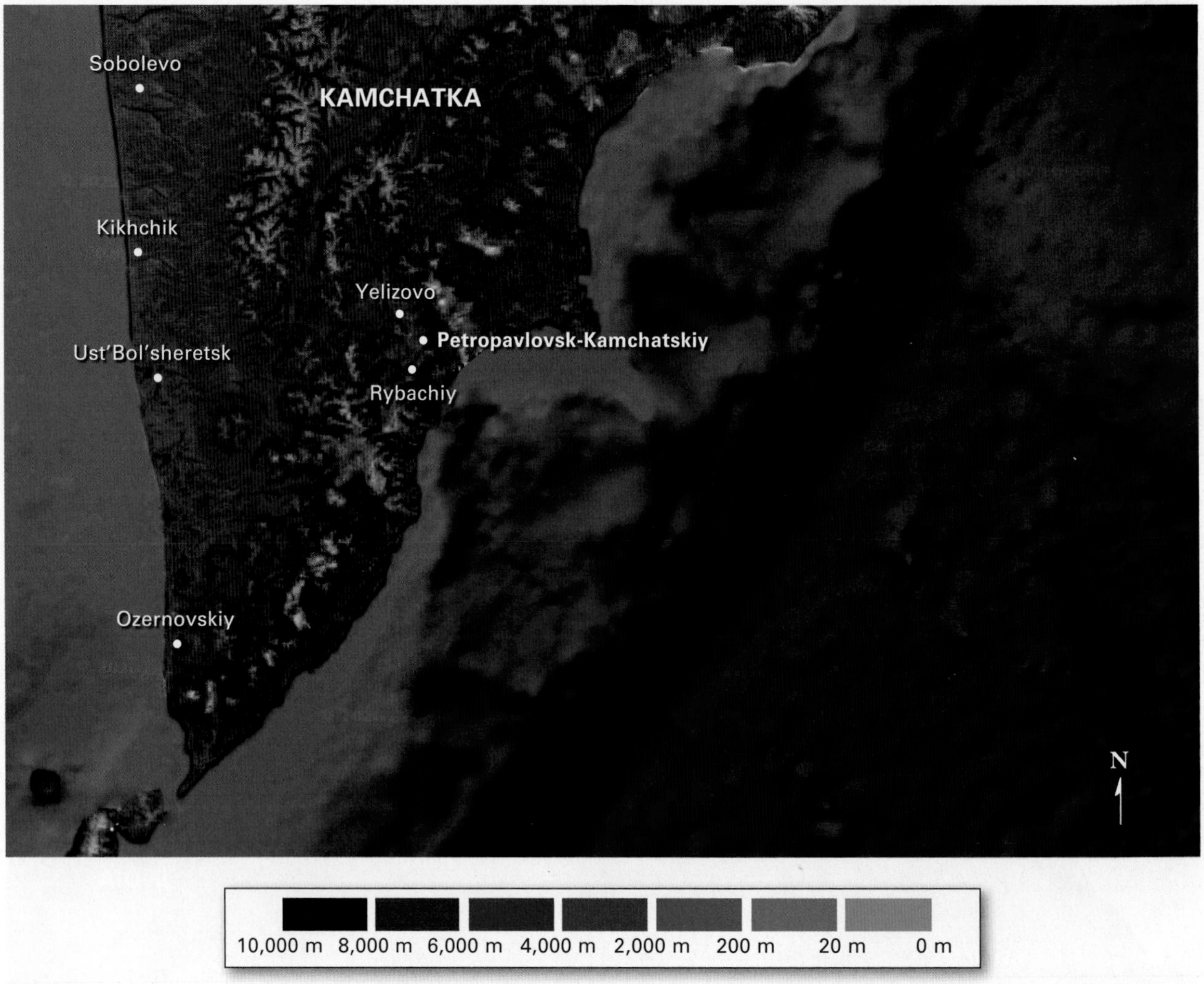

KAMCHATKA

Sobolevo

Kikhchik

Yelizovo

Petropavlovsk-Kamchatskiy

Ust'Bol'sheretsk

Rybachiy

Ozernovskiy

N

| 10,000 m | 8,000 m | 6,000 m | 4,000 m | 2,000 m | 200 m | 20 m | 0 m |

(b) The west coast of India, on the Arabian Sea, is a _____ continental margin. How can you tell?

_____

_____

_____

_____

(c) The east coast of the Kamchatka peninsula is a _____ continental margin. How can you tell?

_____

_____

_____

_____

14.5 Ocean Circulation at the Surface

Remember that the world's geographic oceans and seas are all connected into one ocean; their connector is a vast network of ocean currents that carry energy, nutrients, and oxygen across the globe. The movement of the upper 200 meters (656 ft) or so of the ocean surface also affects the climate of the whole Earth. The interaction between the atmosphere (layer of gases above the Earth's surface) and these surface movements of the ocean is vital not only for our climate, but for all life on the planet.

In this section, you will investigate how the ocean is divided into layers with invisible boundaries and how the movement of water occurs at the surface. In the next section, you will examine how seawater moves vertically, connecting the surface water to the deeper ocean.

14.5.1 Shallow versus Deep Layers

Anyone who has ever been in a large body of water like a lake or ocean knows that as the water gets deeper, it gets darker and colder. The upper part of the ocean—the upper 200 meters—where most of the light from the Sun penetrates, and where the energy from the atmosphere is transferred into the ocean, is known as the **surface layer**. The water molecules in this layer are constantly mixed and moved around by the wind and storms, and are heated by sunlight. Beneath the surface layer is the **thermocline**, stretching from 200 m to about 1,000 m in depth—in this layer, temperatures undergo dramatic changes (**FIG. 14.6**). As you go deeper into the ocean, it gets darker and colder because most of the sunlight has been reflected and absorbed by the water near the surface—this lower layer is called the **deep layer**. The water molecules in the deeper waters also move more slowly with increasing depth, as the energy from the atmosphere above is absorbed by the surface layer of water molecules, leaving less for the water molecules below.

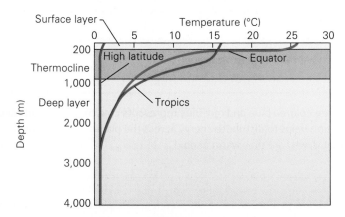

FIGURE 14.6 Layers of the ocean: Surface versus deep water. The colored lines show how temperature changes with depth at the different latitudes.

14.5.2 Atmospheric Circulation and the Ocean Surface

Water at the surface of the ocean is moved by the wind, generating surface currents. The frictional forces between the wind's moving air molecules and the ocean water molecules transfer energy from the atmosphere into the ocean. So it's not surprising that the dominant direction that the winds blow—called the prevailing winds—determines the direction in which the surface currents flow.

Name: _____   Section: _____

Course: _____   Date: _____

Let's investigate the link between winds and surface ocean currents. The figure below is an idealized map of the prevailing winds across the globe. Notice that these winds are separated by latitude. You will investigate this further in the chapters on atmospheric circulation.

(a) On the map, use arrows to draw in the direction that surface ocean currents are likely to flow in each of the oceans shown.

(b) Now, compare your answers above with the map of the surface ocean in **FIGURE 14.7**. In what ways, if any, does your sketch differ from the map?

_____

_____

What pattern or shape appears to define the movement of ocean currents?

_____

(c) Notice that the ocean currents in Figure 14.7 are colored blue and red. Blue represents cold ocean currents and red represents warm ocean currents. This circulation helps to distribute energy across the planet. Explain why the cold and warm currents occur where they do—that is, why are they warm or cold.

_____

_____

_____

(d) What do you think might be the effect on existing ocean currents if there was a change in either the strength or direction of the prevailing winds? Explain.

_____

_____

_____

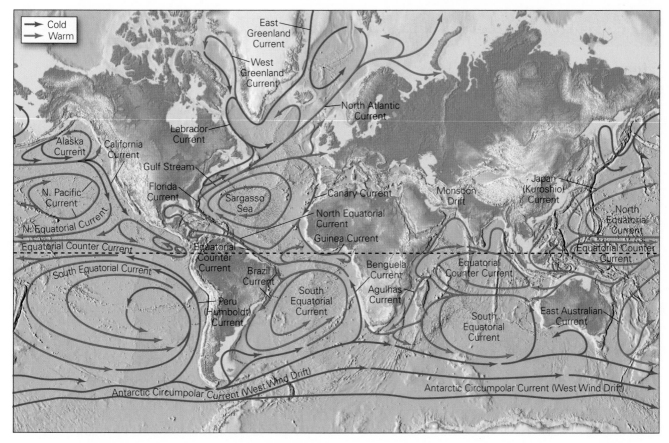

FIGURE 14.7 Map of ocean currents in the Pacific Ocean. Red arrows indicate warm currents, and blue indicate cold currents.

14.5.3 Coriolis Effect and Oceanic Gyres

If you look at the current patterns in Figure 14.7, they make circular loops, called **gyres**, around the ocean basins—clockwise in the northern hemisphere and counterclockwise in the southern hemisphere. Why is this? As we have seen, the surface ocean currents closely mirror the direction of the winds just above the ocean. We associate this with friction between the winds and ocean surface. However, we left out two important factors that affect surface circulation: the Coriolis effect and the pressure-gradient force.

We saw that the winds above all the ocean basins in the northern hemisphere flow clockwise around large, atmospheric high-pressure centers. The surface water, pushed along by the wind, would flow in the same direction. But, due to the Earth's rotation, the water also experiences a force that causes it to curve toward the center of the gyre, perpendicular to the wind currents. This is called the **Coriolis effect** (or Coriolis force). As water moves in this direction, it effectively "piles up" in the center of the gyre. Because of the pile-up, the ocean surface is slightly higher in the center of the gyre than it is closer to the coasts. This pile-up leads to a second force, called the pressure-gradient force, which acts outwardly, opposite the Coriolis effect. With the water piled up in the gyre, these two forces essentially balance, so the currents flow with the wind direction. The shape of the ocean basin (or configuration of the coastline) also affects where the currents ultimately flow. In Exercise 14.6, let's look more closely at how the Coriolis effect causes ocean currents to curve.

Name: _____   Section: _____
Course: _____   Date: _____

As you know, the Earth rotates to the east (remember that the Sun rises in the east and sets in the west!). Based on rotation alone, no matter where you are on the planet, the planet rotates to the east. But if we follow a moving object as it crosses over this moving planet, observers in the northern hemisphere sees objects deflected to the right of their path, while the deflection appears to be to the left for those in the southern hemisphere.

   You can prove this to yourself with a tennis ball and a marker and a bit of dexterity (or get a classmate to help you out).

(a) Draw a circle around the middle of the ball to represent the equator. Write "N" at one pole and "S" at the other pole. Now, look down at the ball so that the N is facing you. Let's say that east is to your left. As you turn the ball to the east, use a marker to draw a line straight down the ball from the N to the equator. You must be sure that your hand is moving in a straight line and not turning with the ball.

   Take a look at the line you drew. Is it straight or curved? _____ If curved, which direction did it curve?

   Now, flip the ball over so that the S is facing you. Turn the ball again to the east while drawing a new line straight from the S to the equator.

(b) Describe the line that you just drew. What shape does it make?

   _____

   _____

   Did you rotate the ball in the same or different direction from step (a)?

   _____

   While drawing, did you move the marker in a straight line or a curve?

   _____

   You have just demonstrated to yourself that as the oceans (your pen) move across the moving Earth (the tennis ball), the oceans appear to curve their paths. This curvature is a result of the rotating Earth.

(c) How would the patterns of ocean currents look different if the Earth was NOT rotating? Explain. (*Hint:* Refer to Exercise 14.5.)

   _____

   _____

   _____

   _____

Now try this out on a map view—a two-dimensional view of the ocean surface. In the box below, first label all the compass directions (N, S, E, W), with north at the top of the box. Then start at the dot and draw an arrow about 1-inch long, traveling toward the top of the box.

(continued)

Name: _____ Section: _____

Course: _____ Date: _____

(d) Let's imagine that we are in the northern hemisphere. In which compass direction will the arrow you drew be deflected? _____ Support your answer by continuing your drawing with another arrow that begins where the first arrow ended, applying the same rule of deflection.

(e) Do this again with another arrow, continuing where you left off with the arrow drawn in (d). And then repeat this step one more time, continuing from your last drawn arrow.

(f) Imagine the path you created was smoother rather than having sharp right angles. (That is, it behaved more akin to actual ocean currents.) If that were the case, what term would you use to describe the pattern you just drew?

_____

(g) Now, try it again in the space below. First label the diagram with the compass directions, and draw an arrow from the dot headed north. Then repeat steps (d) and (e) from above, but this time draw the deflections as they'd appear if you were observing ocean currents in the southern hemisphere.

In the above activity, we simulated how the ocean currents in the open ocean would be deflected as a result of Earth's rotation. But why does deflection also happen near the coasts of landmasses? Whenever a current encounters a landmass, such as a continent, it can be deflected either left or right along the coast, or both. The shape of the coastline likely influences which way the current travels. Once the current is bent from the coast, then the rotation of the Earth determines the flow direction across the ocean basin.

(h) Look at Figure 14.7. In the Atlantic Ocean, the South Equatorial Current flows up the coast of Brazil and eventually into the North Atlantic. Why do you think the water flows like this?

_____

_____

_____

_____

_____

Why do you think the Brazil Current flows in the direction shown in Figure 14.7?

_____

_____

_____

_____

14.5.4 Deviations from the "Norm": When Ocean Currents Change

Our observation of the circulation of surface ocean currents so far has been in terms of the "idealized" pattern. However, the patterns we think of as "normal" do not always occur. Because it is the global winds that drive the surface currents in the ocean, any change in these winds has an effect on the surface ocean. Change the surface ocean currents, and that in turn affects the weather across much of the planet for months. Undoubtedly you have heard of the term *El Niño* in the news or science media. It is less likely that you have heard of *La Niña*. Both of these events are blamed for severe changes in weather such as droughts and floods, although neither of them is a storm.

FIGURE 14.8A illustrates the "normal" conditions in the equatorial regions of the Pacific Ocean, which is the presence of cold ocean currents flowing along the Eastern Pacific Ocean and warmer waters in the Western Pacific Ocean. **El Niño**—or more precisely the El Niño–Southern Oscillation (ENSO)—is the occurrence of unusually warm water in the Eastern Pacific Ocean (**FIG. 14.8B**). **La Niña**, meanwhile, is an extreme version of "normal" conditions where the water off the coast of the Americas (primarily Peru and Ecuador) is extremely cold. In both cases, the change in the ocean currents along the equatorial Pacific Ocean affects the weather mainly along the Pacific Ocean, but they can also have impacts further away, such as in the eastern United States.

The causes of these events are not fully understood, but appear to be related to the weakening or strengthening of the winds that move the equatorial surface currents. Under "normal" conditions in the Pacific Ocean, the northeast and southeast trade winds blow toward and then westward along the equator, generating the

FIGURE 14.8 La Niña and El Niño.

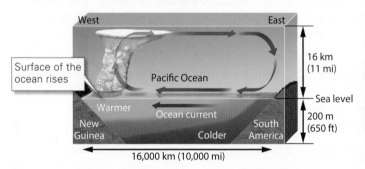

(a) Normal circulation in the equatorial Pacific. During La Niña, this circulation strengthens.

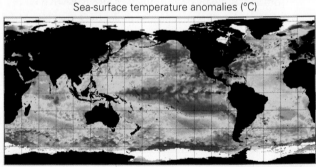

(b) Sea-surface temperature anomalies during La Niña.

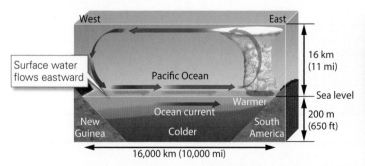

(c) During El Niño, the flow of surface water in the equatorial Pacific changes.

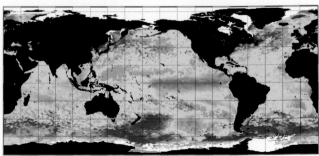

(d) Sea-surface temperature anomalies during El Niño.

North and South Equatorial Currents. These winds cause warm water to be pushed away from the Americas and to pile up in the western Pacific near Australia and Indonesia. This removal of warm surface waters along the eastern Pacific Ocean causes colder ocean water to flow in to replace it. This colder water flows along the coast of North America and South America, coming either from the surface (as in the Peru current) or from the deep waters below (called **upwelling**). This upwelling brings nutrient-rich waters to the surface, which in turn profoundly increases the abundance of marine life in those waters. In later chapters on atmospheric circulation, we will investigate the interactions between the ocean and the atmosphere that have dramatic effects on the weather and drive the climate of the planet.

14.6 Thermohaline Circulation

While the wind is the main driving force that moves the ocean at the surface, this water doesn't always remain at the surface. When the density of seawater increases, either because it gets colder or saltier, it sinks to the deeper layer of the oceans. As it sinks, it slowly moves within deep ocean currents that hug the ocean floor. In places where this water mixes with less dense waters, its density decreases and so it rises to the surface to join the surface ocean currents. This link between the surface and deeper waters is vital for life on the planet; it transports nutrients and oxygen vital to ocean life throughout the entire depth of the ocean, and much of that oxygen is also released into the atmosphere. These currents control our climate by distributing heat (energy) around the globe, which also helps to keep our planet habitable. This movement of water from the surface to the depths forms a great conveyor belt that is illustrated in **FIGURE 14.9**. This density-driven circulation, which begins in the North Atlantic, is called **thermohaline circulation** because the two factors that control the density of seawater are heat (*thermo-*) and salinity (*-haline*). **Salinity** is the measure of the concentration of salt in water. In this section, you will investigate and observe these effects on density, first looking at temperature in Exercises 14.7 and 14.8, and then salinity in Exercise 14.9.

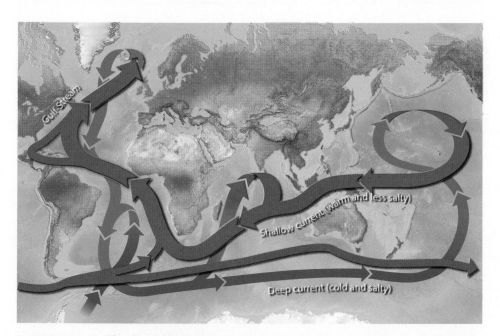

FIGURE 14.9 The thermohaline circulation results in a global-scale conveyor belt that circulates water throughout the entire ocean system. Because of this circulation, the ocean mixes entirely in a 1,500-year period.

14.6.1 Temperature Effects on Density

The intensity of solar energy—or *insolation*—that the surface of the Earth receives varies with latitude. While the equator receives the most direct sunlight as well as consistent daylight hours, the high latitudes, especially those above the Arctic and Antarctic circles (60°N and S latitudes, respectively) receive significantly less intense solar energy. While this variation in solar energy affects sea surface temperatures, the temperature of seawater also varies with depth. Most of the sunlight that reaches the ocean is reflected and absorbed at the surface; therefore, with increasing depth the waters get darker and colder. This unequal distribution of solar energy not only affects where life is most abundant in the ocean (nearer to the surface), but also the density of seawater.

EXERCISE 14.7 **Observing the Effects of Temperature on Density: An Experiment**

Name: _____   Section: _____
Course: _____   Date: _____

Your instructor has provided you with water, food coloring, graduated cylinders, ice, a hot plate, and test tubes. (*Note:* If this exercise is available and being conducted online, the setup, materials, and directions may differ. See the instructions in the online lab for how to conduct this exercise.)

Step 1: If hot water is not available in your lab room, heat up approximately 250 ml of tap water on a hot plate. You want the water to be hot but not boiling.

Step 2: Fill a graduated cylinder approximately two-thirds full with *cold* tap water.

Step 3: Add 2 drops of food coloring to a test tube. Fill the test tube half full with *hot* water. Then, slowly pour the hot water solution from the test tube into the graduated cylinder filled with cold water.

   (a) Describe what you observe:

   _____

   _____

   _____

Step 4: Empty the graduated cylinder and rinse the test tube. Place 2 drops of food coloring in the test tube and fill it with cold tap water. Place the test tube in a beaker of ice water and let it cool for several minutes.

Step 5: Fill a second graduated cylinder two-thirds full with *hot* water. Slowly pour the *chilled* test tube water into the graduated cylinder.

   (b) Record your observations:

   _____

   _____

   _____

Step 6: Clean all items and return them to the instructor.

   (c) Summarize the result of your experiment (that is, describe what happened):

   _____

   _____

   _____

   _____

(continued)

Observing the Effects of Temperature on Density: An Experiment (*continued*)

Name: _____   Section: _____
Course: _____   Date: _____

(d) Based on your experiment, given the same amount of water, which water will be more dense, warm or cold? Explain using evidence from your experiment:

_____

_____

_____

(e) In terms of the density equation (density = mass ÷ volume), which factor is changing in this experiment, mass or volume? _____

EXERCISE 14.8 **How Does Density Vary by Latitude?**

Name: _____   Section: _____
Course: _____   Date: _____

You demonstrated how density varies with temperature in the previous experiment. Now, let's observe how density of the ocean surface varies with latitude. Look at the graph in **FIGURE 14.10A**, which shows the sea surface temperature at various latitudes.

(a) At which latitude(s) do you expect surface water to be most dense? Give evidence to support your choice.

_____

_____

_____

_____

(b) Where would you expect the least dense surface water to be located? Explain.

_____

_____

_____

_____

FIGURE 14.10B shows the difference in sea surface temperature during the summer and winter seasons.

(c) Explain why there is a seasonal change in water temperature.

_____

_____

_____

_____

(continued)

Name: _____   Section: _____

Course: _____   Date: _____

(d) Why might this change be greater at the mid-latitudes (~40°N and S) than at the polar regions?

_____

_____

_____

(e) During which season is the surface water at the mid-latitudes most dense? _____

FIGURE 14.10 Ocean
temperatures by latitude.

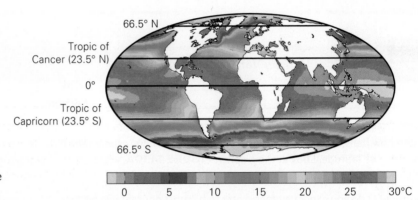

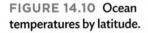

(a) Annual average sea-surface
temperatures.

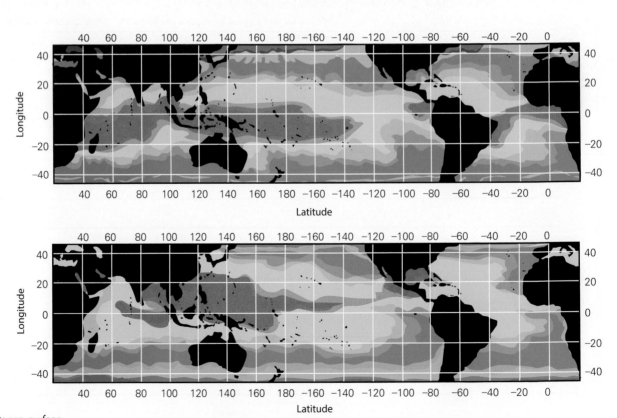

(b) Average sea-surface
temperatures by season: January
(top) and July (bottom).

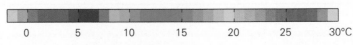

14.6.2 Salinity Effects on Density

The amount of dissolved substances (or salts) in water is known as **salinity**. Seawater has an average salinity of 35‰ (‰ is parts per thousand). This means that in 1,000 g of seawater, there are 35 g of salt (**FIG. 14.11**). In other words, if you evaporated 1,000 g of seawater, then there would be 35 g of salt crystals left behind. But the actual salinity of individual areas of the ocean is not the same because the amount of salt that is in seawater depends not only on how much salt enters the ocean, but also on the processes of the hydrologic cycle—primarily evaporation and dilution processes—that control the amount of freshwater in the ocean.

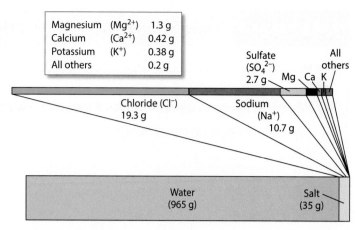

FIGURE 14.11 **The average chemical composition of seawater, in grams per liter.**

EXERCISE 14.9 | **Observing the Effects of Salinity on Density**

Name: _____ Section: _____

Course: _____ Date: _____

For this experiment your instructor has provided you with salt, freshwater, graduated cylinders, a mass balance, food coloring, and a stopwatch. (*Note:* If this exercise is available and being conducted online, the setup, materials, and directions may differ. See the instructions in the online lab for how to conduct this exercise.) You will be making two saltwater solutions (A and B), and determining their relative densities with an experiment that investigates the effects of salinity on the density of seawater.

Before you begin the experiment, form a hypothesis about what you think happens to the density of water as salinity changes, and explain your prediction. Remember that density = mass ÷ volume; in other words, density changes (increases or decreases) when either the mass or volume changes.

_____

_____

_____

(a) What is the volume of saltwater **solution A** that your instructor has asked you to make? _____ ml

How many grams of salt do you need to make solution A? _____ g

Of water? _____ g

Show your calculations in the space below:

(continued)

Name: _____   Section: _____

Course: _____   Date: _____

(b) What is the volume of saltwater **solution B** that your instructor has asked you to make? _____ ml

How many grams of salt do you need to make solution B? _____ g

Of water? _____ g

Show your calculations in the space below:

Check your answers with your instructor before you continue.

Step 1: Fill a graduated cylinder two-thirds full with freshwater. Place a rubber band or other marker at the top of the column of water so that you always use the same amount of water for each trial.

Step 2: Place 2 drops of food coloring in a test tube and fill the test tube half full with **solution A**.

Step 3: When you're ready, slowly pour the test tube contents into the graduated cylinder and time how long it takes for solution A to first reach the bottom of the cylinder. Record this time (Trial 1) in the table below.

Step 4: Rinse out the cylinder and refill with freshwater to the marker you placed in Step 1. Again, put food coloring and solution A in a test tube and repeat step 3, recording the time of this trial (Trial 2) in the table.

Step 5: Repeat steps 1 through 4 for **solution B** and record the times of these two trials in the table.

Step 6: Calculate the average of the two trials for solution A and for solution B and write your answers in the table.

Data table for salinity-density experiment trials.

	Solution A	Solution B
Trial 1 time (seconds)		
Trial 2 time (seconds)		
Average of Trials 1 and 2 (seconds)		

(c) Summarize the results of your experiment:

_____

_____

_____

(continued)

Name: _____   Section: _____

Course: _____   Date: _____

(d) Based on the results of your experiment, which solution is denser? Explain your choice.

_____

_____

(e) Circle your answer from the choices given: The higher the salinity of seawater, the (higher/lower) the seawater density.

(f) Was your initial hypothesis correct? Explain:

_____

_____

As you observed in the salinity-density experiment, the salinity of seawater has an effect on the density of seawater. Now we need to answer *why* the salinity varies in the oceans. As we have seen in the previous section, the temperature of seawater can vary, and with that change comes a difference in how much salt water can hold. In general, warm water can hold more dissolved salt than cold water. Other factors that affect salinity include the mixing of surface waters due to currents and individual storm events. But for the surface ocean in general, processes of the hydrologic cycle control the amount of freshwater in the oceans, thereby controlling surface salinity over long time spans. Processes that increase salinity, such as evaporation and the formation of sea ice, remove liquid water molecules from the ocean, leaving the salts behind. Processes of dilution—precipitation, runoff from rivers and melting glaciers, and the melting of sea ice—cause salinity to decrease as freshwater is added to the ocean.

In Exercise 14.10, we will investigate how salinity varies with depth in the ocean, at different latitudes, and at the same latitude in different ocean basins.

EXERCISE 14.10   How Does Salinity Vary in the Ocean?

Name: _____   Section: _____

Course: _____   Date: _____

Use the graph of average global ocean salinity versus depth shown at right to answer the following questions.

(a) Describe the pattern or trend of salinity in the surface layer (top 200 meters).

_____

_____

_____

_____

_____

Salinity variation in the ocean. (The *halocline* is the layer in the ocean in which salinity levels change most dramatically with depth.)

(continued)

Name: _____   Section: _____

Course: _____   Date: _____

(b) What might be the cause of this pattern?

_____

_____

(c) Describe the pattern or trend of salinity in the deeper waters (below the halocline).

_____

_____

(d) What is a possible explanation for the trend you observe in deeper waters?

_____

_____

Now let's compare the surface salinity at various latitudes in the Atlantic Ocean, as shown in the map below.

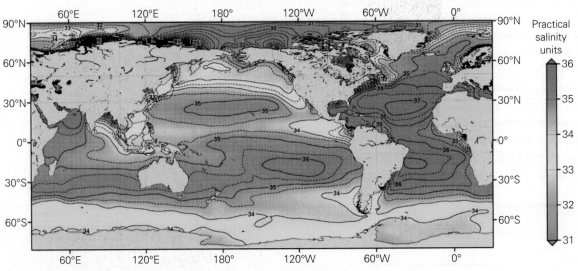

Global sea surface salinity values averaged from 1955 to the present. [Units shown are Practical Salinity Units (PSU), a measure of salinity based on electrical conductivity. Higher PSU values equal higher salinity.]

(e) Compare surface salinity at the high latitudes (polar regions) to the surface salinity in the mid-latitudes.

_____

_____

Explain the reason for any differences observed.

_____

_____

_____

(continued)

Name: _____   Section: _____

Course: _____   Date: _____

Lastly, compare the surface salinities in the North Pacific Ocean to those in the North Atlantic Ocean in the map above.

(f) Which ocean has the highest surface salinities? _____ Explain why that ocean has higher salinity values.

_____

_____

_____

_____

(g) Based on your experiments and observations in this section, explain why global thermohaline circulation begins in the North Atlantic Ocean.

_____

_____

_____

_____

? What Do You Think As the average temperature of the oceans continues to increase, and more freshwater enters the ocean, scientists and global leaders are interested in the potential impacts of these changes. Why should we care about these changes? There are numerous reasons—some occurring now and others in the near future—including but not limited to:

- Changes in climate that would lead to warmer summers and colder winters. This could have enormous impact on the physical conditions people experience during these times of the year.
- Increased storm severity.
- Coastal flooding with rising sea levels, and the resulting displacement of millions of people.
- Negative impacts on nutrient and heat distribution in the ocean.

The last impact has more severe consequences than it may seem at first. Life in the ocean is sensitive to changes in nutrient levels and temperature. For example, without the right mixture of nutrients and temperatures, ocean creatures that are the basis of the food web will perish, and those predators who depend on them will also suffer. Declining stocks of fish and increasing spread of disease with rising water temperatures are concerns for fisheries and the economies that depend on their success.

On a separate sheet of paper, identify one potential immediate impact and one potential longer-term impact of warming oceans where you live or go to school. Explain your answers.

Name: _____   **Section:** _____

Course: _____   **Date:** _____

Exploring Earth Science Using Google Earth

1. Visit **digital.wwnorton.com/labmanualearthsci**
2. Go to the **Geotours** tile to download Google Earth Pro and the accompanying Geotours exercises file.

Expand the Geotour14 folder in Google Earth by clicking the triangle to the left of the folder icon. The folder contains lines crossing an **active margin** (nearby active plate boundary) and a **passive margin** (not nearby active plate boundary) with placemarks keyed to questions that investigate features along each transect.

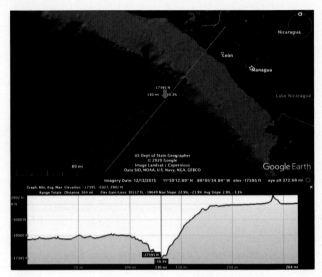

(a) Check and double-click the Active Margin line to fly to the SW coast of Nicaragua. Right-click (or Control-click on a Mac) on the line and select Show Elevation Profile, and then check the (a) placemark to highlight the deepest feature present on the profile. Move your mouse along the Elevation Profile and watch the corresponding red arrow with elevation values in the viewer window. Knowing that this is an active plate boundary, what does this narrow, deep feature likely represent?

_____

(b) Check and double-click the (b) placemark, and identify the corresponding feature on the Elevation Profile. Is this section of the profile a **continental shelf** (submerged continental crust with a gentle, shelf-like slope), **continental slope** (relatively steep transition between continental and oceanic crust), or **abyssal plain** (deepest regions of the ocean underlain by oceanic crust)?

_____

(c) Close the Elevation Profile by clicking the "X" on the right side of the window. Check and double-click the Passive Margin line to fly to the east coast of the United States. Display the Elevation Profile, and then check the (c) placemark. Is placemark (c) highlighting the continental shelf, continental slope, or the abyssal plain?

_____

(d) Check and double-click the Bermuda placemark. Bermuda is mostly composed of ~33 Ma volcanic rocks surrounded by ~110-120 Ma abyssal plain rocks. Did Bermuda likely form at the mid-ocean ridge or as a result of some form of plume activity? How can you tell?

_____

Coasts and Energy

Coastlines come in a wide variety of types, including rocky, sandy, and mangrove swamps.

15.1 Introduction

Hundreds of millions of people live along North America's Atlantic, Gulf of Mexico, Pacific, and Great Lakes shorelines, and millions more vacation in these places every year. Recent hurricanes such as Harvey, Irma, and Maria, all from 2017, remind us that shorelines are the most dynamic places on Earth, able to change dramatically in a few hours or days in ways that can endanger lives and property. Shoreline landscapes are unique places where the relatively stable geosphere interacts with the much more rapidly changing hydrosphere and atmosphere as wind generates waves that crash onshore, eroding and transporting materials. Humans will always live along shorelines, but we must do so intelligently to lessen the impact of these rapid changes, and we must plan wisely as global climate change increases the risk of future catastrophic events. This chapter examines the nature of shorelines, the processes that create and destroy their distinctive landforms, the effects of human activity on those processes, and the disastrous events to which shorelines are uniquely subjected.

15.2 Factors Controlling Shoreline Formation and Evolution

There are many types of shorelines: examples range from rocky cliffs, gentle sandy beaches, untouched coral reefs, marshes, and mangrove forests to heavily urbanized coastlines. As mentioned above, shoreline landscapes can change rapidly. The nature and rapidity of the changes are controlled by four important factors:

- The material of which the shoreline is made
- Weather and climate
- Tidal range
- Tectonic activity

15.2.1 Shoreline Materials

The type of shoreline material of which the shoreline is made is one of the most important factors that controls the effectiveness of erosional and depositional processes and, therefore, shoreline stability. Not all shorelines are sandy beaches—some are made of corals, rocks or boulders, shells, marshes or mangrove forests, or human-made piers or jetties, and have little or no sand (**FIG. 15.1**). Some, like the southern end of Manhattan Island in New York City, parts of San Francisco, and other harbor cities, have been completely modified by human construction. And even when there are beaches, they can come in many colors, depending on the materials from which the sand is derived (**FIG. 15.2**).

15.2.2 Weather and Climate

Waves are the dominant force in shoreline erosion and deposition, and the height and velocity of waves in a region are controlled by weather factors such as the types of weather systems present, the strength and direction of winds, and the frequency of storms. Winds also control how loose materials will be moved along the shoreline in a process called *longshore drift* (discussed in Section 15.3.3), and they can have devastating effects on human-made parts of the shoreline.

shoreline at West Quoddy Head in Maine.

(b) Sandy and rocky shoreline, Hawaii.

ef along the Hawaiian shoreline.

(d) Salt marsh along the coast of Cape Cod, Massachusetts.

e forest in northeastern Brazil.

(f) A seawall built at the southern tip of Manhattan (New York City).

FIGURE 15.2 Beach sand comes in many colors.

(a) A sandy beach in Oregon.

(b) A white-sand beach in the Dominican Republic.

(c) A black-sand beach in Maui, Hawaii.

(d) A green-sand beach on the island of Hawaii.

(e) A pink-sand beach in Indonesia.

(f) A red-sand beach in Maui, Hawaii.

EXERCISE 15.1 Erodibility and Stability of Shoreline Materials

Name: ______________________________ Section: ____________
Course: ______________________________ Date: ____________

Review the types of shorelines shown in Figure 15.1 and answer the following questions:

(a) Which of these shorelines do you think would be the most difficult to erode? Which would be the easiest? Explain your choices in a few sentences.

____
____
____
____

(b) Compare the bedrock cliffs in Figure 15.1a and 15.1b. What factors will determine how resistant to erosion these cliffs are?

____
____
____
____

(c) Compare the human-made shoreline in Figure 15.1f with the natural shorelines in the other photographs of the figure. Identify one natural shoreline that may be stronger than the human-made shoreline in part (f) and one that may be weaker. Explain your reasoning for each choice.

____
____
____
____

(d) Indicate with arrows the wind directions that generate the waves shown in Figure 15.1b. Explain your reasoning.

____
____
____
____

(e) Which shoreline do you think would be most damaged by high winds and high waves? Explain why you made your choice.

____
____
____
____

Climate change has a profound effect on shorelines. During an ice age, as we've seen, glaciers advance across the continents, locking enormous amounts of water in ice that would otherwise have been in the ocean, causing sea level to drop. Today, global warming is having the opposite effect, melting the Greenland and Antarctic ice caps as well as most mountain glaciers, causing sea level to rise. And global warming isn't just adding meltwater: as the oceans get warmer, the water in the upper 700 m expands, a phenomenon called *thermal expansion*. In the past 25 years, thermal expansion has accounted for about 50% of the total global rise in sea level.

EXERCISE 15.2 **Effects of Climate Change on Shorelines**

Name: _____   Section: _____
Course: _____   Date: _____

(a) What would be the effect on the world's **oceans** if continental glaciers worldwide expanded by 10%? Explain.

_____

_____

_____

(b) What would be the effect on the world's **shorelines** if continental glaciers worldwide expanded by 10%?

_____

_____

_____

(c) Conversely, what would be the effect on the world's oceans and shorelines if continental glaciers shrunk by 10%?

_____

_____

_____

(d) About 2 million years ago, much of northern North America, Europe, and Asia were covered with continental glaciers. In what way was the location of the world's shorelines at that time different from that of today's shorelines? Explain your reasoning.

_____

_____

_____

15.2.3 Tidal Range

The gravitational attraction of the Sun and Moon causes the tides—the rising and falling of water along the shoreline—that occurs twice a day in most coastal areas. **Tidal range** (the total change between high and low tide) is typically a few feet, but it is much greater in some places because of coastal geometry. The Bay of Fundy, between the Canadian provinces of Nova Scotia and New Brunswick, has the highest

FIGURE 15.3 Tidal range at Mont-Saint-Michel, France.

(a) At low tide.

(b) At high tide.

tidal range worldwide, as much as 16 meters (about 53 feet). Mont-Saint-Michel in France ranks "only" fourth, at 14 meters (about 46 feet), but as **FIGURE 15.3** shows, the result is dramatic. The tides move enormous amounts of sediment all along the shoreline, and fluvial processes extend across exposed tidal flats at low tide. As we shall see later, a major factor in determining how much damage coastal storms will cause when they come onshore is whether they arrive at high tide or low tide.

15.2.4 Tectonic Activity

Tectonic activity creates ocean basins, enlarges or shrinks them, and may uplift or lower the land along the coast. When an ocean widens due to seafloor spreading, its water must occupy the larger basin and sea level drops. Conversely, when an ocean closes due to subduction, its basin becomes smaller and as a result sea level rises. In addition, when submarine volcanoes build their cones above sea level, they create new lands (**FIG. 15.4**); such lands include heavily populated island nations such as Japan, Indonesia, and the Philippines, as well as the state of Hawaii. These islands are in a constant state of change, as waves erode the land that the new lavas build.

FIGURE 15.4 Growth of the island of Hawaii by addition of lava from Kilauea to the shoreline.

(a) Red-hot lava from the volcano entering the ocean.

(b) New land created by lava flows in the 1940s.

Name: _____   Section: _____

Course: _____   Date: _____

(a) What effect will continued seafloor spreading in the Atlantic Ocean have on East coast and Gulf Coast sea level? Explain your reasoning.

_____

_____

_____

(b) What effect would partial closing of the Atlantic Ocean have on sea level? Explain.

_____

_____

_____

_____

15.2.5 Emergent and Submergent Shorelines

Short-term fluctuations in sea level occur sporadically during major storms and twice daily with the tidal cycle, but long-term sea-level changes are caused by tectonic activity and climate change. Geologists commonly group shorelines into two categories based on how they respond to these long-term changes:

1. Along **submergent shorelines**, the land sinks or sea level rises. The shoreline appears to be drowned, with an irregular coastline, prominent bays, and abundant islands, marshes, and lagoons.
2. Along **emergent shorelines**, the land rises or sea level drops. Emergent shorelines are typically straight and bounded by steep cliffs; where tectonic uplift has occurred, remnants of former shoreline features may be found well above sea level.

The following exercises will give you some practice in recognizing different types of shorelines, the processes that create them, and the ways in which they have changed over time.

EXERCISE 15.4   **Recognizing Emergent and Submergent Shorelines**

Name: _____   Section: _____
Course: _____   Date: _____

Emergent shorelines look very different from submergent shorelines. Examine the following topographic maps showing part of the Atlantic coast in Maine (**FIG. 15.5**) and part of the Pacific coast in California (**FIG. 15.6**). One of these coasts is a typical submergent shoreline, the other a classic emergent shoreline. Apply your geologic reasoning to determine which is which.

(continued)

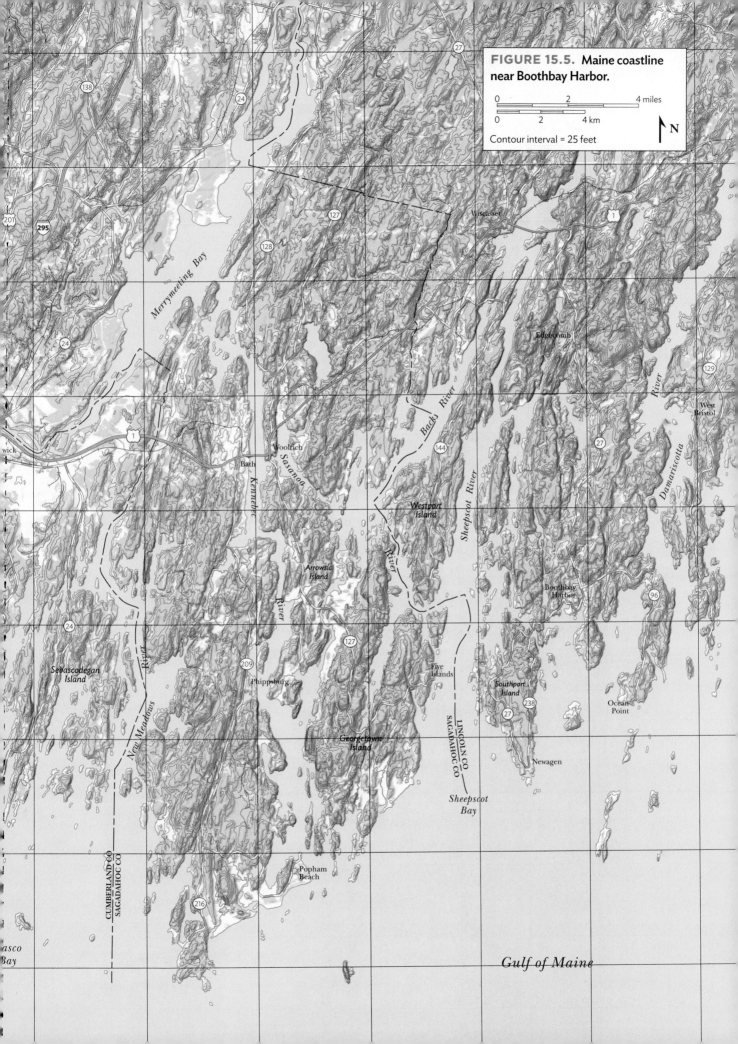

FIGURE 15.5. Maine coastline near Boothbay Harbor.

0   2   4 miles

0   2   4 km

Contour interval = 25 feet

N

FIGURE 15.6 California coastline south of Half Moon Bay.

Contour interval = 40 feet

N

Name: _____   **Section:** _____
Course: _____   **Date:** _____

(a) How do the shapes of the two shorelines differ?

_____

_____

_____

(b) Which of these shorelines is emergent and which is submergent? Explain your reasoning

_____

_____

_____

_____

EXERCISE 15.5   Measuring Sea-Level (and Lake-Level) Change

Name: _____   **Section:** _____
Course: _____   **Date:** _____

The map below shows details of the California coast.

Portion of the Dos Pueblos Canyon quadrangle of California.

(continued)

Name: _____ Section: _____

Course: _____ Date: _____

(a) Describe this shoreline in your own words.

_____

_____

(b) Sketch topographic profiles along lines A–B and C–D using the graph paper provided at the end of this chapter.

(c) What evidence shows that a change in sea level has taken place here?

_____

_____

(d) Based on the map and profile, is this an emergent or submergent shoreline? By how much has sea level changed? Explain your reasoning.

_____

_____

_____

_____

_____

15.3 Wave Action

In this section, we will learn more about how a shoreline changes. As we've seen, waves are the dominant agents of shoreline erosion and deposition. They erode coastal materials when they strike the shore, and deposit loose materials to form beaches and other landforms. Waves also generate currents that parallel the shoreline and build landforms offshore.

A few basic principles explain how waves erode and deposit materials along shorelines and move sediment directly, forming landforms such as barrier islands and spits.

- Shorelines are in a constant state of conflict between destructive coastal erosion, which removes material, and the constructional processes of wave deposition, lava flows, and coral reef development, which add material.
- Waves are generated by the interaction between wind and the surface waters of oceans and large lakes.
- The kinetic energy of waves causes erosion and re-deposition of *unconsolidated sediment.*
- Like streams and glaciers, waves carry loose sediment that abrades the bases of solid bedrock cliffs. Waves move sediment back and forth across the tidal zone (the area between the high tide and low tide lines), undercutting the cliff. When support of its base is undermined, the cliff collapses. Waves then erode the resulting rubble, exposing the base of a new cliff, and the cycle repeats. In this way, shoreline cliffs gradually retreat inland.

FIGURE 15.7 Mechanics of wave action offshore and near the shoreline.

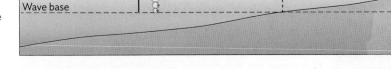

Within a deep-ocean wave, water molecules move in a circular path. The radius of the circle decreases with depth.

Wavelength | ½ wavelength

Wave base

In the *breaker zone*, the wave base intersects the seafloor; here waves lose symmetry and "break."

- Wind itself also moves sediment in shoreline environments, forming coastal sand dunes.
- Shoreline currents redistribute sediment to produce barrier islands, spits, and other landforms.
- We will look first at the basics of wave mechanics and then at how waves produce shoreline landforms.

15.3.1 How Waves Form

Waves are formed by the friction generated when wind blows across the surface of an ocean or lake. The symmetric shape of waves offshore shows how the kinetic energy of the wind is transferred to the water. Offshore, this energy moves water molecules in circular paths (**FIG. 15.7**, left side). Each molecule passes some energy on to the other molecules it contacts, but some energy is lost with each contact. These losses limit the depth of wave action to approximately half of the wavelength, a depth referred to as the **wave base**. When water is shallower than the wave base, as on the right side of Figure 15.7, the seafloor or lake bottom interferes with the orbiting water molecules, causing waves to lose symmetry and "break" onshore. The pileup of water in this breaker zone increases the waves' kinetic energy, enhancing their ability to erode the shoreline. **FIGURE 15.8** shows a breaker zone along the coast in California. The wind that generated these waves blew over thousands of miles of ocean uninterrupted by land or trees; these conditions are the most favorable condition for wave formation.

15.3.2 Wave Characteristics

In deeper parts of the ocean where the wave base is not in contact with the seafloor and the water molecules move in circular orbits, the waves are referred to as deep-water waves, or **swells**. As the waves approach shore and the water depth is shallower, the wave base comes into contact with the seafloor or "feels bottom," becoming a transitional wave. As the wave base touches the seafloor, the orbits of the water molecules flatten and some of the wave energy is transferred to the seafloor. This increase in friction between the water and the seafloor causes the base of the wave to slow down, but the top of the wave continues to move forward. As the wave slows down, the wavelength decreases while the wave height increases, forming a shallow-water wave that ultimately crashes forward as a breaking wave and forms surf.

We can determine where the deep-water waves become shallow-water waves if we know the wavelength (distance between the crests of two consecutive waves). The minimum water depth (D) where the wave base comes in contact with the seafloor is equal to half the wavelength (L), or simply

$$D = \frac{L}{2}$$

FIGURE 15.8 Waves striking the shoreline at a breaker zone in California at low tide.

For deep-water waves, the water depth is greater than ½ *L*; for transitional waves, the water depth is equal to ½ *L*; and for shallow-water waves the water depth is between ¹⁄₂₀ *L* and ½ *L*. Visually, we can see this transition by looking at the sea surface and noting where the wave height begins to increase, the wavelength shortens, and the wave eventually breaks.

EXERCISE 15.6   **Determining When Waves Will "Feel Bottom"**

Name: _____   Section: _____
Course: _____   Date: _____

(a) On the image below, label deep-water waves, transitional waves, and shallow-water waves based on your observations of the waves in the image. Use a ruler to assist you in measuring wavelengths.

Northwest coast of Oahu, Hawaii

(b) Justify your choices for the locations of the three wave types:

_____

_____

_____

(c) Why might it be important to distinguish between deep- and shallow-water waves for:

Swimmers and surfers?

_____

_____

Harbor pilots, sea captains, or naval architects?

_____

_____

_____

15.3.3 Refraction, Longshore Drift, and the Transportation of Coastal Sediment

If you look carefully at waves as they approach a shoreline, you will notice that most waves do not wash up parallel to the beach; rather, they come in at an angle. As the deep-water waves approach land, they enter shallower water, or run into obstacles, and as you investigated in the previous section, this causes part of the wave to lose energy to the seafloor (or obstacle) and slow down. This slowing causes the wave crest to bend and become more parallel to the shoreline. This process is referred to as wave **refraction**. Even with refraction, though, most waves still strike a shoreline at an oblique (slanted) angle, and thus carry sediment onto the beach along that angled path. As a wave recedes, gravity pulls it straight down the slope of the beach along a path that is perpendicular to the shoreline. In areas where there is an abundant supply of sediment, longshore currents move sand and silt parallel to the shoreline in a process called **longshore drift**. Over many such zigzag cycles, sediments gradually move along the beach as *beach drift*—in what looks like a straight-line path to someone who hasn't been watching closely (**FIG. 15.9**).

Similarly, oblique waves generate currents just offshore that parallel the shoreline. These currents are responsible for moving large volumes of sediment and for building distinctive elongate landforms.

In order to manage this movement of sediment along shorelines and maintain current coastal shapes, beaches, and harbor entrances, humans have built structures along the coasts that interrupt longshore drift—either to limit erosion or promote deposition of sediment. We will examine these features in section 15.5. However, many of these approaches, while fulfilling their purpose of moving locations of erosion and deposition, also have negative impacts—overall, all approaches tend to increase rates of erosion or focus the erosion in other areas. In the following experiment, you will investigate and observe how waves approach shore and how the shape of the shoreline and man-made structures impact the location of sediment deposition and erosion.

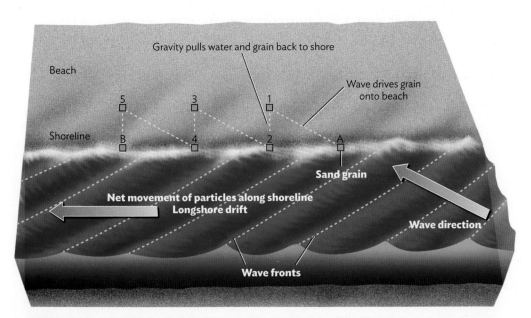

FIGURE 15.9 Mechanics of longshore drift. Grain A eventually moves to Point B by a complex zigzag path shown by the dashed lines. A wave drives the grain (A) up the beach, in the direction that the wave is moving, to Point 1. Then the water is pulled back downslope to the shoreline, carrying the grain with it to Point 2. As these processes are repeated, the grain is transported parallel to the shoreline.

Name: _____ **Section:** _____
Course: _____ **Date:** _____

Your instructor has provided you with materials to simulate different coastal environments: a sandy beach, a rocky coastline, and an inlet to a harbor. For each set-up you will generate waves with the tools provided (fan, paddles, etc.) while observing the movement of the wave crests as the waves approach shore and any obstacles. You'll also observe the movement of sand by wave action. (*Note:* If this exercise is available and being conducted online, the setup, materials, and directions may differ. See the instructions in the online lab for how to conduct this exercise.)

Before you begin, predict how the waves will interact with each of the coastal features provided.

_____

_____

_____

Step 1: When you are ready, use the plastic paddle to move the water gently back and forth, generating waves, and move the paddle only between the marks on the side of the box. Keep the size of your waves consistent. Make waves for about 5 minutes or for as long as it takes for the sand to be moved by your waves. During this time, observe the way that the wave crests move onto the shore (for example, straight up, at a low angle, or at an oblique angle), and where sand is removed (eroded) and deposited by the waves.

Step 2: Next, in the box below, draw a sketch of the shoreline (a simple line or curve illustrating where the land and water are will do).

Step 3: On the shoreline you just made in Step 2, draw the wave patterns you observed in your simulation by using some wavy lines labeled "wave crests." If you observed wave refraction, draw the wave crests bending or use an arrow to show which way the wave crests bent.

Step 4: On your drawing, write "E" where sand was eroded and "D" where sand was deposited.

Step 5: Repeat Steps 1 through 4 for the other two shoreline simulations. For the last simulation, draw and label the jetties (the man-made structures built perpendicular to the shoreline) at Step 2 as well.

Step 6: Finally, for each simulation, evaluate whether or not your predictions matched your observations during the experiment. Write your answers on the lines below each sketch. If you conducted additional simulations, use a separate piece of paper.

Simulation 1

_____

_____

_____

(continued)

Name: _____ Section: _____
Course: _____ Date: _____

Simulation 2

_____
_____
_____
_____

Simulation 3

_____
_____
_____
_____

You have now observed how waves move sediment onshore, offshore, and parallel to shore. In your simulations, your waves were made at consistent speed and direction. However, the strength and direction of the wind often changes, and with those changes come different-sized waves of varying speeds and frequencies. The waves do not always approach the shore at the same angles and short-term events like storms can cause severe changes in the distribution of sediment along a coast.

Let's take a look at a satellite view of the southwest coast of Florida, where a pair of jetties have been constructed along the sides of a harbor entrance:

(continued)

Name: _____ Section: _____
Course: _____ Date: _____

Longshore currents that carry sediment parallel to the shore can form sandbars that block the entrance to a harbor. Jetties are constructed in order to protect the harbor entrance from being closed off by sand deposition. Take a close look at the shape of the beach on each side of the jetties. North is to the top of the image.

Inlet along the southwest coast of Florida.

(a) Compare and contrast the size of the beaches on each side of the jetties and provide an explanation for their differences:

_____

_____

_____

(b) Which way do you think the longshore current is flowing along this coast? Explain how you know.

_____

_____

_____

(c) Why do you think that the jetty on the south side of the harbor entrance is angled? How does this relate to the waves as they come toward shore?

_____

_____

_____

(d) Why do you think that jetties are constructed in pairs? In other words, why not just build one on either the north or south side of the harbor?

_____

_____

_____

15.3.4 Erosional and Depositional Features

The removal and transportation of rock and sediment along a coastline, and the redeposition of shoreline materials, produces distinctive shoreline features and causes coastal landscapes to change in predictable ways. *Emergent coasts* are generally dominated by erosion, *submergent coasts* are generally dominated by deposition. In this section you will first investigate features resulting from coastal erosion, and then characteristic depositional features.

15.3.4a Coastal Erosion   FIGURE 15.10 shows examples of the distinctive features formed by coastal erosion. Coastal erosion is most rapid in locations where land extends out into the ocean, called *headlands*, because this position allows waves to attack the land from nearly any direction. Conversely, coastal erosion is slowest in deep, low areas of coastal land called *embayments*, where wave energy is diffused along a broad stretch of coastline.

Coastal erosion produces two types of landforms. The first type forms at the shoreline. As waves drive loose sediment across shorelines underlain by bedrock, the sediment scours a flat surface, called a **wave-cut bench** (Fig. 15.10a). It then cuts into the base of bedrock cliffs to form a **wave-cut notch** (Fig. 15.10b, c). Where

FIGURE 15.10 Erosional features of bedrock shorelines.

(a) A wave-cut bench at the foot of the cliffs at Étretat, France.

(b) A wave-cut notch along the Hawaiian coast.

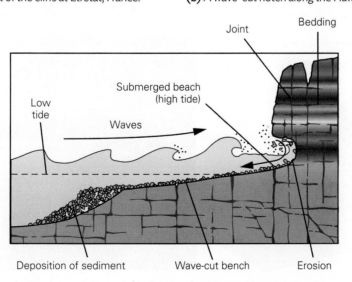

(c) Coastal erosion undercuts a sea cliff, producing a wave-cut notch and bench.

FIGURE 15.11 Hazards from coastal erosion.

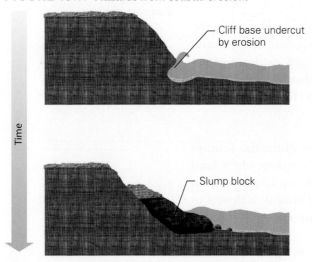

Time

(a) A wave-cut notch in a cliff composed of unconsolidated sediment can lead to slumping.

(b) Coastal-bluff erosion caused this house in Destin, Florida, to collapse.

the coastal cliffs are made of loose sediment, the waves eat into the cliff and pick up sediment that is added to the abrasive material. In either case, the wave-cut notch eventually undermines the cliff, causing slumping or rockfall, and the location of the cliff moves away from the shoreline, in toward land.

Where humans have built homes on coastal cliffs, the results of coastal erosion can be disastrous, particularly where the cliffs are made of loose sediment or glacial deposits, as is the case along the Atlantic coast (**FIG. 15.11**). Over long periods, wave-cut benches become wider as the coastal cliffs migrate landward, or retreat. In tectonically active areas, old wave-cut benches may have been uplifted several meters above present sea level, where a newer bench is being carved today (**FIG. 15.12**). By dating the uplifted benches, geologists can estimate the amount and rate of tectonic uplift.

FIGURE 15.12 Uplifted wave-cut benches.

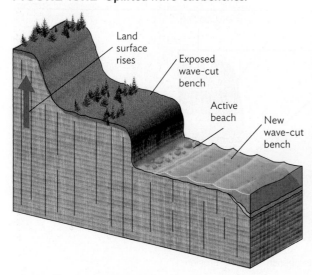

(a) As the land rises, the bench becomes a terrace, and a new wave-cut bench forms.

(b) An example from the California coast.

FIGURE 15.13 A sea arch on the coast of Hawaii.

The second type of landform is found a short distance offshore. When a bedrock cliff migrates landward, it does not do so at the same rate everywhere. Waves may erode zones of weak rock quickly, isolating stronger material, and then start to cut into that material as well. Eventually, a **sea arch**—named for its distinctive shape—is formed (**FIG. 15.13**). When further erosion removes the support for the arch, it collapses, leaving an isolated piece of the bedrock called a **sea stack** (**FIG. 15.14A**). Morro Rock (**FIG. 15.14B**), off the coast of California, is a spectacularly beautiful sea stack. Sea stacks along a coastline mark the former position of the bedrock cliffs, letting us measure the amount of cliff retreat.

15.3.4.b Depositional Features Prominent depositional features develop where there is an abundant supply of sand along shorelines. These features range from continuous sandbars that extend for miles along the coast to small isolated beaches. The Gulf of Mexico and Atlantic coastal plains are underlain by easily eroded unconsolidated sediments, and many places with familiar names display classic depositional features: Cape Cod, Cape Hatteras, the Outer Banks of North Carolina, the eastern Louisiana coast, Padre Island.

FIGURE 15.14 **Creation of a sea stack.**

(a) Formation of sea arches and stacks by erosion at a bedrock headland.

(b) Morro Rock in California, a classic sea stack.

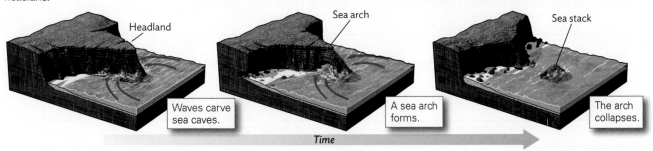

EXERCISE 15.8 Erosional Processes and Shoreline Landforms

Name: ______________________ **Section:** __________
Course: ______________________ **Date:** __________

Refer to the photos specified to answer the following questions.

(a) How did the large blocks in Figure 15.10b get into the breaker zone? What is their eventual fate?

____

____

____

(b) How did Morro Rock (Fig. 15.14b) become isolated from the bedrock shoreline?

____

____

____

(c) What is the eventual fate of the sea arch in Figure 15.13?

____

____

____

Examine the shoreline in **FIGURE 15.15**. Point Sur and False Sur are the same kind of feature and record the multistage development of this part of the California shoreline.

(d) Given its size and steepness, is Point Sur made of resistant bedrock like granite or a softer rock like limestone? Explain.

____

____

____

(e) What material or type of material is the area between the Point Sur lighthouse and the California Sea Otter Game Refuge probably made of? Explain your reasoning.

____

____

____

____

(f) Suggest a sequence of events by which the Point Sur shoreline could have formed.

____

____

____

____

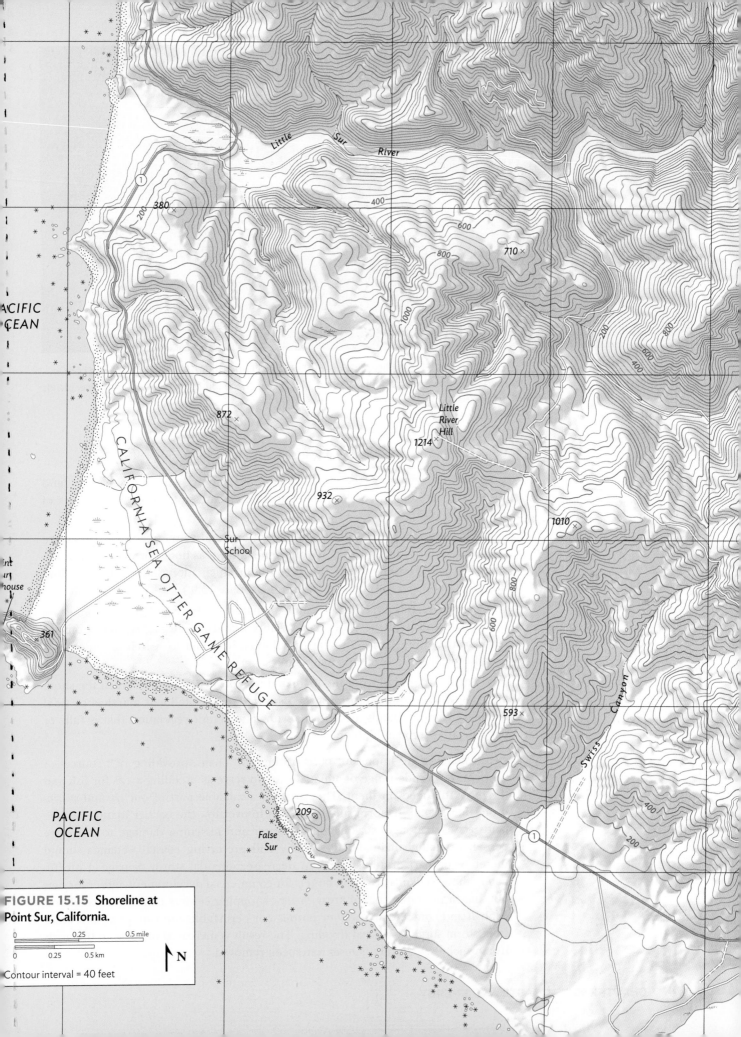

PACIFIC OCEAN

ACIFIC OCEAN

Little Sur River

CALIFORNIA SEA OTTER GAME REFUGE

Sur School

Little River Hill

380 ×

710 ×

872 ×

932 ×

1214 ×

1010

593 ×

209

False Sur

nt ur house

361

Swiss Canyon

FIGURE 15.15 Shoreline at
Point Sur, California.

0   0.25   0.5 mile

0   0.25   0.5 km

↑ N

Contour interval = 40 feet

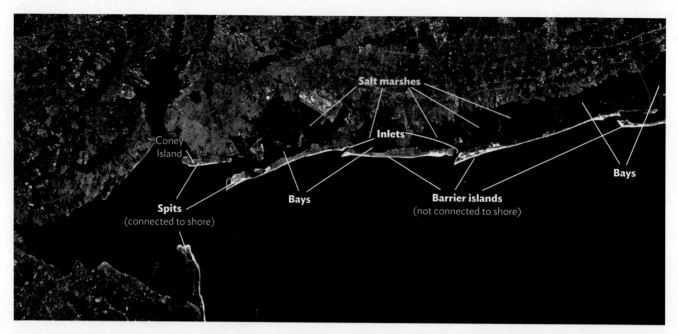

FIGURE 15.16 Satellite view of southwestern Long Island, New York, showing common shoreline depositional features. White areas on the south shores of spits and barrier islands are beaches.

Common shoreline depositional features include the following (**FIG. 15.16**):

- **Beaches**, which are the most common depositional features, consist of sand (or coarser sediment), coral and shell fragments, and so forth—whatever type of sediment is available.

- **Spits** are elongate sandbars attached at one end to the mainland. Some are straight, like those in Figure 15.16, but some, known as hooks or *recurved spits*, are curved sharply. Hooks curve in the direction in which a longshore current is flowing. (*Hint:* This fact will come in handy for Exercise 15.9.)

- **Barrier islands** are elongate sandbars that lie offshore and are not connected to the mainland (e.g., the Outer Banks, Padre Island). Their name comes from the fact that they were barriers to early explorers, who had to search for inlets that would allow them to reach the mainland.

- Sediment eroded from the mainland is deposited in bays between the shore and the barrier islands, forming marshy wetlands called **salt marshes** or **tidal marshes**. These wetlands are covered with salt-tolerant vegetation that is fully or partially submerged at high tide.

Salt marshes are important parts of the food chain, providing rich sources of nutrients for a wide range of aquatic life, and serve as breeding areas for fish and the birds that feed on them. They are also a valuable part of our natural storm-protection system. If storm surge manages to overflow the barrier island, the wetlands act as a sponge, soaking up the water and lessening damage to the more densely inhabited mainland. Protecting and preserving coastal wetlands should therefore be part of any strategy to decrease potential storm damage. Unfortunately, these wetlands are often targets for commercial development, either for new housing in desirable coastal areas or new shopping centers catering to residents. Asphalt and concrete are neither porous nor permeable and cannot absorb water from storm surge or coastal flooding. The result is increased damage to the areas from which the natural defenses have been removed.

Name: _____   Section: _____

Course: _____   Date: _____

(a) Does the shoreline in the map below appear to be emergent or submergent? Explain.

_____

_____

(b) Identify and label the following depositional landforms on the map: spit, hook, barrier island, and beach.

(c) What evidence is there that sediment redistribution is taking place *landward* of the barrier island as well as on the barrier islands and spits that protect the mainland?

_____

_____

(d) The rapid movement of sand by longshore drift could block access to the mainland by closing gaps in barrier islands and between spits. What steps can be taken to prevent further erosion?

_____

_____

(e) Indicate the dominant direction(s) of longshore drift. Explain your reasoning.

_____

_____

Barrier islands along the Texas Gulf Coast, southwest of Galveston.

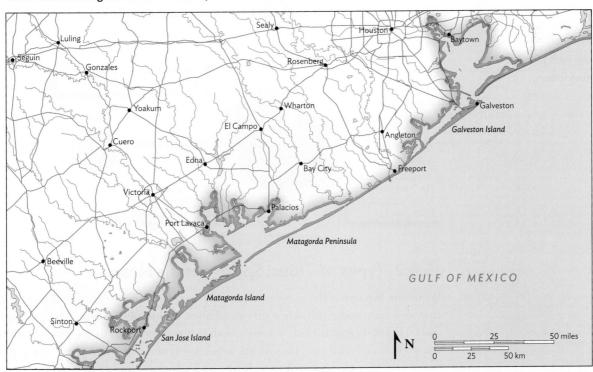

15.4 Tides and Tidal Forces

As we observe the height of water levels along the coast over a period of approximately 25 hours, we see that it rises and falls in a cycle, forming the **tides**. This change in water heights is the movement of the ocean in the form of the largest waves on Earth—true tidal waves. Their wavelength is half the circumference of the Earth, or just over 20,000 km (approximately 12,500 miles)! Tides are important for the organisms that live along coastal waters, as this movement transports sediment and nutrients. They are also important for navigation and as a source of renewable energy. In this section, we will investigate the source of these waves, their patterns, and the harnessing of tidal energy.

15.4.1 Investigating the Forces That Cause Tides

The primary source of the tide-generating force on the oceans is our nearest neighbor in space—the Moon. The Moon tugs on our planet as a result of the force of gravity, and likewise the Earth tugs on the Moon. Although the solid Earth also experiences this gravitational attraction, it is the fluid nature of the ocean that causes it to flow outward toward the Moon, forming a tidal bulge on the spot directly below the moon (**FIG. 15.17**, right side). There is a comparable bulge in the ocean that occurs on the opposite side to the one facing the Moon, which is a result of the outward force (*centrifugal force*) exerted as our planet spins on its axis (Figure 15.17, left side). The centrifugal force is the secondary component of the tide-generating force. These two bulges form the crests of the tidal waves, while the areas where the ocean level is lowest form the troughs. When the crest of the wave is at a particular location on Earth, that location experiences **high tide**; when the trough of the wave is at the location, the location experiences **low tide**. The total change in the height of the ocean surface between high and low tide is called the **tidal range,** or stated mathematically:

High-tide height − Low-tide height = tidal range

FIGURE 15.17 Tide-generating forces.

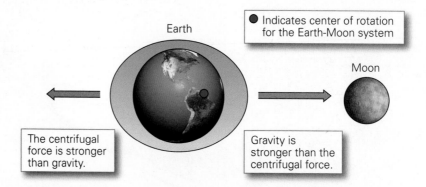

Earth

● Indicates center of rotation for the Earth-Moon system

Moon

The centrifugal force is stronger than gravity.

Gravity is stronger than the centrifugal force.

15.4.2 Types of Tides: Spring versus Neap

While the Moon has the greatest effect on our tides, there are other objects in space that also exert some influence on ocean tides, including the Sun. The gravitational force (F_G) that objects exert on one another is proportional to the distance between the objects (d^2) and the mass of the objects (m_1 and m_2) multiplied by the gravitational constant (G), according to Newton's Universal Law of Gravitation:

$$F_G \propto \frac{Gm_1 m_2}{d^2}$$

While the Sun is the most massive object in our Solar System, it is also more than 150 million kilometers (approximately 93 million miles) away. On the other hand, the Moon—while significantly smaller in mass—is much closer to Earth, at a mere 384,000 kilometers (approximately 239,000 miles) away, on average. It therefore has a greater effect on our tides than the Sun.

Since the tides are primarily dependent on the Moon, we need to look at the motions of the Moon to understand why the location and time of tides on the Earth change each day. Recall that the location on Earth that is directly below the Moon experiences high tide. Every 24 hours the Earth completes a rotation (spinning on its axis) beneath the Moon; so, the location on Earth that lies beneath the Moon changes over that time span. At the same time, the Moon is orbiting the Earth, changing its position about 13.19 degrees each day (for a total of 360 degrees in one complete orbit). These motions result in the arrival of high and low tides at different times at each location on Earth every day. Because we know the timing of these movements, the predicted times of high and low tides can be determined.

FIGURE 15.18 illustrates the various positions of the Moon as it orbits the Earth over a lunar month (about 28 days). As the Moon orbits the Earth, the amount of the illuminated side of the Moon (the daytime side) that is visible from the Earth changes, causing the *phases* of the Moon. If we first observe the position of the Earth, Moon, and Sun during a New Moon phase, we on Earth cannot see the illuminated side of the Moon. During this phase, the Moon tugs the ocean in the same direction that the Sun is tugging on the ocean, so the gravitational forces of the Moon (the **lunar tides**) and the Sun (the **solar tides**) are added together. This also occurs two weeks later when the Moon is at the Full Moon phase, on the opposite side of the Earth. This additional force generated by the Sun causes the greatest bulge in the oceans, or the greatest high tides. As a result, the low tides are at their lowest height,

FIGURE 15.18 Spring and neap tides.

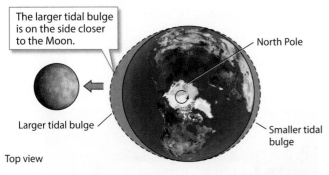

The larger tidal bulge is on the side closer to the Moon.

North Pole

Larger tidal bulge

Smaller tidal bulge

Top view

(a) The larger (sublunar) tidal bulge always faces the Moon, and the smaller (secondary) tidal bulge is always on the opposite side of the Earth.

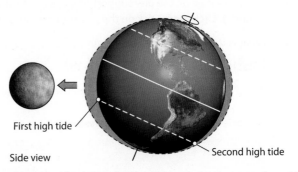

First high tide

Side view

Second high tide

(b) Viewed from the side, the sublunar bulge does not align with the equator.

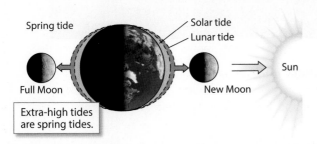

Spring tide

Solar tide
Lunar tide

Sun

Full Moon

New Moon

Extra-high tides are spring tides.

(c) When the Sun is aligned with the Moon, stronger, higher tides, called spring tides, result.

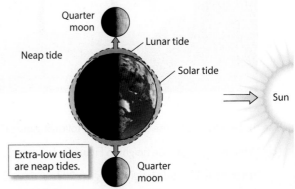

Quarter moon

Lunar tide

Neap tide

Solar tide

Sun

Extra-low tides are neap tides.

Quarter moon

(d) When the Sun is at right angles to the Moon, weaker, lower tides, called neap tides, result.

since water is being tugged even more toward the high tide crests, leaving less in the low tide troughs. When this occurs, it is referred to as the **spring** tides.

So what occurs the other two weeks of the month? In between the spring tides occur the **neap** tides. At the quarter moon phases, the gravitational forces of the Moon and the Sun are opposed to one another, so that the Sun subtracts from the lunar tides. The result is that the high tides are at their lowest height and the low tides are not as low; in other words, the tidal range is smallest. In Exercise 15.10, you will practice with tidal gauges and tidal graphs to see these changes.

EXERCISE 15.10 **Comparing Spring and Neap Tides**

Name: _____   Section: _____
Course: _____   Date: _____

Water levels at a pier during one tidal day.

Use the tidal gauge on the pier support (it looks like a ruler) to answer the following questions (remember to include proper units). Each longer tick on the pier support measures a meter, and each shorter tick is half a meter.

(a) What is the high-tide height in Figure 1a? _____

What is the low-tide height in Figure 1b? _____

Based on your two measurements, calculate the tidal range: _____

Water levels at the same pier one week later.

Use these measurements to answer the following questions (remember to include units):

(b) What is the high-tide height in Figure 2a? _____

What is the low-tide height in Figure 2b? _____

Based on your two measurements, calculate the tidal range: _____

(continued)

EXERCISE 15.10 Comparing Spring and Neap Tides (*continued*)

Name: ______________________ Section: __________
Course: ______________________ Date: __________

(c) Compare your results; which figure likely represents the change in tides during new and full moon phases? __________

During quarter moon phases? __________

Explain your choices.

____

____

Many water levels around the world are measured using buoys and satellites. When plotted on a graph, it is clear that the tides truly are waves, with high tide occurring at the crest, and low tide at the trough of the wave. Use the water level graph shown below to identify the spring and neap tides.

On which dates did the spring tides occur? Explain your selection.

____

____

On which dates did the neap tides occur? Explain your selection.

____

____

Ocean water levels from Portland, Maine.

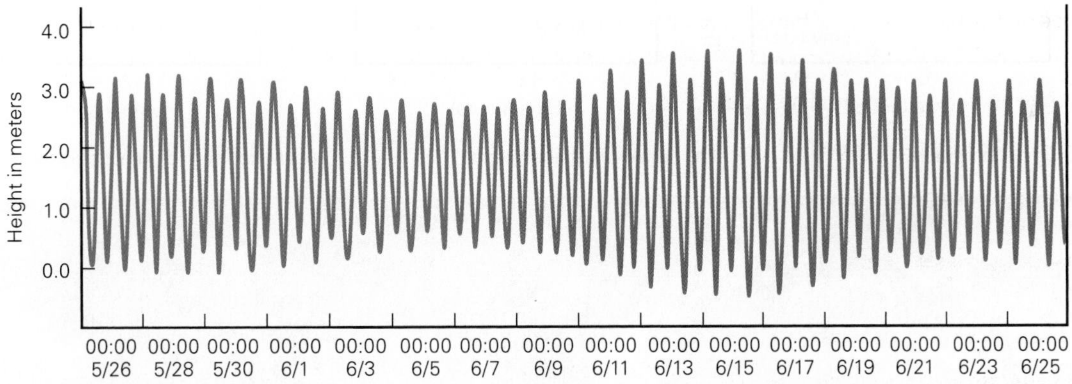

15.4.3 Tidal Patterns

We are familiar with a "day" on Earth; this time (24 hours) is a solar day, the time it takes one location on Earth to make one rotation back to the same position beneath the Sun. For a tidal day, it takes approximately 24 hours and 50 minutes for the Earth to rotate so that the same location is beneath the Moon. These day-to-day discrepancies in the relative positions and angles between the Earth and the Moon impact tidal heights and times.

Another factor that affects the pattern of tides across the globe is the shape of the ocean basin—specifically, where the continents are located and the shape of the shoreline. Although the majority of the world's coastlines experience two high and two low tides each day, as the Earth rotates toward the east, the tidal bulge that moves westward is interrupted by the continents, resulting in more complex patterns. For example, when the tidal bulges reach wide continental margins, the tidal impact can be amplified, whereas smaller margins or islands can minimize the tidal impact. This means that to observers on the coastline, the four tides may not appear as two uniform highs and two uniform lows, but rather a more diverse mix that could include some very large tides and some very small tides.

There are three types of tidal patterns that occur in the ocean basins, all distinguished by the number of high and low tides that take place in a tidal day, and the relative equality of the consecutive high or low tide heights. **FIGURE 15.19** illustrates the pattern of these tides as recorded graphically.

- **Diurnal** tidal patterns: within a tidal day there is one high and one low tide, with the high from one day to the next being of comparable heights.
- **Semidiurnal** tidal patterns: two high tides and two low tides occur within a tidal day, with each consecutive set of high or low tides occurring with approximately equal heights (on the order of only a meter or few feet difference).
- **Mixed** tidal patterns: when there is a significant difference between the heights of consecutive high or low tides, the tidal pattern is referred to as mixed. These can be **mixed diurnal** or **mixed semidiurnal** tidal patterns.

FIGURE 15.19 **Tidal pattern types depicted graphically and their typical locations.**

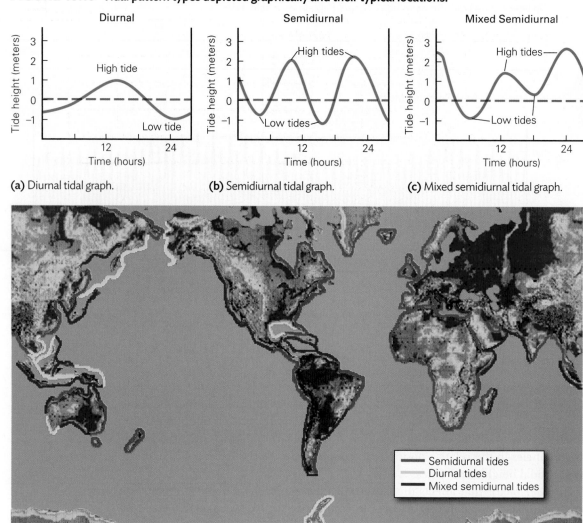

(a) Diurnal tidal graph.

(b) Semidiurnal tidal graph.

(c) Mixed semidiurnal tidal graph.

(d) Map of the ocean basins where the different tidal patterns occur.

Name: _____   Section: _____

Course: _____   Date: _____

You will use the tidal data given in the table below to plot a tidal graph for one tidal day. There are three tides to plot, representing three geographic areas. Plot the data points for one station, and then connect the points with a smooth line or curve in one color. Then do the same for the other two locations, selecting a different color for each location (locations 2 and 3). Include a key that indicates the colors used for each location.

Tidal data for one tidal day at three locations

Location 1	
Time (GMT)	Height (meters)
0:00	0.1
4:00	2.9
8:00	3.0
12:00	−0.1
16:00	2.1
20:00	3.1
0:00	0.5

Location 2	
Time (GMT)	Height (meters)
0:00	2.3
4:00	0.9
8:00	1.5
12:00	2.6
16:00	0.6
20:00	0.3
0:00	2.1

Location 3	
Time (GMT)	Height (meters)
0:00	0.1
4:00	0.0
8:00	0.1
12:00	0.3
16:00	0.5
20:00	0.5
0:00	0.2

Data source: https://tidesandcurrents.noaa.gov/waterlevels

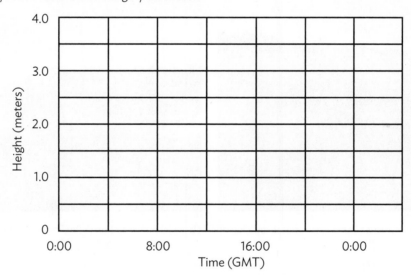

(a) What type of tidal pattern do you observe for Location 1? _____

    Location 2? _____   Location 3? _____

(b) Compare the patterns for Locations 1 and 2. How are they the same?

_____

    How are they different?

_____

(c) What is the tidal pattern for Portland, Maine, in the Ocean Water Levels graph at the bottom of Exercise 15.10?

_____

15.4.4 Tidal Currents and Harnessing Tidal Energy

As you have learned in the previous sections, the energy from the ocean has an impact on shaping the coastline, as rocks and sediment are eroded, transported, and deposited elsewhere. The energy in a tide has the power to move sediments along a coastline, via tidal currents. As the ocean surface rises during high tide, a **flood current** is formed that moves water onshore. Likewise, as the tide recedes during low tide, an **ebb current** forms. These currents not only shape the coastline; they also bring essential nutrients to shallow coastal waters and transport nutrients to the deeper waters. This process represents an essential element for the survival of many marine organisms.

The ebb and flow of the tides is a constant motion that humans have recognized as a source of renewable energy. Scientists and engineers have been developing technologies to capture the energy stored in a tide and turn that energy into electricity to supply power stations. Present technologies include underwater turbines and tidal dams (or barrages) constructed across bays and estuaries (**FIG. 15.20**). As the flood and ebb currents flow through these structures, they turn a turbine that transforms the tidal energy into electricity, which is stored for future use.

FIGURE 15.20 **Examples of tidal power station styles.**

(a) The Pennamaquan River estuary, Maine, is a proposed location of a tidal power barrage.

(b) Tidal power station in the Rance River estuary, Brittany, France.

15.5 Human Interaction and Interference with Coastal Processes

People invest hundreds of thousands of dollars (or more) in shoreline homes, and they want to use those homes and enjoy the local beaches for a long time. But a single storm can change a shoreline in just a few hours. The disconnect between what we want and how nature works has led us to use three expensive coastal management strategies: (1) building seawalls to prevent coastal erosion; (2) replacing eroded beach sand artificially; and (3) trapping sand moved by longshore drift. These strategies sometimes work, but they may also interfere with shoreline processes in such a way as to create new problems. We'll look first at the strategies, then at the problems.

FIGURE 15.21 Examples of seawall construction.

(a) A section of the Galveston, Texas, seawall.

(b) Concrete blocks of a seawall.

15.5.1 Seawalls

To protect areas ravaged frequently by intense coastal erosion, some communities choose to armor the shoreline with **seawalls** made of concrete, blocks of loose rock, or similar materials. These structures are designed to break the force of the waves and prevent further shoreline erosion. The 1900 hurricane that devastated Galveston, Texas, was a wake-up call for that community, and today the barrier-island city is protected from storm surge by an extensive seawall (**FIG. 15.21A**). Seawalls can be made with different designs and different materials, such as the concrete blocks shown in **FIGURE 15.21B**.

15.5.2 Beach Nourishment

A single storm can erode vast amounts of sand from an unprotected beach, as shown in **FIGURE 15.22A**. The most common remedy is to do in a short time what it would take nature decades to do: replace the eroded beach by dredging sand from offshore, pumping it onto the beach, and spreading it out with bulldozers (**FIG. 15.22B**) This process is known as **beach nourishment**, a common strategy along beaches in popular tourist areas such as Florida, New Jersey, and California.

FIGURE 15.22 Beach erosion and nourishment.

(a) Effect of erosion along a sandy shore.

(b) Beach nourishment: sand pumped from offshore replaces a beach whose sand has been eroded away.

FIGURE 15.23 Aerial view showing how jetties are designed to trap sediment moved by longshore drift.

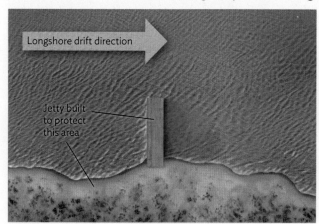

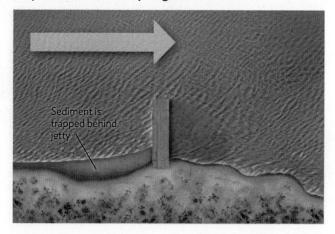

15.5.3 Jetties

Communities on barrier islands often use their knowledge of the longshore drift that built those islands to try to preserve their beaches. They build structures perpendicular to the shoreline in an attempt to trap moving sand and prevent its loss to downdrift areas (**FIG. 15.23**). These structures, called **jetties** (or *groynes*), are effective in preserving the targeted beach, but often have unintended consequences that cause problems for other beaches.

15.5.4 Unintended Consequences of Shoreline Erosion Management

Beach management practices sometimes backfire because the results of building a seawall or jetty were not thought through fully. The most common problem is that while the shoreline is protected in one area, the seawall or jetty concentrates erosion in different areas, creating problems where there were none previously. For example, waves crashing against a seawall may remove the sand that otherwise would have accumulated naturally along the shoreline, doing exactly the opposite of what was intended. Or, as shown in **FIGURE 15.24**, sand trapped on the updrift side of a jetty is no longer available to replenish the beach naturally on the downdrift side. The beach is preserved in one place, as intended, but eroded in another. Exercise 15.12 gives you some experience in recognizing these common problems.

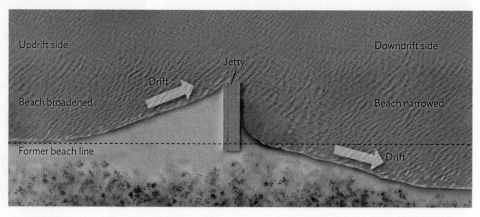

FIGURE 15.24 Aerial view of shoreline showing potential negative effect of jetty construction.

Name: _____   Section: _____
Course: _____   Date: _____

(a) The two photographs below show seawalls built to prevent shoreline erosion and retreat on the west coast (top) and the Gulf Coast (bottom) of North America. They have successfully protected the shoreline, but what negative effects of these seawalls can you observe?

_____

_____

_____

(continued)

Name: _____   Section: _____

Course: _____   Date: _____

(b) The following photograph shows a jetty field and a seawall protecting one side of the hook at North Avenue Beach, Chicago, Illinois.
- Draw an arrow to indicate the direction of longshore drift.
- Describe the width of the beach in the areas between the jetties. Suggest an explanation for this pattern.

_____

_____

_____

(c) The photograph to the right shows a series of jetties (indicated by the red arrows) built to protect the barrier island at Westhampton, New York. Why is the beach narrower south of the inlet than in the area where the jetties were built? What long-term problem will the area in the foreground experience?

_____

_____

_____

_____

_____

_____

Δrge areas of Houston, Texas, were flooded by more than 50 inches of rain Hurricane Harvey.

(b) Storm surge damage in Biloxi, Mississippi, from Hurricane Katrina.

15.5.5 Shore Stabilization on a Dynamic Planet

People living along shorelines are all too familiar with the dangers associated with coastal areas. Graphic images of the violence and devastation caused by coastal storms and tsunamis bring that message to those living inland. Hurricanes (called typhoons in Asia) inundate coastal regions with rain (**FIG. 15.25**)—we'll investigate how these hurricanes form and cause coastal damage further in Chapter 18. And tsunamis, as we saw in Chapter 7, strike suddenly and with a force that overwhelms our coastal defenses. In contrast, the gradual rise of sea level is almost imperceptible, but it will eventually disrupt society more than any single catastrophic event (**FIG. 15.26**). This, too, we'll explore in more depth in Chapter 19. But first, in Exercise 15.13, you'll assess what impact sea-level rise might have on different coastal regions within the United States.

FIGURE 15.26 Potential effects of sea-level rise on New Orleans.

Current sea level

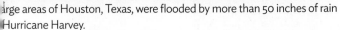

5-foot sea-level rise
88% flooded

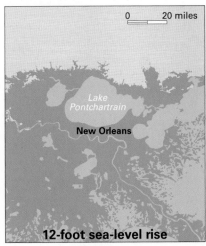

12-foot sea-level rise
98% flooded

Name: _____

Course: _____

Section: _____

Date: _____

? **What Do You Think** Half of the population of the United States lives within 50 miles (80 km) of a shoreline. Disasters caused by recent tsunamis, hurricanes, and typhoons worldwide are making insurance companies reexamine the risks and reevaluate their premiums for insurance along shorelines—or consider whether they should even offer insurance in some areas. They rely on geologists' expertise, and you have been contacted by a company for recommendations about whether there should be different policies or rates for the Atlantic and Gulf coasts than for the Pacific coast. Your staff has constructed shoreline profiles (shown on the next page) for six coastal communities from the two regions, shown on the map below. *Your job is to outline the factors that control potential damage to coastal properties in these two regions.* Here are some questions to consider:

- How would continued global climate change affect risk in each region? How far inland would a 5-foot sea-level rise shift the shoreline? A 25-foot rise?
- How would sea-level rise affect the vulnerability of coastal regions to hurricane or tsunami damage?
- Do the two regions have the same risk of damage by tsunamis and hurricanes?
- What factors determine the amount of potential damage in each of the regions?

On a separate sheet of paper, provide your report and recommendations for each region.

(continued)

Name: _____   Section: _____
Course: _____   Date: _____

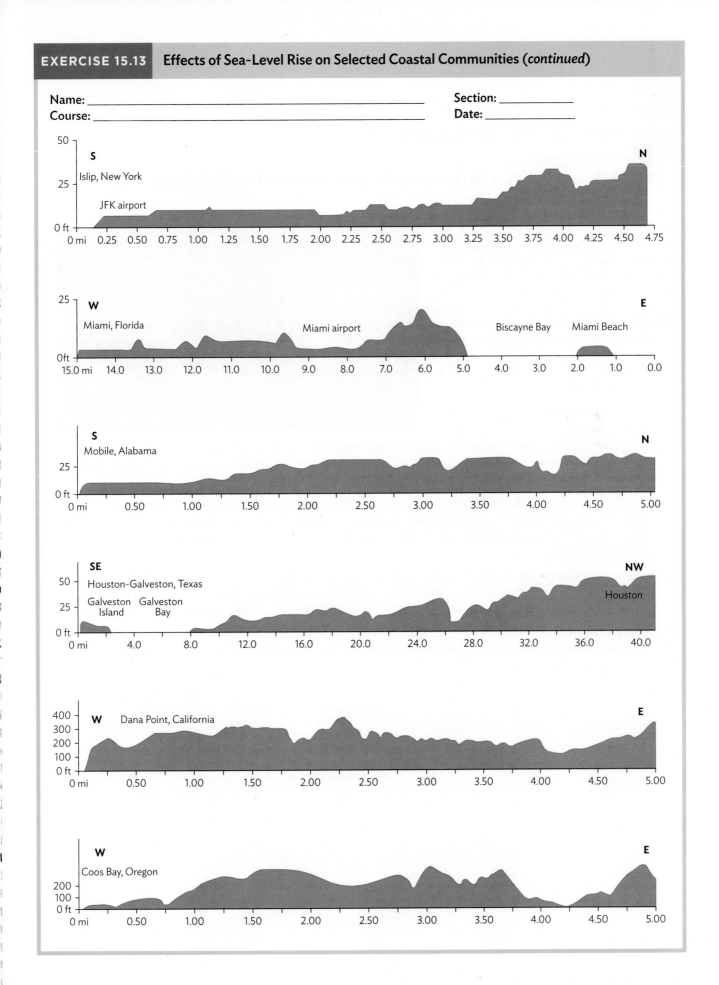

Name: _____   Section: _____

Course: _____   Date: _____

Exploring Earth Science Using Google Earth

1. Visit **digital.wwnorton.com/labmanualearthsci**
2. Go to the **Geotours** tile to download Google Earth Pro and the accompanying Geotours exercises file.

Expand the Geotour15 folder in Google Earth by clicking the triangle to the left of the folder icon. The folder contains placemarks keyed to questions that investigate coastal landforms and coastal processes.

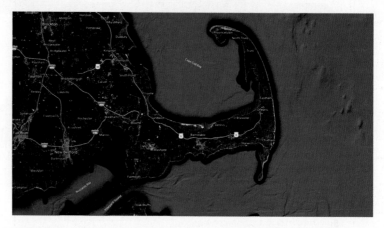

(a) Check and double-click the Cape Cod placemark to fly to the arm-shaped coastline of Cape Cod, MA. Here, the remains of an E-W-oriented glacial moraine form the "muscle" of the "arm," and a N-S glacial moraine creates most of the N-S "forearm." Check and double-click placemark (a) to zoom to the spit being created at the curled "fist" near Provincetown, MA. Watch the animated GIF image in the placemark, or click on the link in the placemark for the Google Earth Engine website for this same location. What is the general direction that the current is transporting sediment (NW, NE, SW, SE)? _____

(b) Check and double-click placemark (b) to zoom to the spit being created at the "elbow" near Chatham, MA. Watch the animated GIF image in the placemark, or click on the link in the placemark for the Google Earth Engine website for this same location. What is the general direction that the current is transporting sediment (NW, NE, SW, SE)? _____

(c) Check and double-click placemark (c) to fly to the barrier island near Chincoteague Bay, MD. Watch the animated GIF image in the placemark, or click on the link in the placemark for the Google Earth Engine website for this same location. Which is more prominent at the placemark location: erosion or deposition? _____

(d) Check and double-click the U.S. East Coast Sea Level Changes overlay, and check the (d) placemarks. If the Greenland and Antarctica ice sheets melt, which placemark city is the best place to own land? _____

Investigating the Earth's Atmosphere and Heat

16

The Earth's atmosphere as seen over the Indian Ocean by the Discovery mission in 1999.

16.1 Introduction

The atmosphere—the layer of air above the Earth's surface—is a rich mixture of gases that protects us from the harsh environment of space and provides the conditions necessary for life to exist. Near the surface of the Earth, in the lower 11 km (7 miles) or so of the atmosphere (**FIG. 16.1**), the exchange of thermal energy (called *heat*) between the solid Earth and the air creates chaotic conditions in the atmosphere. This ever-changing state of the atmosphere is called **weather** and includes properties such as air temperature, pressure, wind, clouds, humidity, and precipitation. These atmospheric conditions at a given location change over short time spans—from seconds to hours—because at different times, any location on the surface of the Earth will receive varying amounts of energy from the Sun, and winds and storms change the character of the weather. In this chapter, we will explore the factors that cause differences in the heating of the Earth's surface. In later chapters, we will investigate the formation of winds, changes in humidity, formation of clouds, precipitation and storms, and long-term averages of all weather in a given region, which we call **climate.**

Before we investigate the causes of weather in more detail, we should distinguish between *heat* and *temperature.* Heat refers to the total amount of thermal energy an

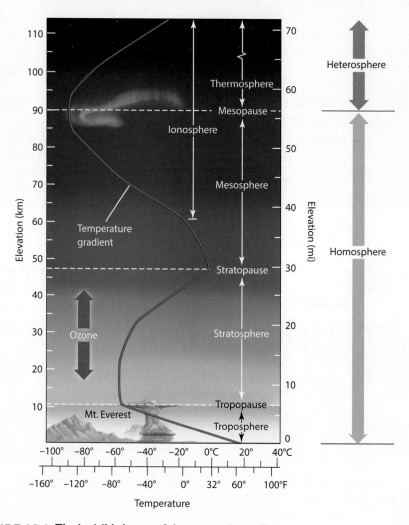

FIGURE 16.1 The invisible layers of the atmosphere. These layers are separated by changes in temperature with increasing height above the surface.

object absorbs or releases. Temperature is one way we measure that amount of heat, typically using a thermometer. Unlike heat, temperature relates to the *average speed* at which atoms or molecules move. Objects with higher temperatures are those in which the atoms or molecules are moving faster, on average. So two objects receiving the same amount of heat may have different temperatures, depending on the average speed of the molecules in each object.

Though we can say that we measure an object to have higher temperature, this object would not necessarily feel hot to our touch—that depends on how that heat is transferred to our hand. An example is the outer edge of our atmosphere, the *thermosphere* (see Fig. 16.1), which stretches 90–1,000 km (56–621 miles) above the Earth's surface and where temperatures can reach 2,480°C (4,500°F)! At this altitude, we approach the vacuum of space and so there are barely any gas molecules present to transfer that heat. The International Space Station (ISS) orbits in this layer; if astronauts did not have protective spacesuits and the safe environment of the ISS, they would feel very, very cold despite the high temperature, because their bodies would radiate heat to space, and the few air molecules would not replace that heat. They would not be able to survive.

16.2 The Transfer of Heat Energy

There are three ways in which heat energy can be transferred between objects: *conduction*, *convection*, and *radiation*. Whenever there is an imbalance (or difference) in temperature between two objects, heat flows between the objects, moving from areas of higher temperature to areas with lower temperature in order to establish a state of equilibrium.

Conduction is the transfer of heat between molecules that are in contact with one another. The heat is transferred through the matter as the molecules collide. An analogy is to think of lined-up billiard balls smacking into one another on a pool table (**FIG. 16.2A**). When you hit the first ball, it bumps into the second one, which then bumps into the third one, and so on. Likewise, the heat moves from the "hot" object toward the "cool" object to equalize heat or temperature. Some materials are good at transferring energy in this way, such as metals. On the other hand, poor conductors—called *insulators*—such as plastics, rocks, and even air, are not good at this method of energy transfer.

FIGURE 16.2 **Examples of heat transfer mechanisms.**

(a) Billiard balls colliding into one another in a sequence. This is analogous to atoms in a solid bumping into their neighbors in a repeated pattern, passing energy through the material through conduction.

(b) A lava lamp is an example of convection. The hot liquid wax rises to the top where it then cools and sinks, to be heated again and repeat the cycle of rising and circulating.

(c) Food heated in a microwave receives radiation emitted in all directions by the oven.

Convection is the transfer of heat by circulation within a fluid, namely liquids and gases—matter that contains loosely connected molecules. An example of convection is a lava lamp (**FIG. 16.2B**). The liquid near the bottom is heated, becomes warm and less dense, and then rises to the top. At the top, it cools down, becomes denser, and then sinks to the bottom where it is reheated. You investigated this energy transfer in the ocean chapters. The atmosphere works in the same way: as air is heated at the surface of the Earth, it rises into the air above, cools, and then sinks. A thunderstorm is a visual example of convection in the atmosphere.

Radiation is energy that needs no connecting molecules to be transferred; rather, it travels in the form of waves (measured by their wavelength, or the distance from the peak of one wave to the peak of the next wave). The Earth receives energy from the Sun—called *solar radiation*—in this way. Another example of radiation is cooking with a microwave oven, where microwaves (long-wavelength radiation) emitted by the oven are absorbed by the food inside the oven (**FIG. 16.2C**).

There are three rules regarding radiation that we need to consider when investigating heat transfer:

- All objects absorb and release radiation.
- The shorter the wavelengths emitted by an object (the closer together the waves), the higher the temperature of the object.
- Hotter objects emit more energy.

In the next section, we'll take a closer look at radiation, as this is the primary way the Earth is heated and the principal control on weather. Bur first, in Exercise 16.1, you'll compare and contrast the different types of energy transfer.

EXERCISE 16.1 **Methods of Energy Transfer**

Name: _____   Section: _____
Course: _____   Date: _____

(a) Look at the figure on the right and answer the following questions:

Which mechanism of heat transfer has occurred? (Specifically, we're referring to the energy given off by the heated objects and the air around them.)

_____

Explain your reasoning.

_____

_____

_____

_____

The figures below illustrate the ways in which heat is transferred.

Give an example of an object or surface of the Earth that you think transfers energy in this way:

_____

(continued)

EXERCISE 16.1 Methods of Energy Transfer (*continued*)

Name: ______________________________ **Section:** __________

Course: ______________________________ **Date:** __________

(b) Which mechanism of heat transfer is illustrated in the figure on the right? __________________

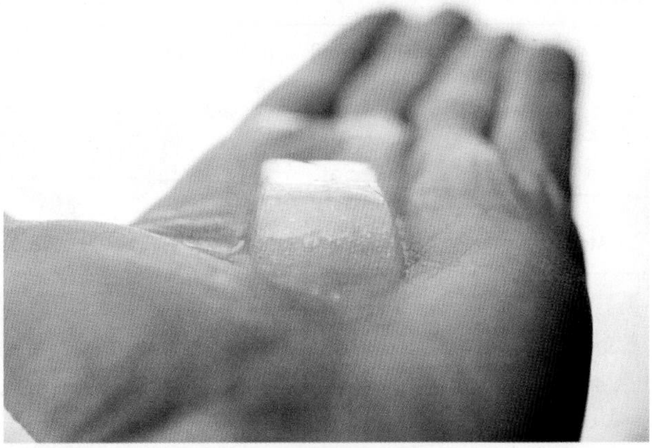

Explain your reasoning.

____

____

____

____

Give an example of an object or surface of the Earth that you think transfers energy in this way:

____

(c) The figure on the right is illustrating what type of heat transfer? ______________________

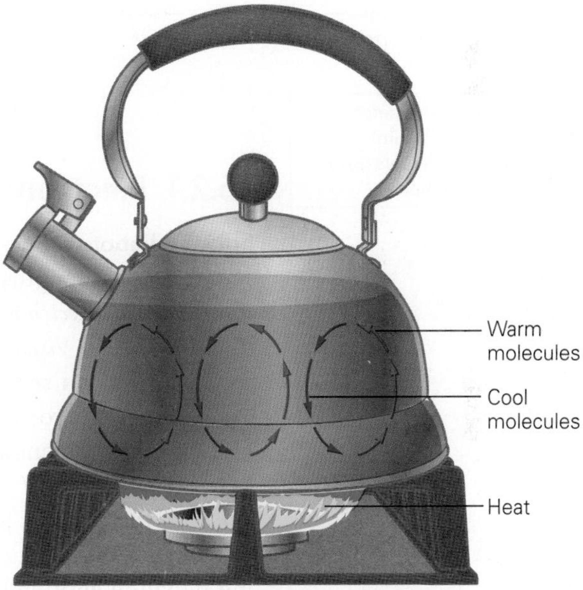

Warm molecules
Cool molecules
Heat

Explain your reasoning.

____

____

____

____

____

____

Give an example of a substance on or above the surface of the Earth that you think transfers energy in this way:

____

(d) Consider the following scenario: You are building a house in a hot climate where you expect to use the air conditioning quite a bit. Where would be the better place to put the air vents in the house, so that the air circulates to keep the house cool: near the floor or near the ceiling? Explain your choice.

____

____

____

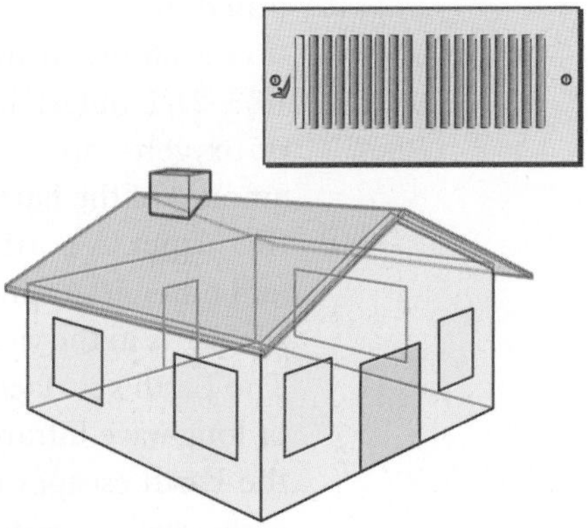

FIGURE 16.3 The electromagnetic spectrum.

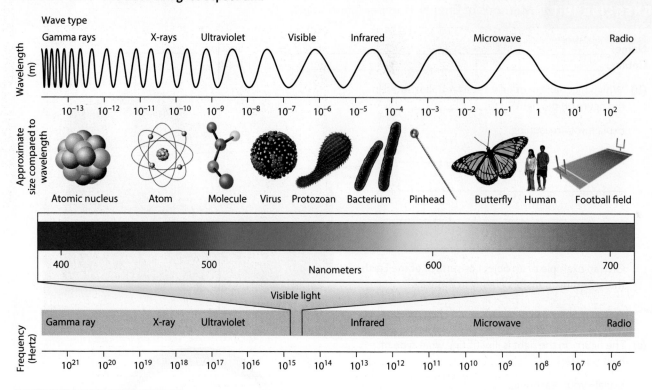

1 meter (m) = 100 centimeters (cm)
1 cm = 10 millimeters (mm)
1 mm = 1,000 micrometers (μm)
1 μm = 1,000 nanometers

16.2.1 Electromagnetic Radiation

Solar radiation is energy that is released by the Sun in the form of *electromagnetic* waves. All wavelengths of radiation coming out from the Sun can be seen on the *electromagnetic spectrum* shown in **FIGURE 16.3**. This is how the surface of the Earth is initially heated. *Visible light* energy coming from the Sun is the only solar radiation wave that we can see naturally, and is made up of all colors (red, orange, yellow, green, blue, indigo, and violet—collectively referred to as ROYGBIV), represented by different wavelengths. Of the solar radiation reaching the Earth, 43% is visible light, 49% is *infrared radiation*, and 7% is *ultraviolet radiation* (UV). The remaining 1% is in the form of gamma rays, X-rays, and radio waves. Radiation that reaches the Earth from the Sun arrives in the form of *shortwave radiation*, whereas the outgoing radiation emitted by the Earth is infrared radiation we cannot see, and is called *longwave radiation* (**FIG. 16.4**).

The transfer of heat between the Earth, its atmosphere, and outer space is illustrated in **FIGURE 16.5**. As solar energy reaches the Earth, much of the UV radiation is absorbed by the ozone layer, which is an invisible layer between 15–35 km (9.3–21.7 miles) in altitude where ozone molecules—composed of three atoms of oxygen—are concentrated. Thanks to this protective layer of gas, only a small amount of the harmful UV energy the Earth receives actually reaches the surface of the planet. A portion of the infrared radiation is absorbed and scattered by clouds and other atmospheric gases. Therefore, most of the energy that reaches the Earth's surface is in the visible part of the electromagnetic spectrum, which allows us to see. The Earth's surface absorbs this radiation and then re-emits the energy in the form of long-wave infrared radiation. Some of the long-wave infrared energy released by the Earth escapes to space, while the rest is trapped in our atmosphere by certain gases, such as water vapor, carbon dioxide, and methane. This trapping of long-wavelength radiation is what helps to keep our planet habitable. This trapping of

FIGURE 16.4 Diagram showing amounts of radiation in terms of the Sun and Earth by wavelength.

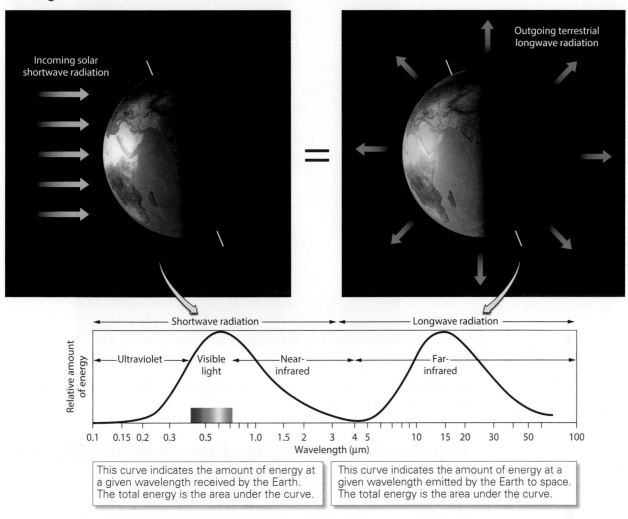

FIGURE 16.5 **The energy balance in the Earth's atmosphere.** The numbers represent arbitrary units of energy and show the relative amounts of energy in each transfer.

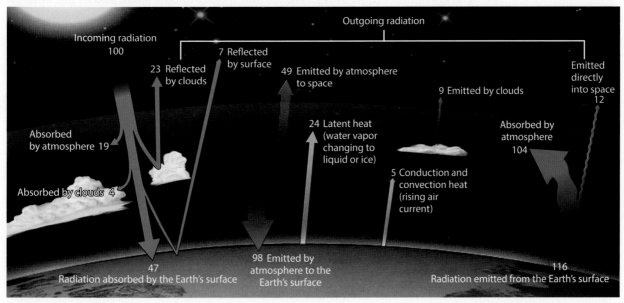

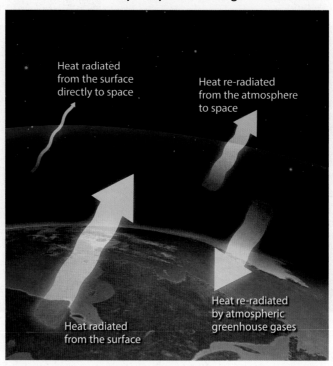

Heat radiated from the surface directly to space

Heat re-radiated from the atmosphere to space

Heat radiated from the surface

Heat re-radiated by atmospheric greenhouse gases

outgoing energy is often referred to as the *greenhouse effect*. **FIGURE 16.6** shows a basic depiction of that effect.

16.3 Earth-Sun Relationships: Seasonal Changes

There are three aspects of how the Sun's radiation reaches the Earth that control our seasons. The primary driver of seasons is related to the tilt of the Earth's axis. Because of the tilt, each hemisphere leans away from the Sun in its respective winter and toward the Sun in summer. This changes the number of hours of daylight. Days are long and nights short during a region's summer, and the opposite is true during winter. This affects the total amount of radiation that the region receives each day. During summer in polar regions, the sun is out 24 hours a day, and in the winter it is continuously dark.

Another factor is the length of the path that the Sun's radiation takes through our atmosphere. When the Sun is low in the sky (as it is in winter) the arriving solar energy travels longer paths, during which its energy can be scattered back into space, so less energy arrives at the ground. In summer, when the Sun is high in the sky, the paths for solar energy are shorter.

The third factor has to do with the fact that the Earth is a sphere. Recall that waves of radiation travel in straight lines from their sources. Because the Earth is so far from the Sun, the Sun's rays all essentially come from the same direction. If the Earth were flat, like the wall illustrated in **FIGURE 16.7A**, all locations on the planet would receive the same amount of solar radiation throughout a year. However, the planet's spherical shape results in different amounts of solar radiation being received at different locations on the Earth's surface as it orbits the Sun. Wherever solar radiation strikes the Earth's surface at a 90° angle, that energy is concentrated in a smaller area than sunlight reaching the surface at any other angle. This is why the equator receives more direct sunlight than the polar regions (**FIG. 16.7B**). A

FIGURE 16.7 Changing angles of incoming solar radiation.

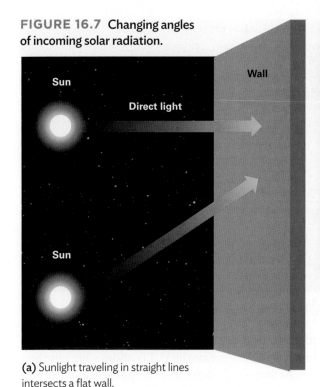

(a) Sunlight traveling in straight lines intersects a flat wall.

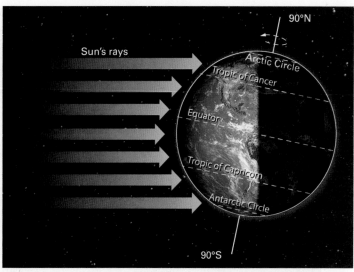

(b) Sunlight strikes different latitudes of the Earth's surface at different angles.

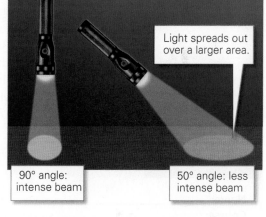

(c) The flashlight analogy for explaining why the intensity of solar heating varies with the angle at which it reaches the Earth.

good analogy for this phenomenon is the different beams produced by a flashlight when it's shined straight down on a tabletop versus when it's shined at an angle. The former produces a more intense and concentrated beam **FIG. 16.7C**).

If the Earth's axis of rotation (the point about which the planet spins) were exactly perpendicular to the plane of the Earth's orbit around the Sun (that is, the Earth stood "straight up" as it spun), the variation in temperature from equator to pole that we just described above would be the same all year everywhere on the Earth's surface. However, the Earth is tilted on its axis of rotation. Currently, this tilt is 23.5° from vertical. As the Earth spins, it remains tilted in the same direction; but, as the planet moves around the Sun, the latitudes on the Earth that receive the highest and lowest amounts of radiation change. Specifically, over the course of a year, the location where the noon sun angle is 90° with the Earth's surface moves slowly between 23.5°N latitude (the Tropic of Cancer) and 23.5° S latitude (the Tropic of Capricorn). The periods when the Sun is directly overhead at these maximum latitudes are referred to as the *solstices*—the June solstice occurs when the Sun is directly overhead of the Tropic of Cancer, and the December solstice occurs when the Sun is directly overhead of the Tropic of Capricorn. The lowest latitude where the sunlight comes in at a right angle is the equator, or 0° latitude, and the periods when this occurs are called the March and September *equinoxes* (**FIG. 16.8**).

The solstices and equinoxes divide the year into seasons: summer, fall, spring, and winter. It's important to note, though, that while the solstices and equinoxes apply to the entire Earth (that is, the December solstice occurs at the same time for every location on the planet), the seasons are relative and can only be used when describing a particular location on the Earth. For example, the June solstice marks the start of the summer season in the northern hemisphere, but it marks the start of the winter season in the southern hemisphere.

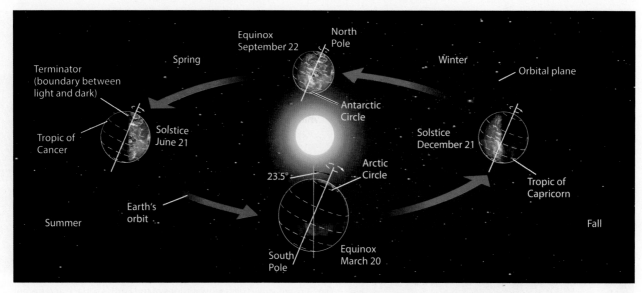

FIGURE 16.8 **The inclined tilt of the Earth's axis of rotation relative to its orbit around the Sun.**

For every 1° of latitude a location is away from the equator on an equinox, or from 23.5° on the solstices, the incoming Sun angle decreases by 1°. Locations where the Sun angle is high (45°−90°) receive more intense sunlight than latitudes where the Sun angle is low (0°−45°). The intensity of sunlight—given equal amounts of radiation—depends on the area over which the sunlight is spread, as illustrated in Figure 16.7C. In turn, the intensity of sunlight controls the length of daylight in a location at a given time, which controls the seasons. In Exercise 16.2, you'll examine the impact of Sun angle on seasonal changes.

EXERCISE 16.2	Recognizing Seasonal Changes in Heat: Sun Angle

Name: _____ **Section:** _____
Course: _____ **Date:** _____

In this exercise, you will observe how changing the Sun angle affects the intensity of solar energy the Earth's surface receives. First, you will investigate how the Sun angle varies by latitude during the same season, and then how the Sun angle varies at the same latitude at different seasons.

The figure below shows two cities in North America that are at different latitudes: Miami, Florida, and Seattle, Washington. The latitude of Miami is approximately 26°N, and Seattle is approximately 48°N. The figure illustrates the position of the Earth (and the cities) during an equinox.

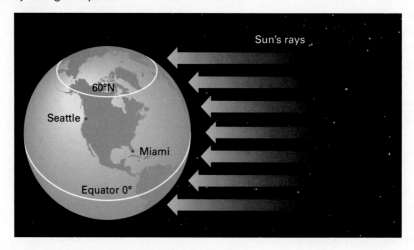

(continued)

Name: _____ **Section:** _____

Course: _____ **Date:** _____

(a) Given the latitudes of each city, calculate the angle at which the noon Sun intersects the surface at each city on the equinox:

Miami: _____

Seattle: _____

(b) Based on your answers in (a), what do you predict the solar intensity will be for each city (high or low)?

Miami: _____

Seattle: _____

(c) Let's test your prediction. The figure directly below shows the Sun above the Earth's surface during an equinox. Draw in sunbeams and measure the amount of the surface the sunlight covers. To do so, using a protractor and a ruler, draw lines on the figure from the Sun to the surface at each city such that the light intersects the surface at the angles you determined in (a). Make each sunbeam 1 cm wide. An example has been drawn for you below using the city of Philadelphia, which is at 40°N.

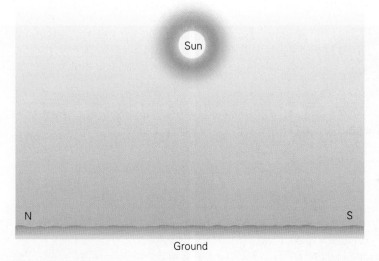

Ground

Example:

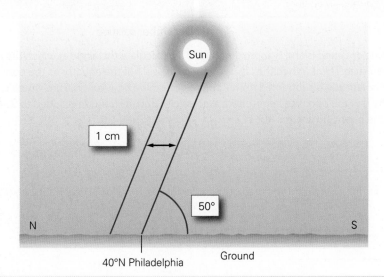

Ground

(continued)

Name: _____   Section: _____

Course: _____   Date: _____

(d) Using the scale of the sunbeam, determine the length of the sunlight spread over the surface at each city:

Miami: _____ cm   Seattle: _____ cm

(e) Given that the amount of energy in each sunbeam is the same, is the intensity of sunlight on the ground greater or less at Miami than it is at Seattle? _____. Does this match your prediction in (b)? Explain.

_____

(f) Summarize the results, indicating the relationship between intensity of sunlight at the surface at different latitudes:

_____

_____

_____

Now let's look at how Sun angle changes at one location over changing seasons. In the figure below, you can see the Earth during the solstices and the location of Grand Island, Nebraska (indicated by the red dot).

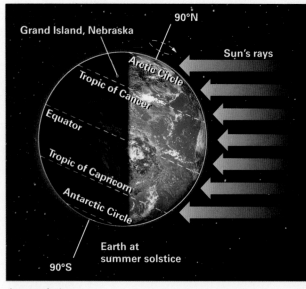

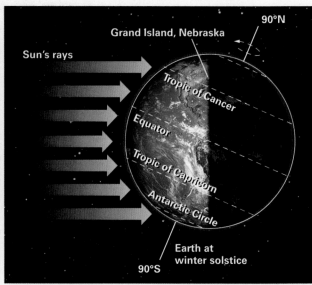

June solstice   December solstice

(g) Given that the latitude of Grand Island is approximately 41°N, calculate the angle at which the noon Sun intersects the surface at each solstice. (*Hint:* When the city is on the same side of the equator as the noon Sun—that is, in the same hemisphere—its latitude is subtracted from 23.5° to determine how far it is from where the Sun strikes the Earth at 90°. When the city is in the opposite hemisphere as the noon Sun, its latitude is added to 23.5° to determine how far it is from where the noon Sun angle is 90°.)

June: _____

December: _____

(h) Based on your answers in (g), what do you predict the solar intensity will be for each solstice (high or low) at this location?

June: _____

December: _____

(*continued*)

Name: _____   **Section:** _____

Course: _____   **Date:** _____

(i) As you did in (c), draw the Sun angle at Grand Island in the figure below for the June solstice (on the left image) and the December solstice (on the image at right), making the beam 1 cm wide for each drawing.

June solstice

December solstice

Ground

Ground

(j) Using the scale of the sunbeam (1 cm), determine the length of the sunlight beam spread over the surface at each solstice:

June: _____ cm

December: _____ cm

(k) Given that the amount of energy in each sunbeam is the same, is the intensity of sunlight greater or less during the June solstice or December solstice? _____ . Does this match your prediction in (h)? Explain.

_____

_____

_____

_____

_____

(l) The average high air temperature in Grand Island in June is 28.4°C (83°F) and 2.7°C (37°F) in December. Part of this has to do with the cold air coming in from Canada during winter. Based on your work in this exercise, what else helps explain the relatively large temperature changes. (*Hint:* Consider how solar intensity relates to the length of daylight.)

_____

_____

_____

_____

16.4 Other Controls on Atmospheric Heating

As you have just investigated, the atmosphere is heated in large part by the longwave infrared radiation that is emitted from the surface of the Earth. In order for the Earth to radiate heat, it first must absorb shortwave solar energy. The amount of solar energy that the surface of the Earth absorbs depends on several factors, some of which you have already explored in this chapter. These factors include, but are not limited to, the wavelength of radiation received, heat capacity of the surface material, reflectiveness of the surface (known as the **albedo**), and the presence or absence of clouds. The Earth also absorbs longwave infrared energy emitted downward by the atmosphere (this is the greenhouse effect). The more energy that the surface absorbs, the more energy the Earth emits to the air above the surface. Solar energy that is reflected, scattered, or absorbed by clouds and gases in the atmosphere doesn't reach the Earth's surface, and so is not available to heat it.

16.4.1 Land or Water?

The amount of energy that a material absorbs before there is a change in temperature is referred to as its **heat capacity**. Specifically, it is the amount of heat required to change the temperature of 1 kg of a substance by 1°C. Water has a high heat capacity compared to land, and that has some profound effects on the amount of longwave energy that the Earth radiates to the atmosphere. This in turn affects the weather in various locations of the Earth, both daily and seasonally, as you'll investigate in Exercise 16.3.

EXERCISE 16.3	**Investigating the Effects of Heating the Atmosphere above Land and Water**

Name: _____   Section: _____

Course: _____   Date: _____

In this investigation, you will be taking temperature readings of land and water, given equal amounts of radiation, and observing how their relative heat capacities affect their temperature and that of the air above. Remember that temperature is the *average speed* of the molecules: higher average speed means higher temperature, and the opposite is also true: lower average speed equals lower temperature.

   With the materials provided by your instructor, set up the experiment as explained below and shown in the diagram. (*Note:* If this exercise is available and being conducted online, the setup, materials, and directions may differ. See the instructions in the online lab for how to conduct this exercise.)

- Place one thermometer in a 250-ml beaker one-third full of water.
- Place the other thermometer in a 250-ml beaker filled one-third with soil or dark sand.
- Place both beneath a heat lamp attached to a ring stand.
- Have a timer ready.

You will record your temperature readings at 1-minute intervals using the table provided on the next page. Then, answer the questions that follow.

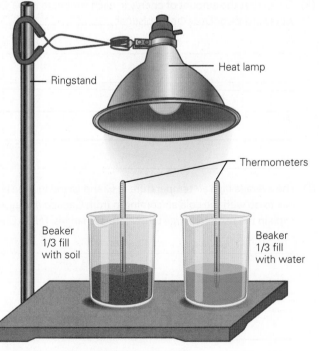

Heat lamp

Ringstand

Thermometers

Beaker 1/3 fill with soil

Beaker 1/3 fill with water

(continued)

Name: _____   Section: _____
Course: _____   Date: _____

(a) Once you have set up the experiment as indicated, make sure that the bulbs of both thermometers are just below the surfaces of the soil (or sand) and the water. Record the starting temperatures of the soil and water and take these in the table. Turn the heat lamp on and start your timer. After 1 minute, take the temperature of the soil and the water and record these in the table. Leaving the heat lamp on, record the temperature of the soil and water every minute for a total of 10 minutes.

Land vs. water heating data

Time (minutes)	Beaker of water temperature (°C)	Beaker of soil (or sand) temperature (°C)	Beaker of moist soil (or sand) temperature (°C)
Start			
1 min			
2 min			
3 min			
4 min			
5 min			
6 min			
7 min			
8 min			
9 min			
10 min			
Total change in temperature (°C)			

(b) After 10 minutes, turn off the lamp and let the soil cool for several minutes. Then, slightly moisten the soil with new water (not the water from part (a)). Run the experiment again with only the damp soil, recording the temperature change at 1-minute intervals, for a total of 10 minutes, and record the results in the table.

(c) Calculate the total change in temperature for each beaker from start to finish and record your answers in the bottom of the table.

(d) Graph your results using the grid below. Include a key for each of the materials tested. Then answer the following questions.

(e) Although the dry soil and water received the same amount of radiation, how did their abilities to change temperature compare?

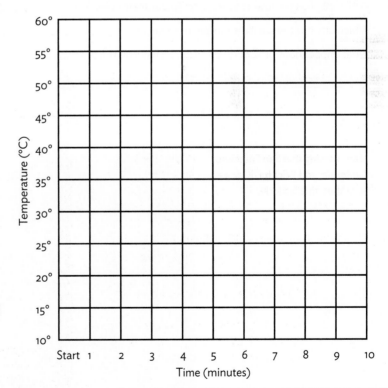

_____

_____

(continued)

Name: _____ Section: _____
Course: _____ Date: _____

(f) Describe how the temperature changed for the damp soil.

_____

_____

Explain why this might be so.

_____

_____

_____

(g) Let's say that you are traveling to the beach on a hot, sunny day. Both the water and the sand are receiving equal amounts of radiation. What would likely feel cooler to your bare feet: the beach sand or the water? Explain.

_____

_____

(h) Now look at the graph on the next page, which shows the air temperature change over a year for two cities located on the same latitude. One city is on the coast, while the other is in the middle of a continent (see map below).

Which city has the highest average temperature in winter? _____

Which city has the lowest average temperature in summer? _____

Which city has the smallest temperature range? _____

What impact does proximity to the ocean have on the seasonal changes in the two cities? Explain your answer.

_____

_____

Map showing that San Francisco, California, and Wichita, Kansas, lie at the same latitude.

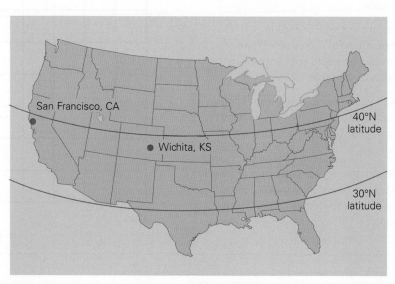

(continued)

Name: _____   Section: _____

Course: _____   Date: _____

Graph of the average high temperature per month for San Francisco, California, and Witchita, Kansas.

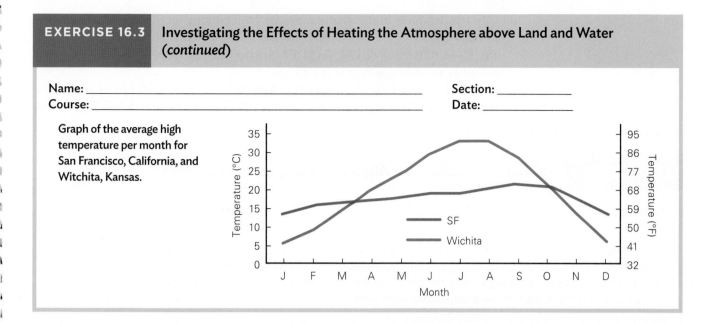

16.4.2 Albedo, Humans, and the Urban Heat Island Effect

The surfaces of the Earth, both natural and human-modified, have different reflective properties, as illustrated in **FIGURE 16.9**. The reflectiveness of a surface, or the *albedo*, affects the amount of radiation that is absorbed at the Earth's surface. Clouds, snow, and ice have high albedos, so they reflect large amounts of solar radiation to space, resulting in cooler air temperatures. Surfaces such as dark rooftops and water have low albedos, and so can absorb more sunlight, which is then radiated to the air above, causing the air temperature to rise.

Although many artificial materials have high albedos (such as metal and concrete), in urban areas heat nevertheless builds up, due to numerous factors, including lack of vegetation and the types of human activity that take place within cities (mass transportation, high energy usage within buildings, etc.). These factors, among others, increase the amount of heat released to the atmosphere. This creates a hot spot compared to the nearby suburban and rural areas. This hot spot is called an **urban heat island**. On warm summer days, the air in urban areas can be $3°C - 4°C$ ($6°F - 8°F$) hotter than in surrounding areas, as shown in **FIGURE 16.10**. For example, in New York

FIGURE 16.9 Albedo of some of the Earth's surfaces.

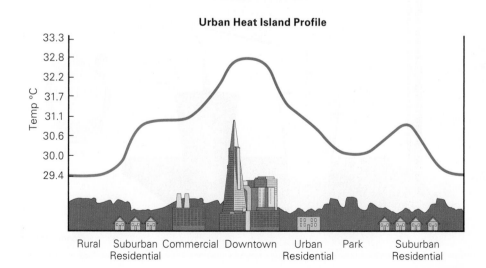

FIGURE 16.10 Chart of an urban heat island profile.

City, temperatures in the Madison Square Mall parking lot during the summer are as high as 48.9°C (120°F) during the day. Tree islands in the lot are only 31.7°C (89°F)—a difference of 17.2°C (31°F)! Nearby wooded areas (in the city parks and suburbs) are as low as 29.4°C (85°F).

In Exercise 16.4, you'll take a closer look at how different surfaces have varying impacts on the heating of the atmosphere, and investigate how urban heat islands form.

EXERCISE 16.4 | **How Do Different Surfaces Affect the Heating of the Atmosphere?**

Name: _____   Section: _____
Course: _____   Date: _____

This activity has two parts. First, you will observe how the albedos of similar objects affect air temperature by conducting an experiment. Then, you will measure outgoing infrared temperatures on your campus or at home (or using data provided by your instructor) to investigate how the albedos of different surfaces (natural and artificial) cause variation in the amount of energy that is absorbed and radiated up to the atmosphere.

Part 1:

Before beginning the experiment, consider the following question:

(a) If it were a sunny summer day, where would you go to cool off in the park: out onto a basketball court or under a tree? _____

Explain your choice. _____

_____

Set up your albedo experiment as shown in the diagram below using the materials provided by your instructor. (*Note:* If this exercise is available and being conducted online, the setup, materials, and directions may differ. See the instructions in the online lab for how to conduct this exercise.)

Step 1: In the table on the next page, record the starting temperature of the air inside each cup or can.

Step 2: Turn on the heat lamp and record the temperature inside each cup or can at 1-minute intervals for a total of 10 minutes.

Step 3: Using the graph provided below the table, make a line graph of your results. Label the temperature axis using your temperature range as a guide. Use a different colored pencil or pattern for each type of cup (or can) and include a legend/key. Then, answer the questions that follow.

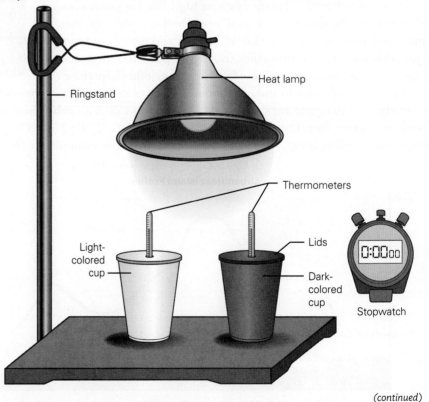

(continued)

Name: _____   Section: _____

Course: _____   Date: _____

Time (minutes)	Dark-colored cup (or can) temperature (°C)	Change in temperature (°C)	Light-colored cup (or can) temperature (°C)	Change in temperature (°C)
Start				
1 min				
2 min				
3 min				
4 min				
5 min				
6 min				
7 min				
8 min				
9 min				
10 min				

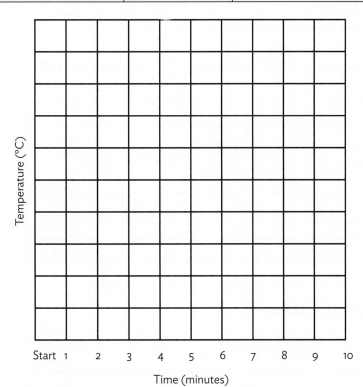

Time (minutes)

(b) Which cup (or can) reached the highest temperature during the experiment? _____

(c) Calculate the change in temperature over *each 1-minute interval* of heating and average these values. Record your answers for the average change in temperature per minute here:

Dark-colored cup or can: _____

Light-colored cup or can: _____

(continued)

Name: _____   Section: _____
Course: _____   Date: _____

(d) Which cup or can heated fastest? Explain why you think this occurred.

_____

_____

Part 2:

Now you will be comparing the heating of different materials to determine how their albedos affect the amount of energy available to heat the air at the surface of the material.

Step 1: Determine how many different types of ground covering exist in the area indicated by your instructor. Then select at least three surfaces to measure. One of these should include a grassy area, and the other two should include artificial materials (such as concrete, asphalt, or brick). Consider taking measurements of these surfaces at spots that are in the shade and also at spots that are exposed to direct sunlight.

Step 2: Once you have selected your surfaces, record them in the table below.

Step 3: To take measurements using the infrared thermometer provided, aim the thermometer at the ground surface, placing it between 1 to 2 inches above the ground. Once the digital display stops changing, record the temperature in the data table below.

Two trials should be run for each surface, and then calculate the average.

Surface type	Sunny or shaded	Trial 1 temperature (°C)	Trial 2 temperature (°C)	Average temperature (°C)

Based on what you observed in your experiment, answer the following questions:

(e) Summarize the results from your experiment:

_____

_____

_____

(f) What might the air temperature be like in a rural area with lots of fields compared to an urban area with lots of pavement? Support your answer with evidence from this experiment.

_____

_____

(continued)

Name: _____ Section: _____

Course: _____ Date: _____

(g) How do the patterns you've observed explain why urban heat islands form?

_____

_____

_____

For the last part of this exercise, use the data provided in the table below or look up the data for three locations given by your instructor, to compare the average daily air temperatures for an urban city and two rural towns to the west and east of the city.

Average daily high temperatures by month for three Nevada locations.

Month	Las Vegas (°C)	Pahrump (°C)	Boulder City (°C)
Jan.	14.4	14.7	11.7
Feb.	16.9	16.7	14.8
March	21.3	20.3	18.5
April	25.7	24.4	23.0
May	31.6	29.6	28.4
June	37.1	34.9	34.4
July	40.1	38.3	37.4
Aug.	38.9	37.4	36.4
Sept.	34.4	33.6	32.2
Oct.	27.0	27.1	25.3
Nov.	19.1	19.6	16.9
Dec.	13.7	14.1	11.8

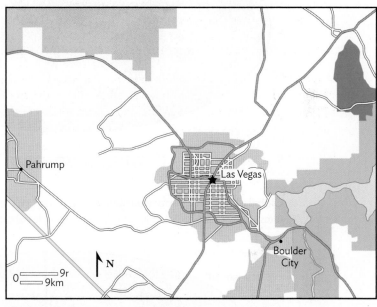

The locations of Las Vegas, Pahrump, and Boulder City, Nevada. Population estimates as of 2017–2018: Las Vegas = 644,644; Pahrump = 36,441; Boulder City = 15,971.

(h) What patterns do you notice when you compare the temperatures of these three locations?

_____

_____

_____

(i) What may be the reason(s) for the patterns you observed?

_____

_____

_____

Name: _____ Section: _____
Course: _____ Date: _____

In this chapter, you have investigated some of the factors that affect the heating of the Earth's surface and the energy that is radiated to the atmosphere above. Like all parts of the Earth System, the atmosphere is a complex system that interacts with other parts of the Earth System. Considering what you have observed and learned about how the atmosphere gets heated, study the maps below, which show global average sea and land surface temperatures. Then answer the questions that follow.

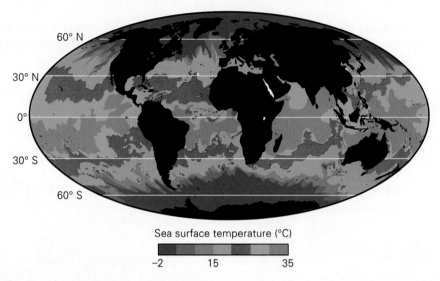

Global average sea surface temperatures. Orange indicates warmer surface temperatures, gray and blue are intermediate values, and purple indicates cold water.

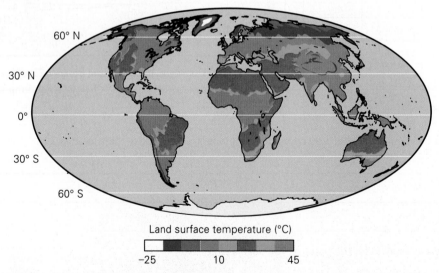

Global average land surface temperature. Gray areas are higher temperature, blue/purple areas are lower temperatures.

(a) What latitudes have the highest sea surface temperature? _____

What latitudes have the highest land surface temperature? _____

(continued)

Name: _____ Section: _____

Course: _____ Date: _____

(b) What factors could be responsible for the temperatures you observed in these hottest areas? Explain your choices.

_____

_____

_____

_____

(c) Do you think the data shown in both parts of the figure above would be different if viewed for individual months within a year, rather than as a yearly average? Explain.

_____

_____

_____

? **What Do You Think** The price of solar panels has finally decreased to a point that is within your friend's budget, and she wants your help in choosing where to place them. To get the most out of the solar panels, the company installing the panels needs to know the optimum angle to set them up so that they will absorb the highest intensity of sunlight. Since the noon Sun angle changes daily, you reason with your friend that you should determine the average yearly angle of the noon Sun at her location.

In the northern hemisphere, in which compass direction will the noon Sun be located (to the north or south)?

_____

What is your starting latitude? (You can use the latitude where your campus is.) _____

What is the noon Sun angle at your latitude during an equinox? _____

Just to check our work, let's calculate the average yearly angle based on the maximum seasonal changes.

What is the noon Sun angle at your latitude during the June solstice? _____

December solstice? _____

What is the midway point, or average yearly angle, based on the maximum and minimum noon Sun angles you just calculated? _____

Of course, the position of the Sun in the sky changes throughout the day. To get the best use out of the panels, your friend would need to track the Sun minute by minute and adjust the angle on the panels. To compensate for the movement, she can buy additional panels set at varying angles or purchase an automated solar tracker. Your friend is not sure she has enough space for more panels and doesn't think she'll like the way they look. On the other hand, solar trackers are often more expensive than buying additional panels and the amount of power a solar tracker uses may take away from its benefits.

Would you recommend that your friend purchase additional panels, a solar tracker, or neither? Explain your answer on a separate sheet of paper.

Name: _____   Section: _____
Course: _____   Date: _____

Exploring Earth Science Using Google Earth

1. Visit **digital.wwnorton.com/labmanualearthsci**
2. Go to the **Geotours** tile to download Google Earth Pro and the accompanying Geotours exercises file.

Expand the Geotour16 folder in Google Earth by clicking the triangle to the left of the folder icon. The folder contains information keyed to questions that investigate the seasonal differences in solar insolation. **Solar insolation** is a measure of how much sunlight has reached the Earth's surface over a specified time interval. Such information not only helps us understand weather and climate patterns, but also informs us about potential areas for solar energy grids.

Google Earth

(a) Check and double-click the Solar Insolation (Jul 2019) overlay to see a visualization of the solar insolation for July 2019. Turn on the Legend overlay to interpret the colors; values range from 0 (dark purple) to 550 (light tan) watts per square meter. Why does the southern hemisphere have lower values than the northern hemisphere?

_____

_____

(b) Based on the data, would you rather develop a solar energy array (a collection of solar panels) for southern California or for Florida? _____

(c) Check and double-click the Solar Insolation (Dec 2019) overlay to see a visualization of the solar insolation for December 2019 [uncheck Solar Insolation (Jul 2019)]. What has changed and why?

_____

_____

_____

(d) Check and double-click on the (d) placemark to fly to Chile (first make sure that Layers > Borders and Labels is turned on). Now select the Solar Insolation (Dec 2019) overlay and make it semi-transparent by clicking on the rectangle at the bottom of the Places panel and then adjusting the slider. Pass your cursor over the green text "Desierto De Atacama Desert." Does this area receive high or low insolation? _____ Can you hypothesize why this might be?

_____

_____

_____

Humidity, Air Pressure, and Wind

Snow kicked up by a tracked vehicle is blown across the Ross Ice Shelf near McMurdo Station, Antarctica.

17.1 Introduction

In today's world, we are accustomed to weather reports that rely heavily on technologies such as weather satellites and radar. The wealth of measurements that are made within the atmosphere both by ground-level instruments and those at high altitude provide us with insights to further our understanding of weather systems and to make accurate predictions of changing weather conditions over short time scales. These day-to-day conditions of the troposphere—temperature, moisture, pressure, humidity, and wind, for example—are all factors of the weather that are measured independently of one another. But they do not operate independently; therefore, in order to understand weather patterns, we need to investigate the interactions between these factors. In this chapter, you will investigate the interactions between air temperature, air pressure, and moisture.

Moisture refers to water in our atmosphere. Water is unique in that it is the only substance that occurs in all three forms of matter at the Earth's surface—solid, liquid, and gas. In its gaseous state, water is referred to as *water vapor* (**FIG. 17.1**). Water can also occur in liquid or solid forms in the atmosphere, depending on the temperature and pressure conditions at a given time and location (**FIG. 17.2**).

17.2 Humidity and Temperature

If you are outside on a hot summer day in the southeastern United States, you will probably notice the clammy feel of moisture on your skin. If you wear glasses and go inside an air-conditioned building, your lenses may fog up. In both cases, you are experiencing *humidity*. Although the gases that make up the air are dominated by nitrogen and oxygen, which remain in relatively constant proportion to one another, the small percentage of other atmospheric gases, whose abundances change over time, have a profound effect on daily atmospheric conditions. These trace gases include water vapor (H_2O), carbon dioxide (CO_2), and

FIGURE 17.1 Water vapor in the atmosphere. Even on this cloudless day on Cape Cod, invisible water vapor molecules are mixed with molecules of other gases.

Nitrogen
Oxygen
Water vapor

FIGURE 17.2 Water in the atmosphere occurs in three forms.

(a) Clouds in the warm sky over Brazil consist of liquid water droplets.

(b) High clouds over the University of Illinois consist of tiny ice crystals.

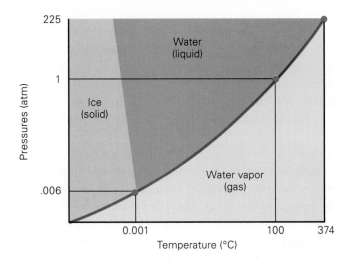

(c) The form water takes in a given location in the atmosphere depends on the temperature and pressure conditions at the location. One atmosphere (1 atm) approximately equals the air pressure at sea level and is equivalent to about 1,013.25 millibars.

methane (CH_4), among others. While you may hear a lot about the effects of CO_2 on atmospheric heating, it is the changing form of water in the atmosphere (and the exchange of energy involved) that has the greatest measurable influence on our weather. When we refer to the **humidity** of the air, we are describing the amount of water vapor present in the air.

17.2.1. Changing States of Water

So why does the amount of water vapor in the air change? As you have investigated in earlier chapters, water moves throughout the Earth System as part of the hydrologic cycle. As the water moves, it may also change form, or state, as illustrated in **FIGURE 17.3.** For a change in state to occur, energy is either absorbed by the water from the surrounding environment or released to the surrounding environment. For example, during **evaporation**, in which liquid changes to gas, the water takes energy from its surroundings; in contrast, during the reverse process of **condensation**, in which gas changes to liquid, the water releases energy to its surroundings.

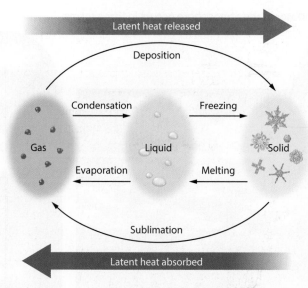

FIGURE 17.3 Changes in the state of water.

This form of energy is referred to as *latent heat* because it is "hidden" energy used to form or break bonds between water molecules such that the water changes form; it does not result in a change in the temperature of the water. While the temperature of the water does not change during this process, the addition or depletion of energy to/from the surrounding air causes a change in the air's temperature. This is why the process of evaporation has a cooling effect on your skin; as the water evaporates from your skin, it is taking heat from your body. On the other hand, as water vapor condenses, or liquid water freezes to form ice, heat is released to the air, causing it to warm.

There are different amounts of heat that are required for each change in state of a given volume of water. In order for the process of **melting** to occur—a change from solid to liquid state—the molecules must become loosely bonded so that they can flow. Thus, energy is required to break bonds between the water molecules in ice. Changing from either liquid to gas (evaporation) or solid to gas (**sublimation**) involves even higher amounts of energy to release individual molecules, so they are not bonded to each other and can move freely. The opposite is true as water condenses from gas to liquid and then freezes into solid ice—less energy is involved in each step as bonds are formed, so the excess energy is released to the surrounding air. It is the exchange of heat as water changes states that drives weather and provides energy to the world's storm systems. The experiment in Exercise 17.1 will help you to understand the nature of latent heat.

EXERCISE 17.1	**Latent Heat and the Changing State of Water**

Name: _____   **Section:** _____
Course: _____   **Date:** _____

In this exercise, you'll record the results of gradual warming of ice water on a hot plate. Your instructor will provide the materials needed to conduct the experiment. (*Note:* If this exercise is available and being conducted online, the setup, materials, and directions may differ. See the instructions in the online lab for how to conduct the exercise.)

Step 1: Fill your beaker approximately two-thirds full with ice and put in enough tap water to just cover the ice.

(continued)

Name: _____   Section: _____

Course: _____   Date: _____

Step 2: Gently stir this ice water mixture with your thermometer for a few minutes and record the temperature. It should be close to 0°C before you continue to the next step. If not, gently stir some more until the mixture's temperature decreases, or you may need to add more ice. Record your starting temperature in the table below.

Step 3: Place the beaker of ice water onto the hot plate. Depending on your particular hot plate, you will set the temperature to around medium-high (usually between 5–7 on a scale of 10). You want it to heat the ice mixture gradually without melting all the ice in less than 5 minutes!

Step 4: Continue to gently stir the ice mixture and record the temperature at 1-minute intervals in the table. When all the ice has melted, note this time on your table and continue to stir and record the water temperature for an *additional* 5 minutes. If you need more time on the table, simply add more rows.

Latent heat experiment data

Time (minutes)	Temperature (°C)	Change in temperature per minute (ΔT/min)
0 (start)		_____
1		
2		
3		
4		
5		
6		
7		
8		
9		
10		
11		
12		
13		
14		
15		

Step 5: When you have finished the experiment, unplug your hot plate and leave your beaker to cool on top. Use Figure 17.3 and your experiment results to complete the following items:

(a) Graph your results on the blank graph on the next page. Use a different colored pencil or patterned line to connect your data points for the temperature before melting and the temperature after melting. Make sure to add appropriate intervals to the y-axis.

(b) Determine the change in temperature over each 1-minute time interval *before* the ice melted completely, and mark this in the third column in the table. Then calculate the average of these temperature changes per minute (add all these changes per minute together and then divide by number of minutes it took for ice to melt): _____ °C/minute

What is the average change in temperature per minute *after* melting? _____ °C/minute

(continued)

Name: _____   Section: _____
Course: _____   Date: _____

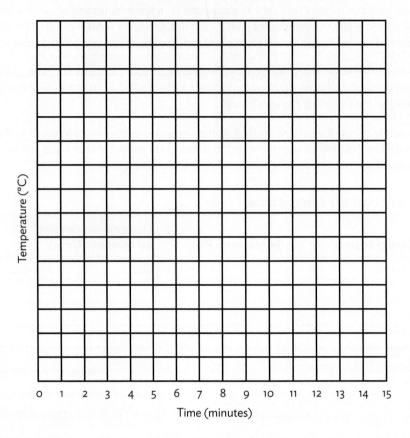

Time (minutes)

(c) Explain why there was a *difference* in the rate of temperature change before and after melting. In other words, what was happening in each case?

_____

_____

_____

(d) If you had continued to heat the water for several more minutes, into which state would the water have changed? _____

(e) Recall from the introductory text before this exercise that evaporation requires more latent heat than melting. Therefore, would it take longer to melt or evaporate the same volume of water?

_____

(f) Sublimation occurs when solid ice turns directly into water vapor, without turning into a visible liquid in between. Would this process require more or less heat than melting or evaporation alone?

_____

_____

Sublimation occurs on a sunny day in winter when a pile of snow gets smaller and smaller without any puddles forming. It can also explain why the ice cubes left in your freezer for long periods of time become misshapen!

(a) Humid air.

(b) Dry air.

17.2.2 Relative Humidity

When referring to the humidity of the air, we are only describing water in its gaseous form, or as vapor. **Absolute humidity** refers to the total mass of water vapor in a given amount of air, recorded as grams per cubic meter (g/m³) or grams per kilogram of air (g/kg). However, a weather report in the media does not describe absolute humidity; what is reported is **relative humidity**, which is a comparison of the percent of water vapor in the air relative to the amount of water vapor that the air can potentially hold at a given air temperature. Just as the ground can be saturated with water, so too can the air. When the actual amount of water vapor present in the air is equal to the amount the air can potentially hold at the current air temperature, the air is at saturation (**FIG. 17.4A, B**). **FIGURE 17.4C** is a graph of water vapor content for 1 kg of air (at saturation) as a function of air temperature.

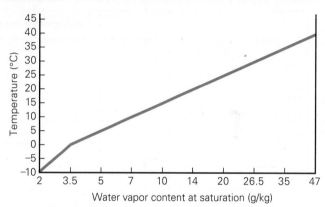

(c) Water vapor content at saturation as a function of air temperature. (*Note:* the x-axis scale is not linear.)

In order to calculate the relative humidity of the air, the air temperature and the amount of water vapor in the air must be measured. Then you can use the following equation to calculate relative humidity:

$$\text{Relative Humidity (\%)} = \left(\frac{\text{water vapor content}}{\text{water vapor saturation}} \right) \times 100$$

For example, if the actual amount of water vapor in the air is measured as 14 g/kg, and the air temperature is 30°C, then:

$$\text{Relative Humidity (\%)} = \frac{14 \text{ g/kg}}{26.5 \text{ g/kg}} \times 100 = 53\%$$

The water vapor at saturation value has been taken from Figure 17.4c, which shows that at 30°C, the air can hold a maximum of 26.5 grams of water vapor for every 1 kg of air. If the air were to cool to 20°C, then relative humidity (RH) would equal 100%, and the air would be saturated with water vapor. Be careful here—this does not necessarily mean that *precipitation* will occur. Precipitation is part of the hydrologic cycle, during which water moves from one reservoir (the atmosphere) to another (the Earth's surface—oceans, land, lakes, glaciers, etc.). If the air were oversaturated with water vapor, the excess vapor molecules would condense or freeze, forming clouds in the atmosphere or dew/frost on the surface, respectively.

EXERCISE 17.2 — Determining Relative and Absolute Humidity

Name: _____   **Section:** _____
Course: _____   **Date:** _____

In the questions below, don't forget to include proper units of measurement in your answers!

(a) Use the graph in Figure 17.4c to help you complete the table below. Some of the table has been filled in for you.

(b) Based on what you recorded in the table, how much water vapor can 1 kg of air hold at 35°C? _____
At 5°C? _____

(c) Based on the general patterns you observe in Fig. 17.4c and the table below, does warm or cool air hold more water vapor? _____

(d) At what temperature (°C) is the air saturated when the water vapor content is 5 g/kg? _____°C

Water vapor content at saturation at various temperatures

Temperature (°C)	Temperature (°F)	Water vapor content at saturation (g/kg)
−10	14	2
0	32	3.5
5	41	
10	50	7
15	59	
20	68	14
25	77	
30	86	26.5
35	95	
40	104	47

As you've observed, the relative humidity of the air changes as air temperature changes. Likewise, relative humidity changes if the amount of water vapor in the air changes. Let's investigate how changing either the air temperature or water vapor content changes relative humidity by doing some calculations using the relative humidity formula and the table from Exercise 17.2.

EXERCISE 17.3 — Investigating How Relative Humidity Changes

Name: _____   **Section:** _____
Course: _____   **Date:** _____

(a) Calculate the relative humidity as air temperature changes:

Air temperature (°C)	Water vapor content (g/kg)	Relative humidity (%)
30	7	_____
20	7	_____
10	7	_____

(continued)

Name: _____   Section: _____
Course: _____   Date: _____

As air temperatures cool, what happens to relative humidity?

_____

_____

(b) Calculate the relative humidity at constant temperature as the water vapor content changes:

Air temperature (°C)	Water vapor content (g/kg)	Relative humidity (%)
20	3.5	_____
20	7	_____
20	14	_____

As the amount of water vapor is reduced, what happens to relative humidity?

_____

_____

(c) Let's say a TV meteorologist has reported that the current relative humidity is 75% and that the air temperature is 20°C. How much water vapor is in the air? _____ g/kg

In order for water to evaporate, there must be room in the atmosphere for water vapor—in other words, the air must be below saturation with respect to water vapor. At 100% relative humidity, the rate of evaporation is equal to the rate of condensation. But, when relative humidity is over 100%, the air is supersaturated with respect to water vapor, and the excess water vapor must condense (or form ice crystals, depending on the air temperature).

(d) If the water vapor content is 20 g/kg, and the air temperature is 30°C, is the air saturated? _____

If the air temperature drops to 20°C, how much excess water vapor is present and must condense? _____ g/kg

(e) In order to cool homes in some places, an evaporative cooler is used (a device that cools air by evaporating liquid water into water vapor). Where do you think an evaporative cooler would work best: in an arid environment in Arizona or a swampy environment in Florida? Explain.

_____

_____

Another way to determine relative humidity is to use an instrument called a *hygrometer*, like the *sling psychrometer* shown in **FIGURE 17.5**, the hygrometer most often used in laboratory classes. This sling psychrometer comprises two thermometers—a *dry-bulb thermometer* and a *wet-bulb thermometer*. Notice that the wet-bulb thermometer has a piece of cotton gauze or cloth covering it, which is dampened during use. The two thermometers are spun in the air for several minutes to both evaporate water from the wet-bulb thermometer and to measure the air temperature with the dry-bulb thermometer. If the air is not saturated (RH < 100%), then evaporation can take place. Recall that evaporation is a cooling process—heat energy is absorbed by liquid water to change it to gaseous form. So, as the water evaporates from the gauze, the temperature on the wet-bulb thermometer drops. The drier the air, the more evaporation can occur, and the greater the difference between the dry and wet bulb temperatures.

FIGURE 17.5 A sling psychrometer.

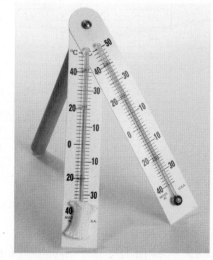

Name: _____   Section: _____

Course: _____   Date: _____

Below is a relative humidity chart that you will use along with your psychrometer measurements to determine the relative humidity inside and outside your lab room. To use the chart, first find the air (dry-bulb) temperature on the vertical axis (rows), then go across the row until you reach the column that represents the difference between the air temperature and wet-bulb temperature, called the *wet-bulb depression*. The numerical value where the row and column intersect is the relative humidity (in %).

For example, let's assume that the air temperature is 20°C and the wet-bulb temperature is 18°C. So, the row you need is 20°C (the air temperature), and the column (the difference between the air temperature and wet-bulb temperature) is 2°C. Therefore, at these temperatures the RH = 82%.

Before taking your measurements, take a look at the row and column headings in the chart, and the relative humidity values.

Relative humidity (values in %)

Number of degrees difference between dry-bulb & wet-bulb temperatures

	0	1	2	3	4	5	6	7	8	9	10	11	12	13	14	15
−20	100	28														
−18	100	40														
−16	100	48														
−14	100	55	11													
−12	100	61	23													
−10	100	66	33													
−8	100	71	41	13												
−6	100	73	48	20	11											
−4	100	77	54	32	20											
−2	100	79	58	37	28	1										
0	100	81	63	45	36	11										
2	100	83	67	51	42	20	6									
4	100	85	70	56	46	27	14									
6	100	86	72	59	51	35	22	10								
8	100	87	74	62	54	39	28	17	6							
10	100	88	76	65	57	43	33	24	13	4						
12	100	88	78	67	60	48	38	28	19	10	2					
14	100	89	79	69	62	50	41	33	25	16	8	1				
16	100	90	80	71	64	54	45	37	29	21	14	7	1			
18	100	91	81	72	66	56	48	40	33	26	19	12	6			
20	100	91	82	74	68	58	51	44	36	30	23	17	11	5		
22	100	92	83	75	69	60	53	46	40	33	27	21	15	10	4	
24	100	92	84	76	70	62	55	49	42	36	30	25	20	14	9	4
26	100	92	85	77	71	64	57	51	45	39	34	28	23	18	13	9
28	100	93	86	78	72	65	59	53	47	42	36	31	26	21	17	12
30	100	93	86	79	73	66	61	55	49	44	39	34	29	25	20	16
32	100	93	86	80	74	68	62	56	51	46	41	36	32	27	22	19
34	100	93	86	81	75	69	63	58	52	48	43	38	34	30	26	22
36	100	94	87	81	76	69	64	59	54	50	44	40	36	32	28	24

Air (dry-bulb) temperature (°C)

(continued)

Name: _____   Section: _____

Course: _____   Date: _____

(a) What two pieces of information do you need to use this chart?

_____

(b) What pattern do you observe when there is *no difference* between air temperature and wet-bulb temperature?

_____

_____

_____

(c) What pattern do you observe as the difference between the air temperature and wet-bulb temperature increases?

_____

_____

_____

Is this consistent with your observations when using the relative humidity calculation in Exercise 17.2? Explain.

_____

_____

_____

(d) If the air temperature is 32°C, and the wet-bulb temperature is 28°C, what is the relative humidity? _____

(e) At what wet-bulb temperature is RH 100% if the air-temperature is 18°C? _____

(f) If a psychrometer is available to you, follow the instructions provided by your instructor to use it to measure the wet-bulb and dry-bulb temperatures in the room and outside the building. If one is not available, your instructor will provide the temperatures for you. Use these temperatures to calculate the RH in the room and outside.

	In room	Outside
Dry-bulb temperature	_____°C	_____°C
Wet-bulb temperature	_____°C	_____°C
Difference between dry and wet temperatures	_____°C	_____°C
Relative humidity	_____%	_____%

17.2.3 Humidity and the Dew Point

The last way that we will describe humidity in this chapter is by determining the **dew point** temperature. You have just determined this temperature for the conditions given in Exercise 17.4e. The dew point is the temperature that the air must cool to, without changing the air pressure, so that the air will be saturated with water vapor. In other words, it is the temperature at which the RH = 100%—the moment **condensation** occurs. Unlike relative humidity, which changes as either air temperature or water vapor content changes, the dew point only depends on the amount of water vapor in the air. It, too, can be determined via psychrometer measurements, as you'll see in Exercise 17.5.

Name: _____ Section: _____
Course: _____ Date: _____

In this exercise, you will make two measurements of the dew point, first using the psychrometer data you collected in Exercise 17.4, then by way of a condensation experiment.

Psychrometer data

Using the data from Exercise 17.4, question (f), refer to the chart below to answer question (a).

Dew point (values in °C)

Number of degrees difference between dry-bulb & wet-bulb temperatures

Air temp	0	1	2	3	4	5	6	7	8	9	10	11	12	13	14	15
−20	−20	−33														
−18	−18	−28														
−16	−16	−24														
−14	−14	−21	−36													
−12	−12	−18	−28													
−10	−10	−14	−22													
−8	−8	−12	−18	−29												
−6	−6	−10	−14	−22												
−4	−4	−7	−12	−17	−29											
−2	−2	−5	−8	−13	−20											
0	0	−3	−6	−9	−15	−24										
2	2	−1	−3	−6	−11	−17										
4	4	1	−1	−4	−7	−11	−19									
6	6	4	1	−1	−4	−7	−13	−21								
8	8	6	3	1	−2	−5	−9	−14								
10	10	8	6	4	1	−2	−5	−9	−14	−28						
12	12	10	8	6	4	1	−2	−5	−9	−16						
14	14	12	11	9	6	4	1	−2	−5	−10	−17					
16	16	14	13	11	9	7	4	1	−1	−6	−10	−17				
18	18	16	15	13	11	9	7	4	2	−2	−5	−10	−19			
20	20	19	17	15	14	12	10	7	4	2	−2	−5	−10	−19		
22	22	21	19	17	16	14	12	10	8	5	3	−1	−5	−10	−19	
24	24	23	21	20	18	16	14	12	10	8	6	2	−1	−5	−10	−18
26	26	25	23	22	20	18	17	15	13	11	9	6	3	0	−4	−9
28	28	27	25	24	22	21	19	17	16	14	11	9	7	4	1	−3
30	30	29	27	26	24	23	21	19	18	16	14	12	10	8	5	1

Air (dry-bulb) temperature (°C) (vertical axis)

(a) What is the dew point inside your lab? _____

And outside? _____

(continued)

Name: _____ Section: _____
Course: _____ Date: _____

Condensation experiment

In this experiment, you'll use the materials provided by your instructor to measure the dew point for the air in your lab and the air outside. (*Note:* If this exercise is available and being conducted online, the setup, materials, and directions may differ. See the instructions in the online lab for how to conduct the exercise.)

(b) Before you begin, form a hypothesis comparing the dew point inside and outside your lab room.

_____

_____

_____

Inside dew point:

Step 1: Without getting any water on the outside of the beakers, place some ice-cold water (no ice) in one beaker and the same amount of room-temperature tap water in the second beaker.

Step 2: After about 1 minute, check for condensation on the outside of the beakers. Place a thermometer in each beaker. (*Note:* If condensation forms on the tap-water beaker, your water is cooler than room temperature and you need warmer water). After the thermometer reading reaches a fairly constant temperature, record the results in the table below.

Step 3: Use more beakers or cups (2 to 6) to create containers of differing temperatures by mixing different amounts of the ice water and tap water. After a minute or two, look for the pair of beakers whose water temperatures are closest to each other but have different results in condensation (one has condensation and the other does not). Record their temperatures and condensation status in the table.

Step 4: Continue mixing the two noncondensing and condensing beakers you selected in Step 3 until you have narrowed the temperature difference between the two to approximately 2°C and record the results in the table. Calculate the average temperature of these two beakers and record this as the "dew point" in the table.

Outside dew point:

Step 1: First determine the outside air temperature—it must be above freezing to conduct this part of the experiment. After confirming that the outside temperature is warm enough, mix up batches of water in the paper or Styrofoam cups so that the water temperatures are between 0°C and the outside temperature, in increments of 5°C, or as indicated by your instructor.

Step 2: Take your batches of water, thermometers, and metal cups outside. Carefully pour the water from each cup into a separate metal cup and observe if moisture condenses on the outside surface or not. Record the temperatures of the water and the results in the table.

Step 3: As you did inside the lab, identify the pair of cups whose water temperatures are closest but have different condensation results. Then mix those different waters to get intermediate temperatures until you can narrow the dew point measurements. Record the results in the table.

Step 4: Return all items to the lab room. Then, examine your results for inside and outside the lab and answer the questions that follow.

(continued)

EXERCISE 17.5 Measuring Dew Point by Two Methods *(continued)*

Name: ______________________ **Section:** __________
Course: ______________________ **Date:** __________

Dew point determination

Inside dew point measurement		**Outside dew point measurement**	
Inside air temperature (°C):		**Outside air temperature (°C):** **Is it foggy? Yes or No**	
Water temperature (°C)	**Condensation? Yes or No**	**Water temperature (°C)**	**Condensation? Yes or No**

(c) How are the results of your experiment for inside the lab and outside the lab similar?

____

____

____

(d) Is the dew point different? If so, where is it lowest? Where is it highest? Explain.

____

____

____

(e) Considering the dew point temperature inside and outside the building, did the building's air conditioning or heating units *add* or *remove* moisture from the outside air when it was brought inside? __________

Explain how you can tell in each case. ______________________

____

____

Compare your condensation results with the **indoor** psychrometer results.

(f) Are the results the same? Yes No (circle your answer)

(g) *Should* you expect the results to be the same? Yes No (circle your answer)

Why or why not?

____

____

____

FIGURE 17.6 Examples of barometers.

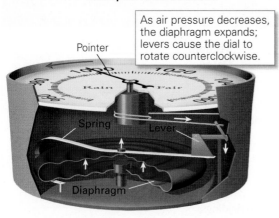

As air pressure decreases, the diaphragm expands; levers cause the dial to rotate counterclockwise.

Pointer

Spring
Lever
Diaphragm

(a) In an aneroid barometer, compression or expansion of a diaphragm moves a dial.

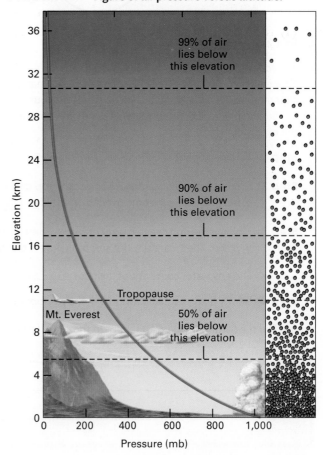

Vacuum

(b) In a traditional mercury barometer, air pressure pushes mercury up a vacuum tube. The greater the pressure, the higher the mercury rises.

760 mm (29.92 in)

Mercury pressure equal to air pressure

Atmospheric pressure

17.3 Atmospheric Pressure

The 17th century saw the invention of the mercury thermometer, and in 1643 Italian physicist Evangelista Torricelli invented the **mercury barometer**, a device used to measure **atmospheric pressure:** the force of the air that is pressing down on an area of the Earth's surface. Over time, other types of barometers were invented, such as the **aneroid barometer** (**FIG. 17.6**). These instruments were not only used by scientists to relate observed weather phenomena with measurements of atmospheric conditions, but they were also used by sailors to detect approaching storms and changing winds—especially important when sailing tall ships across the vast reaches of the open oceans!

If you have ever flown in an airplane, traveled up a mountain, or even been a passenger in an elevator going up a tall building, then you probably noticed the change in air pressure affecting your ears. That's because the air inside our bodies is in balance with the force of the air outside us, the force of the atmospheric pressure. At sea level, the air presses down on us with a weight of 14.7 pounds on every square inch, or a pressure of about 1,013 millibars (mb). But as you travel to higher altitudes, the air pressure decreases rapidly, as illustrated in **FIGURE 17.7**. In a weather report, the air pressure values reported across the United States are the barometric pressures for multiple cities that are not all at the same altitude. For example, New York City is 10 m (33 ft) above sea level, while Denver is more than one hundred times that distance above sea level at 1,609 m (5,280 ft). In order to compare the weather in various locations and predict weather changes, the air pressures at the different cities are "corrected" (or adjusted) as if they were recorded at sea level.

FIGURE 17.7 Figure of air pressure versus altitude.

99% of air lies below this elevation

90% of air lies below this elevation

Tropopause

Mt. Everest

50% of air lies below this elevation

Elevation (km)

Pressure (mb)

| EXERCISE 17.6 | Investigating Atmospheric Pressure |

Name: _____   Section: _____

Course: _____   Date: _____

Part 1

If a barometer is available to you, record the atmospheric pressure inside and outside your lab room. If no instrument is available, obtain the data from your instructor.

(a) What is the pressure inside your lab? _____

What is the pressure outside your lab? _____

Are they the same or different? Suggest factors that might affect the two readings:

_____

_____

_____

(b) Based on the diagram below showing two parcels of air at the Earth's surface, label each as either high or low pressure. Note that volume is constant in both cases; only the number of molecules varies.

Part 2

The following experiment is another way to investigate the effects of air pressure in two locations: inside and outside a cup. Your instructor has provided you with the needed materials. In order to contain any escaping water, you will conduct the experiment over a beaker or bucket, or whatever your instructor has provided. (*Note:* If this exercise is available and being conducted online, the setup, materials, and directions may differ. See the instructions in the online lab for how to conduct the exercise.)

Step 1: Fill a cup about two-thirds full with tap water.

Step 2: Place the piece of cardboard across the top of the cup. The cardboard must be large enough to cover the entire top of the cup with some overlap on all sides. Place one hand flat but firmly on top of the cardboard.

Step 3: Holding the cup over a beaker or bucket, pick up the cup with your free hand and continue to hold the cardboard in place. Then, carefully but smoothly, invert the cup.

Did any water come out of the cup? _____

Step 4: Slowly take your hand away from the cardboard.

(c) Did the cardboard stay "attached" to the cup? If not, why not? If yes, explain how that can be.

_____

_____

_____

17.4 Modeling Vertical Air Movements

The atmosphere is a fluid that, like the oceans, circulates globally, redistributing energy across the planet. What drives the rising and sinking of parcels of air are differences in their densities relative to the surrounding air. You are probably familiar with the idea that warm air rises. It does so because as air warms, it expands and becomes less dense than the surrounding cooler air. On the other hand, as air cools, it becomes denser than the surrounding air, and so sinks.

Temperature change within a rising or sinking parcel of air occurs as a result of the changing speed of the molecules within the air parcel. The temperature change is called **adiabatic temperature change** when it occurs without the addition or loss of heat or mass in the air parcel. As a parcel of air rises, it expands due to the decrease in the surrounding air pressure. This expansion causes the air molecules to slow down, and so the air cools as it rises. The parcel will stop rising once its temperature is equal to the surrounding air. When a parcel of air becomes denser than the surrounding air (generally because its temperature decreases), it sinks towards the surface. As it sinks, the motion of the air molecules increases, causing the air to warm.

EXERCISE 17.7 **Rising and Sinking Parcels of Air**

Name: _____ Section: _____
Course: _____ Date: _____

You will use the diagram on the right, which is a side view of the atmosphere above the Earth's surface, to answer the following questions:

(a) Assuming that both locations are in the troposphere, at which location (X or Z) will the air temperature be higher? Explain. (*Hint:* Refer back to Chapter 16.)

_____

_____

(continued)

Name: _____   Section: _____
Course: _____   Date: _____

At which location will the air pressure be higher? Explain.

_____

_____

_____

(b) Assume a parcel of air at location X is less dense than the surrounding air, and a parcel of air at location Z is denser than the surrounding air. Given these conditions, complete the diagram on the previous page by drawing arrows up from or down to the surface at the two locations (that is, the vertical motion of the air).

(c) Does the water vapor in a parcel of *sinking* air condense to make clouds? _____ . Does the relative humidity of the parcel of air increase or decrease? _____ . Explain your answers.

_____

_____

_____

(d) Does the water vapor in a parcel of *rising* air condense to make clouds? _____ . Does the relative humidity of the parcel of air increase or decrease? _____ . Explain your answers.

_____

_____

_____

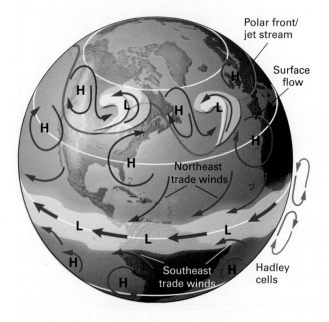

FIGURE 17.8 Global wind circulation pattern. Note that this is an idealized pattern, averaging out the typical movement of air on a large scale such as latitude.

17.5 Atmospheric Pressure and the Horizontal Movement of Air

Low-pressure systems (called *cyclones*) form where near-surface air heats and becomes less dense. High-pressure systems (called *anticyclones*) form where near-surface air cools and becomes denser. Some cyclones and anticyclones are semi-permanent in that they don't move for long periods of time, while others, associated with storms, move about the Earth. Their movement is related to forces in the jet stream. We call the horizontal movement of air *wind*. On a global scale, the direction the dominant (*prevailing*) winds blow depends on latitude and hemisphere. These global wind patterns are formed by a combination of low- and high-pressure systems at major belts of latitude (**FIG. 17.8**) and by the rotation of the Earth. Let's look more closely at the formation of wind before we investigate global wind patterns and their relationship to atmospheric moisture and precipitation.

17.5.1. Driving Forces of Wind

The horizontal movement of air across the Earth's surface requires a force. This force is called the **pressure gradient force** (PGF), which states that in order for the air to flow, there must be a change in air pressure between two locations. Think of it like the Earth's surface, where higher pressure values are like hills, and lower pressure values are like valleys. Just like water flows downhill (or down elevation), so too does the air—it moves from areas of high pressure to areas of low pressure. Where there is no difference in pressure, there is no force and no wind. On the other hand, the greater the pressure difference, the steeper the gradient and the faster the wind blows. Just as you investigated in the topographic maps chapter, where contour lines of equal elevation illustrate the surface topography, on weather maps lines of equal sea-level air pressure—called **isobars**—are drawn that allow us to label high and low pressure centers and to determine general wind speed and direction. Just as with elevation contour lines, the spacing of the contours indicates relative speed. Where the isobars are close together, there is a greater rate of change in air pressure and faster winds. If the PGF were the only force, the air would move in a straight line, crossing the isobars at right angles.

EXERCISE 17.8   **Pressure Gradient Force and Winds**

Name: _____   Section: _____
Course: _____   Date: _____

On the diagram of isobars shown below:

(a) Label an area of high pressure with "H" and low pressure with "L".

(b) Draw an arrow indicating the direction the wind will blow due to PGF only.

(c) Label where the wind blows "faster" and "slower" in the appropriate places on the diagram.

The pressure gradient is not the only force that controls the wind, however. Two other factors must be considered at the surface: the rotation of the Earth (which affects wind direction) and friction (which influences the wind speed).

As the Earth rotates, the **Coriolis effect** causes the wind direction to deflect, just as it does with ocean currents. Just as with the oceans, the direction in which the winds are deflected from their course depends on which hemisphere you're in.

For example, look at Figure 17.8; the trade winds in both hemispheres are blowing toward the equator, from higher to lower pressure. However, in the northern hemisphere they are deflected to the right (coming from the northeast), while in the southern hemisphere they are deflected to the left (coming from the southeast). Meteorologists describe winds based on the direction they come from; so the Northeast trade winds come from the northeast, and vice versa.

As the air parcels accelerate to a critical speed, the Coriolis force becomes equal and opposite to the pressure gradient force—that is, an arrow indicating the Coriolis force would be 180 degrees from an arrow indicating the pressure gradient force. When this happens, the air has attained geostrophic balance, and the wind that results, the *geostrophic wind*, flows parallel to isobars.

EXERCISE 17.9   **PGF and the Coriolis Effect**

Name: _____   Section: _____
Course: _____   Date: _____

Perform the following steps on the three surface pressure maps below:

(a) Label the cyclone with "L" and anticyclone with "H."

(b) First, considering PGF only, draw an arrow in each diagram illustrating a direction the wind blows. Label this pencil line or pen color "PGF."

(c) Using a different-colored pencil or pen, draw an arrow at right angles to your PGF arrow (right or left, depending on hemisphere) and label it "CE" for Coriolis effect. Then, using the same-colored pencil or pen, draw an arrow at 180 degrees at your PGF arrow and label it "GB" for geostrophic balance. This reflects the ultimate impact of the Coriolis effect.

(d) Finally, using a third color, draw curved arrows that are a combination of the PGF and CE arrows (that is, that connect the PGF and CE arrows). These are the actual wind directions considering these two forces.

| Northern Hemisphere | Southern Hemisphere | Northern Hemisphere |

Based on the markings you've made on your maps, answer the following questions:

(e) In the northern hemisphere, which way do the surface winds circulate around a cyclone: clockwise or counterclockwise? _____

(f) In the southern hemisphere, which way do the surface winds circulate around a cyclone: clockwise or counterclockwise? _____

As the wind moves across the Earth's surface, it encounters obstacles, such as vegetation, buildings, and mountains. The force of friction causes the wind to slow down. At high altitudes, there is essentially no friction and the winds blow at very fast speeds, up to 400 kph (250 mph)! The belt of strongest winds are called the jet streams. These winds help to "steer" pressure systems and other weather phenomena like hurricanes, which you will investigate in the next chapter.

Name: _____   Section: _____

Course: _____   Date: _____

The map below depicts the continental United States labeled with lines of equal sea-level pressure (isobars) at the surface.

Surface barometric pressure map of the United States and southern Canada.

(a) On the map, label the low pressure centers/areas with a capital "L" and the high pressure centers with a capital "H."

(b) Using arrows, draw the wind directions on the map around the H and L pressure centers and a few in other areas of your choosing. Consider all three factors that affect wind (PGF, Coriolis effect, and friction) when drawing your arrows.

(c) Using geographic locations in your answers, where are the winds fastest? _____

And the slowest? _____

(d) Find Chicago, Illinois, on the map. What direction is the wind coming from on this date? _____

(e) Find Seattle, Washington, on the map. What direction is the wind coming from on this date? _____

(f) On this date, which has the faster winds, Chicago or Seattle? _____. How can you tell?

_____

_____

(g) The prevailing winds for most of the United States are called the westerlies; as their name suggests, these pressure systems move more or less eastward. Considering the pressure map you have made here, predict the winds for where your school is (include general wind speed—faster or slower—and direction.)

_____

_____

17.6 Modeling Global Patterns

Many Earth Science textbooks provide a diagram that is a model of the idealized global pattern of winds, such as Figure 17.8. It is idealized because, as you have been learning in your course, there are always exceptions to the "big picture" when conditions are normalized. However, this generalized global pattern provides us with information to help us predict large-scale changes in winds, especially the regional or local conditions that we receive in our weather forecasts. These global belts are driven by the unequal heating of the Earth's surface, and so are separated from one another by belts of latitude. Broadly speaking, at the equator hot air rises, forming low-pressure systems at the surface and the trade winds. At high altitudes, this excess heat flows away from the equator toward the higher latitudes, where the air cools and sinks to the surface, forming the sub-tropical high-pressure zone and the westerlies. The colder air at the polar regions flows toward the equator but clashes with the air flowing away from the mid-latitudes. However, this is an ideal pattern; in reality, as these prevailing winds flow they are also interrupted by the formation of high- and low-pressure systems at the surface, and by the jet streams. In the majority of the continental U.S., this results in weather systems moving more or less from west to east across the country.

EXERCISE 17.11   **Why Do Deserts Exist Where They Do?**

Name: _____   Section: _____

Course: _____   Date: _____

? What Do You Think When you look at a satellite image of the Earth like the one below, great swaths of brown areas are visible. These are the arid regions of the landmasses: deserts. But why are they in these locations? You have the clues to that answer, as you just investigated the causes!

Satellite view of the Earth showing the vegetated and arid lands.

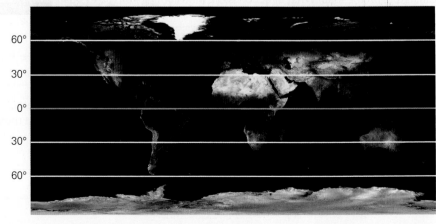

(continued)

Name: _____   Section: _____
Course: _____   Date: _____

Explain why deserts, like the Sahara in North Africa, exist in the subtropical regions rather than in the equatorial regions.

_____

_____

Antarctica, while mostly white rather than brown in the satellite view, is also an arid region—one of the driest continents on the planet. Why do you think it is so dry?

_____

_____

GEOTOURS EXERCISE 17   **Investigating Air Pressure and Wind Circulation Concepts**

Name: _____   Section: _____
Course: _____   Date: _____

Exploring Earth Science Using Google Earth

1. Visit **digital.wwnorton.com/labmanualearthsci**
2. Go to the **Geotours** tile to download Google Earth Pro and the accompanying Geotours exercises file.

Expand the Geotour17 folder in Google Earth by clicking the triangle to the left of the folder icon. The folder contains an animation and static overlay that show wind circulation as it relates to air pressure highs and lows.

(a) Check and double-click the Air Pressure & Wind Circulation overlay to fly to a point above the U.S. The labeled white lines connect points of equal barometric pressure (isobars in units of inHg; inches of mercury), the colors show wind velocity ranging from ~0 km/hr (blue/purple) up to ~70 km/hr (pink), and the animated lines in the placemark indicate wind direction (*velocity is not scaled*). Which direction do lows (L) rotate: clockwise or counterclockwise? (*Note:* Use the animated image.) _____

(b) Which direction do highs (H) rotate: clockwise or counter-clockwise? (*Note:* Use the animated image.) _____

(c) Based on the wind velocity data provided for the coast of the United States, is it better to place wind turbines just offshore or on land? _____

(d) Describe a general relationship between the spacing of the lines of equal barometric pressure and the wind velocity. (*Note:* Look at the pink areas.)

_____

_____

(e) Using the data provided, from what general direction is the wind that is affecting Illinois coming? _____ Would you expect it to be cooling or warming Illinois? _____

Weather Patterns

18

Hurricanes are among the largest and most destructive weather phenomena on our planet. Hurricane Irma, pictured here, caused severe damage and loss of life in the Southeast United States and the Caribbean in 2017.

MATERIALS NEEDED

- Pencil
- Colored pencils or pens
- Metric ruler
- Calculator

18.1. Introduction

For centuries, modern humans have used many methods to predict the weather—from weather folklore, superstitions, and personal observations to measuring atmospheric changes using the new tools of the Renaissance era. While many of these methods have been replaced with mathematical equations and modern technology, including radiosondes and satellite sensors, society today still relies on dependable weather forecasts for their economic and safety impacts. Such forecasts tell us when to plant and harvest crops and when to protect them from a freeze; what kind of traffic and air travel conditions we can expect; when to plan outdoor activities; and how to prepare for natural disasters such as hurricanes, to name just a few examples.

Despite the advances in our understanding of the atmosphere and development of new technologies for measuring it, the reliability of weather forecasts remains only as good as the quantity and quality of weather data and the mathematics used in developing computer modeling programs. Weather reports are based on current atmospheric measurements and statistical analyses of past weather events. Given a set of weather conditions, meteorologists use computer programs to create a model of the probability of certain changes occurring over a few hours to several days. As you probably have experienced, sometimes these predictions turn out to be inaccurate. In some parts of the country, the weather is more consistent than others and that makes the predictions more reliable. But a weather forecast is just that: a prediction of the future. Any change in a single atmospheric condition can have profound impacts on the actual outcome of the weather. Unlike climate, which is a long-term average of all weather for an area (usually calculated at a minimum of 10 years), weather is constantly changing. Therefore, weather conditions are continuously monitored, and weather reports and weather maps are updated frequently.

In this chapter, you will investigate how weather maps are created and how we can use them to make observations about current conditions and make simple predications about future changes.

18.2 Weather Patterns from Contour Maps

When meteorologists investigate the weather, they are collecting data on all the measurable aspects of the atmosphere—at the surface as well as at high altitude. Maps that contour surface temperature and barometric pressure data help us to create a picture that we can interpret; that is, they make it easier to "see" the weather.

18.2.1 Isotherms

If you have ever looked at a weather map on TV, over the internet, or in a newspaper, you would see that the map is divided into color-filled areas representing different temperatures, such as the one shown in **FIGURE 18.1**. In this example, temperatures in the 80s are colored pink while temperatures in the 70s are colored orange. The **isotherm**—the line of constant or equal temperature—is the line that divides the pink from the orange. As such, all areas between the 70-degree isotherm and the 80-degree isotherm have temperatures between 70 and 80 degrees Fahrenheit.

These maps give us a broad view of temperatures across a region, such as the United States. This makes it much easier to view large-scale patterns than by looking at the raw data from weather stations. For example, in Figure 18.1 you can see where in the country the coolest and warmest temperatures are located. And if you have information about wind patterns for a region, you can also use weather maps to predict the change in temperatures over the following few days for any location on the map.

In Exercise 18.1, you will use the surface temperature data from weather stations across the United States to create a similar map.

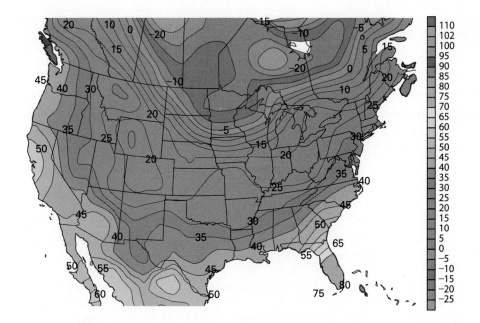

FIGURE 18.1 **Surface temperature map of the United States.** Units are in degrees Fahrenheit.

EXERCISE 18.1 | Creating and Analyzing Isotherm Maps

Name: _____ Section: _____

Course: _____ Date: _____

In Chapter 9, you contoured surface elevations to create topographic maps. In this exercise, you will contour surface temperatures to create isotherm maps. The data you will use are surface temperatures across the United States, collected by the National Oceanic and Atmospheric Administration (NOAA), which includes the National Weather Service (NWS). Note that here we are using degrees Fahrenheit, the temperature unit used for surface temperatures in the United States.

Using the data on the map below, draw contour lines of equal temperatures (*isotherms*) every 10°F. An example has been drawn for you for a 40°F isotherm. Label each isotherm with its temperature value. Use different colored pencils to shade in the areas between isotherms to represent the temperature value, and make a key to accompany your color choices.

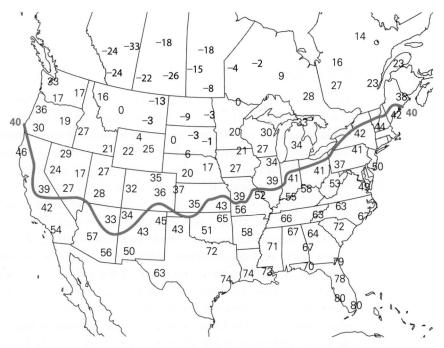

(continued)

Name: _____   Section: _____

Course: _____   Date: _____

(a) Looking at the temperature values on the map, at what time of year (approximately) do you think the data were likely recorded? Explain.

_____

_____

_____

(b) Based on the temperature patterns that you observe, how do the temperatures in the central United States compare with the temperatures in the west and east? Explain why you think these patterns are present.

_____

_____

_____

(c) Label the location where the greatest change in temperature occurs over the shortest distance (that is, the steepest temperature gradient), with the word "steep." How does this temperature gradient compare with the temperature at your campus location?

_____

_____

_____

(d) Considering that the prevailing winds generally move weather systems from west to east across most of the United States, what change in temperature do you predict for the east coast in the next 24 hours? _____

And for your campus location? _____

18.2.2 Isobars

Often even more informative than temperature maps are weather maps that show the sea-level barometric pressure—these maps can tell us how intense the changes in weather may be and what to expect in terms of clouds, winds, and precipitation. These weather maps are made by contouring areas of constant or equal sea-level pressure using **isobars**. As you look at **FIGURE 18.2A** you will notice areas that are labeled with either a capital "L" or "H" where the isobars form closed loops or circles. These are areas where you will find the lowest and highest pressures, respectively. Generally speaking, areas of low pressure are where precipitation or storms are likely to occur, whereas high-pressure centers generally experience sunny and dry conditions. You can also get a relative idea of wind speed by looking at the spacing of the isobars; where they are closer together, the pressure gradient is steepest and so windier conditions are expected. For example, in Figure 18.2a, Denver is experiencing faster winds than is Chicago. When the isobars form notable dips and rises, meteorologists refer to these as *troughs* if the pressure is lowest in the center, and *ridges* if the pressure is highest in the center (**FIG. 18.2B**). Troughs and ridges tend to include similar weather conditions to low and high pressure centers, respectively.

FIGURE 18.2 US weather maps showing barometric pressure values using isobars.

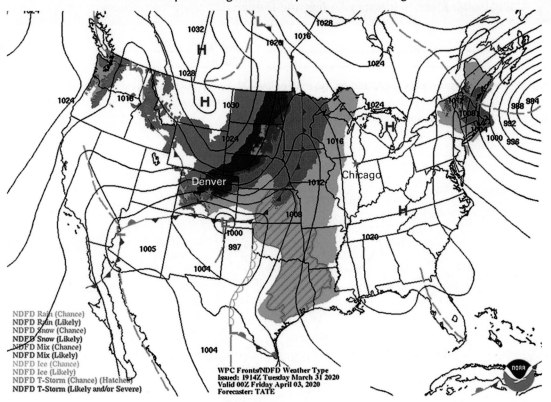

(a) A map showing low- and high-pressure centers, precipitation, and fronts (discussed in the next section).

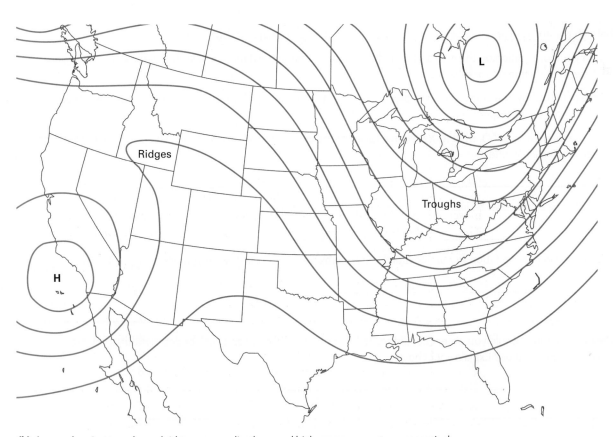

(b) A map showing troughs and ridges surrounding low- and high-pressure centers, respectively.

Name: _____ Section: _____

Course: _____ Date: _____

Examine the sea-level air pressure measurements plotted on the map below. Draw and label isobars on the map every 4 millibars (mb) beginning with 1008 mb and ending with 1040 mb. A 1020 mb isobar has been drawn for you as an example. Then, study your map and answer the questions that follow.

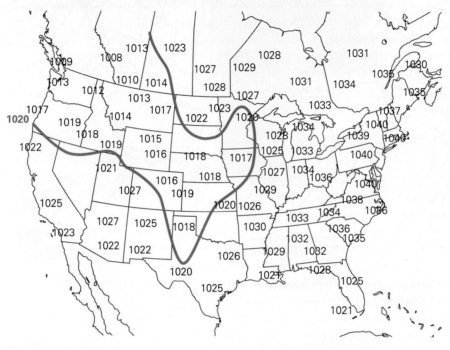

(a) Label all high- and low-pressure centers on the map using the appropriate symbols and colors.

(b) Plot your campus location on the map and then draw arrows showing the direction of surface winds for your region (the Midwest or southeast, for example).

(c) Label the map to show the areas where the pressure gradient is steepest and where it is gentlest by writing "steep" and "gentle" in those areas, respectively.

(d) Based on your answers in (c), what is the relative speed of the wind in the areas labeled as "steep" compared with the areas you labeled as "gentle"?

_____

(e) What might the cloud cover be like for your location on this date? Explain.

_____

(f) Would you expect precipitation at your location? Why or why not? If you can't tell at your location, explain why.

_____

_____

(g) In the next 24 hours, would you expect conditions to be improving (skies clearing up) or deteriorating (getting stormier) at your location? Explain your choice.

_____

_____

18.3 Air Masses and Fronts

A vast body of air that has a relatively uniform temperature and humidity across the same altitude is called an **air mass**. Air masses can cover distances of thousands of kilometers, and originate either over land that has little topography or over vast areas of water. Air masses originating over land develop lower humidity than those forming over the oceans. The temperature of an air mass is determined by the latitude from which the air mass originates. Those masses forming at the high latitudes, such as the polar regions, will have cooler temperatures, while those forming at the low latitudes or equator will be warmer. As these air masses move across the Earth's surface, they bring changing temperatures and moisture to the areas they cover. Where two air masses meet, the boundary is called a **front**. While air masses do not typically appear in your daily weather forecast, fronts are shown on weather maps and are used by meteorologists to predict changes in weather, particularly storm conditions.

18.3.1 Air Masses and Associated Weather Changes

To predict the changes in temperature and humidity of an area, meteorologists use a naming system to define different types of air masses. The names of the air masses are a combination of two letters: the first represents the source of the moisture (maritime or continental) and the second letter represents the temperature (tropical or polar). In other words, air masses are warm or cold, and moist or dry. Note that the moisture letter is lower case, while the temperature letter is capitalized. For example, the air masses originating in Canada would typically be labeled as "cP" (continental Polar) meaning that the air is "dry" and "cool," respectively. Meanwhile, an air mass originating over the tropical oceans would be labeled "mT" (maritime Tropical), and would be moist and warm. There are a few additional air mass names, but they are not common—some occur only in one location (the Antarctic or equator) or only form during one season, such as arctic air masses in the winter months.

By locating these air masses and measuring the direction in which they are heading, we can make predictions about the expected changes in weather for a particular area. **TABLE 18.1** is a brief list of the expected humidity and temperature within each air mass. Although we are focusing here on horizontal movement of the atmosphere, keep in mind that air can also rise or descend depending on several factors: changes in temperature and pressure; the presence of topographic features such as mountains that an air mass may have to rise over; or the collision of one air mass with another that has different characteristics, as we will investigate in the next section on fronts.

TABLE 18.1 Air-mass classification based on humidity and temperature and their related regions

Symbol	Air mass	Humidity	Temperature	Source region	US area affected
mT	Maritime tropical	Wet	Warm	Over water at low latitudes	Southwest, southeast
mP	Maritime polar	Wet	Cold	Over water at high latitudes	Northwest, northeast
cT	Continental tropical	Dry	Warm	Over land at low latitudes	Southwest
cP	Continental polar	Dry	Cold	Over land at high latitudes	Midwest to east
cA	Continental arctic	Dry	Very cold/frigid	Over land at high latitudes	North, winter only

Name: _____ Section: _____

Course: _____ Date: _____

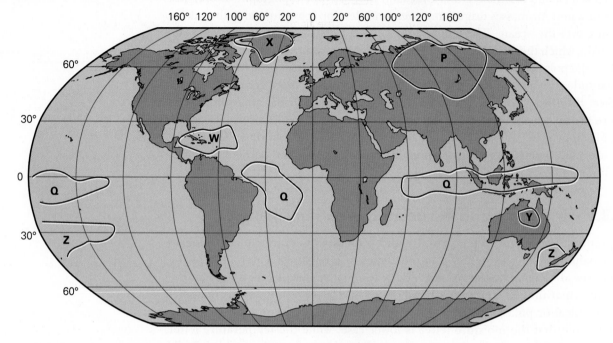

(a) On the map above, label the air mass in each lettered location as either continental (c) or maritime (m), and as equatorial (E), tropical (T), polar (P), Arctic (A), or Antarctic (AA).

P: _____   W: _____

Q: _____   X: _____

Y: _____   Z: _____

(b) What kind of temperature and humidity would you expect in North Africa? Explain.

_____

_____

_____

(c) What kind of temperature and humidity would you expect in Greenland (X)? Explain.

_____

_____

_____

(d) Imagine that an mT air mass moves onto a continent. What changes in the weather will this air mass bring to the landmass?

_____

_____

_____

(continued)

Name: _____   Section: _____

Course: _____   Date: _____

What happens to the temperature and humidity of the air mass after it moves over the continent?

_____

_____

_____

(e) Which air mass dominates North America east of the Rocky Mountains during the winter months? _____

During the summer months? _____

How do these air masses impact the weather during each of these seasons? Be specific and justify your answer.

_____

_____

_____

18.3.2 Fronts and Weather Predictions

In Chapter 2 you investigated the movement of tectonic plates across the Earth's surface and related the features that formed at the surface to the interactions between two plates. A similar relationship can be made when observing the motions of neighboring masses of air. However, while the boundaries between tectonic plates can be located with fairly high precision, that is not the case with air masses. A weather front is not a precise boundary; rather, it is a transitional zone occurring where a new air mass moves into an area and replaces the original air mass. As with tectonic plates, density plays a role in the interaction between air masses. In the case of the atmosphere, this is primarily driven by temperature differences. Warm air is less dense than cool air, and so will always be the mass that is forced upward when two air masses interact.

FIGURE 18.3 illustrates the four typical situations that drive the movement of air masses and force the less dense air upwards. In Figure 18.3a, the differential heating of the surface causes the warmer, less dense air to rise. If an obstacle blocks the horizontal motion of the air, then the air is forced to rise up and over the obstacle, as shown in Figure 18.3b. Air masses traveling in opposite directions converge over land, as illustrated in Figure 18.3c, which is a common scenario for the peninsula of Florida in mid-summer, where an air mass from the Gulf of Mexico moves eastward and collides with a westward-moving air mass from the Atlantic Ocean. And in Figure 18.3d, two air masses collide and the colder, denser air acts as a wedge to force the warm air aloft. The boundary between these two air masses is a front, which is where we turn our attention now.

Look back at the weather map in Figure 18.2. This map contains colorful lines, each with different symbols. These symbols represent the different types of fronts that can occur where air masses meet (**FIG. 18.4**), and each front is associated with a set of expected weather conditions (**TABLE 18.2**). Each type of front is named based on the properties of the air mass that is moving in to replace the "old" air mass. For example, a "cold front" means that the air that is moving into an area is significantly cooler in temperature than the current air mass. Even without the symbols on the

FIGURE 18.3 Forces that force warm air aloft.

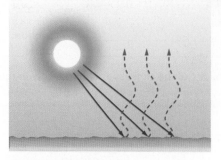

(a) Surface heating.

(b) Mountains.

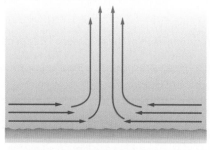

(c) Widespread rising air due to convergence (flowing together) of surface air.

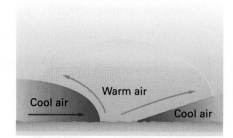

(d) Uplift along weather fronts.

map, you can generally locate the position of a front by determining where a temperature boundary exists between two air masses—a front is especially likely to exist on the warm air side of a steep temperature gradient. However, because there are many factors that control temperature at the surface, often it is the higher-altitude temperatures that make the location of fronts more clear.

Table 18.2. Fronts and their symbols, method of formation, and expected weather changes. The symbols point in the direction that the front is moving.

Front name	Map symbol	How air moves	Expected outcome
Cold		Cooler air replaces warmer air	Thunderstorms, intense but brief precipitation, winds typically moving northwest to southeast, and increased wind speed, followed by higher pressure, clearing skies, westerly (east-moving) wind, and cooler temperatures as front passes. Typically takes only a few hours to pass through.
Warm		Warmer air replaces cooler air	High clouds increasing in thickness to deeper clouds; light to moderate precipitation; and wind switching from west-moving to north-moving as front passes. Front passes over a 12-24 hour period.
Occluded		Colder air overtakes a warm front	Precipitation and winds similar to a warm front, except that when occluded front passes, air becomes colder and shifts to northerly (south-moving) winds.
Stationary		Cold air flows parallel to front, so front doesn't move	Rain, snow, or freezing rain over a location for a long time, leading to hazards such as flooding.

Note that the predicted weather changes are based on years of weather data and calculations of the average, or "normal," conditions that occur. The actual changes at a given time will depend on the conditions of the atmosphere at that moment, which may differ from the calculated average outcome.

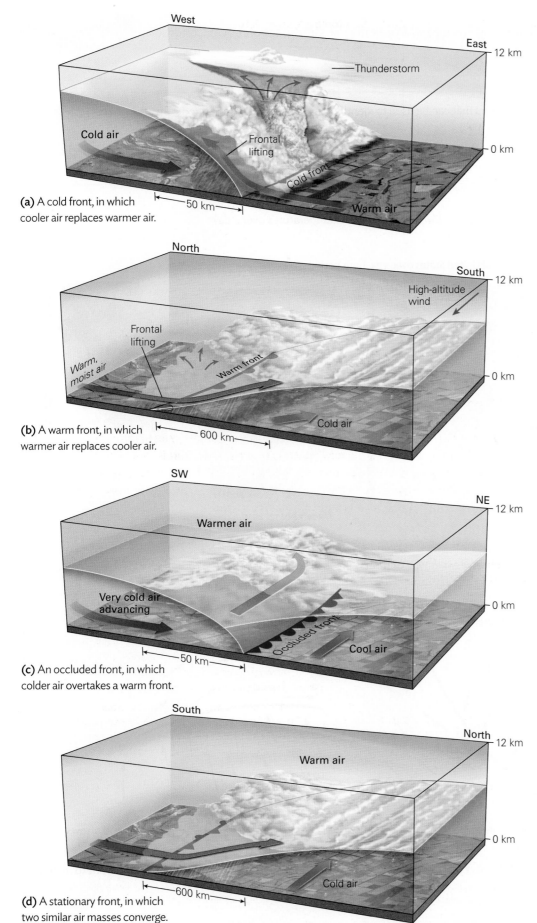

FIGURE 18.4
The four types of fronts.

(a) A cold front, in which cooler air replaces warmer air.

(b) A warm front, in which warmer air replaces cooler air.

(c) An occluded front, in which colder air overtakes a warm front.

(d) A stationary front, in which two similar air masses converge.

Name: _____   Section: _____
Course: _____   Date: _____

As there are several weather factors that can be used to identify the location and types of fronts on a weather map, let's take a look at a few of them individually.

(a) Use the isotherm map you created in Exercise 18.1 to locate the likely position of a front. Explain your location choice.

_____

Now, look at the time series images below, which show the temperatures over 2 days.

Surface temperatures in central United States over a 24-hour period

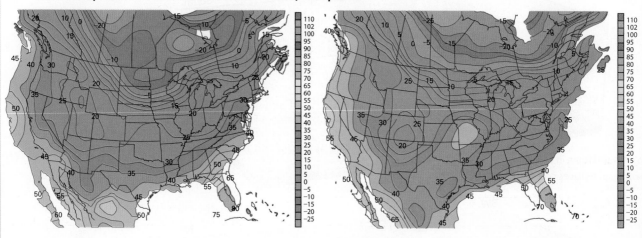

December 11, 2019   December 12, 2019

(b) Based on the changes you observe, what type of front is present in the upper Midwest (around Minnesota)—cold, warm, stationary, occluded? _____ Draw the position of this front on both images and use the correct symbol to represent it.

Now, look at the two images below, which show the location of two fronts, and the wind arrows on each side.

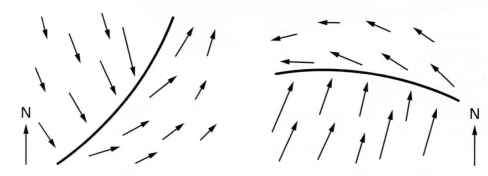

(c) For each front, place the appropriate symbol on the line to indicate the type of front indicated by the wind arrows. (*Hint:* Review Table 18.2.)

(d) On the map at the top of the next page, draw wind arrows on each side of the five fronts and label each front to match the symbols given.

(continued)

Name: _____   Section: _____
Course: _____   Date: _____

Map of the central and eastern portion of the United States with the location of five fronts shown.

(e) Explain how the occluded front formed.

_____

_____

_____

(f) Predict the change in weather in the next day or two for:
 i. Arkansas _____
 ii. New York _____
 iii. Ohio _____
 iv. Iowa _____

18.4 Using Station Models to Determine Current Weather and Predict Changes

Meteorologists collect a lot of information about current weather conditions. If they were to attempt to place all of that information on one map, it would be illegible. Instead, they have devised a code system—called a **station model**—to represent all of the pertinent weather conditions recorded at weather stations, and from these models, anyone can interpret the weather over the given area and produce weather forecasts.

An example of a station model is shown in **FIGURE 18.5** along with a key to the typical symbols used and their meanings. In some cases, not all of the weather elements are included, but typically air temperature, air pressure, cloud cover, precipitation, and wind speed and direction are given.

In Exercise 18.5, you will use station models to identify current weather conditions, make your own station model, and develop a weather forecast.

FIGURE 18.5 A weather station model and a legend of common weather station symbols with their meanings.

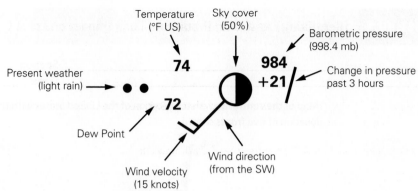

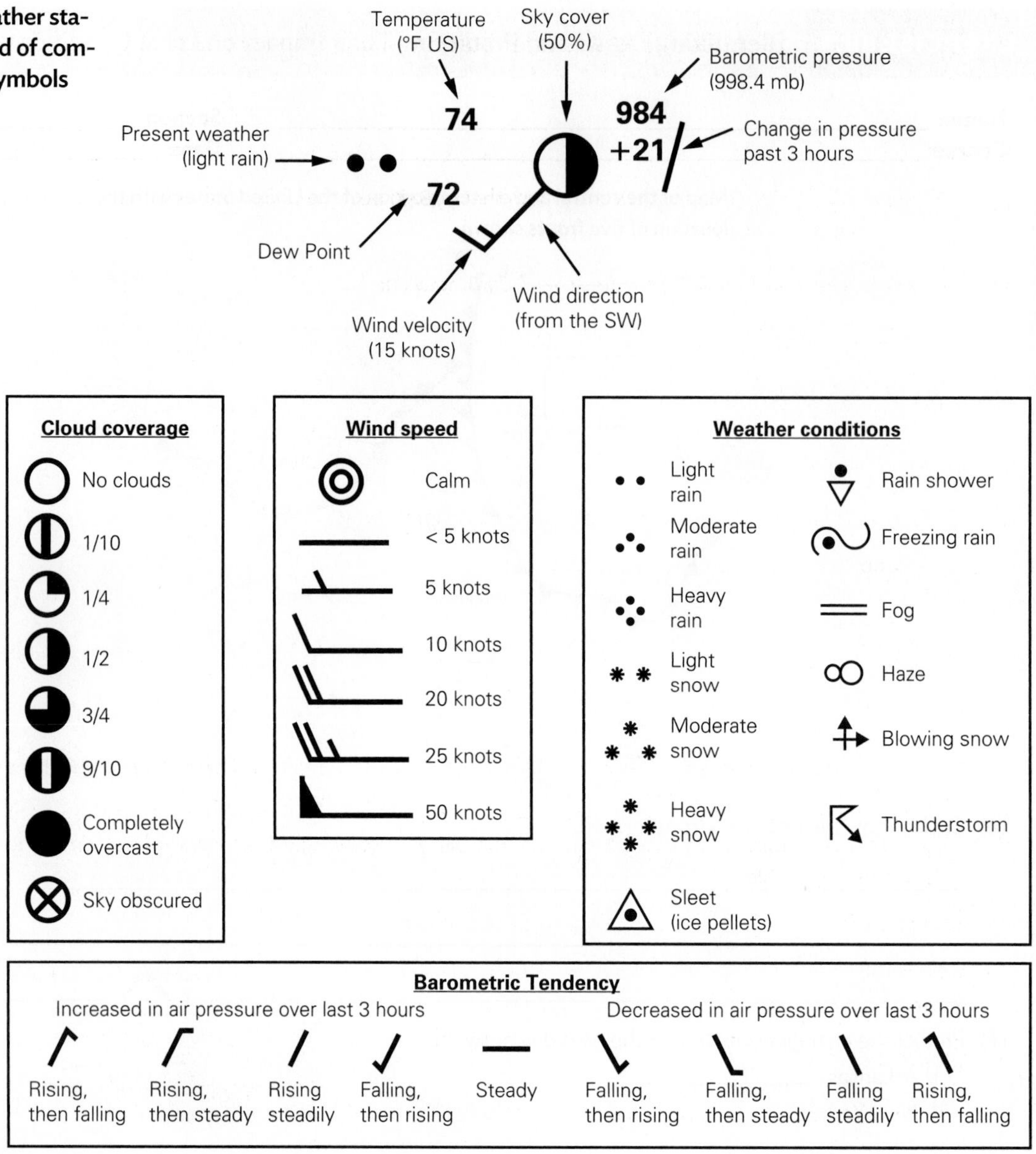

Temperature (°F US)
Sky cover (50%)
Barometric pressure (998.4 mb)
Change in pressure past 3 hours
Present weather (light rain)
Dew Point
Wind velocity (15 knots)
Wind direction (from the SW)

74
72
984
+21

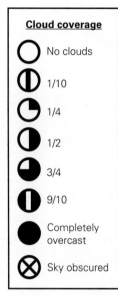

Cloud coverage

◯	No clouds
◐	1/10
◔	1/4
◑	1/2
◕	3/4
◑	9/10
●	Completely overcast
⊗	Sky obscured

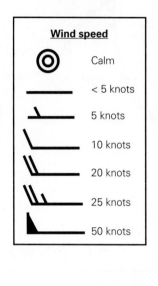

Wind speed

◎	Calm
—	< 5 knots
	5 knots
	10 knots
	20 knots
	25 knots
	50 knots

Weather conditions

Symbol	Meaning	Symbol	Meaning
• •	Light rain	▽	Rain shower
•• •	Moderate rain	∿	Freezing rain
•• • •	Heavy rain	=	Fog
* *	Light snow	∞	Haze
* *	Moderate snow	✦	Blowing snow
** **	Heavy snow	�	Thunderstorm
△	Sleet (ice pellets)		

Barometric Tendency

Increased in air pressure over last 3 hours   Decreased in air pressure over last 3 hours

| Rising, then falling | Rising, then steady | Rising steadily | Falling, then rising | Steady | Falling, then rising | Falling, then steady | Falling steadily | Rising, then falling |

EXERCISE 18.5   Understanding Weather Station Models

Name: _____   Section: _____
Course: _____   Date: _____

Before you begin deciphering weather station models, you'll need some practice converting the abbreviated barometric sea-level pressure reported on the model to the full (actual) barometric pressure. Barometric pressure is recorded to the nearest tenth of a millibar, but on the station model you will observe that there are no decimal places shown. To convert from a station model to the sea level pressure, use the following formula: If the coded number is >500, add a 9 in front of the pressure value shown on the model and move the decimal one place to the left, so that 988 becomes 998.8. If the coded number is <500 add a 10 in front and move the decimal one place to the left, so that 123 becomes 1012.3 mb. *Note:* Acceptable barometric pressures are typically between 940 mb and 1060 mb. Standard sea level barometric pressure is 1013.2 mb.

Take the example station model given in Fig. 18.5 where the barometric pressure is shown as "984."

Step 1: In order to bring that value closest to 1000 mb, you would add a 9 to the front so that 984 becomes 9984 (which is 9984.0).

(continued)

Name: _____   Section: _____
Course: _____   Date: _____

Step 2: Move the decimal one place to the left so that 9984.0 becomes 998.4 mb, which is the measured barometric pressure at the surface.

(a) Convert the following model pressures to full barometric pressures:

922   _____ mb

107   _____ mb

247   _____ mb

998   _____ mb

(b) Convert these barometric pressures to station model pressures:

1002.4 mb   _____

911.8 mb   _____

986.1 mb   _____

1016.5 mb   _____

You are now ready to decode a station model, using Figure 18.5 and the practice you just completed as a guide.

(c) Identify the weather conditions that are given by the following simplified weather station model used by the National Weather Service. Some of the data given on a complete station model is not shown. Remember to include proper units where applicable.

Temperature: _____

Dew point: _____

Sea-level pressure: _____

Wind: _____

Weather: _____

Sky cover: _____

Pressure trend: _____

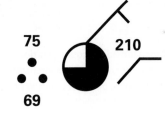

(d) Create a weather station model. If tools or the internet are available to you, determine the weather conditions outside your lab or at your current location. If they not available, and if directed to by your instructor, use the following weather conditions. Complete the weather station model using the circle drawn below:
  i. A southwest wind is blowing 20 knots.
  ii. Barometric pressure is 1016.2 and falling.
  iii. The temperature is 80 and falling.
  iv. The dew point is 75.
  v. The sky is approximately 90% overcast.
  vi. It is lightly raining.

Name: _____   Section: _____
Course: _____   Date: _____

Now that you are an expert at identifying pressure centers, fronts, and reading station models, you will apply your knowledge to interpreting the weather map below and making weather predictions. Remember that these individual weather factors are related! For example, high pressure typically means drier conditions, while low pressure relates to clouds or storms. If you only have station models or only isobars on your map, you can still interpret the weather.

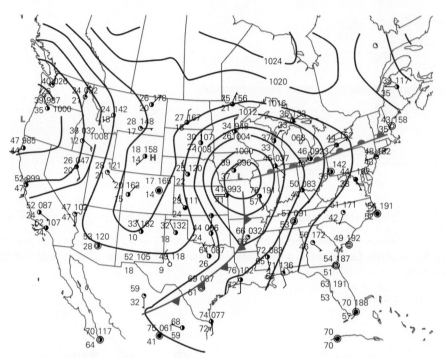

Weather map including isobars, station models, fronts, and precipitation. The echo intensity scale is an indicator of the rate or intensity of precipitation indicated by radar, where higher values indicate more or higher intensity precipitation in general.

(a) What types of fronts are shown on the map? _____

(b) In which direction is the front that passes through Arkansas moving? _____

What type of air mass lies to the northwest of this front? _____

(c) What type of air mass is east of the front that runs through Arkansas? _____

(d) What are the thin dark-blue lines indicating? _____

What is the contour interval of these lines? _____

(e) Find the station model in southwestern Louisiana.
What is the temperature, dewpoint temperature, cloud cover, sea-level pressure, wind speed, wind direction, and weather at this station?

_____

_____

Now find the station model in northern Nevada. What type of weather is occurring at this station? _____

(continued)

Name: _____   Section: _____

Course: _____   Date: _____

(f) Find the center of low pressure in northern Missouri. Winds around this pressure center blow in a (**clockwise/counterclockwise**) direction. (Circle your answer). Winds are blowing (**in toward/out from**) the low pressure system. (Circle your answer).

(g) Which station has the higher relative humidity on this date: Lake Charles, Louisiana (southwest Louisiana), or El Paso, Texas (western tip of Texas)? _____

   Explain how you determined your answer:

   _____

   _____

   _____

(h) Based on the station reports, what type of weather is occurring just north of the low pressure center?

   _____

(i) Draw an isotherm on the map defining the location where the temperature is 32°F.

(j) Where is the heaviest rainfall occurring in the United States at the time of this map? _____

18.5 Tropical Cyclones

When episodes of weather occur that cause deteriorating conditions—which can include strong winds and heavy precipitation—we consider these to be *storms*. When a storm system consisting of many individual thundertorms originates and intensifies over tropical waters, it forms a tropical cyclone. Weak tropical cyclones are called tropical storms, and more intense tropical cyclones are called hurricanes, typhoons, or cyclones, depending on where over the tropical oceans they occur.

Not all tropical storms will develop into hurricanes. In order to classify the level of storm development, scientists use a system based on sustained wind speeds. As an example, let's use **FIGURE 18.6** to follow the development of a hurricane. When an area of low pressure develops over the tropical oceans in the Atlantic basin (a *tropical disturbance*), producing sustained winds between 23 and 39 mph, this weak tropical cyclone is termed a **tropical depression**. On a weather map, the location of this storm is designated by an open circle (Fig. 18.6a). If the barometric pressure continues to drop, the rising air brings more moisture and energy to higher altitudes where condensation increases, more energy is released to the surrounding atmosphere, and the sustained wind speeds increase beyond 39 mph. At this point the storm is reclassified as a **tropical storm**, and the location of the storm's center is plotted on a weather map using an open circle with curved barbs, representing the circulation within the storm (Fig. 18.6b). As the storm intensifies, barometric pressure continues to decrease and sustained wind speeds increase to 74 mph or above, and the strong tropical cyclone is now a **hurricane.** The open symbol for a tropical storm is filled in to represent this new classification (Fig. 18.6c). Hurricanes are further divided into categories based on sustained wind speed, which

FIGURE 18.6 The progressive development of a tropical cyclone.

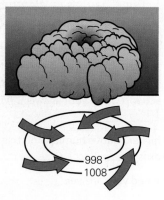

(a) Schematic view of a tropical depression and its weather map symbol.

(b) Tropical storm features and its weather map symbol.

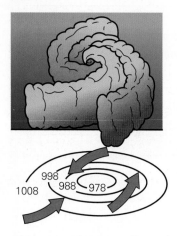

(c) Hurricane features and its symbol.

relate to the potential property damage they may cause. The system used is called the Saffir-Simpson Hurricane Wind Scale, which rates the hurricanes on a scale from 1 to 5 (**TABLE 18.3**).

The names of hurricanes are maintained and updated by an international committee of the World Meteorological Organization. The names for the eastern Pacific and the Atlantic basins are listed in alphabetical order, with the first named storm of the tropical cyclone season beginning with A, such as "Albert" or "Anne," and with names alternating between male and female. Names are applied to tropical cyclones once they reach tropical storm status and retain the given name throughout the life of the storm. Although the lists are repeated every 6 years, if any hurricane was particularly strong (that is, costly and deadly), the name may be retired and a new name will be added to the list to

TABLE 18.3 The Saffir-Simpson scale for hurricane intensity

Category	Sustained wind speed	Types of damage resulting from wind
1	74–95 mph 64–82 knots 199–153 km/h	**Dangerous winds produce some damage:** Well-constructed frame homes could have damage to roof shingles, vinyl siding, and gutters. Large branches of trees snap, and shallowly rooted trees may topple. Some power outages occur.
2	96–110 mph 83–95 knots 154–177 km/h	**Extremely dangerous winds cause extensive damage:** Well-constructed frame homes sustain major roof and siding damage. Many shallowly rooted trees are snapped or uprooted, blocking roads. Near-total power outages occur.
3	111–129 mph 113–136 knots 178–208 km/h	**Devastating damage occurs:** Well-built frame homes may incur major damage or removal of roof decking. Large trees snap or are uprooted. Numerous roads become blocked. Electricity and water are unavailable for days to weeks.
4	130–156 mph 113–136 knots 209–251 km/h	**Catastrophic damage occurs:** Well-built homes sustain severe damage. Most trees snap or are uprooted, and most power lines are downed. Debris isolates residential areas, and power outages last weeks to months. The area becomes temporarily uninhabitable.
5	≥157 mph ≥137 knots ≥252 km/h	**Total catastrophic damage occurs:** Most homes are destroyed. Only strongly reinforced buildings remain standing. Debris isolates large areas, and power outages last for weeks to months. Most of the area remains uninhabitable for weeks or months.

FIGURE 18.7 Formation of storm surge and its effects.

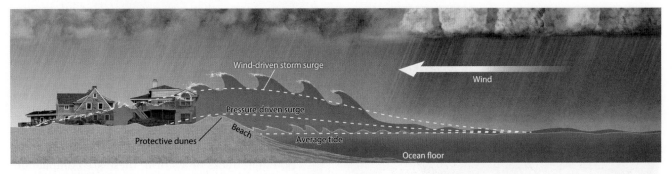

(a) Wind and low atmospheric pressure combine to produce a bulge of water.

(b) Storm surge accompanying a hurricane floods coastal areas and leaves behind devastation.

replace the deleted name. Examples of names that have been retired are Katrina (US, 2005), Sandy (US, 2012), Harvey (US, 2017), Irma and Maria (Caribbean, 2017), and Michael and Florence (US, 2018).

While we have focused on sustained wind speed as the criterion for dividing tropical cyclones into categories of intensity, the primary damage that is associated with such storms is flooding due to intense precipitation and **storm surge**. Storm surge occurs when winds blow onshore, pushing the surface of the ocean toward land, as illustrated in **FIGURE 18.7**. Low-lying coastal communities are at the greatest risk from storm surge, but depending on the topography, the waters may surge inland for several miles, flooding roads, homes, and businesses. Storm surge erodes beaches and undermines the support for homes and roads. Particularly devastating storm surge was generated by Hurricane Katrina in 2005, which produced flooding of over 25 feet in parts of Louisiana (especially in and around New Orleans) and Mississippi. Hurricane Sandy in 2012 caused a storm surge of 14 feet in New York City and inundated other areas in the northeast. And in August and September of 2017, hurricanes Irma and Maria caused intensive storm surge and flooding in the American southeast, wreaking destruction in Louisiana and Florida, and devastating Puerto Rico. Flooding also occurs inland due to extreme rainfall, as happened in 2017 in Houston during Hurricane Harvey, well after it weakened back to tropical storm intensity.

So how can science assist in protecting lives and property from the effects of severe tropical cyclones? Advances in satellite technology have given scientists vast amounts of information about these storms at many levels within the atmosphere.

FIGURE 18.8 Tracking the paths of hurricanes.

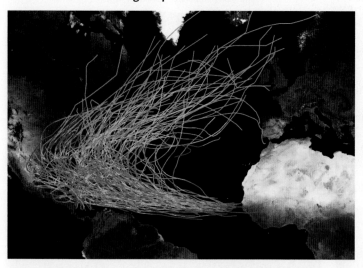

(a) Historic category 4 and 5 hurricane tracks in the Atlantic basin over the past 150 years.

(b) Track of Hurricane Jeanne in 2004 shows the northward movement and loop east of the Bahamas, bringing it to Florida and the US east coast.

NOAA also uses aircraft to gather firsthand data from storms whenever possible. This information helps forecasters at the National Hurricane Center to make predictions about the path of the storm and its intensity. **FIGURE 18.8A** shows the tracks of the strongest tropical cyclones in the Atlantic basin over the past 150 years. Many factors determine the strength of the storm and the path that it may take, including conditions at upper levels of the atmosphere. When you see the reports of the potential tracks of a storm, these are compilations of data and computer simulations taken from the NWS and weather services internationally. The plotted storm tracks focus on the center of the storm, but the impacts of the storm may be felt hundreds of miles away from the center. People may make the mistake of thinking that they are out of the path of the storm simply because they are not in the direct path of the storm's center, only to find later that they are now in the storm's path. As we have seen in the past, changing conditions can steer the storm in several possible directions, and storms that have passed one area may loop back, as shown in **FIGURE 18.8B**.

To assist communities in their preparations when a tropical cyclone's approach to land is more certain, the NWS has designated particular storm conditions that are used as guides to issue alerts at specific times. When hurricane conditions are *possible* in an area, a **hurricane watch** is issued. This watch is issued 48 hours in advance of the onset of tropical-storm-force winds. When hurricane conditions are *expected* in an area, then a **hurricane warning** is issued 36 hours in advance of the arrival of tropical-storm-force winds.

In Exercise 18.7, you will use data recorded during the development of a tropical cyclone to make observations and predictions.

EXERCISE 18.7	Investigating Tropical Cyclones

Name: _____   Section: _____

Course: _____   Date: _____

The table on the next page includes data from the National Hurricane Center that you will use to make simple measurements and predictions regarding a storm and its path. Note that the data in the table do not represent the entire history of this storm. The original wind data have been converted from knots to miles per hour (mph) for this exercise.

Use a pencil to lightly mark the location of the severe tropical storm on the map provided on the next page. You will need to approximate the locations to the best of your ability and give the map scale. *At first, make each point an open circle (the symbol for a tropical depression.)* Draw a line connecting each location on the chart. The first three data points have been plotted for you.

(continued)

Name: _____   Section: _____
Course: _____   Date: _____

Partial tracking data for a storm event in the Atlantic basin

Date (month/day)	Time*	Latitude (°N)	Longitude (°W)	Pressure (mb)	Wind speed (mph)
10/06	18:00	17.8	86.6	1006	29
10/07	00:00	18.1	86.9	1004	29
10/07	06:00	18.4	86.8	1004	34
10/07	12:00	18.8	86.4	1003	40
10/07	18:00	19.1	85.7	999	52
10/08	00:00	19.7	85.5	996	57
10/08	06:00	20.2	85.4	984	69
10/08	12:00	20.9	85.1	982	75
10/08	18:00	21.7	85.1	977	86
10/09	00:00	22.7	85.2	971	98
10/09	06:00	23.7	85.8	973	98
10/09	12:00	24.6	86.2	968	104
10/09	18:00	25.6	86.4	961	115

*Time is given in 24-hour format where 00:00 to 11:59 is morning (A.M.) and 12:00 to 23:59 is afternoon/evening (P.M.).
Data source: NHC. These data are a part of the full tracking record from Hurricane Michael, October 7–11, 2018.

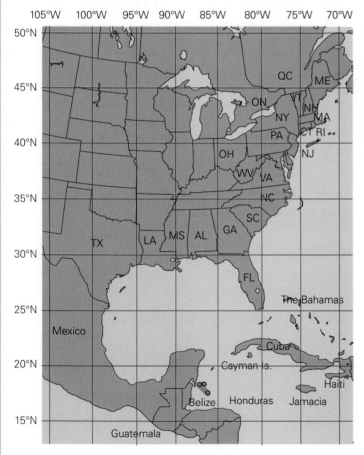

(a) If this is the second tropical storm of the season in the Atlantic basin, and the first was given a male name, use the National Hurricane Center naming system to give this storm a name. (*Hint:* Recall the system used to name hurricanes discussed earlier).

The name I have given this tropical storm is

_____

(b) Based on the data in the table, when did the tropical depression officially reach tropical storm status? Explain how you know.

Date: _____   Time: _____

_____

_____

_____

_____

From the point at which the depression became a tropical storm onward, change the location marks on your chart to the correct symbol for a tropical storm.

(continued)

Name: _____   **Section:** _____
Course: _____   **Date:** _____

(c) Based on the data in the table, when did this storm officially reach hurricane status? Explain how you know.

     Date: _____   Time: _____

     _____

     _____

     From the point at which the tropical storm became a hurricane onward, change the location marks on your chart to the correct symbol for a hurricane.

(d) What was the maximum wind speed recorded during the dates shown? _____ What does the table on the previous page suggest about the relationship between wind speed and air pressure? _____

     _____

(e) What is the average speed at which this storm moved as a hurricane? _____Show your calculations below.

(f) Based on its latest movement and average speed, which cities/states should have received watches and/or warnings? Explain.

     _____

     _____

     _____

(g) When and where do you predict the hurricane to make landfall? Explain your choices.

     _____

     _____

     _____

(h) If the storm keeps its intensity, what category hurricane will it be when it makes landfall? Explain.

     _____

     _____

     How long (approximately) did it take for the hurricane to strengthen from a category 1 to this category? _____

(i) Once the storm makes landfall, what is likely to happen to the conditions of the storm (consider wind, pressure, etc.)? Explain why.

     _____

     _____

(j) Ultimately, in what direction do you predict this storm to travel as it diminishes? Why? (*Hint:* Look at Figure 18.8 for historical trends.)

     _____

     _____

18.6 Variations from the "Norm": The El Niño-Southern Oscillation (ENSO)

So far, we have investigated short-term changes in weather: short-lived storms and tropical cyclones. These weather events happen over relatively short time spans of a day to a week. These types of weather conditions form patterns that develop during particular seasons of the year, and so can be predicted with some accuracy. But what about changes that occur on time scales between 1 and 7 years, last for several months and are irregular and therefore less predictable?

As we've seen, the energy exchange that occurs between the ocean and the atmosphere helps to drive our weather and our climate. In the tropical eastern Pacific Ocean, periodic changes between below-normal and above-normal sea surface temperatures cause alternating wet and dry conditions on each side of the Pacific Ocean basin for months at a time. These irregular periodic variations are called the El Niño-Southern Oscillation (ENSO). (Chapter 14 examines this phenomenon in the context of ocean currents.) Between 1 and 7 years, the atmospheric circulation in the region—known as the *Walker circulation*—increases or decreases, which causes a change in the flow of warm surface ocean waters across the equatorial Pacific. Under "normal" conditions—or ENSO-neutral conditions—the atmospheric circulation winds cause the warmer surface waters to "pile up" in the western Pacific Ocean (**FIG. 18.9A, B**). When this atmospheric circulation increases, the waters in the eastern Pacific are cooler than normal, forming the La Niña phase. During El Niño, the atmospheric circulation weakens, and warmer surface waters move back toward the eastern side of the Pacific Ocean. This brings extra moisture to the west coast of the Americas and starving Australia and Indonesia of their usual moisture, creating drought conditions (**FIG. 18.9C, D**).

One way to recognize the formation and presence of an ENSO event is to observe the sea surface temperatures (SST) in the eastern Pacific—a region referred to as Niño 3.4 (**FIG. 18.10**). When the three-month running average of SST in this region is $\geq 0.5°C$ warmer than normal (a positive temperature anomaly), it is an El Niño event; and when it is $\geq 0.5°C$ cooler than normal (a negative temperature anomaly), it is a La Niña event (**FIG. 18.11**).

When El Niño events occur, the warmer surface waters in the eastern Pacific do not allow the deep ocean waters to well up, so nutrient levels in the surface ocean decrease, as does photosynthetic productivity. This effect can be seen from space! Using satellite data, we can see the amount of photosynthesizing organisms in the surface ocean (*phytoplankton*), as shown in **FIGURE 18.12**. In this image, billions upon billions of phytoplankton stain the surface oceans a deep green as cold, nutrient-rich waters well up toward the surface. The yellow-boxed area in the figure is region Niño 3.4. The level of phytoplankton bloom in this region indicates that these are "normal" or ENSO-neutral conditions. During El Niño events, warmer waters (higher than normal SST) in the eastern Pacific inhibit upwelling and as a result there are noticeably fewer blooms than usual in Niño 3.4. You'll explore this for yourself in Exercise 18.8.

While ENSO events occur in the Pacific Ocean, they disrupt weather beyond the Pacific Rim, and their effects can be felt globally. While the changes in weather beyond the Pacific Rim may be minimal in the short term, the floods, droughts, fires, and changes in the productivity of the oceans can have social and economic impacts that can far outlast these short-lived weather changes.

FIGURE 18.9 The formation of ENSO conditions.

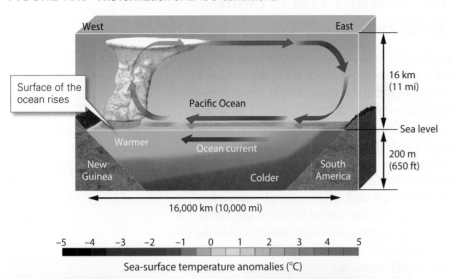

(a) Normal Walker circulation. During La Niña, this circulation strengthens.

Sea-surface temperature anomalies (°C)

| −5 | −4 | −3 | −2 | −1 | 0 | 1 | 2 | 3 | 4 | 5 |

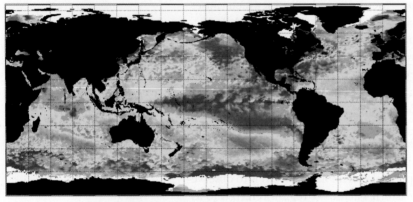

(b) Sea-surface temperature anomalies during La Niña.

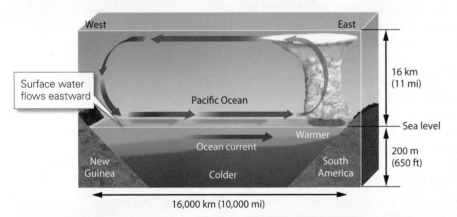

(c) During El Niño, the Walker circulation weakens or reverses.

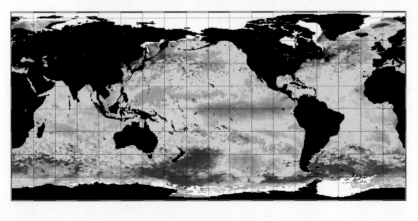

(d) Sea-surface temperature anomalies during El Niño.

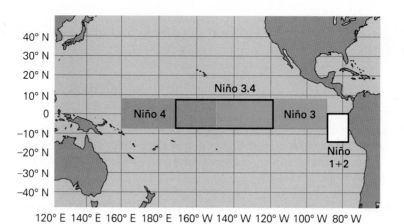

FIGURE 18.10 The location of Niño 3.4 region in the eastern Pacific Ocean (5°N to 5°S latitude, 120° to 170° W longitude).

FIGURE 18.11 Temperature anomalies provide signals as to when El Niño and La Niña events may occur.

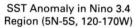

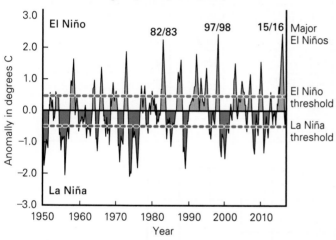

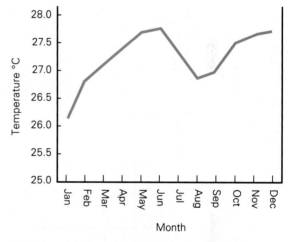

(a) The thresholds for El Niño and La Niña conditions in region Niño 3.4 and past events.

(b) The 2017 SST as an example of normal conditions in the region.

FIGURE 18.12 ENSO-neutral ("normal") chlorophyll concentrations (top image) in the sea surface around the globe, measured by satellite in 2013. The area highlighted in yellow is region Niño 3.4. Lower image is SST for the same region and time frame.

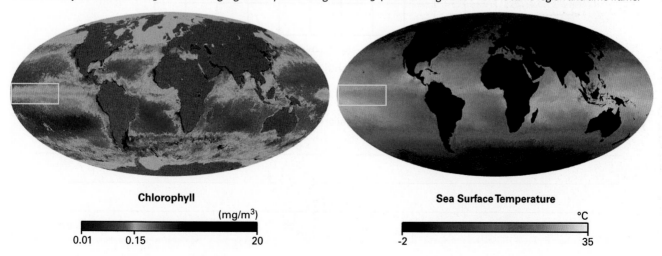

Short-term Changes in "Normal" Conditions and Atmosphere–Ocean Interactions: Is El Niño Phenomenon Real?

Name: _____   Section: _____

Course: _____   Date: _____

The table below contains sea surface temperature measurements taken in the equatorial Pacific Ocean in region Niño 3.4, from 2018 to 2019. Use the data to plot the average temperatures for each month shown in the table below. Use one color or pattern to represent the 2018 data, and a different color or pattern to represent the 2019 data. Be sure to connect your data points with a smooth line and include a legend.

Average monthly equatorial sea surface temperatures in the Pacific Ocean in the region Niño 3.4 from 2018 through 2019.

Year	Month	Average temperature (°C)	Year	Month	Average temperature (°C)	Year	Month	Average temperature (°C)
2018	Jan	26.2	2018	Sept	27.0	2019	May	28.2
2018	Feb	26.8	2018	Oct	27.5	2019	June	27.8
2018	Mar	27.1	2018	Nov	27.6	2019	July	27.2
2018	Apr	27.4	2018	Dec	27.7	2019	Aug	26.5
2018	May	27.7	2019	Jan	27.6	2019	Sept	26.4
2018	June	27.8	2019	Feb	27.8	2019	Oct	27.1
2018	July	27.3	2019	Mar	28.2	2019	Nov	27.3
2018	Aug	26.9	2019	Apr	28.4	2019	Dec	N/A

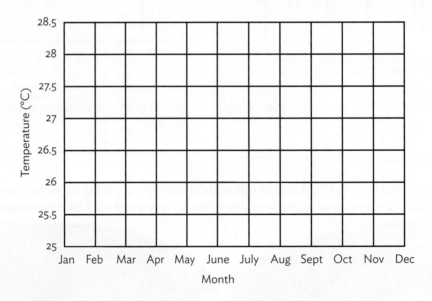

Compare your graph with the 2017 graph in Figure 18.12b.

(a) How are the temperature patterns in the two graphs similar? How are they different?

_____

_____

_____

(continued)

Name: _____   Section: _____

Course: _____   Date: _____

(b) If the 2017 data represent the "normal" patterns of this region, how would you describe the 2018–19 data pattern?

_____

_____

To investigate the SST patterns in 2018–19 and determine if these conditions were normal or if an El Niño event occurred, plot the 3-month-average consecutive temperature anomalies that are given in the table below on the graph on the next page. Use the two reference lines shown in Figure 18.11a to remind you of the temperature anomalies that signal the onset of El Niño and La Niña events.

The 3-month-average temperature anomalies in region Niño 3.4, 2018–2019

	Three-month period	Average temperature (°C)	Anomalies (°C)
2018	Dec–Jan–Feb	25.72	−0.87
	Jan–Feb–Mar	26.02	−0.76
	Feb–Mar–Apr	26.6	−0.60
	Mar–Apr–May	27.18	−0.41
	Apr–May–June	27.61	−0.13
	May–June–July	27.64	0.06
	June–July–Aug	27.38	0.11
	July–Aug–Sept	27.18	0.20
	Aug–Sept–Oct	27.25	0.43
	Sept–Oct–Nov	27.47	0.70
	Oct–Nov–Dec	27.57	0.85
	Nov–Dec–Jan	27.44	0.82

	Three-month period	Average temperature (°C)	Anomalies (°C)
2019	Dec–Jan–Feb	27.40	0.81
	Jan–Feb–Mar	27.60	0.83
	Feb–Mar–Apr	28.02	0.82
	Mar–Apr–May	28.36	0.76
	Apr–May–June	28.38	0.64
	May–June–July	28.11	0.53
	June–July–Aug	27.58	0.31
	July–Aug–Sept	27.11	0.12
	Aug–Sept–Oct	26.97	0.15
	Sept–Oct–Nov	27.07	0.30
	Oct–Nov–Dec	27.18	0.46
	Nov–Dec–Jan	27.17	0.55

(c) Based on your graph, and keeping in mind the requirements for identifying an El Niño event, was there any time during 2018–2019 that would be recognized as an El Niño event? If so, when did it begin and end? Use evidence to support your answer.

_____

_____

(d) Was there any time when there was a La Niña event? If so, indicate when and explain how you know.

_____

_____

_____

(continued)

Name: _____   Section: _____
Course: _____   Date: _____

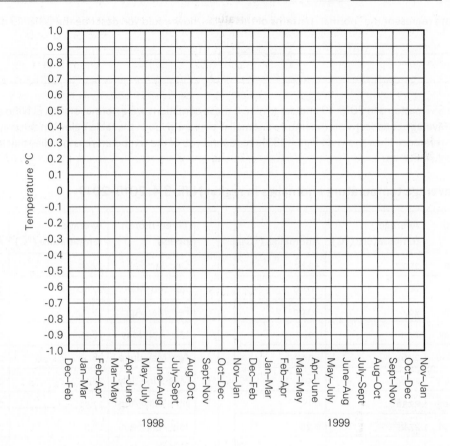

For comparison, you will now use the chlorophyll and sea-surface temperature data shown in the satellite images on the next page . Recall that chlorophyll levels relate directly to the amount of phytoplankton in the surface waters.

(e) Compare the yellow-boxed area in the 2015 images of chlorophyll and sea-surface temperature with the 2013 satellite images (Fig. 18.12). Does the concentration of chlorophyll along the Niño 3.4 region of the equatorial Pacific (the yellow-boxed area) in 2015 indicate high or low levels of nutrients? Support your answer with evidence.

_____

_____

_____

(f) Does the 2015 concentration of chlorophyll along this region of the equatorial Pacific indicate normal/neutral conditions or El Niño conditions? Explain.

_____

_____

(g) Look at the yellow-boxed area in the images from 2018 and compare it to the same area in the 2013 and 2015 images. Does the concentration of chlorophyll along this area of the equatorial Pacific in 2018 indicate high or low levels of nutrients? _____

(continued)

Name: _____   Section: _____

Course: _____   Date: _____

Global chlorophyll concentrations and sea-surface temperature (region Niño 3.4 is the boxed area on the images), December 2015 and 2018

December 2015

Chlorophyll (mg/m³)
0.01   0.15   20

Sea Surface Temperature °C
-2   35

December 2018

Chlorophyll (mg/m³)
0.01   0.15   20

Sea Surface Temperature °C
-2   35

(*continued*)

Name: _____   Section: _____

Course: _____   Date: _____

(h) Does the 2018 chlorophyll concentration along this region of the equatorial Pacific indicate normal/neutral conditions or El Niño conditions? Explain.

_____

_____

_____

_____

(i) Summarize the relationship you observe between SST and chlorophyll concentrations along the Niño 3.4 region of the equatorial Pacific during non-El Niño years.

_____

_____

_____

_____

(j) Summarize the relationship you observe between SST, El Niño events, and chlorophyll concentrations along the Niño 3.4 region of the equatorial Pacific.

_____

_____

_____

_____

_____

EXERCISE 18.9   Why Are El Niño Events Such a Big Deal?

Name: _____   Section: _____

Course: _____   Date: _____

? What Do You Think Considering the connections between the atmosphere and the oceans that drive our weather, what changes related to El Niño events do you think affect humans across the globe, beyond the Pacific Rim? What about changes in your local weather? Explain your choices on a separate sheet of paper, and consider changes in oceans, weather, economics, and environmental/health conditions, for example.

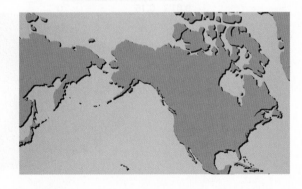

Name: _____   Section: _____

Course: _____   Date: _____

Exploring Earth Science Using Google Earth

1. Visit **digital.wwnorton.com/labmanualearthsci**
2. Go to the **Geotours** tile to download Google Earth Pro and the accompanying Geotours exercises file.

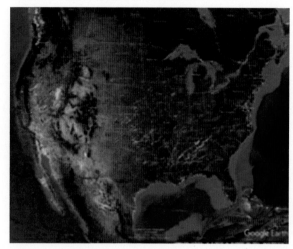

Expand the Geotour18 folder in Google Earth by clicking the triangle to the left of the folder icon. The folder contains a sub-folder with tornado tracks for the years 2010–2018 colored by season: Mar–May ~spring = green; Jun–Aug ~summer = red; Sep–Nov ~fall = purple; and Dec–Feb ~winter = blue. In this lab, we examined how hurricanes form, move, and impact the locations where they land. Now we'll examine the movement and seasonal occurrences of another severe weather hazard, tornadoes, which can form from particularly large storms called supercell thunderstorms.

(a) Check and double-click the 2010–2018 US Tornado tracks folder to fly to a position overlooking the United States. What is the general trend of the tornado paths (NW to SE, SW to NE, S to N, or W to E)? _____

(b) Which state experienced the greatest number of tornadoes during this period, Montana or Indiana? _____

(c) West Virginia has relatively few tornadoes during this period relative to the adjacent states. Suggest a reason to explain this occurrence (*Hint:* Look at the geography of the majority of the state).

_____

(d) Use the colors to compare the geographic distribution of tornadoes by season. In which season do the northern states experience the most tornadoes? _____

(e) Which geographic region of the United States experiences the most tornadoes in the winter? _____

Climate

19

Climate change leading to warming global temperatures has accelerated the melting and calving of large glaciers, as seen here in Alaska. This, in turn, has contributed to rises in sea level, which can have enormous consequences for coastal populations around the Earth.

19.1 Introduction

Hurray! You're headed on vacation, so now it's time to pack. But how do you figure out what you might need? You check the weather report, but you know weather reports can vary from day to day, especially over several days in advance. What you really need to know is the climate for your destination so that you have all the items you might need over your long stay. In the media, the terms *weather* and *climate* are often used as if they are interchangeable. However, as you have learned through your investigations on the atmosphere and weather, there is a distinction between the two with respect to time frame. So as you pack for your trip, remember to use the climate of the area to pack your suitcase, and use the weather reports to determine what you will wear each day.

As with many aspects of the Earth System, there are several factors that work simultaneously to control a region's climate. Among these factors are latitude, elevation, long-term atmospheric circulation, ocean circulation patterns, and proximity to oceans, mountains, and other surface features. In this chapter, we will examine how weather and climate differ, and how climates vary by location. Then we will investigate the forces that cause changes in climate and the resulting impacts of those changes.

19.2 Contrasting Weather and Climate

We've seen earlier that forecasting the weather involves taking a series of measurements of atmospheric conditions and then using computer models and statistics to determine the most likely weather outcomes given those sets of conditions. Since the factors controlling weather may vary each day and seasonally, short-term patterns are the most difficult to forecast. For example, FIGURE 19.1 is a plot of the high temperatures in Las Vegas, Nevada, for the month of January. As you can see, the high temperature can vary widely from day to day. However, despite that variability, there is an overall trend of increasing temperature from the beginning of the month to the end. So while predicting the high temperature each day is hard to do accurately, you could more confidently predict that the temperature will increase overall over that month, based on past seasonal records.

In Exercise 19.1, you will examine how weather and climate patterns differ for a region by graphing both long- and short-term data and comparing the results.

FIGURE 19.1 The high temperature for Las Vegas, Nevada, each day for the month of January 2019.

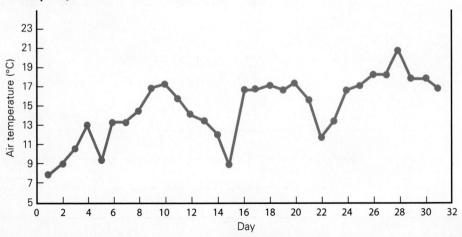

Name: _____   Section: _____

Course: _____   Date: _____

In this exercise, you will use the weather data diagram in Figure 19.1 and the data in the tables below to determine the trends in the high air temperatures for Las Vegas, Nevada, over two different time scales. Note that to get the most accurate and meaningful picture of a region's weather and climate patterns, we would ideally measure all the relevant air conditions (temperature, precipitation, etc.). But for the sake of time, in this exercise we will focus only on the air temperature.

The table below provides the average monthly high temperature for Las Vegas over the course of a year. Use the data in the table to plot the temperature data on the graph. Connect your data points with a line.

Monthly average high temperatures in Las Vegas, Nevada, in 2019

Month	High temp °C	Month	High temp °C
Jan	14.4	July	40.1
Feb	16.9	Aug	38.9
Mar	21.3	Sept	34.4
Apr	25.7	Oct	27.0
May	31.6	Nov	19.1
June	37.1	Dec	13.7

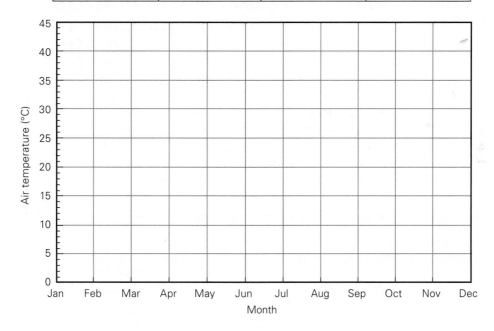

(a) Using the averages given, calculate the average high air temperature (in °C) for the year. _____

(b) Compare your graph with Figure 19.1. How do the temperature trends we obvserve change when we use monthly averages rather than individual dates?.

_____

_____

_____

(continued)

Name: _____   Section: _____

Course: _____   Date: _____

This next table (below) provides the average annual high temperature for Las Vegas over the course of a decade. Plot the data in the table on the graph below. Connect your data points with a line.

Yearly average high temperatures for Las Vegas, Nevada, 2010–2019

Year	High temp °C	Year	High temp °C
2010	26.4	2015	27.9
2011	26.4	2016	27.8
2012	27.7	2017	28.2
2013	27.0	2018	27.9
2014	28.1	2019	26.7

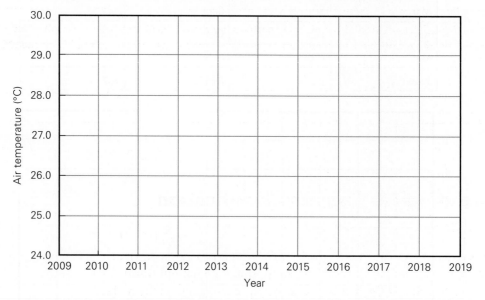

(c) Calculate the 10-year average high temperature for Las Vegas (in °C). _____

(d) How does the pattern or trend of these data on the graph (the smoothness of the line) compare with the single year average you plotted above? (*Note:* Pay close attention to the vertical scales on the graphs).

_____

_____

_____

(e) Note the difference in the vertical (*y*-axis) scale between the two graphs in this exercise. If you were to plot the data from the second table onto a graph using the vertical scale from the first graph, how would the smoothness of the line change?

_____

_____

_____

(continued)

Name: _____   Section: _____

Course: _____   Date: _____

(f) What can you conclude about the variability of weather versus climate for a region?

_____

_____

_____

(g) If you had more data and could calculate an average over a longer time, how would you expect the appearance of the lines you drew in the graphs to change?

_____

_____

_____

(h) Which controls (such as global wind and air-mass patterns, Earth's orbital process, and atmospheric conditions) are most likely to have the greatest impact on the average weather conditions over the course of a decade? On the daily averages?

_____

_____

19.3 Köppen-Geiger Climate Classification System

As you have observed, scientists like to categorize things. This process makes it easier for us to see patterns and predict results when we study the natural world. So far in this course you have investigated classification systems for minerals, rocks, plate boundaries, and other aspects of the Earth System. Likewise, a classification system has been developed for identifying climates in different regions of the world. Why is this important? Other than helping you pack the right clothes and supplies for your travels, classifying climates helps us choose the crops that we plant, the ones that are most likely to be successful under local conditions. It also helps us to choose the materials with which we construct local buildings and roadways, the ones best suited to provide durability, protection, and convenience under the given conditions.

Several classification systems for climate have been utilized over the years. The one that is used most frequently today is the Köppen-Geiger climate classification system (KGCC), originally developed by the German botanist and climatologist Wladamir Köppen in the early 1900s and later updated by Rudolph Geiger in 1961. This system divides climate into five different "groups," as shown in **TABLE 19.1**, each one designated by a capital letter, A through E. These groups can then be divided into "types" based on seasonal precipitation levels and the length of time over which precipitation occurs—the type is indicated by the second letter in the legend. Subtypes are the third letter in the legend and are based on the temperature of the region—whether it is hot, warm, or cold.

TABLE 19.1 The Köppen-Geiger climate classification

Group	Type	Subtype	Description
A			**TROPICAL:** Temperature of the coldest month is 18°C or higher.
	f		Rainforest: Precipitation in driest month is at least 6 cm.
	m		Monsoonal: A short dry season, with precipitation in the driest month of < 60 mm, but ≥ [100 − (R/25) mm].[1]
	w		Savannah: Well-defined winter dry season, with precipitation in the driest month of < 60 mm or < [100 − (R/25) mm].
B[2]			**ARID:** Either ≥ 70% of the annual precipitation falls in the summer half of the year and R < [20T + 280] mm, or ≥ 70% of the annual precipitation falls in the winter half of the year and R < 20T mm. *Alternatively,* neither half of the year has ≥ 70% of annual precipitation and R < [20T + 140] mm.
	W		Desert: R is < one-half of the upper limit for classification as a B type.
	s		Steppe: R < the upper limit for classification as a B type, but is > 50% of that amount.
		h	*Hot:* T ≥ 18°C
		k	*Cold:* T < 18°C
C			**TEMPERATE:** Temperature of the warmest month ≥ 10°C, and temperature of the coldest month < 18°C but > −3°C.
	s		Dry summer (Mediterranean): Precipitation in the driest month of the summer half of the year is < 30 mm and is < 33% of the amount in the wettest month of the winter half.
	w		Dry winter: Precipitation in the driest month of the winter half of the year < 10% of the amount in the wettest month of the summer half.
	f		No dry season: Precipitation is more evenly distributed throughout the year; criteria for neither s nor w satisfied.
		a	*Hot summer:* Temperature of the warmest month is ≥ 22°C.
		b	*Warm summer:* Temperature of each of the four warmest months is ≥ 10°C, but the warmest month is < 22°C.
		c	*Cold summer:* Temperature of one to three months is ≥ 10°C, but the warmest month is < 22°C.
D			**COLD:** Temperature of the warmest month is ≥ 10°C, and temperature of the coldest month is ≤ −3°C.
	s		Same as for Group C
	w		Same as for Group C
	f		Same as for Group C
		a	Same as for Group C
		b	Same as for Group C
		c	Same as for Group C
		d	*Very cold winter:* temperature of the coldest month is < −38°C (d designation is then used instead of a, b, or c).
E			**POLAR:** Temperature of the warmest month is < 10°C.
	T		Tundra: Temperature of the warmest month is > 0°C but < 10°C.
	F		Frost (Ice cap): Temperature of the warmest month is ≤ 0°C, and ice covers most land.

[1]In the formulas given, R refers to the annual rainfall in millimeters; T refers to the average annual temperature in degrees centigrade. The summer half of the year is defined as the months April–September for the northern hemisphere and October–March for the southern hemisphere.
[2]Any climate that satisfies criteria for designation as a B type is classified as such, regardless of its other characteristics.

Name: _____   Section: _____

Course: _____   Date: _____

In this exercise, you'll refer to Table 19.1 to identify the climate of a city. [*Hint:* Make sure to check *all* possibilities (A-E) and read the footnotes in Table 19.1 before determining the climate designation. You can't just go by temperature!]

Average temperature and precipitation total each month over one year for Honolulu, Hawaii

	Average temperature (°C)	Rainfall (mm)
Jan	22.4	113
Feb	22.3	80
Mar	22.8	89
Apr	23.4	60
May	24.7	43
June	25.5	26
July	26.2	32
Aug	26.1	29
Sept	26.0	30
Oct	25.5	66
Nov	24.3	93
Dec	23.0	109

(a) Calculate the average annual temperature and precipitation for Honolulu, given the data in the table above.

Average temperature (°C) _____

Average precipitation (mm) _____

(b) Based on the KGCC, what is the climate group, type, and subtype (if applicable) for this city?

_____

(c) How would increasing only the temperature affect the KGCC classification for this area?

_____

_____

_____

And increasing only precipitation?

_____

_____

_____

Name: _____   Section: _____

Course: _____   Date: _____

For this exercise, temperature and precipitation data are given for four cities in Peru, in South America. From these data, you will determine the KGCC for each city and then discuss the factors that lead to the different climates in these cities. Use the figures below in your investigation.

The map of Peru below has four lettered locations representing the cities from which weather data were obtained. The white arrow indicates the prevailing wind direction.

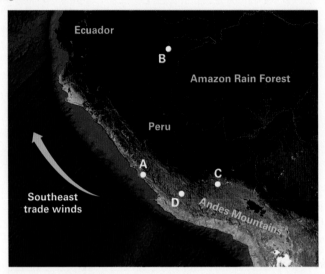

The four graphs below show temperature (red) and precipitation (blue) data over one year for the four lettered cities. (*Note:* These diagrams aren't necessarily in order from Location A–D).

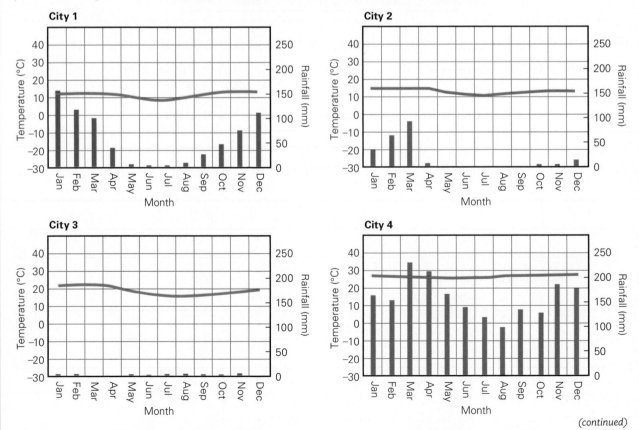

(continued)

Name: _____ Section: _____

Course: _____ Date: _____

(a) Which city (or cities) experiences the highest average annual rainfall? _____

The lowest average annual rainfall? _____

Which city experiences the highest average annual temperature? _____

The lowest average annual temperature? _____

(b) Determine the KGCC for each diagram, using Table 19.1 to assist you.

City 1 _____

City 2 _____

City 3 _____

City 4 _____

(c) Match the lettered locations in the first figure with each diagram.

Location A = City _____

Location B = City _____

Location C = City _____

Location D = City _____

Explain your choices from above.

_____

_____

_____

(d) What are the possible reasons for the different climates at the four locations? That is, how are the controlling factors of climate—surface features and winds, for example—different in these four places?

_____

_____

_____

19.4 Changes in Climate over Short Time Scales

Climate measurements have only been taken systematically since the 1800s. To determine patterns and identify changing atmospheric conditions over several hundred to a few thousand years (which is considered short-term in the context of the Earth's history), we need information beyond direct measurement records. Scientists have recognized that core samples taken from trees can provide information that can be used to interpret past atmospheric conditions over these longer time scales. This information is termed *proxy data*, meaning that it is a substitute for direct measurements. You're likely familiar with the concept that we can determine the age of a tree by counting its rings, which form whenever a new layer of bark is added

FIGURE 19.2 Analyses of tree rings provide clues to past climate conditions.

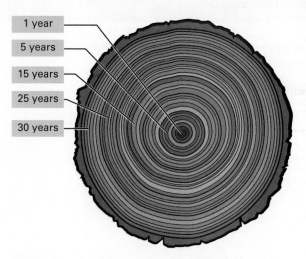

1 year
5 years
15 years
25 years
30 years

(a) Circular cross section of a tree showing alternating dark and light rings of bark growth (years).

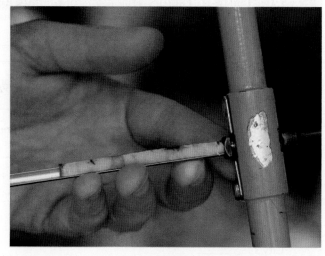

(b) A core sample from a tree that will be analyzed with a microscope.

(**FIG. 19.2**). For example, the supporting center of the tree (known as *heartwood* or *pith*) is the oldest part, representing the first year of growth from a seed into a young tree. As you move outward from the center, the rings become younger and younger. Take a close look at the tree rings in Figure 19.2a. You will see that there are alternating dark-colored rings and light-colored rings. Each pair of dark and light rings represents one year of growth, with the darker layer made up of small cells of wood formed during summer and fall (when growing conditions are poor), and the lighter rings made up of larger cells formed during spring (when conditions allow for faster growth). Favorable growing conditions—warmer and wetter—also increase the width of the rings, while poor growing conditions—cooler and drier—result in narrower rings. However, temperature and moisture are not the only factors that affect tree growth; changes in levels of CO_2 affect photosynthesis, crowding of nearby trees can limit the amount of water and sunshine, infestation by insects and disease can be devastating to tree growth, and events such as lightning strikes, erosion, or forest fires can also leave their marks on trees.

Although scientists use microscopes to more accurately count tree rings and measure their thicknesses and shapes, you can get a general idea of climate patterns over hundreds to thousands of years by using images of tree core samples. In Exercise 19.4, you will examine tree ring cores to determine the growing conditions in one area.

EXERCISE 19.4   Tree Ring Cores as Indicators of Local Climate

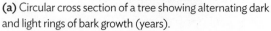

Name: _____   Section: _____
Course: _____   Date: _____

The illustrations in the figure on the next page are reproductions of cores taken from trees grown in the same area. As such, we can assume that they were subjected to similar growing conditions. Core 1 was taken from a living tree in the forest; Core 2 is from a downed tree; and Core 3 is from timber that is a part of the structure of a log cabin at the edge of the forest. You will correlate these cores and, based on the data you collect through your observations, determine the climate conditions for this area over the time period shown.

(continued)

Name: _____   Section: _____

Course: _____   Date: _____

Three tree cores from a forest in the Pacific Northwest.

(a) Look at the Core 1 and Core 2 tree rings. Determine where the tree ring patterns are the same in each core. Draw lines connecting Cores 1 and 2 where the tree ring patterns are the same.

(b) Repeat this process with Core 2 and Core 3.

(c) Given that Core 1 was taken from a tree in 2019, and based on your tree ring correlations above, determine the year when the trees of Cores 2 and 3 died.

Core 2 _____   Core 3 _____

(d) What is the age of each tree?

Core 1 _____   Core 2 _____   Core 3 _____

(e) Were there any seasons of drought? _____ Explain.

_____

_____

_____

(f) Were there any extended periods of poor growing conditions (2 years or more)? _____   If so, how many such periods? _____

(g) List the years of extended wet periods.

_____

(h) Briefly describe how the climate changed in this area over the time period that these trees represent.

_____

_____

_____

(i) If carbon dioxide levels are rising in the atmosphere, and temperatures and precipitation are increasing as a result, what differences would you expect to observe between the tree rings you investigated and those that will form over the next century?

_____

_____

_____

19.5 Climate Change over Long Time Scales

To go back further in time, we need to use other proxy data that span time scales longer than a few thousand years. To look at changes in climate over tens to hundreds of thousands of years, scientists use information that is stored inside glacial ice. Recall that when the ice formed, it was originally snow, which captured the water, air, and aerosols from the atmosphere at the time it fell. Like tree rings, each season of growth (snow fall) and season of snow melt leaves a "ring" or band in a core sample of the ice that can be used to determine the age of the sample. The chemistry of the ice and the contents of the air bubbles it contains can be analyzed to determine the atmospheric conditions at the time it formed, and so are proxies of past climate when observed over long time scales. Also like the rings of a tree, the layers of ice can be counted, though this method is more limited in the data it can collect. The farther down the core you go, the older and more compacted the ice gets, making the distinction between layers difficult to identify. But since each year's snowfall has different properties, scientists can use various methods, such as analyzing isotopes or electrical conductivity of the ice, to separate layers and provide age data.

Ice cores have been recovered from the ice sheets in Greenland and Antarctica, as well as from glaciers around the globe. To date, the oldest ice core records go back 800,000 years and were recovered from East Antarctica. **FIGURE 19.3** is an example of an ice core recovered from Greenland. Note the different shades of gray in the core. The lightest bands represent summer, while the darker gray are winter layers. The darkest bands contain debris, such as dust, ash, or human-generated pollutants.

Studying the composition of the ice and air bubbles allows scientists to better understand how the atmosphere, hydrosphere, geosphere, cryosphere, and biosphere interact. As global temperatures rise, the race is on to collect as many ice cores as possible so that, with information from these climate proxies, climate models can be validated and a clearer picture of our climate future can be made.

FIGURE 19.3 Photograph of a section of an ice core sampled from a glacier in Greenland. In this example, the top of the core is to the left.

EXERCISE 19.5 | Going Back in Time: Investigating Ice Core

Name: _____ Section: _____
Course: _____ Date: _____

Use Figure 19.3 or the physical ice core your instructor has provided to complete this exercise.

(a) How many years are represented in this ice core sample? _____

Compare your answer with another student's. Are they the same or different? How did you determine which layers you were grouping together as a year?

_____

_____

(continued)

Name: _____   Section: _____

Course: _____   Date: _____

(b) Compare the individual years (pairs of light and dark bands); how do they change in appearance over time?

_____

_____

(c) What do the differences between the years represented in the ice core indicate about the amount of precipitation (low, high), melting, and any geologic or other events that occurred during this time span?

_____

_____

(d) Briefly describe the climate history represented by your ice core sample, based on the observations you made.

_____

_____

_____

Now that you have honed your ice core observation skills, let's use some chemistry to assist in climate reconstruction. Specifically, in Exercise 19.6, you'll determine temperature changes using the isotopes of hydrogen (which, alongside oxygen, make up the water molecules in the ice). Hydrogen has two stable isotopes: *protium*, which is your typical hydrogen consisting of one proton and zero neutrons, and *deuterium*, the rare variety of hydrogen that contains one proton and one neutron in its nucleus, making it the "heavy" hydrogen isotope. These two characteristics of deuterium—being rare and heavy—make it an excellent proxy for temperature. To understand this, you need to think about the hydrologic cycle. As water changes form and moves through the cycle, bonds between molecules are formed and broken, resulting, respectively, in the absorption and release of energy. When water evaporates from oceans and lakes, it is "easier" (meaning it takes less energy) to evaporate water containing the lightweight protium, thereby leaving more of the water made of heavy deuterium behind in liquid form. And conversely, because deuterium-rich water is heavier, it more easily leaves the atmosphere in the form of precipitation than the protium-rich water. Furthermore, when the air temperature is colder, the deuterium does not make it to the poles, as it is "rained" out of the atmosphere at lower latitudes and remains in the oceans. During warmer periods, meanwhile, there is more energy available in the atmosphere to transport deuterium-rich water to higher latitudes, where it can then precipitate out as rain or snow.

The amount of deuterium in an ice core is measured and compared with a standard (the ocean), which is assigned a value of zero. The result is that the deuterium concentration in ice is expressed as a negative number. The colder the air temperature, the less deuterium in the ice, and the more negative the value; the warmer the temperature, the more deuterium in the ice, and the less negative the deuterium concentration.

Name: _____   Section: _____
Course: _____   Date: _____

You are able to sample an ice core that is 20,000 years old. The deuterium concentrations from the ice core samples are given in the table below. Plot the data on the graph, noting the labeling and scale of both axes.

Concentrations of deuterium recovered from ice core samples over the past 20,000 years . The deuterium concentrations of the ice core samples are given in the table below as parts per thousand (0/00).

Age of sample (Years before present)	Deuterium (o/oo)	Age of sample (Years before present)	Deuterium (o/oo)
1,000	−396	11,000	−396
2,000	−394	12,000	−394
3,000	−395	13,000	−410
4,000	−393	14,000	−413
5,000	−397	15,000	−412
6,000	−394	16,000	−418
7,000	−398	17,000	−420
8,000	−403	18,000	−432
9,000	−400	19,000	−438
10,000	−409	20,000	−441

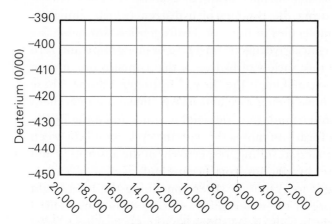

(a) Using the deuterium concentrations you plotted on the graph, when were temperatures warmer? _____

Cooler? _____

Explain your choices.

_____

_____

_____

(continued)

Name: _____   Section: _____

Course: _____   Date: _____

(b) Are there any temperature cycles you can identify? (A cycle length is from one warm/cold peak to the next.) _____

   If so, how many years does each cycle last? _____

(c) Is there an overall trend toward warmer or cooler conditions from 20,000 years ago to the present? Explain.

   _____

   _____

   _____

   _____

(d) Based on your answer in (c), what is the likely effect on global sea level if this trend continues? (*Hint*: Consider the impact on water currently trapped in glacial ice.)

   _____

   _____

   _____

   _____

19.6 What Is the Human Impact on Climate?

While nearly all scientists agree that oceanic and atmospheric data indicate a global change in the Earth's climate, the political debate regarding the human impact on that change persists. You are already familiar from Chapter 16 with how the temperature of the atmosphere can change when its chemical composition changes—that is, when the volume of *greenhouse gases* that absorb infrared radiation is altered. Of all the greenhouse gases, carbon dioxide is the easiest to measure over time. While there are natural sources that produce CO_2, these emissions are balanced by systems that remove the carbon from the atmosphere, known as *sinks*, such as the ocean, photosynthesizing organisms, and geologic processes like burial and lithification. When humans remove vegetation for agriculture and urbanization, and burn vegetation and fossil fuels, we disrupt that natural carbon balance.

While single measurements of the atmospheric concentrations of CO_2 cannot tell us if the source of the CO_2 is natural or anthropogenic (human-caused), we can tell that the majority of the CO_2 in the atmosphere since the industrial revolution is a result of human activity by studying the isotopes of carbon and levels of oxygen in the atmosphere. Fossil fuels—gas, oil, and coal, which formed over hundreds of millions of years from ancient plants and microorganisms—contain no carbon-14, unlike living plant material (recall that the half-life of C-14 is 5,700 years). Fossil fuels are also depleted in carbon-13 but enriched in carbon-12. Therefore, the decrease in C-14 and C-13 in atmospheric CO_2 is related to the burning of fossil fuels.

Name: _____ Section: _____
Course: _____ Date: _____

You will use the figures below to investigate changes in atmospheric carbon dioxide levels and global average temperature, and to understand what they indicate about the role of humans in changing our climate.

The concentration of CO₂ over the past 800,000 years, as recovered from the gas bubbles in ice cores. Units are in parts per million (ppm).

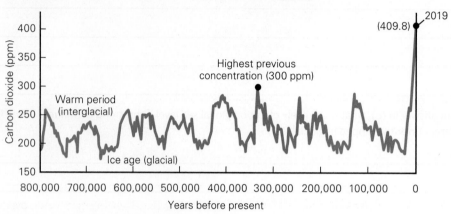

(a) What is the highest concentration of CO₂ on this graph? _____

(b) Has the concentration of CO₂ ever been higher in the past 800,000 years than it was in 2019? _____

(c) How does the rate of change in the concentration of atmospheric CO₂ in the last century or so compare with that in the more distant past?

_____

_____

_____

Now that you have an idea of what has been happening in the atmosphere with respect to this greenhouse gas, let's take a look at temperature changes over the same time scale using the data in the figure below. The temperature data in this figure come from a range of sources, including historical measurement records and from various proxy data. The figure gives the change in temperature—or temperature anomaly—compared with an average.

Change in temperature (temperature anomaly) over the last 800,000 years from proxy and historical data.

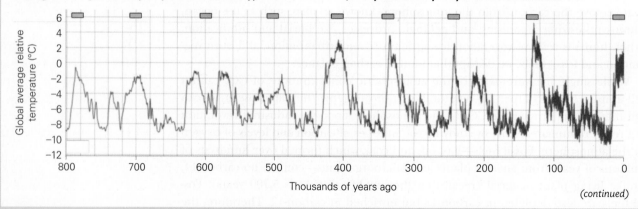

(continued)

Name: _____   Section: _____
Course: _____   Date: _____

(d) Compare the present temperature to past temperature—is it hotter today than it ever has been in the past? Explain.

_____

_____

_____

_____

(e) How does the temperature trend today compare with those of the past?

_____

_____

_____

_____

(f) Investigate the relationship between air temperature and the amount of CO_2 in the atmosphere over time. Is there a correlation between these two graphs? Explain.

_____

_____

_____

_____

(g) What are some natural and anthropogenic (human-caused) reasons that the amount of atmospheric CO_2 has been increasing?

_____

_____

_____

_____

(h) What happens as the amount of CO_2 continues to increase but the natural CO_2 sinks cannot keep up?

_____

_____

_____

_____

19.7 Tropical Storms and Climate

When we look at global climate trends, we often see changes that may seem small and trivial. For example, the global change in temperature is often reported at about 1°C since 1880. So why do we care about such small changes? The key is to consider the vast volume of material that is heating up (the oceans, atmosphere, and land) and the vast amount of heat required to collectively raise the global temperature by just a degree. The so-called Little Ice Age in the late Holocene, where a dramatic series of advances and retreats of mountain glaciers and ice sheets occurred, resulted from only a 1° to 2°C drop in temperature, while a 5°C drop 20,000 years ago resulted in vast areas of North America being buried under a massive ice sheet.

As we've seen, our planet is currently experiencing, on average, an increase in temperature. This increase, particularly in the oceans and atmosphere, raises concerns about its impact on future tropical storms. This is because the more heat that is absorbed into the oceans, the more energy there is to produce more storms and to increase the intensity of these storms.

The formation of tropical cyclones requires a minimum sea surface temperature of 26.5°C, along with favorable sea level pressure, latent heat flux, and vertical wind shear. The optimum locations for storm conditions in the Atlantic basin are shown in **FIGURE 19.4**.

FIGURE 19.4 Tropical cyclone tracks in the Pacific basin (data 1949–2017) and the Atlantic basin (data 1851–2017).

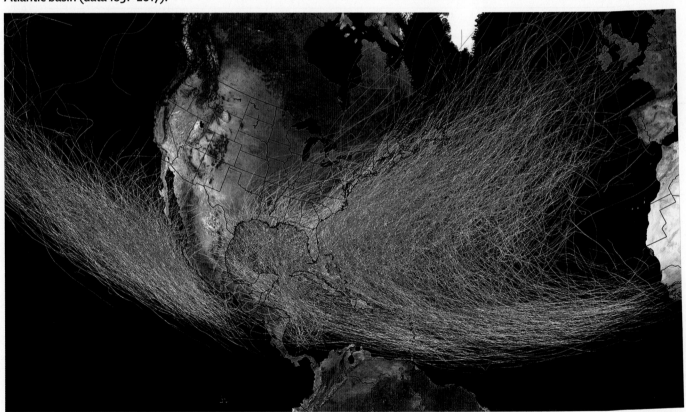

—— Tropical and subtropical storms   —— Hurricanes   —— Major hurricanes   —— Depressions, extratropical, disturbance, low

EXERCISE 19.8 Investigating the Connections between Tropical Cyclones and Climate Change

Name: ______________________ **Section:** __________

Course: ______________________ **Date:** __________

Use the figures below and what you have learned about tropical cyclones and ocean and atmospheric circulation for this exercise.

The average sea-surface temperature (SST) anomaly for the oceans globally between 1880 and 2019, relative to the global average between 1950 and 2005.

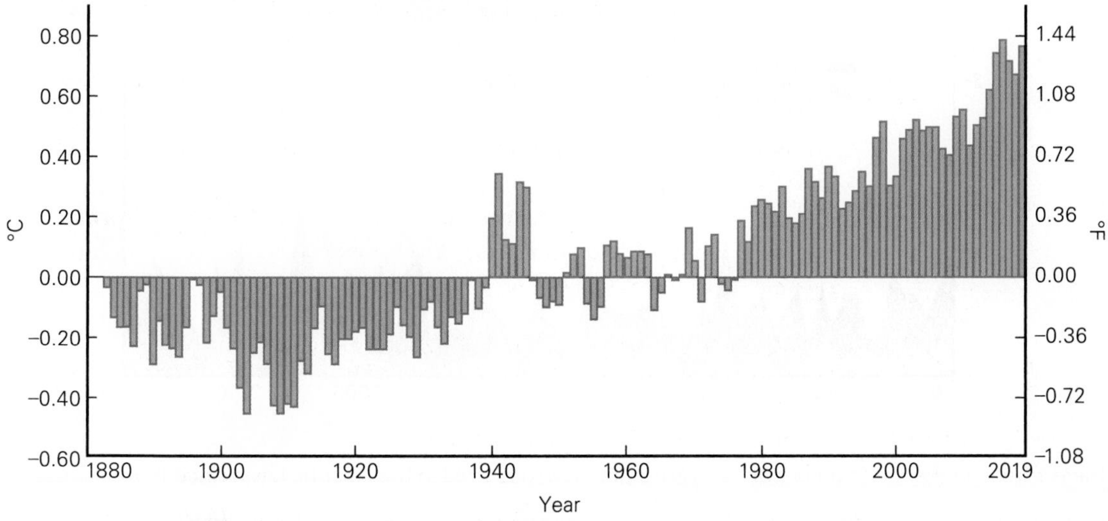

(a) Based on the above graph, what is the trend in global SST over the past 40 years?

____

____

____

(b) From what you have learned about tropical cyclones in Chapter 18, would you predict this trend to result in more or fewer tropical cyclones? Explain.

____

____

____

____

(c) A graph of tropical cyclone counts is shown on the next page. The blue portion is the total count of all hurricanes between the years 1950 and 2019. How do the data in this graph compare with the SST data in the first figure over the same time period?

____

____

____

____

(continued)

Name: _____   Section: _____

Course: _____   Date: _____

Counts of Atlantic basin tropical cyclones from 1950 to 2019. The blue region is for all hurricanes recorded during this time period. The red region represents hurricane categories 3 and higher.

North Atlantic 1950–2019
Average period: 1981-2010

(d) Is there a trend in the total number of hurricanes that have occurred in the Atlantic basin since 1950? _____

(e) Does there appear to be a direct "one-to-one" relationship between SST and tropical cyclone frequency (the number of hurricanes each year)? _____

If so, why would this be the case?

_____

_____

_____

Another relationship that meteorologists need to consider in predicting future tropical cyclone occurrences is the effect of SST on storm intensity, or strength.

(f) What kind of change in SST would you expect to cause an increase in tropical cyclone wind speeds? Explain.

_____

_____

_____

_____

(g) How might an increase in SST affect precipitation? Why?

_____

_____

_____

(continued)

Name: _____ **Section:** _____

Course: _____ **Date:** _____

Look at the second figure again and note the number of major hurricanes that occurred in the North Atlantic basin since 1950.

(h) Has the number of major hurricanes increased, decreased, or stayed the same? _____

(i) Compare this figure with the global SST anomalies. How does the relationship between global SST and hurricane strength compare with what you observed between SST and frequency of hurricanes?

_____

_____

_____

_____

_____

(j) In 2019, during the Atlantic basin hurricane season, there were several storms that were located in higher latitudes than normal. Why might this be? (*Hint:* Consider what you have learned about the conditions necessary for hurricane development.)

_____

_____

_____

_____

(k) Why is it difficult to correlate sea surface temperature with tropical cyclone characteristics and therefore make accurate and consistent predictions?

_____

_____

_____

_____

(l) In the meantime, how can humans reduce the impacts of tropical cyclones?

_____

_____

_____

_____

_____

_____

(continued)

Name: _____

Course: _____

Section: _____

Date: _____

? What Do You Think Unlike the Antarctic, which is a continent buried under ice, the Arctic is primarily ocean. During the winter months, the freshwater molecules in the ocean freeze to form sea ice. While historically sea ice remains intact during the summer seasons in the Arctic, the graphic below illustrates how the sea ice formed each winter has been decreasing over the past few decades, such that, come summer, the Arctic is becoming more and more ice-free. Another difference from the Antarctic is that the Arctic is one of the most biologically productive areas of the oceans. There are many organisms that inhabit this region, some of which remain year-round, both on land and in the sea, such as tiny plants, plankton, walruses, migratory and sea birds, whales, fish, seals, and polar bears, just to name a few. Indigenous peoples in the Arctic today still rely on fishing, hunting, and herding for their survival.

In terms of climate, recall from the chapter on ocean circulation that it is the cold waters of the North Atlantic that drive the global ocean conveyor; therefore, changes in the ocean temperatures here can have drastic global consequences. We have evidence such temperature changes have occurred in the Earth's past, likely resulting in the largest mass extinction in the planet's history at the end of the Permian period.

The maximum winter sea ice extent in March 1985 and 2018. Less than 1% of sea ice is older than 4 years. In other words, more is melting in summer, and less is forming in winter.

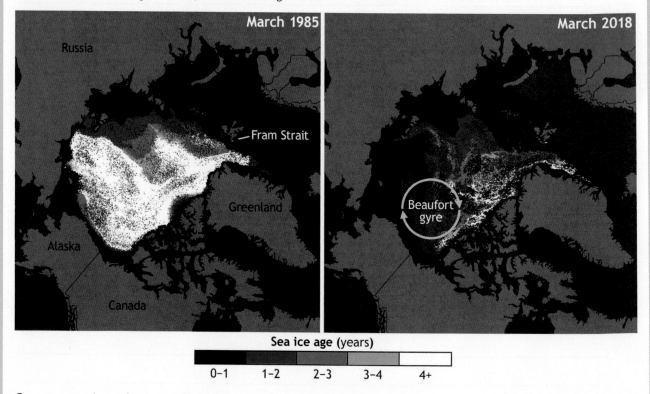

On a separate sheet of paper, explain what you think may be short-term benefits of decreasing sea ice in the Arctic, as well as what you think could be negative long-term (decades to centuries) effects of these changes. Consider both the ecological and human impacts.

Name: _____   Section: _____

Course: _____   Date: _____

Exploring Earth Science Using Google Earth

1. Visit **digital.wwnorton.com/labmanualearthsci**
2. Go to the **Geotours** tile to download Google Earth Pro and the accompanying Geotours exercises file.

Expand the Geotour19 folder in Google Earth by clicking the triangle to the left of the folder icon. The folder contains placemarks keyed to questions that document the effects of global climate change.

(a) Check and double-click the overlays dated Jan 31, 2002 (the dark spots on the ice are pools of water), Feb 23, 2002, and Mar 07, 2002 (the light blue is reformed thin ice after the main ice shelf broke apart) in sequential order to view overlays of historical imagery of the Larsen B Ice Shelf in Antarctica. Much like the "canary in the coal mine," the Larsen B Ice Shelf has responded to global warming in dramatic fashion (which surprised scientists). What happened?

_____

_____

(b) Uncheck the (a) overlays. Now check and double-click the (b) placemark to fly to the Amazon rain forest in Brazil. Rain forests absorb CO_2 and help regulate the impact that excess CO_2 in the atmosphere from human activities exerts on our climate. Watch the animated GIF image in the placemark, or click on the link in the placemark balloon for the Google Earth Engine website for this same region. Estimate the percentage (1%, 15%, or 75%) of land in the field of view that has experienced deforestation from 1984 to 2012. _____

(c) Check and double-click the (c) placemark to fly to the Aral Sea in Central Asia to study what has happened as rivers that once flowed into the Aral Sea have been diverted for agricultural irrigation in the face of climate change. Watch the animated GIF image in the placemark, or click on the link in the placemark balloon for the Google Earth Engine website for this same region. Over this time period, what has happened to the size/surface area of the Aral Sea?

_____

_____

Investigating the
Properties of the Universe

The Milky Way Galaxy moves over CARMA Array Radio Observatory in the White Mountains of California.

20.1 Introduction

Our understanding of the *observable universe* has changed greatly since the time of the ancient Greeks, whose early models of the Universe contained only a handful of planets, the Sun, the Moon, and a fixed sphere of stars. Nevertheless, their observations—achieved with only their eyes—laid the foundation for early scientific thought. As time progressed, the invention of new instruments that enabled astronomers to determine distances by measuring angles led to the development of new mathematics: geometry and trigonometry. By the 1600s C.E., the first optical telescopes enabled astronomers to take a closer look at the "fuzzy" objects in the heavens. While these early telescopes could not provide the clarity of today's lenses, their use represented a systematic attempt to study the cosmos. Today, astronomical investigations are advanced through the use of other types of telescopes—long-wavelength (radio and infrared) and short-wavelength (ultraviolet, x-ray, and gamma)—and space probes. What was once invisible to us is now visible, depending on the wavelength of radiation detected.

Remember that all forms of electromagnetic (EM) radiation (**FIG. 20.1**) travel in the form of waves, and that a **wavelength** is the distance between crests in a wave. The number of wave crests that pass a fixed point every second is the wave's **frequency**. All wavelengths of radiation travel at the same speed, 300,000 kilometers (186,000 miles) each second, or the speed of light. A relationship exists between wavelength (w) and frequency (f) as given in the equation

$$f = c \div w$$

where c is the speed of light. For example, visible light has wavelengths between 400 and 700 nanometers (a nanometer equals a thousand-millionth of a meter), which is higher than that of infrared and lower than that of ultraviolet (UV) radiation. The frequency of visible light is therefore higher than that of infrared radiation, but lower than that of UV radiation.

In this chapter, you will investigate the formation of the Universe, as we understand it today, using the properties of radiation obtained by the study of stars.

20.2 Space, Time, Matter, and Motion

To understand the origins of the Universe, we need to observe all aspects of space: the number, types, and characteristics of celestial (or space) objects and what is between the objects. Unlike the ancient Greeks, who had no tools for determining distances to celestial objects beyond the Moon, modern astronomers have access to a variety of tools that allow them to determine not only what is beyond the Earth, but also to determine the distance, age, and motion of those celestial objects. As astronomers began measuring the distances to these objects they discovered that the numbers involved were far larger than any that had been measured previously. To simplify these truly astronomical numbers, new units of measurement were created. Three of the most widely used units for distance in astronomy are the astronomical unit (AU), light-year, and parsec. The **astronomical unit (AU)** represents the average distance from the center of the Earth to the center of the Sun, which is 150 million km (93 million mi), or 1 AU. A **light-year** is the distance that light travels through space in one year, 9.5 trillion km (5.9 trillion mi), and is the unit you are most likely to come across as an amateur astronomer. The **parsec (pc)** is equivalent to approximately 3.3 light-years (about 31 trillion km or 19.5 trillion mi), so it is very useful for objects that are vastly distant from our Solar System. We will look at the basis for this unit later in the chapter.

The tools that scientists use to determine distances in space include radar, parallax, the brightness of stars, and Doppler shifts (to be discussed in the next section).

FIGURE 20.1 **The electromagnetic spectrum.**

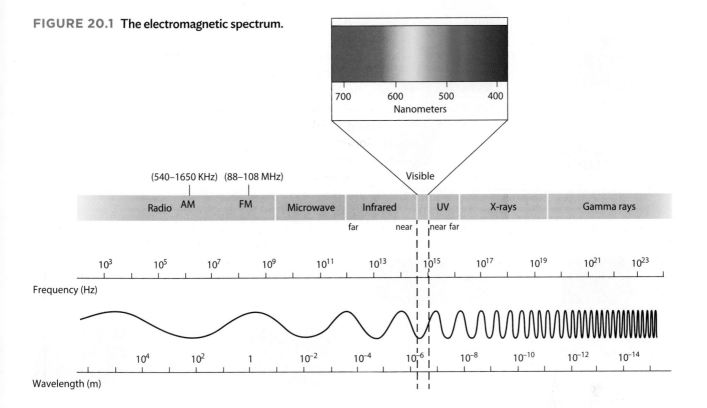

For the purposes of this chapter, we will focus on the tools that utilize information from visible light.

Since early studies of space, observers have been able to calculate the distances between stars using **parallax**, the apparent displacement of an object in the foreground against a background of distant objects. This is what you experience when you hold out your thumb, and alternate closing one eye and then the other (**FIG. 20.2**). Your thumb appears to change its position relative to the background. This apparent motion depends on the observer's position (line of sight) and the distance to the object. For example, you can see your thumb moving against the background of a room or trees in your backyard, but if you were to attempt to see its displacement against the Eiffel Tower (when viewed from anywhere other than Paris), you would not observe parallax, as the angle between the two objects is too small to observe with your eye. One objection by the ancient Greeks to the premise that the Earth moved in the "heavens" was that they did not observe parallax, and so concluded that the Earth was not in motion about any other object. Gaileo Galilei later proved that the reason no parallax was observed was that the stars are very far away.

While distance is a limitation of using parallax for measurements to distant stars, astronomers discovered that the Earth's motions around the Sun provide the right distance for detecting the slight parallax that closer stars display relative to more distant stars. By using trigonometry, the distance to stars that are less than 500 light-years away—a relatively close distance—can then be calculated. This parallax is the basis for defining a parsec, which is the distance that an imaginary star would have to be from Earth for its parallax to equal one *arc-second* ($\frac{1}{360}$ of a degree) if an observer were to move from a point at the Sun's center to a point on the Earth's surface, 1 AU away.

To calculate the distance to more distant stars, astronomers use the stars' brightness and the shifts in the frequency and wavelength of light they emit; you will investigate these methods in more detail in the next section.

Before you dive into these topics, Exercise 20.1 will give you some experience contextualizing the scale of distance and time in the known Universe.

FIGURE 20.2 Geometric parallax.

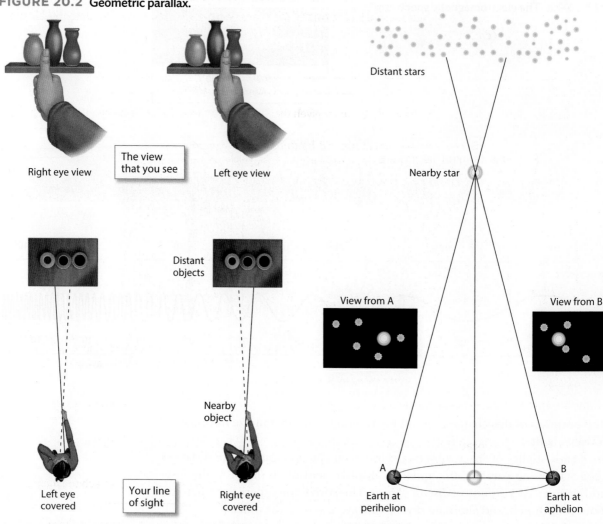

(a) You can observe geometric parallax by holding your thumb out at arm's length. When you close one eye and then the other, your thumb covers different distant objects.

(b) A nearby star shifts its position relative to the background stars when viewed from opposite sides of the Earth's orbit.

EXERCISE 20.1 | **A Sense of Scale: The Relationship of Distance and Time in the Observable Universe**

Name: _____   Section: _____

Course: _____   Date: _____

(a) If Jupiter is about 5.2 AU (astronomical units) from the Sun, what is the distance in AU from the Earth to Jupiter given the configuration shown in the figure on the right? _____

Diagram of the Sun, Earth, and Jupiter: First alignment.

Not to scale

Sun

E

J

(continued)

Name: _____ Section: _____

Course: _____ Date: _____

(b) What is the distance in AU from the Earth to Jupiter given the relationship shown in the figure below? _____

Diagram of the Sun, Earth, and Jupiter: Second alignment.

(c) You're looking through a telescope one night and see the planet Saturn, which is at an average distance of 7 AU away from the Earth. You are able to view Saturn because the light emitted from the Sun 9.6 AU away is reflected off Saturn's outer atmosphere and then travels back to you on the Earth.

How long did it take light from the Sun to reach Saturn? _____ seconds. Convert that to minutes and seconds: _____.

Ignoring changes in location as these planets move in space, how long did it take the light from Saturn to reach the Earth? _____ seconds. Convert that to minutes and seconds: _____.

(d) If you were aboard the Voyager 2 spacecraft traveling at 15 km/s, how long would it take you to reach the most distant object known to orbit our Sun—the planetoid Sedna—at a distance of 90 AU? _____

(e) The Voyager 1 spacecraft was launched on September 5, 1977. Traveling at approximately 17 km/s, it reached interstellar space in 2012.

Assuming Voyager 1 is still traveling at 17 km/s through space, how many kilometers will it have traveled by September 5, 2027 (exactly 50 years from its launch)? _____

How many light-years? _____

(f) After the Sun, the next closest star to the Earth is Proxima Centauri, one of three stars in the Alpha Centauri system. Proxima Centauri is 4.2 light-years from the Earth.

How far is that in kilometers? _____

In parsecs? _____

(g) If time were the only consideration, would it be possible today for a person to travel to other star systems within a human lifetime, if traveling aboard a spacecraft at 15 km/s? Why or why not?

_____

_____

_____

_____

_____

20.3 Origins of the Universe

How and when did the Universe begin? What existed before? As in other aspects of science, many hypotheses have been offered over time. The one that has most successfully supported the available evidence and observations, and has thus become a widely accepted theory, is the "Big Bang." While it may be the idea that is best supported by data, it continues to be tested (and adjusted) as more observations are made. Advances in telescope and space-detection technologies, and continuing experiments in physics, will undoubtedly fuel further cosmological debates in the future. In this section, we will explore the concept of the Big Bang—what it states, the evidence that exists to support it, and new discoveries since its elevation as a theory in 1964 that have yet to be explained.

20.3.1 The Big Bang Theory

The foundation of the Big Bang theory was first proposed in 1927, when astronomer Georges Lemaître, building on observations and mathematical calculations made by other researchers in prior years, proposed that the Universe was expanding. Universal expansion was confirmed two years later by astronomer Edwin Hubble, who observed that the distance between other **galaxies** (groups of stars, planets, and interplanetary matter) and our own was indeed increasing, and that the greater the distance between the galaxies, the faster the motion. These conclusions were based on observations of changes in the frequency and wavelength of light from other galaxies. As the distance between two objects increases, both the frequency and wavelength of the light traveling between them change. This change is referred to as the **Doppler effect**, after Austrian physicist Christian Doppler (1803–1853), who first explained it in another context. A shift toward a higher frequency (shorter wavelength) happens when the object moves toward the observer, and a shift toward a lower frequency (and longer wavelength) happen when the object moves away from the observer. Applied to visible light, the results are a shift in color—a light source moving toward an observer on the Earth exhibits **blue shift** (toward shorter wavelengths), while a light source moving away from an observer on the Earth exhibits **red shift** (toward longer wavelengths) (**FIG. 20.3A**). Additionally, the faster the object travels, the greater the *Doppler shift*.

Astronomers use the Doppler shift to determine the movements of celestial objects. To detect blue or red shifts, astronomers need to compare the spectrum received from a celestial object with the spectrum produced by a stationary source of light, such as the Sun. The black absorption lines within the spectrum are the wavelengths of light that are absorbed by the gas(es) in the object. As **FIGURE 20.3B** shows, the absorption lines in the spectrum from a distant galaxy have shifted toward the red end of the spectrum, relative to equivalent absorption lines displayed on the spectrum of light from the Sun. That indicates that the galaxy is moving away from us. Nearly all other galaxies we can observe exhibit a red shift.

Given these observations, Lemaître then proposed that if the Universe is expanding, that means at one point it must have been much more contracted—in fact, he determined that all matter within the Universe could be traced back to a single origin point in time (and space). This origin was a very hot and dense single point of minute particles, light, and energy in space, which came to be known as a **singularity**. Current estimates place the origin of the Universe at 13.8 billion years ago. For reasons that are still not known, this singularity rapidly expanded outward, in what we now call the "Big Bang." As the Universe continued to expand and take up more space, it cooled, and atoms began to form. The motions of free electrons that

FIGURE 20.3 The Doppler effect for light.

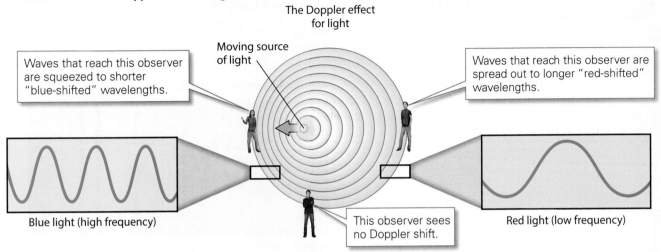

The Doppler effect for light

Moving source of light

Waves that reach this observer are squeezed to shorter "blue-shifted" wavelengths.

Waves that reach this observer are spread out to longer "red-shifted" wavelengths.

Blue light (high frequency)

This observer sees no Doppler shift.

Red light (low frequency)

(a) If a light source moves toward an observer, the radiation detected by the observer will be shifted toward shorter wavelengths (blue-shifted). If the light source moves away, the radiation will be shifted to longer wavelengths (red-shifted).

were not yet bound to atoms scattered and absorbed the particles of light (*photons*) so that the Universe at this time was dark and opaque. As more atoms clumped together, forming "clouds" of hot gas and particles (**nebulae**), there were fewer

Sun

Distant galaxy

(b) The shift in wavelengths between radiation from a distant source, such as a galaxy or a star, and from the Sun can be used to determine the velocity of the source's movement toward or away from the Earth.

free electrons whizzing about, meaning photons were able to move around with a lower chance of interacting with another particle and could spread out. As a result, the Universe became transparent.

You will begin investigating the evidence for the Big Bang in Exercise 20.2 using only information about visible light emitted by stars. Later, you will look at other forms of radiation and how they relate to the idea of an expanding Universe.

EXERCISE 20.2 **Visible Light as Evidence of Expansion**

Name: _____ Section: _____
Course: _____ Date: _____

The visible part of the electromagnetic radiation spectrum is shown in the figure below. Use this figure and the previous information to answer the questions in this exercise.

(a) What color would you expect to see at 450 nm? _____

At 575 nm? _____

At 650 nm? _____

Wavelengths of visible light and their corresponding colors.

700   600   500   400
Nanometers

(continued)

Name: _____ Section: _____
Course: _____ Date: _____

Look at the spectra of light emitted by three stars of identical composition within our galaxy, labeled Star A, B, and C in the figures here. Star A is a star that does not appear to be moving away from or toward the Earth.

Absorption spectra of Star A.

Violet   Blue   Green   Yellow Orange Red

Absorption spectra of Star B.

Violet   Blue   Green   Yellow Orange Red

Absorption spectra of Star C.

Violet   Blue   Green   Yellow Orange Red

(b) How do the black lines in the spectrum for Star A differ from the black lines in the spectrum for Star B?

_____

(c) Do you think that Star B is moving toward or away from the Earth? Explain.

_____

_____

(d) What is the apparent motion of Star C? Support your answer with evidence.

_____

_____

(e) In comparison to Star A, which star is moving faster, Star B or Star C? Use data to justify your answer.

_____

_____

20.3.2 An Expanding Universe: Seeing Beyond the Visible

Although early astronomy was limited to only what could be seen in the visible part of the electromagnetic (EM) spectrum, scientists were still able to gather an incredible amount of information about stars, including their temperature, composition, distance, and relative motion. Advances in space telescope technology have since allowed scientists to detect wavelengths within the EM spectrum beyond the visible part. With these new telescopes, some of what had appeared to be dark and empty regions of space were suddenly revealed to be full of celestial objects.

If the Universe has been expanding since its formation, and energy has spread through the Universe in all directions and places equally, astronomers hypothesize that the wavelengths of thermal radiation originating from the Big Bang would become so long that they could be detected in microwave and radio wave frequencies once they reach the Earth. This theory was confirmed in 1964 by American radio astronomers Robert Wilson and Arno Penzias, who detected what they thought was "noise" affecting their radio telescope. They were later awarded the Nobel Prize for this "accidental" discovery of what is called Cosmic Microwave Background radiation, or CMB.

EXERCISE 20.3 **Cosmic Microwave Background Radiation and an Expanding Universe**

Name: _____ Section: _____

Course: _____ Date: _____

Your instructor has provided you with a balloon to represent the Universe. The balloon is marked with a waveform representing a light wave. (*Note:* If this exercise is available and being conducted online, the setup, materials, and directions may differ. See the instructions in the online lab for how to conduct the exercise.)

Step 1: On a separate sheet of paper, draw or sketch the balloon and what the waveform looks like now, before you inflate the balloon:

Step 2: Slowly expand your "Universe" by inflating the balloon until it has partially, but not fully, expanded. Draw the way the balloon and waveform appear now.

Step 3: Inflate the balloon fully and sketch the way the balloon and waveform appear.

(a) Describe the change in the *wavelength* of the light wave on your balloon once the balloon expanded (was inflated).

_____

(b) What change in the *frequency* of the light wave occurred after the balloon expanded?

_____

(c) Do the results of your experiment agree or disagree with the Doppler effect? Explain.

_____

_____

(d) If infrared radiation that was present at the formation of the Universe was affected in this way—that is, by an expanding and cooling Universe—would it experience a blue shift or red shift?

_____

(e) Into what wavelength on the EM spectrum would the radiation move first?

_____

(f) Do you think the discovery of the CMB radiation supports the idea of an expanding Universe—that is, one that was once smaller and contracted? Explain.

_____

_____

20.4 Dark Matter and Dark Energy

While the evidence of red shift and the discovery of the cosmic background radiation elevated the Big Bang hypothesis into a theory, there are some observations that are yet to be explained by the theory as it currently stands. For example, in an expanding Universe, all matter and energy are expected to be spread out equally. But in 2001, an exceptionally large supercluster of stars was discovered. What forces could account for this clumping together of stars? Astronomers have proposed the existence of undiscovered massive objects existing in space (*dark matter*) that could exert a large enough gravitational pull to have clumped these stars—and other matter—together.

Another reason astronomers have proposed the existence of dark matter is that much of the known, observable Universe is "hidden" to our usual methods of investigation. When we determine the total mass of objects known in space and study the motions of objects in systems of stars, there seems to be "missing" matter—that is, the mass we calculate does not add up to the total mass of the Universe. This missing matter has been referred to as dark matter because it does not emit any light that is part of the EM spectrum. While the nature of dark matter is unclear—what it is, how it formed, what role it played in the formation of the Universe, and so on—we can

---

EXERCISE 20.4   **Where's the Matter?**

Name: _____   Section: _____

Course: _____   Date: _____

This exercise demonstrates one way in which we can think of dark matter. Use the materials provided by your instructor. (*Note:* If this exercise is available and being conducted online, the setup, materials, and directions may differ. See the instructions in the online lab for how to conduct the exercise.)

First weigh a stack of loose paper plates. Then weigh a stack of the same number of paper plates that have been glued together.

(a) What is the mass of the loose paper plates? _____

(b) What is the mass of the glued paper plates? _____

(c) Compare the two masses. What do you think is causing the differences observed?

_____

(d) Take a loose paper plate and place it on top of the wooden dowel or a handle so that it is balanced. Gently spin the plate and note its movement: does it wobble, spin flat, tilt to one side, etc.? Describe the movement.

_____

_____

(e) Repeat (d) with the glued paper plates. Describe the movement as you spin the plates.

_____

_____

(f) Why do you think the motions of the paper plates in (d) and (e) differ?

_____

_____

surmise that it is present and predict where it may exist. The most difficult part of understanding dark matter is that we are dependent upon indirect measurements, as with all aspects of astronomy where we don't have direct measurements. We cannot physically pull a galaxy apart to see what hidden matter lies inside.

We have observed that the Universe is expanding, but questions remain: Why it is still expanding? Will the expansion ever cease? Will it reverse? If gravity is a force that results in objects moving toward one another, we would assume that at some point in time the expansion of the Universe would slow. However, in 1998, two teams of researchers led by Saul Perlmutter and Brian Schmidt published their findings on the study of distant exploded stars (supernovae), which showed that the expansion of the Universe is not slowing, but rather is accelerating. (The Nobel Prize for Physics was awarded to these team leaders and Adam Riess in 2011 for their discovery.) This raises yet another question: *Why* is the expansion accelerating? Currently, the explanation for the continued and increased rate of expansion is attributed to another force that counteracts the pull of gravity by repelling objects and pushing them away from each other. This force of energy is termed *dark energy* and exists everywhere throughout space—including the vacuum between objects.

20.5 Celestial Objects

Most of space—an estimated 80%–90%—is dark, cold, and empty of physical material that we can observe with our current technology. The sparsely distributed "stuff" that fills up the void is referred to as astronomical (or celestial) objects and bodies. In addition to the galaxies mentioned earlier, the celestial objects most likely familiar to you are:

- *Stars*—objects in which the fusion of atomic nuclei occur extensively, producing vast amounts of energy; our Sun is a star.
- *Planets*—large bodies that orbit stars and shine only by light reflected from the star.
- *Planetesimals*—tiny, solid pieces of rock and metal that can accumulate to form a planet.
- *Moons*—less-massive satellites orbiting a more massive object.
- *Asteroids*—small, irregularly shaped rocky bodies left over from planetary formation or produced by the collision of planetesimals.
- *Comets*—objects composed of ice and dust that orbit the Sun on highly elliptical paths.

However, these are not the only observable celestial objects; other notable objects include (but are not limited to):

- *Kuiper belt objects*—small, icy objects that orbit the Sun beyond the orbit of the planet Neptune; they are not affected by Neptune nor do they cross its orbital path.
- *Nebulae*—clouds of gas or dust in space.
- *Star clusters*—groups of stars that are gravitationally bound for some length of time and share a common origin; they can be open (each component easily seen through a telescope) or globular (several thousand to a million stars in a spherical system).
- *Dwarf planets*—bodies with characteristics similar to those of a planet, but that are smaller than a planet and share their orbits with other small bodies.
- *Black holes*—objects in space that are so dense that their escape velocity (the velocity required to escape the gravitational force of an object) exceeds the speed of light; a singularity in space.

Because much of our understanding of what the Universe is composed of and how it forms comes from the study of galaxies and stars, we will focus on these two objects for now. You will investigate the other objects within our Solar System in later chapters.

20.5.1 Stars and Galaxies

When you look up into the sky at night, especially if you are in a location with very little light pollution, the most likely objects that you will see—aside from the Moon and perhaps a planet or two—are stars and galaxies. Without the aid of a telescope or a star-gazing application on your mobile device, it is very difficult to discern whether you are looking at an individual star or a system of millions of stars forming a galaxy. If all the stars visible from the Earth were brighter or closer, the nighttime sky would be lit up like our daytime sky, which is of course illuminated by our Solar System's star, the Sun. The Sun is just one star out of hundreds of billions in our galaxy, the *Milky Way*. Dwarf galaxies contain about 10 million stars while a giant galaxy may consist of a thousand billion stars.

Estimates of the total number of galaxies that humans have detected in the observable Universe reached 200 million just a few years ago, but recent findings from the Hubble Space Telescope suggest that the actual number could be 10 times greater! In that patch of nighttime sky that you observe, just how many galaxies are there? We can use just one grain of rice as a guide. Although very small, if you held a grain of rice at arm's length up to the nighttime sky, the diameter of sky it covers would be several light-years, and it would likely block the light from at least 100 billion stars. You might be asking yourself, if there are billions of galaxies containing billions of stars, how did they form? That's a great question that astronomers are fairly confident they have figured out, but are continually investigating. We will try to answer that question by looking at the life cycle of stars.

20.5.2 Stellar Classification and Evolution

The light that we see from stars and that is reflected off of other celestial objects has provided countless amounts of information that has expanded our knowledge of the Universe. You have investigated radiation in terms of the formation of the Universe, and now we will look at how that radiation provides us with indirect measurements of the temperature, composition, and life cycle of stars.

You might be asking yourself how we can tell the temperature of a star simply from the light that it emits. An object that could absorb all the radiation it receives would be considered a **blackbody**. Though no existing objects are perfect blackbodies, objects in space, such as the Earth and the Sun, act much like blackbodies. The blackbody object heats up to a certain temperature, and then re-radiates the energy back to space as *blackbody radiation*, with a distinctive spectrum defined by the maximum wavelength it contains. To relate radiation to temperature, experimental physicists determined the character of the radiation emitted from blackbodies at various temperatures, as shown by physicist Max Planck's radiation curves in **FIGURE 20.4**. Planck recognized the fundamental relationship between wavelength and temperature shown by the curves in the figure, and from this relationship he developed three laws governing the behavior of blackbody radiation:

- A blackbody emits radiation at all wavelengths, but in varying amounts.
- Hotter objects give off more energy at every wavelength than cooler objects.
- As the temperature of an object increases, the wavelength of radiation emitted becomes shorter (that is, the apparent color of an object is an indication of the temperature of the object).

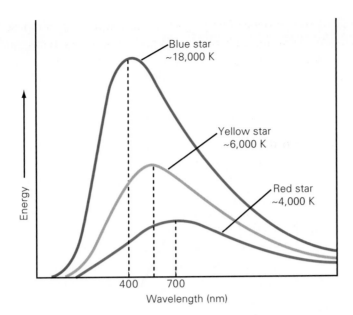

FIGURE 20.4 Plank's radiation curve for three temperatures.

Building on Plank's radiation laws, Wien's law (formulated by physicist Wilheim Wien) states that the radiation spectra emitted by objects of different temperatures peaks at different wavelengths. Hotter objects emit most of their radiation at shorter wavelengths and so appear bluer, while cooler objects emit more of their radiation at longer wavelengths and appear redder. Resulting from these findings is the following formula, known as Wien's law, that relates the temperature of a black body and the wavelength at which it emits most of its radiation:

$$T = \frac{3 \times 10^7}{\lambda}$$

where T = temperature (in Kelvin) and λ = wavelength (in angstroms). (An Angstrom is equal to one hundred-millionth of a centimeter, or 0.1 nm.)

For astronomers, this means that by attaching a light meter and specific filters to a telescope, the energy distribution of a star may be observed and plotted as a blackbody curve. Then, by using the peak wavelength of a blackbody (from the blackbody curve) and inserting it into Wien's law, the temperature of the star may be determined.

Soon, methods of classifying stars were developed based on temperature, such as the one in **TABLE 20.1**.

Table 20.1 **Spectral Classification of Stars**

Spectral class	Approximate surface temperature (Kelvin)*	Color	Example
O	$\geq$ 33,000	Electric blue	Mintaka (09)
B	10,000 – 30,000	Blue	Rigel (B8)
A	7,500 – 10,000	White	Vega (A0)
F	6,000 – 7,500	Yellow-white	Canopus (F0)
G	5,200 – 6,000	Yellow	Sun (G2)
K	3,700 – 5,200	Orange	Arcturus (K2)
M	$\leq$ 3,700	Red	Betelgeuse (M2)

* To convert a Kelvin temperature to degrees Celsius, subtract 273.15 degrees from the Kelvin temperature.

Name: _____   Section: _____
Course: _____   Date: _____

Use the figure on the right and Wien's law to record peak wavelength and temperature in the table for each star. Note that 1 nm = 10 Å. Then, use the table to determine the color of each star and record your answer.

Blackbody radiation curves for 5 stars.

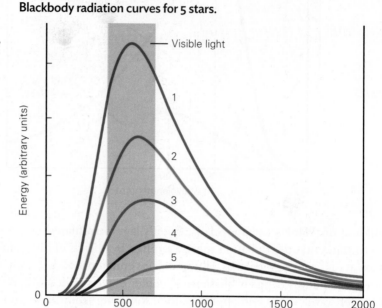

Blackbody data table

Star #	Peak wavelength (nm)	Peak wavelength (angstroms)	Temperature (Kelvin)	Color
1				
2				
3				
4				
5				

Once astronomers had data about the temperature of stars, their color, their masses and their **luminosity**—how bright a star appears compared to the Sun based on the amount of energy it emits—they began to investigate what relationship, if any, existed among these features. Two astronomers, Ejnar Hertzsprung and Henry Russell, independently examined the relationship between luminosity and surface temperature. They both found that there was indeed a relationship between these variables; once combined, their studies resulted in the development of the *Hertzsprung-Russell diagram* (**H-R diagram**) (**FIG. 20.5**). Later studies led to the addition of **absolute magnitude**, the brightness of a star if all stars were seen from the same distance. (Comparatively, **apparent magnitude** is how bright a star appears to us on the Earth, without adjusting for distance.) Notice on the diagram that absolute magnitude values can be positive (dim stars) or negative (bright stars). What you see in the diagram is that stars all lie within discrete regions of the plot: hot, dim

FIGURE 20.5 An H-R diagram shows the relationship between luminosity and temperature (represented by spectral class). Most stars, including the Sun, fall into the main sequence.

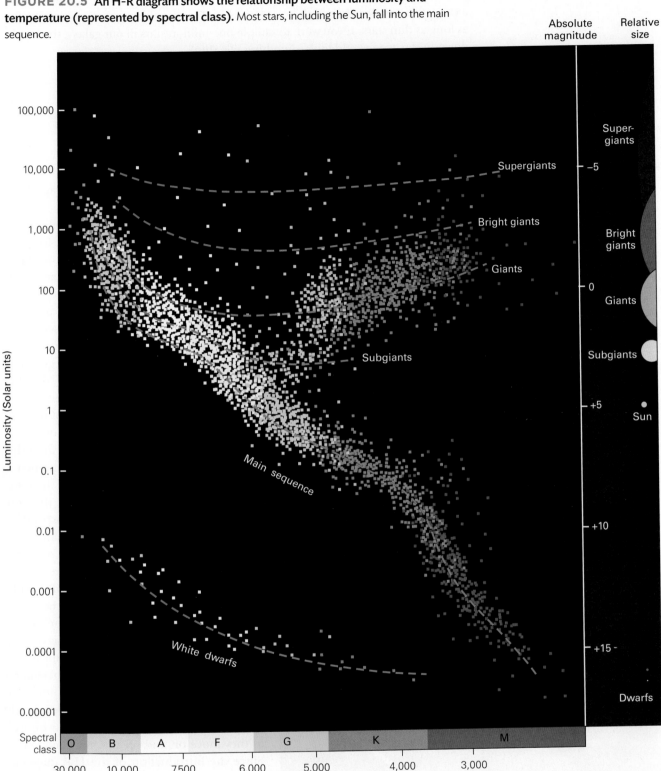

stars called white dwarfs lie in the left corner, while cool, bright stars (giants and supergiants) plot in the top right. Most of the stars on the diagram lie along a band that stretches diagonally, forming the main sequence.

The distribution of stars in the diagram is also a result of the mass of the star. The more massive stars have more fuel and so shine brighter than low-mass stars. But just like large, gas-guzzling vehicles that burn through their fuel more quickly than a car with a smaller engine, the more massive, brighter-shining stars won't burn for as long as dim stars. If you were to sample one million stars in our galaxy, the breakdown in star type would look something like this:

- 900,000 main sequence stars
- 96,000 white dwarfs
- 4,000 giants
- 1 supergiant (maybe!)

Note that not all types of stars are plotted on this diagram. There are many varieties of stars that have been subdivided from these main star classifications because a star follows an evolutionary path, and will move from one region of the diagram to another in its lifetime.

Currently, scientists think that the formation and evolution of stars follows the following sequence: First a *nebula* collapses under the force of gravity, forming "clumps" of clouds that begin to heat up. Once the gravitational crush on these clumps is sufficient to force the nuclei of hydrogen atoms together, **nuclear fusion** occurs, which creates helium atoms and releases energy up to infrared wavelengths. Once fusion in the core of the young star produces enough outward gas pressure to counteract the inward force of gravity, the star becomes a stable main sequence star, emitting visible light. It is estimated that most stars spend approximately 90% of their lives in the main sequence, before they leave to become dwarfs or giants.

The timing of when a star leaves the main sequence, and how the star "dies," is dependent on its initial mass (**FIG. 20.6**). When massive stars (those several times more massive than an average star) run out of usable hydrogen, the core of the star collapses under the force of gravity and the outer layers of gas expand and cool, forming a *supergiant*. The core of the supergiant, as it collapses, heats up until the chemical reactions inside become so unstable that it explodes in what is termed a *supernova*. The gas cloud is dispersed into space, forming a nebula, and the core either becomes a *black hole* or a *neutron star*. For average stars, like our Sun, they too will undergo expansion and cooling, but there is not sufficient mass to expand to the size of a supergiant; instead, they transform into giants. The core of a giant eventually collapses, forming a white dwarf, and the gaseous outer layers of the star drift away to form a nebula. Eventually, the nebula dissipates into space and becomes sufficiently transparent that the white dwarf becomes visible. Low-mass stars burn their fuel much more slowly than high- or medium-mass stars, and at lower temperatures. They appear red throughout their lives until all that remains of the core is a white dwarf star.

Using the H-R diagram, we are able to track the life cycle of stars and predict their futures as they move on and off the main sequence. Since the introduction of the H-R diagram, astronomers have combined the information provided by blackbody curves and mathematical equations to determine the distance to stars and estimate their ages. But that topic is beyond the scope of this course.

FIGURE 20.6 Life cycles of stars with different masses.

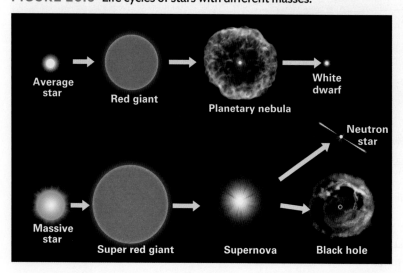

Name: _____ Section: _____
Course: _____ Date: _____

The H-R diagram.

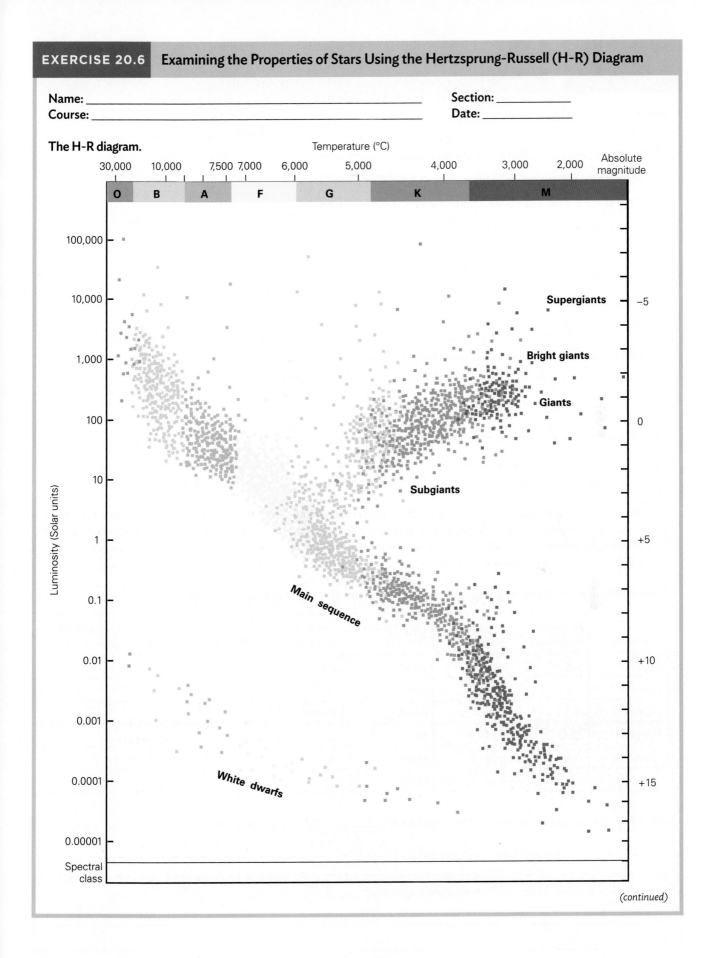

(continued)

Name: _____ Section: _____

Course: _____ Date: _____

(a) Plot the location of the Sun on the diagram on the previous page:

Absolute magnitude = 4.8   Temperature ≅ 5,800 K

What spectral class is our Sun? _____

(b) How does the luminosity of red giant stars compare with the luminosity of white dwarf stars?

_____

(c) What is the spectral class for the most massive stars along the main sequence? _____

Explain your reasoning.

_____

(d) Which stars have high luminosities and low temperatures? _____

(e) On which region of the diagram would an M-class star plot if it was 100 times brighter than the Sun? _____

(f) Which spectral class will leave the main sequence first? Explain.

_____

_____

(g) Which spectral class will leave the main sequence second? _____ Third? _____ Last? _____

The table below contains data for six of the brightest stars that can be seen from the northern hemisphere. Plot the position of each of these stars on the H-R diagram on the previous page using a different colored pen or pencil for each star. You will fill in the last two columns in the table below based on the stars' temperatures and positions on the diagram—check both to make sure you plotted the stars correctly!

Stellar data for six of the brightest stars observed from the northern hemisphere.

Star name	Absolute magnitude	Apparent magnitude	Temperature (Kelvin)	Distance (light-years)	Spectral class	Type of star (white dwarf, giant, etc.)
Sirius A	1.45	−1.46	9,900	8.6		
Canopus	−2.5	−0.72	2,800	74		
Rigel	−8.1	0.12	11,000	~1400		
Procyon	2.6	0.38	6,600	11.4		
Aldebaran	−0.3	0.85	4,100	60		
Vega	0.6	0.03	9,900	25		

(h) Which star is brightest as observed from the Earth? _____. Why do you think this is so?

_____

(i) Compare Sirius A and Vega. Although their temperatures are the same, why is Sirius A apparently brighter than Vega, even though Sirius A has a positive absolute magnitude?

_____

20.5.3 Galaxies and Their Classification Schemes

In the 18th century, Sir William Herschel referred to "fuzzy" objects that he observed through his telescope as *nebulae* because he could not discern any individual objects within. Based on his observations, he made an attempt to classify these nebulae by their brightness, form, and size. By the 19th century, his son Sir John Herschel extended his father's classification system by distinguishing between objects that were *galactic nebulae* (today called galaxies) whose appearance varied from *non-galactic nebulae* (the nebula of today's terminology). By the 1920s, advances in optical telescopes and the use of **spectroscopy** improved our ability to develop classification schemes. Spectroscopy is the process whereby light (EM radiation) is split into distinct wavelengths, such as passing white light through a prism, which produces a signature spectrum for each star or galaxy, like the absorption spectra you analyzed in Exercise 20.2.

In the 1920s, Edwin Hubble recognized that individual stars could be identified within *galactic nebulae* and that they were actually composed of groups of stars that existed outside our own galaxy. By 1925, Hubble had developed his own galaxy classification system (**FIG. 20.7**), in which he distinguished between three types of galaxies—elliptical, spiral, irregular—based on their shape, the age of their stars, and the amount of interstellar matter they contain. **Elliptical** galaxies are circular to lens-shaped, and contain primarily old stars and little interstellar matter. **Spiral** galaxies have a flat, rotating disc of spiraling arms that contain abundant young stars, gas, and dust surrounding a central bulge. A special kind of spiral galaxy where there are only two arms that radiate from opposite "ends" of the elongated center is termed a **barred spiral**. **Irregular** galaxies are those that have no regular shape and do not fall into one of the other galaxy categories.

FIGURE 20.7 Hubble's classification of galaxies.

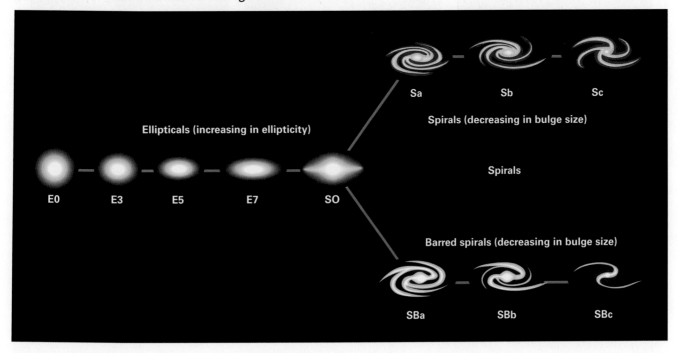

Name: _____   Section: _____
Course: _____   Date: _____

Shown below are six galaxies viewed from the Hubble Space Telescope. Note that the much smaller objects in many of these views are primarily other galaxies.

(a) First, group the galaxies into four categories that you determine (ignore Hubble's categories for now) based on appearance (brightness, form, size, etc.) and fill in Table A.

Table A: Student classification system of galaxies.

Category	Galaxy number(s) from the figure	Characteristics used to define the category
I		
II		
III		
IV		

(continued)

EXERCISE 20.7 What's in a Name? Classifying Galaxies (*continued*)

Name: ______________________ Section: ____________

Course: ______________________ Date: ____________

(b) Now group the same galaxies using Hubble's classification scheme by placing each in their proper location in Table B.

Table B: Classification data using Hubble's system.

Hubble's categories	Galaxy number(s) from the figure	Characteristics
Elliptical		
Spiral		
Spiral barred		
Irregular		

(c) Compare your first classification (appearance) to the second (Hubble). Which groupings aligned between the two tables? Which were different?

____

____

____

____

____

There are challenges we face in trying to classify galaxies. For one thing, recall that we are viewing other galaxies from the perspective of our own. The Hubble Space Telescope (HST) captures images as it orbits the Earth, and sometimes the Earth is in the way, so that the HST cannot be aimed where astronomers may like at any given time. Another challenge is illustrated by the figure below, which shows two spiral galaxies 500,000 light-years apart viewed at different angles.

(d) What challenge to properly classifying both galaxies emerges with this figure?

____

____

____

____

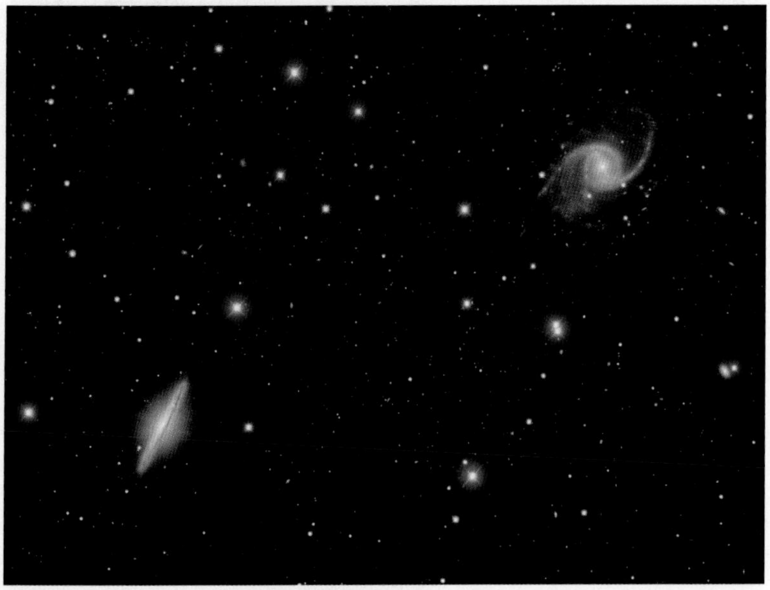

(e) What perspective or angle for capturing images of other galaxies do you think is most useful for classifying them? Why?

____

____

____

____

Two spiral galaxies 140 million light-years from the Earth. NGC 5905 (left) and NGC 5908 (right).

20.5.4 Star Clusters

Stars that appear to have a common origin, but are not clumped together into a galaxy, instead form **star clusters**. Unlike galaxies, the gravitational force holding star clusters together is weaker, so much so that it may only last for a relatively limited time (whereas a galaxy will hold together until or unless another force rips it apart). Astronomers divide star clusters into two distinct types: *globular* and *open*. Globular clusters (**FIG. 20.8A**) are roughly spherical in shape and consist of a few thousand to more than a million stars; very few of their stars are considered young. These clusters commonly are found near the *halo* of galaxies—the glowing area near the central bulge. Located in the disk and spiraling arms of galaxies, **open clusters** contain only hundreds to perhaps thousands of young stars that are loosely arranged. The Pleiades—also called the Seven Sisters or Subaru—is a familiar example (**FIG. 20.8B**) that can be seen from nearly anywhere on Earth.

FIGURE 20.8 **Examples of star-cluster features.**

(a) Globular cluster known as M13.

(b) Open star cluster M45, known as the Pleiades.

Identifying the Locations within the Milky Way Galaxy and Properties of Cluster Types

Name: _____ Section: _____

Course: _____ Date: _____

A series of Hubble Space Telescope images of star clusters: NGC 299 (top left); Messier 2/NGC 7089 (top right); Messier 28 (bottom left); Westerlund 1 (bottom center); NGC 362 (bottom right).

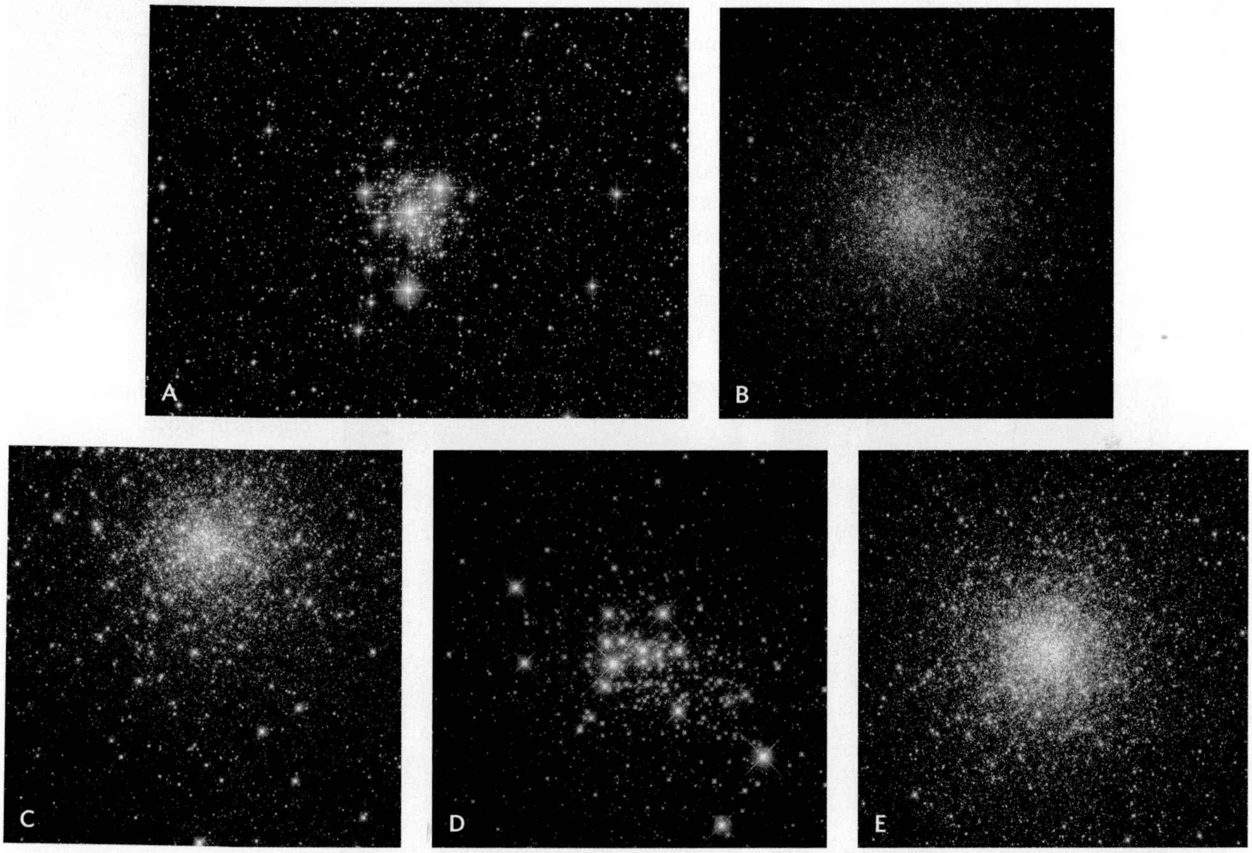

(a) Based on what you have learned about the luminosity, apparent brightness, temperature, and distance of stars, fill in the table below with information about the star clusters pictured above. In the last column, provide the justification for the classification of each star cluster.

Data table for star cluster classification

Cluster ID letter	Type of cluster	Evidence for the classification
A		
B		
C		
D		
E		

(b) Suggest a possible reason why globular clusters cannot be easily seen with your eyes.

_____

(continued)

Name: _____   Section: _____

Course: _____   Date: _____

? What Do You Think The Hubble Space Telescope (HST) was launched into Earth orbit on April 24, 1990, aboard the space shuttle Discovery. Since then, it has captured more than 1.4 million images, from our own "backyard" of the Solar System to the oldest regions of the observable Universe billions of light-years away. The HST primarily captures images in the visible part of the EM spectrum, but can also detect light in the near-infrared range. A limitation of the HST's detectable wavelength range is that by primarily detecting visible light, much of the matter that lies behind large objects like galaxies and nebula (where new stars are forming) is obscured by the particles of dust that absorb light. The near-infrared cameras aboard the HST help to "see through" these dust clouds only somewhat.

The left image shows objects in space obscured by dust particles. The right image shows the clearer view obtained by the Hubble Space Telescope's near-infrared cameras.

The HST is nearing the end of its time in space—it has lasted years beyond its initial estimates. It will no longer receive upgrades; instead, new telescopes will serve as successors. One such telescope is the James Webb Space Telescope (JWST), with a launch date of 2021. The table below is a comparison of the HST and the JWST.

Comparisons of the features of the Hubble Space Telescope and the James Webb Space Telescope.

	HST	JWST
Launch & servicing	NASA Space Shuttle (for both)	European Space Agency (ESA) rocket, no service by shuttle
Orbit	Earth, altitude 570 km (~354 mi)	Sun at the Earth–Sun L2 Lagrange point (1.5 million km/~1 million mi from Earth)
Wavelengths	UV to near-infrared (0.1 to 2.5 microns, or 100 to 2,500 nm)	Visible to infrared (0.6 to 28 microns, or 600 to 28,000 nm)
Mirror/collection size	2.4 m	6.5 m (15× the area covered by the HST)
"Age" of galaxies detected (as a substitute for distance)	Toddler	Babies

(continued)

Name: _____   Section: _____

Course: _____   Date: _____

The figures below illustrate the orbital locations for both telescopes.

The location of the James Webb Space Telescope in comparison with the Hubble Space Telescope (not to scale).

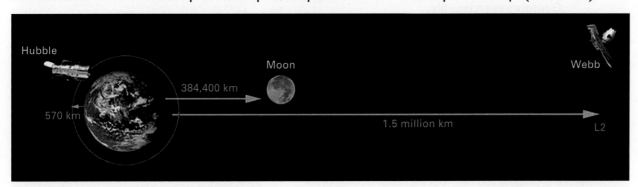

(c) What are the advantages of having more sensitive filters to capture light from near- and far-infrared radiation rather than primarily visible light and near-infrared only?

_____

_____

_____

_____

_____

_____

_____

Lagrange points for the James Webb Space Telescope.

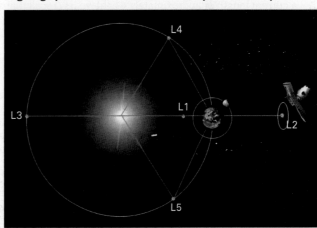

(d) What do you think are some advantages and disadvantages of the JWST's orbital location?

_____

_____

_____

_____

(e) Why might being able to collect images from a greater distance be an advantage in astronomers' study of the formation of the Universe?

_____

_____

_____

Name: _____   Section: _____

Course: _____   Date: _____

Exploring Earth Science Using Google Earth

1. Visit **digital.wwnorton.com/labmanualearthsci**
2. Go to the **Geotours** tile to download Google Earth Pro and the accompanying Geotours exercises file.

Select Google Sky by clicking 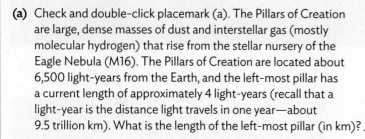 and then selecting Sky. Turn on Sky Database > Imagery in the Layers panel. Expand the Geotour20 folder in Google Earth by clicking the triangle to the left of the folder icon. The folder contains placemarks keyed to questions that explore various celestial features in our Universe. *Please click* and then *select Earth to return to Google Earth after you finish this Geotour.*

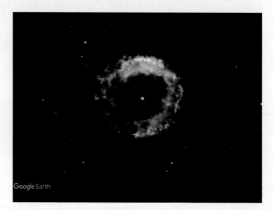

(a) Check and double-click placemark (a). The Pillars of Creation are large, dense masses of dust and interstellar gas (mostly molecular hydrogen) that rise from the stellar nursery of the Eagle Nebula (M16). The Pillars of Creation are located about 6,500 light-years from the Earth, and the left-most pillar has a current length of approximately 4 light-years (recall that a light-year is the distance light travels in one year—about 9.5 trillion km). What is the length of the left-most pillar (in km)? _____

(b) Check and double-click placemark (b) in turn. The Crab Nebula represents a violent supernova explosion of a high-mass star (greater than 8 times the mass of our Sun). In contrast, intermediate- to low-mass stars (0.8 times to 8 times the mass of our Sun) will form smaller **planetary nebulas** like the Little Ghost Nebula. Which of these two features will our star (the Sun) form? _____

(c) Check and double-click placemark (c). This image is of a spiral galaxy (M51) that resembles what our Milky Way might look like if viewed from outside the galaxy. Note how the curved spiral arms develop around the more quickly rotating central cluster of stars. Looking at the spiral arms from this viewpoint, in what direction is this galaxy rotating: clockwise or counter-clockwise? _____

Properties of the
Solar System

21

Saturn and its moon Titan, as captured by NASA's *Cassini* spacecraft.

21.1 Introduction

A sudden explosion of a star sends out a shockwave of energy that collides with existing nebulae and produces a new **nebula**—a swirling cloud of gas or dust in space. The force of gravity causes the nebula to contract and spin, and pulls the particles of hot gas and dust together. A prominent mass eventually forms at the nebula's center—a *protostar*—with an encircling disc of rocky and icy particles. Hydrogen fusion in the core of the protostar becomes stable, emitting light in all forms, producing a star. A similar process acts on the materials in the disc around the star—gravity causes them to clump together, growing from specks to blocks to planetesimals to protoplanets and, for some of these protoplanets, to planets. Repeated collisions and condensation of material in the orbiting disc forms moons and other rocky and icy bodies. A new planetary system is born.

This is one possible story for the formation of our Solar System and its contents, an explanation known as the **condensation theory** (**FIG. 21.1**). In this chapter, we'll examine the properties of those objects, including our Sun, the planets, and the "other stuff" of the Solar System, such as dwarf planets, moons, asteroids, and comets. First, though, we need to begin by exploring the laws that govern the motions of these objects within the Solar System.

FIGURE 21.1 **The processes in the condensation theory.**

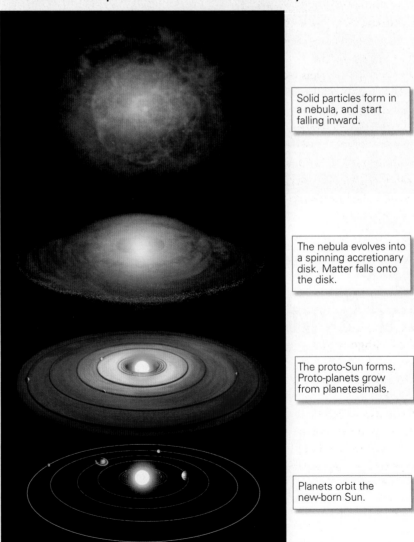

Solid particles form in a nebula, and start falling inward.

The nebula evolves into a spinning accretionary disk. Matter falls onto the disk.

The proto-Sun forms. Proto-planets grow from planetesimals.

Planets orbit the new-born Sun.

21.2 Motions in the Solar System

Our understanding of the celestial objects that exist in our Solar System and how they move has radically evolved since the observations of early scholars. The ancient Greeks proposed that the Earth was at the center of the Solar System, an idea that persisted into the 1500s C.E. Even after the model developed by Copernicus in 1543 placed the Sun in its proper location at the center of the system, predictions of planetary positions were still inaccurate. Danish mathematician Tycho Brahe, for example, still maintained that the Earth could be in the center while other objects orbited the Sun. Although Brahe spent years collecting detailed observations of the changing positions of the planets, he failed to rectify his model to match the data he collected. It was only after his death that his young apprentice Johannes Kepler—a brilliant mathematician—was able to mathematically work out the motions of the planets, for which he published three mathematical laws. Not only did Kepler prove with his mathematics that the planets orbit the Sun rather than the Earth, but he maintained that they did so in paths that were not perfect circles. These were ideas that were not welcomed by the political and religious powers of the time, but laid the foundation for the works of Galileo Galilei and Isaac Newton, and those who followed.

Kepler's three laws of planetary motion can be summarized as follows:

- *First law*: the orbit of each planet is an ellipse, with the Sun as one focus (**FIG. 21.2A**).
- *Second law*: As a planet moves along its orbit it sweeps out equal areas during equal intervals of time, so that a planet's speed along its orbit changes over a year: faster when it is closer to the Sun (at *perihelion*) and slower when it is farthest from the Sun (at *aphelion*) (**FIG. 21.2B**).
- *Third law*: The orbital period of a planet is longer for planets that are farther from the Sun (**FIG. 21.2C**), and a planet's orbital period can be calculated if we know its distance from the Sun.

You will examine each of these laws in Exercise 21.1.

FIGURE 21.2 Kepler's laws of planetary motion. (The eccentricity of the ellipses shown is exaggerated.)

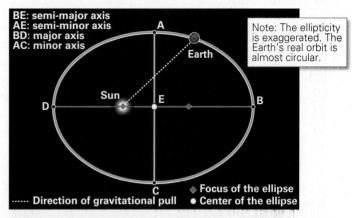

BE: semi-major axis
AE: semi-minor axis
BD: major axis
AC: minor axis

Note: The ellipticity is exaggerated. The Earth's real orbit is almost circular.

····· Direction of gravitational pull
◆ Focus of the ellipse
● Center of the ellipse

(a) Kepler's first law: The orbit of every planet is an ellipse, with the Sun at one of the ellipse's two foci.

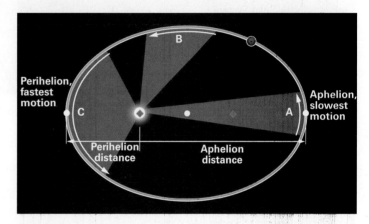

Perihelion, fastest motion

Aphelion, slowest motion

Perihelion distance

Aphelion distance

(b) Kepler's second law: As a planet orbits the Sun, an imaginary line joining that planet and the Sun sweeps out equal areas (labeled A, B, and C) during equal intervals of time. As a result, planets move fastest at perihelion, when they are close to the Sun, and slowest at aphelion, when they are farthest away.

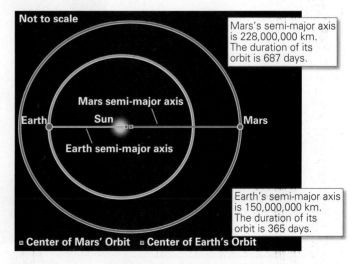

Not to scale

Mars's semi-major axis is 228,000,000 km. The duration of its orbit is 687 days.

Mars semi-major axis
Earth semi-major axis
Earth   Sun   Mars

Earth's semi-major axis is 150,000,000 km. The duration of its orbit is 365 days.

■ Center of Mars' Orbit   ■ Center of Earth's Orbit

(c) Kepler's third law: The larger the semi-major axis of a planet's orbit (a measure of the planet's distance from the Sun), the longer the planet's orbital period, the time it takes the planet to orbit the Sun.

Name: _____   Section: _____

Course: _____   Date: _____

Part 1: Kepler's first law

Eccentricity in astronomy refers to the degree of deviation an orbit exhibits from perfect circularity. A circle has an eccentricity of zero, and the greater the deviation an orbit exhibits, the closer the number will be to 1.

(a) Based on the orbits shown in the figure below, which planet's orbit has each eccentricity?

    0.007 _____   0.056 _____   0.206 _____

Schematic diagram of the orbits of planets Mercury, Venus, and Saturn.

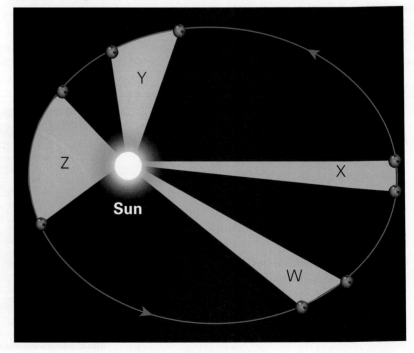

Mercury   Venus   Saturn

Part 2: Kepler's second law

The figure below shows the orbital path of a fictitious planet around the Sun.

(b) Fill in the blank: The Sun is one _____ about which the planet orbits.

(c) Did the *time* it took for the planet to move along each of the shaded segments of the orbit (W, X, Y, and Z) vary or was it the same? Explain your answer.

_____

_____

_____

_____

_____

_____

_____

_____

_____

_____

Schematic orbit of a planet about the Sun with points W, X, Y, and Z representing equal areas of the ellipse covered during orbit.

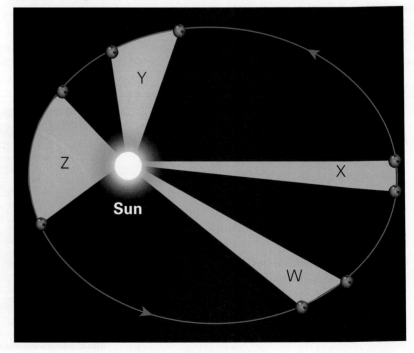

(continued)

Name: _____ Section: _____
Course: _____ Date: _____

(d) Did the *distance* the planet traveled along each of the segments of the orbit (W to Z) vary or was it the same? Explain, and if it varied, rank the segments from least to greatest distance traveled.

_____

_____

_____

_____

Ranking (least) _____ _____ _____ _____ (greatest)

(e) Did the *speed* (velocity) the planet traveled along each of the segments of the orbit (W to Z) vary or was it the same? Explain, and if it varied, rank the segments from least to greatest speed of travel.

_____

_____

_____

_____

Ranking (least) _____ _____ _____ _____ (greatest)

Part 3: Kepler's third law

Kepler's third law is a mathematical calculation, which we can simplify as $P^2 = a^3$. If you know the orbital period of a planet (P = the time it takes to make one revolution about the Sun), then you can determine the distance the planet is from the Sun (a = the semi-major axis of the planet's orbit) (see Figure 21.2c). Likewise, if you know a planet's distance from the Sun, you can calculate the time it takes to make one revolution.

(f) Using the semi-major axis data for the planets in our Solar System from the table below, determine their orbital periods. In this exercise, the orbital period will be expressed in Earth years.

Semi-major axis distances for the Solar System planets.

Planet	Mercury	Venus	Earth	Mars	Jupiter	Saturn	Uranus	Neptune
Semi-major axis (AU)	0.3871	0.7233	1.000	1.5273	5.2028	9.5388	19.1914	30.0611
Orbital period (yr)								

(g) How many times greater is the orbital period of Jupiter compared to the Earth? _____

(h) How many times greater is the orbital period of Neptune compared to Jupiter? _____

In our search of exoplanets—planets outside our solar system—we can detect the presence of a planet orbiting its host star by the dimming of the star's light as the planet passes in front (using the Kepler Space Telescope, for example). Because we cannot see the planet itself, we don't know its distance from its star.

(i) Using Kepler's third law of planetary motion, determine the distance (in AU) an exoplanet is located from its star if we know its orbital period is 4.599 years. _____

21.3 The Sun: The Center of the Solar System

Our current age of scientific discovery has produced vast amounts of new information about the Sun, aided by missions that allow us to observe it more closely from the ground, via satellites orbiting the Earth and via probes orbiting the Sun. Through these investigations, we've learned much about the Sun's features:

- *Layers*—The Sun consists of three layers: the core, the radiative zone, and the convective zone. It's surrounded by a solar atmosphere that also has three layers: the photosphere, the chromosphere, and the corona (**FIG. 21.3**).
- *Temperature*—The various layers of the Sun are not constant in temperature. The core, where fusion occurs, can reach temperatures of 15 million degrees C, but it can also drop to 3.5 million degrees C. Moving outward from the core, the temperature does not decrease continuously to the corona. Temperatures in the photosphere and chromosphere are estimated to be a mere 5,500°C and 4,500°C, respectively. The corona, meanwhile, can be several hundreds of times hotter than the layers below, ranging in temperature from 1 million to 10 million degrees C.
- *Sunspots*—These are areas of the Sun's surface that appear darker because of the cooler temperature (3,600°C/6,500°F) in that location (**FIG. 21.4**).
- *Magnetically active areas*—Areas of the Sun where magnetic fields are stronger than surrounding areas (**FIG. 21.5**) can produce solar eruptions like flares, prominences, and coronal mass ejections (CMEs).
  - *Solar flares and prominences*—These are sudden (minutes to tens of minutes long), local explosions of energy near a sunspot in the corona, resulting from the twisting of magnetic fields generated by the Sun's electrically charged gases. These link the interior layers to the corona.

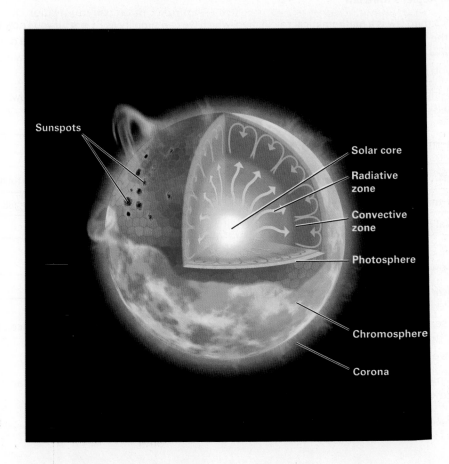

FIGURE 21.3 A model of the Sun's layers and prominent features.

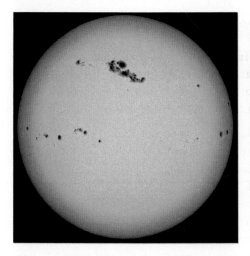

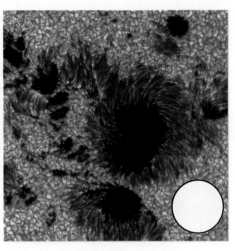

FIGURE 21.4 Sunspots appear as dark patches on the Sun's surface.

(a) At any given time, sunspots cover 0%–0.4% of the area of the Sun.

(b) A close-up shows that a sunspot has a cooler, darker central area and a warmer outer area. The circle represents the Earth, for scale.

- *Coronal mass ejections (CME)*—These are large eruptions of individual clouds of gas and energetic matter into space from the outermost regions of the corona, which can travel at speeds up to 2,000 km/s (~1,240 mi/s) (**FIG. 21.6**). Fortunately, only those emitted from the Sun's east limb can impact the Earth directly.
- *Solar wind*—This consists of a stream of particles that have enough energy to escape the Sun's gravity and flow outward into space.

Our interest regarding the Sun is not simply one of scientific curiosity. Solar activity affects us on Earth, from impacting our climate to producing magnetic storms that impair the functioning of our satellites, aviation equipment, radio communication, and electric power grids.

While the Sun is the focus about which objects in our Solar System orbit, the Sun too has motion. Like the planets, the Sun has a north and south pole and rotates on

FIGURE 21.5 Magnetically active regions of the Sun can be represented by a bar magnet, where electrically charged gases create magnetic lines that flow between oppositely charged areas, resulting in eruptions at the surface.

FIGURE 21.6 **Coronal mass ejection (CME) images taken by the SOHO's LASCO C2 and C3 instruments (SOHO = NASA/ESA's Solar and Heliospheric Observatory).** The white circles in the center of the images are the location of the Sun. The dark disc is a blocking filter on the instruments so that the material ejected from the Sun can be viewed. Both images are of the same event viewed with two different light filters.

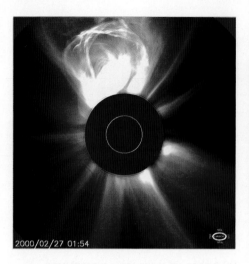
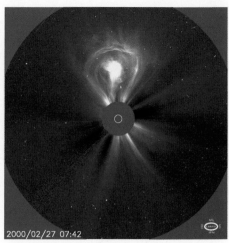

this axis. In the 1850s, astronomer Richard Carrington calculated the Sun's rotation period by observing the motion of low-latitude sunspots across the Sun's surface. From his observations, he determined that the Sun rotates on its axis once every 27.28 Earth days. Rotation rates based on observable features on the Sun's outer layers are referred to today as Carrington rotations.

However, determining the Sun's rotation using Carrington rotations is a limited method because features may not last long; in fact, sunspots appear to come and go in cycles, called sunspot or solar cycles, as shown in **FIGURE 21.7**. Unlike the rotation of the planets in our system, the Sun exhibits what is called *differential rotation*, meaning that its rotation period varies by latitude: at the poles its rotation period is 35 Earth days, while at the equator it takes 25 days to complete one rotation. Why this differential rotation occurs for the Sun is still a debate among astronomers today.

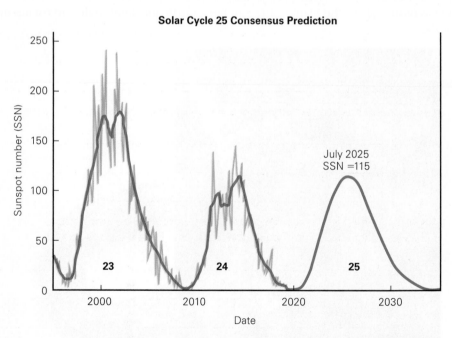

FIGURE 21.7 **The number of sunspots varies with time in a cycle of approximately 11 years.** Solar cycles are numbered, as shown. The solid red line is a best-fit curve of the actual data (in orange). The blue curve is the prediction for cycle 25.

EXERCISE 21.2 Comparing Solar Cycles and Coronal Mass Ejections

Name: ______________________ **Section:** __________
Course: ______________________ **Date:** __________

Use the figure below to complete this exercise.

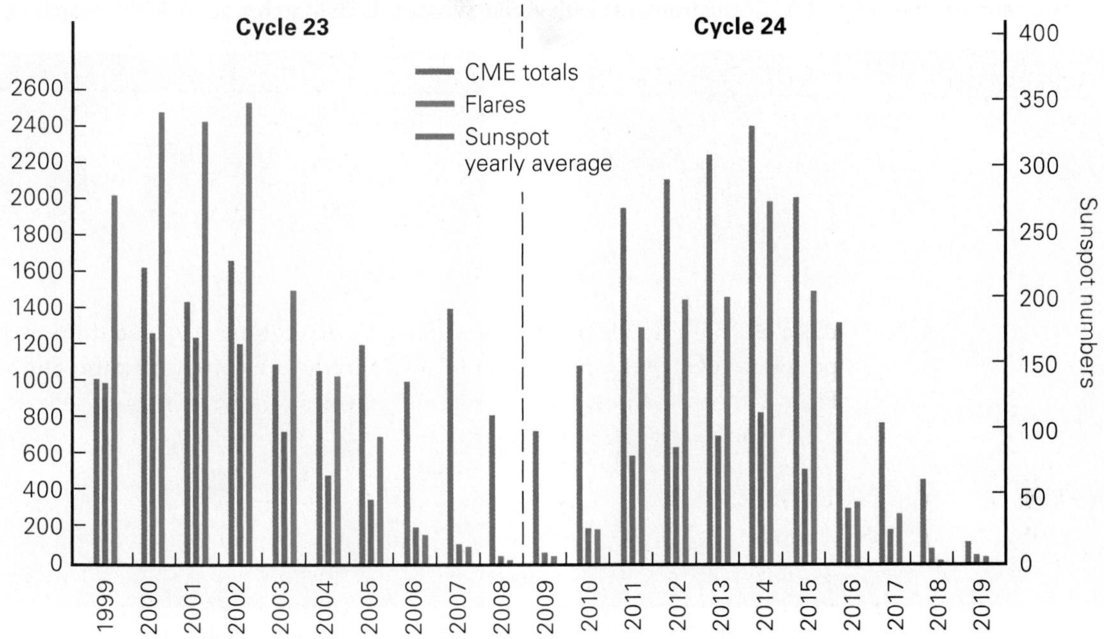

Average yearly sunspot activity (sunspot numbers), solar flares (all types), and CME from 1999 to 2019. This time range covers most of solar (sunspot) cycles 23 and 24, with cycle 24 beginning at the end of 2008 as indicated by the dashed line.

(a) How well does the CME activity follow the solar cycle? Explain.

____

____

____

(b) Do the maximums and the minimums of the three phenomena happen at about the same time? Explain your answer.

____

____

____

(c) Based only on this example, do you think that sunspots give scientists clues as to how and when the next CME may happen so that society can have advance warning? Explain why or why not.

____

____

____

____

Name: _____   Section: _____
Course: _____   Date: _____

Use the figures below and the transparency located at the end of the manual to complete this exercise.

Images of the Sun from the SOHO LASCO instruments with visible sunspot. Left: March 7, 2020; Right: March 14, 2020.

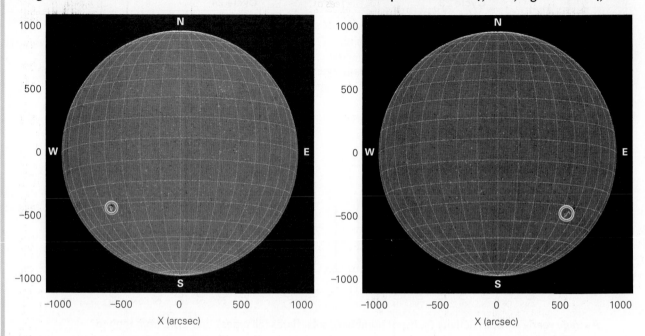

Estimate the Sun's rotation using the motion of the sunspot observed over one week (7 Earth days).

(a) Using a metric ruler, measure the distance (in cm) from the western edge of the Sun to the sunspot in the figure on the left above. _____

(b) Measure the distance (in cm) from the western edge of the Sun to the sunspot in the figure on the right above. _____

(c) Subtract your answer to question (a) from your answer to question (b). _____

This is the distance the sunspot moved in one week.

(d) Using a metric ruler, measure the distance (in cm) from the western limb of the Sun to the eastern limb and double your answer (we need to include the other "side" of the Sun) _____. This is the distance around the Sun in one rotation (the Sun's circumference).

(e) What fraction of the total distance did the sunspot travel in one week? _____

(f) How many days would it take the sunspot to go all the way around (assuming it would last that long)? This is the Sun's rotation period. _____

You can use this formula to assist you:

$$\text{Sun's rotation period} = \text{sunspot travel time} \times \frac{\text{Sun's circumference}}{\text{distance traveled by sunspot}}$$

You can now calculate the Sun's rotation at the sunspot's latitude using the Sun's angular velocity and the template printed on the transparency.

(g) Lay the transparency over the images above and determine the latitude at which the sunspot traveled: _____

(continued)

Name: _____   Section: _____

Course: _____   Date: _____

(h) Calculate the Sun's angular velocity = _____ degrees per day

$$Sun's\ angular\ velocity = \frac{degrees\ of\ longitude\ the\ sunspot\ traveled}{sunspot\ travel\ time}$$

(i) Calculate the Sun's rotation period. _____

$$Sun's\ rotation\ rate = \frac{360\ degrees}{Sun's\ angular\ velocity}$$

(j) How do your two methods of determining the Sun's rotation period compare (the estimate versus the angular velocity calculation)?

_____

_____

_____

(k) Do your answers fit within the range of known solar rotation periods from the poles to the equator? If not, why not?

_____

_____

_____

21.4 Properties of the Planets

Today astronomers recognize many more objects in the Solar System than were known in ancient times, and we are still discovering new information about celestial bodies that have been long known, including the planets. As we explore the Solar System, we encounter observations that no longer fit with what had been well-accepted definitions, which can force us to rethink what we name celestial bodies and come up with new definitions. Such was the case in 2006 when, after long (and still ongoing) debate, the International Astronomical Union (IAU) developed a new category of space object: the dwarf planet. (We'll examine dwarf planets in more detail later in this chapter.) According to the IAU, to be classified as a **planet**, an object must orbit the Sun, have enough mass that it forms a nearly round shape, and not share its orbital plane with any other object. Because of this revised definition, former planet Pluto, which hasn't cleared its orbit of other objects, was reclassified as a dwarf planet, leaving eight planets in our Solar System. These are split into two groups: the **terrestrial planets** of the inner Solar System (Mercury, Venus, Earth, and Mars), which are closer to the Sun and have solid surfaces; and the **Jovian planets** of the outer Solar System (Jupiter, Saturn, Uranus, and Neptune), which are farther from the Sun and are primarily composed of gases. Exercises 21.4, 21.5, 21.6, and 21.7 explore the properties of the planets and Pluto.

Name: _____   Section: _____
Course: _____   Date: _____

Tape or staple 10 pages of letter-sized paper together lengthwise, so that you have a long sheet stretching about 280 cm in length. Make a scale model of the planet positions in the Solar System using a meter stick provided by your instructor. The scale of the Solar System is reduced such that the Sun is 0.02 cm (1/100 in) in diameter (roughly the size of the pointy end of a pin).

   Place the meter stick on the paper running lengthwise, and make a mark for each planet at the scaled distances in the table below.

Planetary distances from the Sun, scaled down approximately 6 trillion times.

Planet	Scaled distance in cm (from 0)	Planet	Scaled distance in cm (from 0)
Mercury	~1	Jupiter	~13
Venus	1.8	Saturn	23.8
Earth	2.5	Uranus	47.8
Mars	3.8	Neptune	75

Now do the same for three additional celestial bodies, using the distances provided in the table to the right:
   • The dwarf planet Pluto
   • The asteroid belt (which contains a ring of many irregularly shaped fragments of solid material, called asteroids, which we discuss later in the chapter)
   • The Kuiper Belt (which contains a ring of many icy objects that are remnants of the Solar System's formation)
For the asteroid belt and Kuiper Belt, mark both ends of the belts' ranges on the paper.

Other Solar System bodies' distances from the Sun, scaled down approximately 6 trillion times.

Other celestial bodies	Scaled distance in cm (from 0)
Pluto	~98.4
Asteroid belt	5 to 10
Kuiper Belt	73 to 248

(a) Do the distances between the planets increase or decrease with increasing distance from the Sun? Is there an exception to the trend?

_____

_____

_____

(b) What separates the terrestrial planets from the Jovian planets? _____

(c) How many times greater is Jupiter's distance from the Sun than the Earth's, on this scale? _____

(d) How does the ring of celestial objects orbiting the Sun in the outer Solar System compare with the ring of objects orbiting the Sun in the inner Solar System?

_____

_____

_____

Name: _____   Section: _____

Course: _____   Date: _____

Use the data in the table below to plot the distance from the Sun versus mass (Earth masses) of the eight planets and Pluto on the graph below. (Note that the distance for Pluto is an average due to its highly eccentric orbit). Use a different colored pencil or pen for each group (terrestrial and Jovian) and for Pluto and make a legend to accompany your color choices.

Distance, mass, diameter, and density for the major planets and Pluto.

Celestial object	Distance (AU)	Mass (Earth masses)	Diameter (km)	Density (k/m³)
Mercury	0.39	0.1	4,879	54.37
Venus	0.72	0.8	12,104	5,243
Earth	1.00	1.0	12,756	5,514
Mars	1.52	0.1	6,792	3,933
Jupiter	5.20	318.0	142,984	1,326
Saturn	9.54	95.0	120,536	687
Uranus	19.22	14.6	51,118	1,271
Neptune	30.06	17.2	49,528	1,638
Pluto	39.50	0.002	2,370	2,095

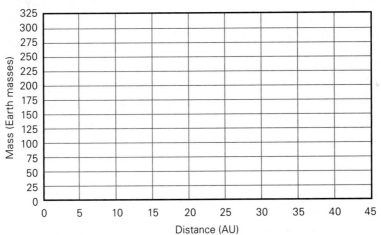

(a) Based on your graph, how do the terrestrial and Jovian planets differ in terms of size?

_____

_____

Does Pluto "fit in" with either of these groups? Explain.

_____

_____

_____

(continued)

EXERCISE 21.5 **Distance, Mass, and Density (*continued*)**

Name: ______________________________ **Section:** ______________
Course: ______________________________ **Date:** ______________

Plot the *distance* versus *density* of each planet and Pluto on the graph below. Use a different colored pencil or pen for each group (terrestrial and Jovian) and for Pluto and make a legend to accompany your color choices.

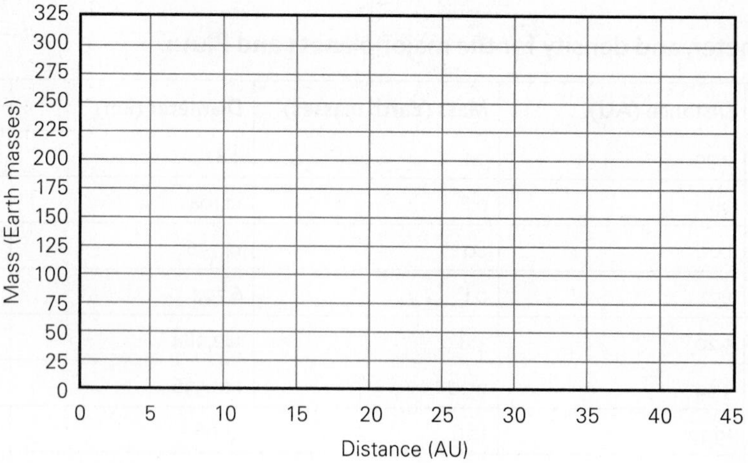

(b) Are the results what you expected? Explain.

____

____

____

____

(c) Why do you think the two graphs you have plotted differ?

____

____

____

____

(d) Does Pluto "fit in" with either of these groups? Explain.

____

____

____

____

Name: _____ **Section:** _____

Course: _____ **Date:** _____

The figure below shows two ways that we can study temperatures on the planets and Pluto: the mean surface temperature and the temperature relative to that on the Earth. The mean surface temperature takes into account the heating of the surface due to absorption by the body's atmosphere.

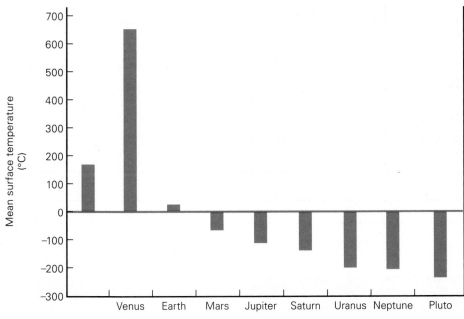

Temperatures of the Solar System planets and Pluto. Top: Mean surface temperature; bottom: The relative amount of radiation each planet receives compared with the Earth.

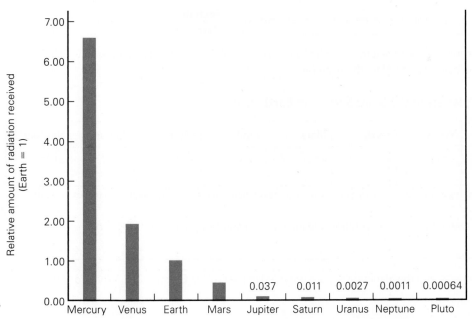

(a) What relationship is there, if any, between the mean surface temperature of the planets and Pluto and the amount of radiation they receive?

_____

_____

(continued)

Name: _____   Section: _____
Course: _____   Date: _____

(b) Look at the data for the planets Mercury and Venus. Approximately how much more radiation does Mercury receive than Venus? _____

Which planet is hotter, Mercury or Venus? _____

Give a possible reason why this is the case. (*Hint:* Recall your investigations of the factors that impact the Earth's heating from previous chapters.)

_____

(c) How does distance from the Sun affect temperatures of the planets and Pluto?

_____

(d) Suggest other factors that govern the temperature of a planet. Explain your answers.

_____

_____

EXERCISE 21.7   Planetary Spin

Name: _____   Section: _____
Course: _____   Date: _____

Other factors astronomers observe are the time it takes each planet to rotate on its axis and the direction in which they spin. Use the values in the table below to complete this exercise.

Rotational periods of the planets in the Solar System in Earth days.

Planet	Mercury	Venus	Earth	Mars	Jupiter	Saturn	Uranus	Neptune
Rotation period (Earth days)*	58.8	−244	1	27.4	0.415	0.445	−0.72	0.673

* The rotation period for each planet is given in Earth days. Negative values denote rotation is in the opposite direction to that of the Earth.

(a) Is there any correlation between the period of rotation and increasing distance from the Sun? Explain.

_____

(b) Compare the rotational periods with the planetary masses provided in the table in Exercise 21.5. What do you observe?

_____

(c) How many times does Jupiter spin on its axis each Earth day? _____

(d) How does the rotation of Venus compare with the orbital period that you calculated in Exercise 21.1?

_____

(e) In which compass direction do Venus and Uranus rotate? _____

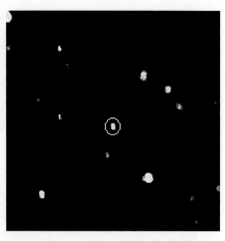

) Pluto (39.5 AU).   **(b)** Makemake (approximately 46 AU).   **(c)** Eris (approximately 68 AU).

21.5 Other Solar System Objects

Our Solar System contains many types of rocky or icy bodies that do not fit under the International Astronomical Union's definition of a planet. To date, these include:

- Five known *dwarf planets*—bodies with characteristics similar to those of a planet except that they have not cleared smaller bodies from their orbital plane (**FIG. 21.8**);
- Approximately 200 confirmed moons (**FIG. 21.9**);
- More than 957,000 *asteroids*—irregularly shaped fragments of solid material residing between the orbits of Mars and Jupiter (**FIG. 21.10A**);
- Approximately 3,615 *comets*—objects composed of ice and dust that orbit the Sun in highly elliptical orbits (**FIG. 21.10B**);

IGURE 21.9 **Examples of moons.**

a) Enceladus (Saturn).   **(b)** Callisto (Jupiter).   **(c)** Io (Jupiter).

FIGURE 21.10 Examples of asteroids, comets, Kuiper Belt objects, and meteoroids.

(a) Asteroid. Asteroids range in size from about 10 m to 530 km in diameter for Vesta, the largest one in the asteroid belt that does not qualify as a dwarf planet. The one pictured here is about 31 km.

(b) Comet. Comet sizes vary widely; the coma cloud can be up to 50,000 km wide, and the tail can extend over 1,000,000 km in length, depending on proximity to the Sun.

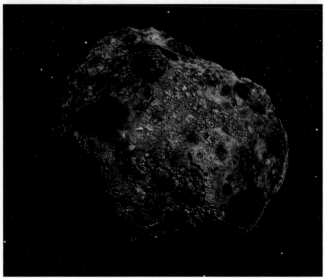

(c) Kuiper Belt object (KBO). KBOs typically range between 20–2,500 km in diameter. The one pictured here is among the smallest yet observed, at just 975 km in diameter.

(d) Meteoroid. Meteoroids range in size from dust to about 1 m in diameter.

(e) When a meteoroid enters the Earth's atmosphere and starts burning up, it is called a meteor.

- Millions of *Kuiper Belt objects*—icy bodies orbiting the Sun between 30–50 AU (**FIG. 21.10C**). About 100,000 of these are over 100 km in diameter;
- Many millions of *meteoroids*—small rocky fragments broken off of asteroids and comets. (**FIG. 21.10D, E**).

Additionally, there are many other possible objects remaining to be discovered in the distant Kuiper Belt and Oort Cloud (a cloud of icy objects that orbits the Sun in the outermost region of the Solar System—reaching a point 1,000 times deeper into space than the outer edge of the Kuiper Belt). **TABLE 21.1** lists the known objects and their characteristic features. It is conceivable that most of these objects are the remnants of the Solar System's formation. The asteroids are likely the leftovers of planet formation

TABLE 21.1 Properties of Solar System objects

Object	Location (AU)	Composition	Shape	Other features
Asteroid	2.2–3.2	Mostly rock	Irregularly shaped; size range from dust to over 500 km (300 mi) in diameter	Reside mainly between Mars and Jupiter.
Comet	30–55 (Kuiper Belt), 2,000–200,000 (Oort Cloud)	Mainly ice, some rock	Nucleus is few km in diameter; coma cloud is up to 50,000 km wide; tail can extend over 1,000,000 km in length	Highly elliptical orbit; approach to Sun warms the comet, producing melting and vaporization and forming both a glowing "coma" around nucleus and two tails: a dust tail and an ion tail of particles blasted off by the solar wind. Originate in the Kuiper Belt or Oort Cloud. May be trapped by a planet's gravity and orbit in the inner Solar System.
Meteoroid	~1.3 (inner Solar System)	Rock/ice	Irregularly shaped	Broken-off pieces of an asteroid or comet; called a *meteor* when it passes through the atmosphere and a *meteorite* when it reaches the Earth's surface.
Kuiper Belt object (KBO)	30–55	Ice and rock	Irregularly shaped; smaller than Pluto (2,300 km/1,430 mi in diameter)	Most do not have eccentric orbits; those that do may be ejected from the Solar System or become short-period comets around Jupiter.
Dwarf planet	Asteroid belt; Kuiper Belt	Rock	Large enough to be round	Share orbits with other objects.

and collisions of planetesimals that did not become large enough to form into terrestrial planets. Comets are the "dirty iceballs" that were blown to the outer system when the Sun ignited. And many of the moons of the Jovian planets may be dwarf planets captured by the Jovian planets' gravities.

In Exercise 21.8, you'll compare two dwarf planets.

EXERCISE 21.8   Dwarf Planets

Name: _____   Section: _____
Course: _____   Date: _____

Pictured below are the dwarf planets Ceres and Haumea

Ceres

Haumea and its two satellites

(continued)

Name: _____   Section: _____
Course: _____   Date: _____

Dwarf planets Ceres and Haumea data

Dwarf planet	Distance (AU)	Orbital period (yrs)*	Rotation period (hours)*	Diameter (km)	Mass (relative)	Satellites	Rings	Eccentricity
Ceres	2.8		9	952	1/14th Pluto's	0	No	0.079
Haumea	43.13		4	1,240	1/14th Earth's	2 confirmed	Yes**	0.198

* in Earth units
** Ring not shown in above image.

(a) Calculate the orbital period of both dwarf planets using Keppler's third law of planetary motion and write your answer in the table.

(b) Within what region of the Solar System is Ceres located? _____ And Haumea? _____

(c) Which of these dwarf planets has a more eccentric orbit? _____

(d) How many times does Haumea rotate in one Earth day? _____ . How does this compare to the rotation speeds of the Jovian planets? _____

(e) How does Haumea's rotation speed impact its shape?

_____

_____

Those objects that are closest to the Earth and are potential impactors on the planet are designated as *Near Earth Objects* (NEOs). NEOs primarily consist of asteroids, comets, and meteoroids. Fortunately, many of these objects collide with Mars, Jupiter, or the Sun before they can impact the Earth. Nonetheless, we know from geologic studies that our planet has been hit by many objects from space over time—frequently in its early years by objects of varying size, and more recently by objects that are dust- to pebble-sized. Impacts of large enough objects can be catastrophic; for example, the 150-km asteroid or comet that struck the Earth about 66 million years ago is thought to have caused the extinction of the dinosaurs. But most smaller objects that reach the Earth are vaporized by our atmosphere. Indeed, an object the size of a compact car is reduced through vaporization and melting in our atmosphere to the size of a grapefruit by the time it reaches the Earth's surface. Most of the material that survives the descent through our atmosphere settles out as dust particles onto the ocean floor.

In Exercise 21.9, you'll examine how the speed and mass of an object affects the impact it would make if it struck our planet.

Name: _____   Section: _____
Course: _____   Date: _____

Suppose an asteroid strikes the Earth. There are two variables that affect the amount of kinetic energy that would be released during impact: the velocity and mass of the impactor. Typical asteroid impact velocities range between 20 and 70 km/s and can have a mass up to 78 billion kg!

(*continued*)

Name: _____   **Section:** _____

Course: _____   **Date:** _____

The amount of kinetic energy released can be calculated using the formula

$$KE = \frac{1}{2}mv^2$$

Where *m* is the mass of the impactor (in kg), *v* is the velocity of the impactor (in m/s), and *KE* is the kinetic energy released on impact (in joules). For reference, 1 joule is approximately the amount of energy required to lift a medium-sized apple (100 g) from the floor onto a desk.

(a) Assume an asteroid weighing in at only 1 kg (the equivalent of a liter of water or a small bag of rice) strikes the Earth at a velocity of 20 km/s. How much kinetic energy is released on impact? _____

(b) Use the same asteroid, but say it struck the Earth at a speed of 50 km/s. How much kinetic energy is released in this collision? _____ .

     What is the effect on the energy released when we change the object's mass, while the velocity remains constant?

(c) How much kinetic energy is released by the impact of a 5 kg asteroid traveling at 20 km/s striking the Earth? _____

     How does this amount compare with your answer in (a)? _____

(d) How much did the KE increase by only changing the velocity of the impactor? _____

(e) How much did the KE increase by only changing the mass of the impactor? _____

Name: _____   **Section:** _____

Course: _____   **Date:** _____

? **What Do You Think** Mathematical predictions have been used in the past to predict the existence of unknown planetary bodies. Neptune was the first planet to be predicted with mathematics before its discovery by telescope in 1846. The Kuiper Belt and Oort Cloud were also predicted to exist using mathematics. What Neptune and these regions of our Solar System have in common is that the orbits of known planets were not "behaving" as expected. It was deduced and later proven via mathematics that there must be another object or region of objects to explain the peculiar orbits of the outer planets. More recently, researchers at Caltech have found mathematical evidence indicating the presence of another planet (Planet X) deep in our Solar System. It is thought this planet could explain the peculiar orbit of some items within the Kuiper belt. Nicknamed "Planet Nine" by the researchers, the planet is predicted to be about the size of Neptune, could be 10 Earth masses, and have an orbital period of 10,000 to 20,000 Earth years.

(a) What information about this "Planet X" would astronomers need to confirm that it is indeed a planet?

_____

(b) What are the challenges in finding this predicted planet? _____

(c) If indeed this planet exists, suggest what its composition would be, given what you have learned about other objects in our Solar System and their composition.

_____

Name: _____   Section: _____

Course: _____   Date: _____

Exploring Earth Science Using Google Earth

1. Visit digital.wwnorton.com/labmanualearthsci
2. Go to the Geotours tile to download Google Earth Pro and the accompanying Geotours exercises file.

Expand the Geotour21 folder in Google Earth by clicking the triangle to the left of the folder icon. Check and double-click the Scaled Solar System folder icon to zoom out to space to see the solar system scaled from Los Angeles, CA (Sun) to New York, NY (Neptune). The placemarks in the folder contain numerical information about original and/or scaled parameters (e.g., radius, orbital distance, distance between objects). Turn on Layers > Borders and Labels to see the state boundaries.

(a) The inner planets of our solar system (Mercury, Venus, Earth, and Mars) can be classified as rocky, terrestrial planets, whereas the remaining outer planets are considered giant, gaseous Jovian planets. Which of the following is correct relative to the scaled Solar System: the terrestrial planets occupy positons across several western U.S. states, the terrestrial planets are located entirely in California, or no Jovian planet lies west of the Mississippi River?

_____

(b) Expand the Scaled Solar System folder and click on any planet placemark to see numerical information about original and/or scaled parameters for the displayed Solar System objects. Assume that you and a friend both are capable of traveling a straight-line path from the Earth to Jupiter. You take a spacecraft that is capable of traveling an average speed of 70,811 km/hr (44,000 mi/hr) to the actual planet, while your friend drives a vehicle at an average speed of 112.7 km/hr (70 mi/hr) along the red line of the scaled Solar System. Determine how many hours the car will travel versus how many days the spacecraft will travel.

_____hours (car) _____days (spacecraft)

(c) When you look at Neptune in a telescope, you are actually looking into the past as the light has to travel from Neptune to your eyes. If the speed of light is -300,000 km/s, how far back into the past are you looking (or put another way, how long does it take light to travel from Neptune to your eyes on Earth)? _____hours

Sun–Earth–Moon Relationships

Solar eclipses, like the one shown here, occur when the disc of a new Moon blocks our view of the Sun.

22.1 Introduction

As we gaze up each day at the sky from the Earth's surface, it appears that the Sun, Moon, and stars travel from the east to the west, rising above and disappearing below the horizon. It is no wonder that the ancient Greeks thought that the Earth was in the center of all objects in the heavens—from this perspective, such a conclusion seems obvious. However, it is wrong. We now know that it is the rotation of the Earth that causes this apparent east-to-west movement of objects across the sky. As the Earth orbits the Sun along the **ecliptic plane** (**FIG. 22.1**), it rotates on its axis at an angle of 23.5° from vertical. This results in periods of day and night whose lengths vary by location on the planet over a year, producing the seasons (**FIG. 22.2**).

Most of the planets in the Solar System orbit the Sun along similar planes to the Earth, and when we describe their motions we do so in reference to this plane. However, not all bodies orbiting the Sun orbit along this plane; as you investigated in Chapter 21—comets, for example, have highly eccentric orbits. Additionally, while most of the planets orbit in nearly circular paths, some do not. The Earth's only natural satellite, the Moon, follows an elliptical orbit about the Earth at an incline

FIGURE 22.1 The Earth's ecliptic plane is the orbit of the Earth around the Sun as viewed from space.

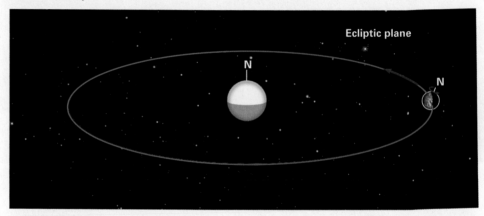

FIGURE 22.2 **The Earth revolves around the Sun in one year.** The tilt of Earth's axis relative to its orbital plane causes seasons.

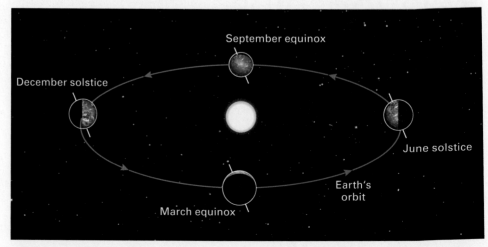

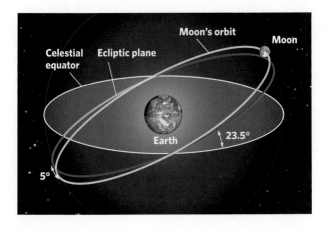

FIGURE 22.3 The Moon's orbit is 5.2° off the ecliptic plane.

of approximately 5 degrees (**FIG. 22.3**). Even though textbooks often illustrate the Moon as orbiting the Earth in a circular pattern and along the plane of the ecliptic (when viewed from above), this is purely for ease of representing this system in two dimensions. In actuality, the Moon does not orbit the Earth in the ecliptic plane, nor at the equator. You will investigate the movements of the Moon and the effects of its alignment with the Sun and the Earth in this chapter.

22.2 Lunar Motions: Sidereal Months

The time it takes the Moon to complete a 360° orbit (one revolution) around the Earth is approximately 27.3 Earth days, a **sidereal** month. As the Moon revolves around the Earth, it does so on a more elliptical path than the Earth completes around the Sun. You'll examine the Moon's orbit in more detail in Exercise 22.1.

EXERCISE 22.1   **The Moon's Motion**

Name: _____   Section: _____

Course: _____   Date: _____

Use the figure on the right to answer these questions. If directed to by your instructor, use a physical model of the Sun-Moon-Earth system instead.

Diagram of the Moon in eight positions as it orbits the Earth, as viewed above the North Pole. The Moon's orbit is drawn as a circle for this exercise.

(continued)

Name: _____ **Section:** _____
Course: _____ **Date:** _____

As you observe the figure (or the model), you will see that a dot has been placed on the Moon. This dot is for reference only, so that you can relate a fixed position on the Moon as the Moon makes its motions around the Earth.

(a) In position 1, is the dot visible to an observer on Earth or on the Sun? (The observer can be anywhere on the Earth or Sun.) _____

(b) As the Moon moves about the Earth to position 2, is the dot visible to an observer on Earth? _____

(c) Look at the Moon at each of the remaining positions (3 through 8 and back to 1): what is the relationship between the dot and an observer on Earth?

_____

_____

_____

(d) From the perspective of an observer on Earth, the Moon does not appear to rotate. Does it actually rotate? Explain.

_____

_____

_____

_____

(e) Is there any time when the dot is visible from the position of an observer on the Sun? Use evidence to support your answer.

_____

_____

_____

_____

(f) If the time it takes the Moon to complete one *revolution* of the Earth (from position 1 all the way back to position 1) is 27.3 days, how many degrees does the Moon move around the Earth each day? _____ degrees/day

(g) How long does it take the Moon to make one complete *rotation* on its axis (for the dot to make a 360° spin)? _____ days

(h) Summarize the relationship between the time it takes the Moon to make one complete rotation (a day on the Moon) and the time it takes to make one complete revolution of the Earth.

_____

_____

_____

_____

_____

22.3 Lunar Phases

Galileo Galilei supported the idea of a Sun-centered system by observing the **phases** of Venus—proving that the planet must orbit between the Earth and the Sun and that the Earth cannot be in the center. What he observed was the changing amount of the illuminated surface of Venus that was visible from the Earth. Recall that only stars emit light—the light that we see from planets and satellites, like our Moon, is the light reflecting off their surfaces. The intensity of the reflected light—or the *albedo*—depends upon the material that makes up the surface of the object, as well as the object's distance from the Sun and other objects. In discovering that objects in our Solar System orbit around the Sun, we also learned that these motions are repeated in a predictable cycle.

We will use the Moon in our example of this cyclical pattern of phases, as illustrated in **FIGURE 22.4**. By convention, the cycle begins at the **new** phase—when none of the illuminated side (i.e., the daytime side) is visible from the Earth. As the Moon moves about the Earth in a counterclockwise motion (as viewed from the North Pole), the amount of the illuminated portion of the Moon that is visible increases, or *waxes*. Therefore, the next phase is termed a **waxing crescent**. When the right half of the Moon is visible from Earth, the Moon is in its **first quarter**. The term *quarter* may seem contrary as we see half of the disc of the Moon lit; but in reality, we only ever see half the Moon at most—the other half of the Moon is not visible from the Earth (remember the previous exercise!), and half of a half is a quarter. A **waxing gibbous** phase follows until the entire daytime side of the Moon is visible from the Earth, or the Moon is **full**. As the Moon continues on the second half of its orbit, the amount of light that is visible from the Earth begins to decrease, or *wane*. Proceeding from a full Moon, the phases are in reverse order: **waning gibbous**, **third** (or last) **quarter**, **waning crescent**, and finally back to new, thereby completing a cycle. These phase terms are applied to any celestial object, such as the phases of Venus.

The time it takes for the Moon to complete one cycle of phases is called a **synodic** (or **lunar**) month, which is approximately 29.5 days. A synodic month is longer than a sidereal month because as the Moon orbits the Earth in a sidereal month, the Earth has moved 30° in its orbit around the Sun, so the Moon must travel more than 360° to get from one new Moon phase to the next (**FIG. 22.5**).

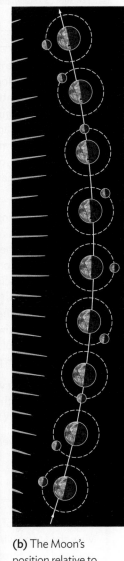

FIGURE 22.4 The Moon as it appears from the Earth through a cycle of phases.

New Moon

Waning Crescent

Last Quarter

Waning Gibbous

Full Moon

Waxing Gibbous

First Quarter

Waxing Crescent

New Moon

(a) The Moon completes its phases in 29.5 days.

(b) The Moon's position relative to the Sun and the Earth determines how much of the illuminated part of the Moon can be seen from the Earth.

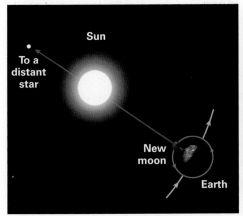
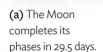
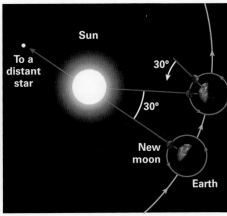

FIGURE 22.5 **The sidereal versus synodic months.** The sidereal month is the time it takes for the Moon to orbit the Earth as measured against a distant star (left). During the synodic month the Earth orbits approximately 30° around the Sun and the Moon orbits approximately 390° (360° + 30°) to align with the Sun again at the next new Moon (right).

Name: _____ Section: _____

Course: _____ Date: _____

The figure on the right is a diagram of the Moon orbiting the Earth. The Moon has been shaded to indicate the daytime and nighttime portions of the Moon's surface. Recall that for simplicity, the Earth, Sun, and Moon are depicted in the same plane, but in reality, the Moon orbits the Sun at an angle to the Earth's ecliptic plane. If directed to by your instructor, use a physical model of the Sun-Moon-Earth system rather than the figure.

The Earth-Moon system as viewed from above the Earth's North Pole.

(a) If you are using the figure rather than a physical model, use a ruler to draw a line across the Moon separating the "side" of the Moon that is visible to an observer on Earth at each numbered position (the side "facing" the Earth) and the side that is not visible. (You can assume your observer is standing at the location directly across from the Moon at each time.) Then, use a pencil to shade in the portion of the Moon that is *not* visible from the Earth at each position.

(b) Imagine that you are floating in space, looking at the Sun-Earth-Moon system in the orientation shown in the figure. As an observer in space, does the shading of the light and dark side of the Moon change at each position? _____

(c) To an observer in space, are lunar phases visible? Explain. _____

_____

_____

For the next set of questions, you will answer from the viewpoint of an observer on the Earth. You may find it helpful to view the Moon while looking from the opposite position. For example, orient the page so that position 5 is right in front of you, and look across the Earth to position 1, where the side facing the Earth is completely dark (shaded in). Do the same for every position by rotating the page.

(d) As the Moon moves from position 1 to position 3, the amount of the illuminated side of the Moon that is visible from the Earth is (circle your answer):   increasing   decreasing

The term for this change is: _____

(e) The number of days that it takes the Moon to move from position 1 to position 3 is _____ , which is _____ % of its orbit.
The side of the Moon that is visible from the Earth at position 3 is (circle your answer):   illuminated   dark

(f) What is the phase name when the Moon is at position 5? _____

What percent of the Moon's orbit is completed at this position? _____

(g) As the Moon moves from position 5 to position 6, the amount of the illuminated side of the Moon that is visible from the Earth is (circle your answer):   increasing   decreasing

The term for this change is: _____

(h) The side of the Moon that is visible from the Earth at position 7 is (circle your answer):   illuminated   dark

(i) How many days will pass before the Moon returns to position 1 from position 7? _____

(continued)

Name: _____   Section: _____
Course: _____   Date: _____

(j) On the diagram below, shade in the circles to represent how the Moon appears to an observer on the Earth at each numbered position from the figure. Write the name of each phase below the sketch. Remember to include terms such as *waning, waxing, first* and *last* (or *third*) quarter as appropriate.

Position: 1   2   3   4

Phase
Name: _____   _____   _____   _____

Position: 5   6   7   8

Phase
Name: _____   _____   _____   _____

(k) If you were an observer standing on the Moon looking at the Earth, what would be the names for the phases of the Earth at each of these numbered positions in the figure (or physical model):

Position 1 _____   Position 3 _____   Position 5 _____   Position 7 _____

(l) Briefly explain why phases (of any celestial object) occur: _____

_____

Just as the Sun appears to rise in the east and set in the west, so too does the Moon. However, since we know that the Moon only moves 13.19 degrees each day around the Earth, it is not really the Moon moving across the sky from horizon to horizon (or 180°) each day and night; rather, it is the rotation of the Earth that makes it appear that the Moon and Sun are moving fully east to west. The Moon actually moves around the Earth counterclockwise (as viewed from the North Pole)—that is, from west to east. We know this because as you observed in the previous exercise on lunar phases, in the northern hemisphere the waxing portion of the Moon begins on the "right" side and moves to the "left." If the Moon orbited clockwise around the Earth, we would see the opposite phase sequence.

While the Sun rises in the morning and sets in the evening, the rising and setting of the Moon can occur during daytime or nighttime. As it turns out, due to the Moon's regular motions, if we know the lunar phase, we can determine the time at which the Moon will rise and set. Conversely, you can also tell the time of day by observing the lunar phase and its position in the sky—rising above the horizon, crossing at its highest point (the meridian), or setting below the horizon.

In Exercise 22.3, you will investigate the relationship between lunar phases and the timing of moonrise and moonset.

Name: _____   Section: _____
Course: _____   Date: _____

The table below contains data for the lunar phases and times of moonrise, the meridian, and moonset for New York during the month of April 2020.

Times of moonrise, meridian crossing, and moonset for New York.

Date	Phase	Moonrise	Meridian	Moonset
Apr 1, 2020	First quarter	11:52 A.M.	7:37 P.M.	3:20 A.M.*
Apr 7, 2020	Full	7:05 P.M.	12:16 A.M.	7:05 A.M.*
Apr 14, 2020	Third quarter	2:06 A.M.	6:46 A.M.	11:28 A.M.
Apr 22, 2020	New	6:14 A.M.	12:43 P.M.	7:22 P.M.
Apr 30, 2020	First quarter	11:48 A.M.	7:22 P.M.	2:48 A.M.*

* = next calendar day

(a) Based on your observations of the data in the table, how does the time of moonrise change over the course of the month? _____

(b) Based on your observations of the data in the table, how does the time of moonset change over the course of the month?_____

(c) Based on your answers in (a) and (b) above, does the Moon at its highest point in the sky appear to move progressively westward or eastward during a lunar cycle?

_____

(d) Can a full Moon be observed from the Earth at noon? Why or why not?

_____

_____

(e) If an observer in New York wanted to view a waning crescent Moon when it was at its highest position in the sky, at approximately what time would it be in that position? _____

(f) Note that there are two first quarter Moon phases in this month. Explain how that can occur.

_____

_____

(g) How can you tell that the Moon orbits the Earth (from a northern hemisphere view) in a counterclockwise direction (that is, the same way that the Earth rotates)? (*Hint:* Consider the order in which the Moon's phases change, per Figure 22.4.)

_____

_____

(h) Given that the "rising" and "setting" of the Moon is a result of the Earth's rotation, should an observer in New York look westward or eastward to see the setting full Moon? Explain your reasoning.

_____

_____

22.4 Eclipses

Eclipses have been shrouded in mystery from the days of early humans, and over the course of history societies have generated many explanations for the phenomenon, including monsters, omens, and curses by gods. Even today eclipses hold us in wonderment for a few simple reasons: eclipses happen infrequently, total eclipses occur even less often, and not all locations on Earth can view the same eclipse simultaneously. As illustrated in **FIGURE 22.6**, an eclipse occurs when either the shadow of the Earth crosses over the full Moon (a **lunar eclipse**) or the disc of the new Moon blocks our view of the Sun (a **solar eclipse**). For any eclipse to occur, these three celestial bodies—the Sun, Moon, and Earth—must be aligned in the same plane. As you have already seen, the Moon does not orbit the Earth in the plane of the ecliptic—it is inclined, and so eclipses do not occur each month. The location of the Moon as it orbits the Earth also limits the frequency of eclipses.

There are two parts to the shadows that form during an eclipse: the darkest, inner part of the shadow, called the **umbra**, and the less-dark, outer part of the shadow, called the **penumbra**. The umbra is darker because the Earth itself blocks any additional sunlight from entering and diluting the shadow. The addition of sunlight

FIGURE 22.6 Eclipses.

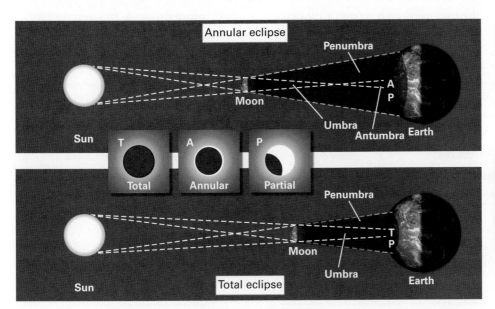

(a) We see a solar eclipse when the Moon blocks the Sun. The distance between the Earth and Moon determines the sizes of the umbra and penumbra (or antumbra, in the case of an annular eclipse), and thus whether a solar eclipse will be viewed from the Earth as partial, total, or annular.

(b) During the total eclipse of the Sun in 2017, visible across the US, observers could see gases streaming from the Sun into space. Millions looked up at the amazing sight when the sky went dark.

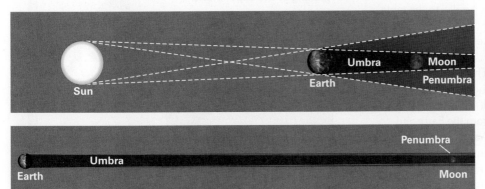

(c) We see a total lunar eclipse when the Moon passes through the Earth's umbra and a partial lunar eclipse when it passes through the penumbra.

(d) The Moon during a total lunar eclipse appears red because light from the Sun passes through the Earth's atmosphere and is bent toward the Moon.

that bends around the Earth on the outer part of the shadow is what causes the penumbra's lighter shading. When the Earth or Moon is covered by the penumbra, an *annular* or partial eclipse occurs. When the umbra crosses the Earth or Moon, it results in a *total* eclipse (Figure 22.6a). Because the umbra track on the Earth is quite small, fewer people get to witness a total solar eclipse than a lunar eclipse.

These differences in the cone-shaped shadows of an eclipse also result in the brightness and colors of the Moon that are observed during lunar eclipses. As the Moon moves into and out of the penumbra, the Moon appears to dim only slightly, if at all. When it enters the darker shadow of the umbra, the darkening is quite noticeable, and once at totality, the color will often turn a reddish-orange shade (Figure 22.6d). The effect that causes the Moon to appear reddish is the same as that which causes our red sunsets: the blue wavelength of sunlight is scattered by Earth's atmosphere, leaving mostly red light, which is bent into Earth's shadow. Clouds, dust from windstorms, ash from forest fires and volcanic eruptions, and other aerosols in the atmosphere can also change the amount of light that is bent into the umbra, creating different appearances of the Moon during an eclipse. The five-point Danjon scale (**TABLE 22.1**), named after French astronomer Andrè-Louis Danjon, ranks total lunar eclipses based on their brightness and other visible characteristics. Anyone can observe these differences—in fact, they are best viewed with the eye, using only a pair of binoculars. You will utilize this scale in Exercise 22.7.

Table 22.1. **The Danjon scale of lunar eclipse brightness and color**

L value	Brightness and color	Other characteristics
0	Very dark	Moon is almost invisible
1	Dark, gray, or brownish	Details distinguishable only with difficulty
2	Deep red or rust-colored	Very dark central shadow, with relatively bright outer umbra
3	Brick red	Bright or yellow rim on umbral shadow (usually)
4	Very bright copper-red or orange	Bluish, very bright rim of umbral shadow

EXERCISE 22.4 **Modeling a Solar Eclipse**

Name: _____ Section: _____
Course: _____ Date: _____

Your instructor has provided you with a model of the Sun-Moon-Earth system. Before you begin, make sure that the light source (the Sun) is parallel to the table so that it shines across the table onto your models of the Earth and Moon. Be careful of the pins holding the Earth and Moon in place!

Turn on the light (the Sun). The Moon should be placed between the Earth and the Sun. Place a piece of blank paper vertically (like a wall) a few inches behind the Earth so that the shadows from the Earth and Moon are cast onto the paper. Adjust the angle of the model to the paper as needed until you can clearly see the shadows. See the figure on the right for reference.

Set-up of model at the beginning of the experiment.

(*Note:* If this exercise is available and being conducted online, the setup, materials, and directions may differ. See the instructions in the online lab for how to conduct the exercise.)

(continued)

Name: _____   **Section:** _____

Course: _____   **Date:** _____

Look at the Sun-Moon-Earth system from the side.

(a) Describe how the Moon's shadow (umbra) is formed and where it falls.

_____

_____

(b) Create a sketch below showing what this setup looks like from above (a top view) including where the Moon's and Earth's shadows fall. Label each object.

(c) Move the Moon to another location in its orbit around the Earth. Is the Moon's shadow cast onto the Earth at this location? Explain.

_____

(d) During what phase of the Moon does a solar eclipse occur? _____

Explain why it occurs during this phase and none of the others.

_____

_____

(e) On Earth, do we see a solar eclipse during the daytime or nighttime? Explain your answer.

_____

_____

(f) A solar eclipse occurs at least twice a year at some location on Earth, but a total solar eclipse occurs only about once every 375 years on average. Give one or two possible reasons why you might not see *total* solar eclipses each month.

_____

_____

(g) Explain why we don't see an eclipse when Venus comes between the Earth and the Sun. (Remember that Galileo observed *phases* of Venus).

_____

_____

Name: _____   Section: _____
Course: _____   Date: _____

The total solar eclipse on August 21, 2017, was visible in a narrow band that swept across the United States from the coast of Oregon to the coast of South Carolina—in other words, from the Northwest to Southeast, as shown in the figure on the right. Use this figure to assist you in completing the exercise.

The path of the umbra (shadow of totality) across the United States during the total solar eclipse of 2017.

(a) Explain why the shadow (umbra) moves in this direction across the United States.

_____

_____

_____

_____

_____

The shadow of totality took only about 91 minutes to cross the United States in a narrow band, which comes out to a speed of approximately 2,013 kph (approximately 1,250 mph). However, this calculation is very simplified. There are some "wrinkles" in making a general calculation of the speed of the umbra.

The Earth is not a flat, 2-D surface; rather, it is spherical. If we measure the circumference of a sphere, then the results will vary by latitude. As a consequence, the rotational speed used in determining the velocity of the umbra will vary by location. Even so, following the steps below, you will prove that a reasonably accurate prediction about how long an eclipse will last can be made.

(b) The eclipse of 2017 crossed the Earth at an average of 39° North latitude. Is the rotational velocity of the Earth at 39° North latitude larger or smaller than at the equator? _____

(c) Calculate the circumference at 39°N. To do so, multiply the circumference of the Earth at the equator by the cosine of the latitude. Yes, this is trigonometry, but we'll help with the hardest calculations: The circumference at the equator = 40,075 km. The cosine of 39 is 0.777.

Circumference at 39°N is _____ km.

(d) What is the rotational velocity at this latitude in kph? (*Hint:* Divide the circumference at 39°N by the number of hours in a day.) _____

How does your answer compare with the value given above (2,013 kph)? _____

(e) Is the Moon's orbit around the Earth a perfect circle? _____

What effect might this have on the revolutionary velocity of the Moon around the Earth (think back to Kepler's second law)? _____

(f) Finally, the Moon's shadow is cast onto a *curved* surface and not a flat one. What effect (if any) do you think this has on the velocity of the shadow as it crosses the face of the Earth?

_____

_____

Name: _____ **Section:** _____

Course: _____ **Date:** _____

For this exercise, you will use the same physical model that you used to model solar eclipses in Exercise 22.4, or additional materials provided by your instructor. Before you begin, be sure that your model is set up as shown in the figure on the right. (*Note:* If this exercise is available and being conducted online, the setup, materials, and directions may differ. See the instructions in the online lab for how to conduct the exercise.)

Turn on the light (the Sun). The Moon should be placed "behind" the Earth so that the Earth is between the Moon and the Sun. Place a piece of blank paper vertically (like a wall) a few inches behind the Moon so that the shadows from the Earth and Moon are cast onto the paper.

Look at the Sun-Moon-Earth system from the side.

Set-up of model at the beginning of the experiment.

(a) Describe how the Earth's shadow is formed and where it falls. _____

_____

(b) Create a sketch below showing what this setup looks like from above (a top view), including where the Moon's and Earth's shadows fall. Label each object.

(c) Move the Moon to another location in its orbit around the Earth. Does the Earth's shadow cross the Moon at this location? Explain.

_____

(d) During what phase of the Moon does a lunar eclipse occur? _____

Explain why it occurs during this phase and none of the others. _____

_____

(e) On Earth, do we see a lunar eclipse during the daytime or nighttime? Explain your answer.

_____

_____

(f) Explain why a lunar eclipse does not occur every month. (*Hint:* Push or pull the pin holding the Moon so that the Moon is not in the same plane as the Earth [it is "above" or "below" the level of the Earth] and rotate the model to observe the change in the shadow's movements against the blank piece of paper.)

_____

_____

Name: _____   Section: _____

Course: _____   Date: _____

The color and brightness of total lunar eclipses vary. In this exercise, you will use the Danjon scale (Table 22.1) to assign the lunar eclipse brightness (L value) for the images of three total eclipses photographed at totality below.

Images of three total lunar eclipses at the moment of totality.

August 28, 2007   September 27, 2015   December 10, 2011

(a) Justify your choices for each figure:

   L value for the 2007 eclipse = _____

   Evidence: _____

   L value for the 2015 eclipse = _____

   Evidence: _____

   L value for the 2011 eclipse = _____

   Evidence: _____

(b) Examine the pattern of coloring on any one of the above images in the figure. Is the Moon image you selected the same color and brightness across the entire image? _____ . Explain your answer.

   _____

   _____

   _____

(c) Compare the images to one other and suggest a reason for the difference in color observed.

   _____

   _____

(d) As you observe in the images and the Danjon scale, not all eclipses are the same level of brightness. Suggest events or conditions that could contribute to a darker eclipse.

   _____

   _____

(continued)

Name: _____   Section: _____
Course: _____   Date: _____

? **What Do You Think** You are living in a Mars colony. Mars has two natural satellites—Deimos and Phobos—that orbit Mars as shown in the figure below. Mars is approximately 228 million km (approximately 134 million mi) from the Sun and a minimum of 54.6 million km (32.8 million mi) from the Earth.

The two Martian moons: Deimos (left) and Phobos (right) orbit Mars.

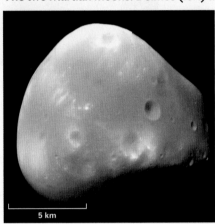

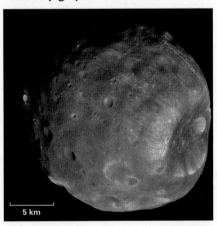

(e) On the Earth you observe phases of the Moon. As a Martian colonist, would you observe phases of the Martian moons? Explain.

_____

_____

_____

_____

Illustration of the orbit of the two Martian moons.

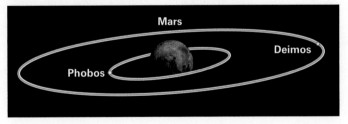

(f) What would you expect the phases of the Earth to look like as viewed from Mars? Explain your reasoning.

_____

_____

_____

_____

_____

_____

Name: _____   Section: _____
Course: _____   Date: _____

Exploring Earth Science Using Google Earth

1. Visit **digital.wwnorton.com/labmanualearthsci**
2. Go to the **Geotours** tile to download Google Earth Pro and the accompanying Geotours exercises file.

Expand the Geotour22 folder in Google Earth by clicking the triangle to the left of the folder icon. The folder contains subfolders for two total solar eclipses that have (or will) cross North America on August 21, 2017, and April 8, 2024, respectively.

(a) What causes a total solar eclipse to occur?

_____

_____

_____

_____

(b) Check and double-click the 08.21.2017 Total Solar Eclipse Path folder. Assuming good weather (and correct timing), any location within the shaded region saw a total solar eclipse. Outside this shaded region, viewers only saw a partial solar eclipse. Check and double-click the Glen Jones Lake placemark in Illinois. What did viewers see at this location: a total or a partial eclipse? _____

(c) Turn on Layers > 3D Buildings, and then check and double-click the St. Louis Gateway Arch placemark in Missouri. Did visitors to the Gateway Arch see a total solar eclipse in 2017? _____

(d) Uncheck the 08.21.2017 Total Solar Eclipse Path folder, and now check and double-click the 04.08.2024 Total Solar Eclipse Path folder. Will visitors to the Gateway Arch see a total solar eclipse in 2024?

_____

(e) Turn off Layers > 3D Buildings, and check and double-click the (e) placemarks. Which university will be in the total solar eclipse path of the moon's umbra? _____

CREDITS

Line Art Permissions

Figure Ex12.4b: Reprinted by permission of the Harris-Galveston Subsidence District

Figure 17.1d: Figure created with CHEMIX School Software, © Arne Standnes. Reprinted by permission of Arne Standnes.

Figure Ex17.10.1: Reprinted by permission of the University of Illinois Department of Atmospheric Science.

Figure 18.1, 18.2b, 18.4.1, and **18.6:** Reprinted by permission of the University of Illinois Department of Atmospheric Science.

Photos

Chapter 1

Page 1: Hackman/Deposit Photos; **p. 2:** Corey Templeton; **p. 7:** NASA; **p. 17 (left):** WilcoUK/Shutterstock; **(right):** The Natural History Museum/Alamy Stock Photo; **p. 20 (top left):** Dr. Stephen Hughes/Coastal Hydraulic Laboratories; **(top right):** National Geographic Image Collection/Alamy Stock Photo; **p. 21 (left):** Courtesy of Stephen Marshak; **(right):** Courtesy of Stephen Marshak; **p. 23:** © 2017 Google Image Landsat/Copernicus.

Chapter 2

Page 25: Buurserstraat386/Dreamstime.com; **p. 29:** NGDC/NOAA; **p. 45:** The Protected Art Archive/Alamy Stock Photo; **p. 47:** © 2017 Google Image Landsat/Copernicus.

Chapter 3

Page 49: Javier Trueba/MSF/Science Source; **p. 54 (top left):** QA International/Science Source; **(top right):** Mark A. Schneider/Science Source; **(bottom left):** Mineral Sciences Department/Smithsonian, National Museum of Natural History; **(bottom right):** Mark A. Schneider/Science Source; **p. 55 (left):** Courtesy of Allan Ludman; **(right):** Courtesy of Allan Ludman; **p. 57 (top left):** Courtesy of Paul Brandes; **(top right):** Courtesy of Paul Brandes; **(bottom left):** Courtesy of Paul Brandes; **(bottom right):** Courtesy of Paul Brandes; **p. 58 (top left):** Kent, Breck P./Animals Animals - Earth Scenes; **(top right):** Epitavi/Alamy Stock Photo; **(middle left):** R. Weller/Cochise College; **(middle center):** Mark A Schneider/Dembinsky Photo Associates/Alamy Stock Photo; **(middle right):** Courtesy of Allan Ludman; **(bottom left):** Millard H. Sharp/Science Source; **(bottom center):** Mark A. Schneider/Science Source; **(bottom right):** 1993 Jeff Scovil; **p. 59 (left):** Coldmoon Photoproject/Shutterstock; **(right):** Biophoto Associates/Science Source; **p. 60 (top to bottom):** R. Weller/Cochise College; Courtesy of Paul Brandes; Courtesy of Paul Brandes; Courtesy of Allan Ludman; Courtesy of Paul Brandes; **p. 70:** © 2017 Google.

Chapter 4

Page 85 (left): Rafael Laguillo/Dreamstime.com; **(center):** Dirk Wiersma/Science Source; **(right):** Les Palenik/Dreamstime.com; **p. 91 (left):** Courtesy of Allan Ludman; **(center):** Albert Copley/Visuals Unlimited, Inc.; **(right):** Courtesy of Allan Ludman; **p. 92 (top left):** Dirk Wiersma/Science Source; **(top right):** Mike McNamee/Science Source; **(middle left):** Courtesy of Stephen Marshak; **(middle right):** albertoblu/Shutterstock; **(bottom):** John Cancalosi/Alamy Stock Photo; **p. 94 (top left):** Courtesy of Stephen Marshak; **(top right):** Marli Bryan Miller, University of Oregon; **(bottom left):** Courtesy of Allan Ludman; **(middle right):** Courtesy of Douglas W. Rankin; **(bottom right):** Courtesy of Allan Ludman; **p. 95 (left to right):** Courtesy of Allan Ludman; Albert Copley/Visuals Unlimited, Inc.; Courtesy of Allan Ludman; Courtesy of Allan Ludman; Courtesy of Stephen Marshak; Courtesy of Stephen Marshak; Courtesy of Allan Ludman; **p. 101 (top left):** geogphotos/Alamy Stock Photo; **(top right):** Lee Rentz/Alamy Stock Photo; **(bottom left):** PjrRocks/Alamy Stock Photo; **(bottom center):** Panther Media GmbH/Alamy Stock Photo; **(bottom right):** Siim Sepp/Alamy Stock Photo; **p. 103 (left):** Marli Bryan Miller, University of Oregon; **(right):** Kavring/Dreamstime.com; **p. 104 (top left):** Siim Sepp/Alamy Stock Photo; **(top right):** Valery Voennyy/Alamy Stock Photo; **(bottom left):** Science Stock Photography/Science Source; **(bottom right):** Artur Mroszczyk/Alamy Stock Photo; **p. 111 (top left):** Ashley B. Staples, courtesy Appalachian State University; **(top right):** R. Weller/Cochise College; **(middle top left):** Courtesy of Allan Ludman; **(middle top right):** Courtesy of Allan Ludman; **(middle bottom left):** R. Weller/Cochise College; **(middle bottom right):** R. Weller/Cochise College; **(bottom left):** Biophoto Associates/Science Source; **(bottom right):** R. Weller/Cochise College; **p. 112 (top left):** Ashley B. Staples, courtesy Appalachian State University; **(top right):** Biophoto Associates/Science Source; **(middle left):** Fokin Oleg/Shutterstock; **(middle right):** R. Weller/Cochise College; **(bottom left):** Dirk Wiersma/Science Source; **(bottom right):** Kurt Hollocher/Geology Department, Union College; **p. 113 (left):** R. Weller/Cochise College; **(right):** Courtesy of Allan Ludman; **p. 117 (top left):** www.sandatlas.org/Shutterstock; **(top right):** Breck P. Kent/Shutterstock; **(middle left):** Courtesy of Allan Ludman; **(middle right):** Courtesy of Stephen Marshak; **(bottom left):** Steve Lowry/Science Source; **(bottom right):** Courtesy of Allan Ludman; **p. 119 (left):** vvoe/Shutterstock; **(center):** Breck P. Kent/Shutterstock; **(right):** Fokin Oleg/Shutterstock; **p. 120:** macrowildlife/Shutterstock; **p. 122 (top left):** SAPhotog/Shutterstock; **(top center):** DeAgostini/Getty Images; **(top right):** TheSoon/Shutterstock; **(middle left):** Steve Vidler/SuperStock; **(middle center):** imageBROKER/Shutterstock; **(middle right):** Shutterstock; **(bottom left):** imageBROKER/Alamy Stock Photo; **(bottom center):** Ian Dagnall/Alamy Stock Photo; **(bottom right):** Michele Burgess/Alamy Stock Photo; **p. 123:** © 2017 CNES/Airbus/© 2017 Google/Image © 2017 Digital Globe.

right): Bob Gibbons/Science Source; **(bottom left):** George Steinmetz; **(bottom right):** Jean-Paul Ferrero/Pantheon/Super Stock; **p. 356:** Sinclair Stammers/Science Source; **p. 361:** Ben Renard-wiart/Dreamstime.com; **p. 363 (left):** Marli Bryan Miller, University of Oregon; **(right):** Global Warming Images/Alamy Stock Photo; **p. 364 (left):** Courtesy of Stephen Marshak; **(right):** Courtesy of Allan Ludman; **p. 365 (top):** Courtesy of Allan Ludman; **(middle):** Ross Dickie/Shutterstock; **(bottom):** Granger; **p. 366:** ©2018 Google; **p. 367 (top):** ©2018 Google; **(bottom):** ©2018 Google; **p. 368:** RADARSAT using MICRODEM; **p. 371:** Marine Geoscience System/National Science Foundation; **p. 376 (top right):** Wolfgang Meier/Corbis/Getty Images; **p. 377 (top):** Patrick Lynch Photography/Shutterstock; **(bottom):** RADARSAT using MICRODEM; **p. 380 (top):** ARCTIC IMAGES/Alamy Stock Photo; **(middle):** Athabasca and Dome Glaciers, 1906, Mary Schaffer/photographer, Whyte Museum of the Canadian Rockies, Mary Schaffer fonds (V527/ng-4); **(bottom):** Dr. Brian Luckman, University of Western Ontario; **p. 381:** Delphotos/Alamy Stock Photo; **p. 382:** ©2018 Google.

Chapter 14
Page 385: kerenby/Alamy Stock Photo; **p. 386:** NASA; **p. 388:** NASA; **p. 390:** Howard Perlman, USGS; globe illustration by Jack Cook, Woods Hole Oceanographic Institution and Adam Nieman; **p. 393:** 2020 Google Earth; **p. 394:** 2020 Google Earth; **p. 400:** NOAA; **p. 404:** NOAA/OSPO; **p. 408:** NOAA; **p. 410:** 2020 Google Earth.

Chapter 15
Page 411 (left): John Kershner/Dreamstimes.com; **(center):** Fever Pitch Productions/Dreamstime.com; **(right):** Jonathlee/Dreamstime.com; **p. 413 (top left):** Courtesy of Allan Ludman; **(top right):** Neil Rabinowitz/Getty Images; **(middle left):** Courtesy of Stephen Marshak; **(middle right):** Courtesy of Stephen Marshak; **(bottom left):** Courtesy of Stephen Marshak; **(bottom right):** MIKE SEGAR/REUTERS/Newscom; **p. 414 (top left):** Don Paulson/Alamy Stock Photo; **(top right):** Lars Christensen/Shutterstock; **(middle left):** Mike Brake/Shutterstock; **(middle right):** Galina Barskaya/Shutterstock; **(bottom left):** Pocholo Calapre/Shutterstock; **(bottom right):** Marisa Estivill/Shutterstock; **p. 417 (top left):** Atlantide Phototravel/Getty Images; **(top right):** RIEGER Bertrand/hemis.fr/Getty Images; **(bottom left):** Roger Ressmeyer/Corbis/VCG/Getty Images; **(bottom right):** USGS; **p. 423:** Courtesy of Allan Ludman; **p. 424:** 2020 Google Earth; **p. 428:** 2020 Google Earth; **429 (left):** Courtesy of Stephen Marshak; **(right):** Education Images/UIG via Getty Images; **p. 430 (top):** AP Photo/ Phil Coale; **(bottom):** Courtesy of Allan Ludman; **p. 431 (top):** RGB Ventures/SuperStock/Alamy Stock Photo; **(bottom):** John Elk III/Alamy Stock Photo; **p. 434:** NASA; **p. 440:** NOAA; **p. 442 (left):** Michael R. Johnson/NOAA; **(right):** olrat/Alamy Stock Photo; **p. 443 (top left):** Ed Darack/Superstock; **(top right):** Gitanna/Shutterstock; **(bottom left):** Steven J. Taylor/Shutterstock; **(bottom right):** AP Photo/The Virginian-Pilot, Mort Fryman; **p. 445 (top):** Dee Golden/Shutterstock; **(bottom):** Jeff R. Clow/Getty Images; **p. 446:** (top) NOAA; **(bottom):** U.S. Army Corps of Engineers; **p. 447 (left):** AP Photo/Gerald Herbert; **(right):** AP Photo John/David Mercer, Pool; **p. 450:** 2020 Google Earth.

Chapter 16
Page 451: NASA; **p. 453 (left):** eddie linssen/Alamy Stock Photo; **(center):** Romans Klevcovs/Alamy Stock Photo; **(right):** Monkey Business Images/Shutterstock; **p. 454:** ksamurkas/Shutterstock; **p. 455 (top):** Bobbo's Pix/Alamy Stock Photo; **p. 467:** John Luke/Getty Images; **p. 472 (top):** MODIS Ocean Science Team/NASA; **(bottom):** Marit Jentoft-Nilsen/Reto Stockli (NASA/GSFC)/NASA Earth Observations; **p. 474:** 2020 Google Earth.

Chapter 17
Page 475: Courtesy of Matthew Olney; **p. 476:** Courtesy of Stephen Marshak; **p. 477 (left):** Courtesy of Stephen Marshak; **(left inset):** Nigel Cattlin/Alamy Stock Photo; **(right):** Jeffrey Frame; **(right inset):** Ted Kinsman/Science Source; **p. 481 (left):** Octavio Campos Salles/Alamy Stock Photo; **(right):** Courtesy of Matthew Olney; **p. 483:** Science History Images/Alamy Stock Photo; **p. 496:** NASA; **p. 497:** 2020 Google Earth.

Chapter 18
Page 499: NOAA/CIRA/Joseph Smith; **p. 503:** US National Weather Service/NOAA; **p. 514:** NOAA/NWS, OCT 1999; **p. 517 (left):** XM Collection/Alamy Stock Photo; **(right):** POOL/REUTERS/Newscom; **p. 518:** National Ocean Service/NOAA; **p. 523 (bottom):** NASA Earth Observatory; **p. 527:** NASA Earth Observatory; **p. 529:** 2020 Google Earth.

Chapter 19
Page 531: Don Mennig/Alamy Stock Photo; **p. 538:** 2020 Google Earth; **p. 540 (right):** Custom Life Science Images/Alamy Stock Photo; **p. 542:** Courtesy of Jessica Olney; **p. 548:** NOAA; **p. 552:** NOAA Climate.gov Data: ARC 2018; **p. 553:** 2020 Google Earth.

Chapter 20
Page 555: Tony Rowell/Alamy Stock Photo; **p. 562:** H. S. Photos/Science Source; **p. 574 (top left):** ESA/Hubble and NASA; **(top center):** ESA/Hubble and NASA; **(top right):** NASA, ESA, S. Beckwith (STScI) and the Hubble Heritage Team (STScI/AURA); **(bottom left):** ESA/Hubble & NASA; **(bottom center):** NASA, ESA, and The Hubble Heritage Team (STScI/AURA); **(bottom right):** ESA/Hubble & NASA; **p. 575:** T.A. Rector/University of Alaska Anchorage, H. Schweiker/WIYN and NOAO/AURA/NSF; **p. 576 (top):** Stocktrek Images, Inc./Alamy Stock Photo; **(bottom):** Robert Gendler; **p. 577 (top left):** ESA/Hubble & NASA; **(top right):** ESA/Hubble & NASA, G. Piotto et al.; **(bottom left):** ESA/Hubble & NASA, J. E. Grindla; **(bottom center):** ESA/Hubble & NASA; **(bottom right):** ESA/Hubble & NASA; **p. 578:** NASA/ESA/M. Livio & Hubble 20th Anniversary Team (STScI); **p. 579:** NASA/Goddard Space Flight Center/JWST; **p. 580:** 2020 Google Earth.

Chapter 21
Page 581: NASA/JPL-Caltech/Space Science Institute; **p. 582:** NASA; **p. 586:** Tim Brown/Science Source; **p. 587 (top left):** Courtesy of the SOHO-MDI consortium. SOHO is a project of international cooperation between ESA and NASA; **(top right):** Observed with the Swedish 1-m Solar Telescope (SST) by the Institute for Solar Physics; **(bottom):** NASA/Stanford Lockheed Institute for Space Research; **p. 588 (left):** SOHO ESA & NASA; **(right):** SOHO ESA & NASA;

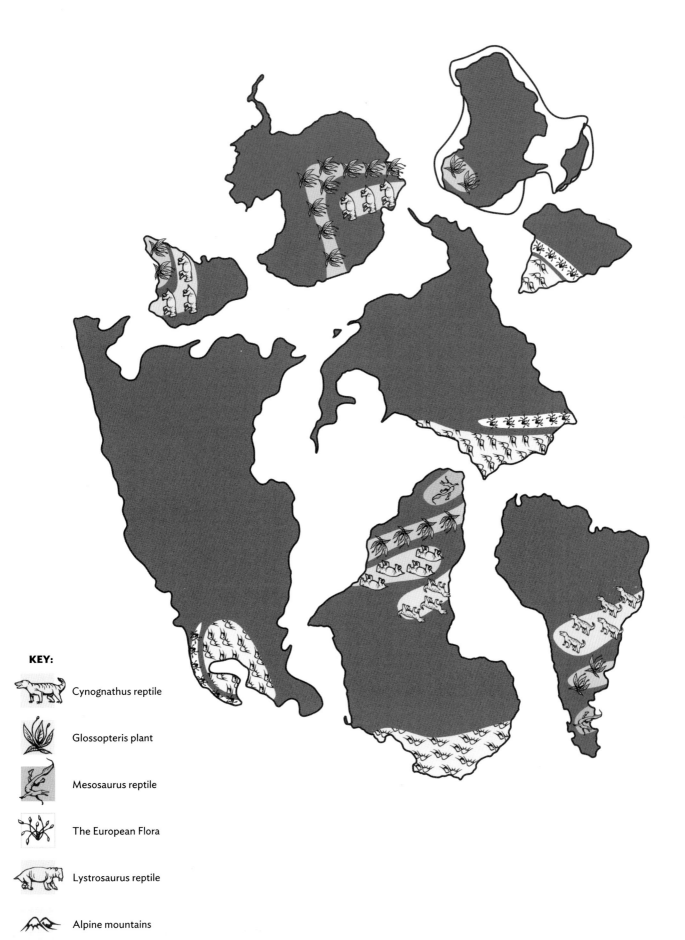

KEY:

Cynognathus reptile

Glossopteris plant

Mesosaurus reptile

The European Flora

Lystrosaurus reptile

Alpine mountains

Block
Diagram 1

1. Cut along solid red lines.
2. Fold along dashed red lines.
3. Tape or glue tabs together
   to create a 3-D block diagram.

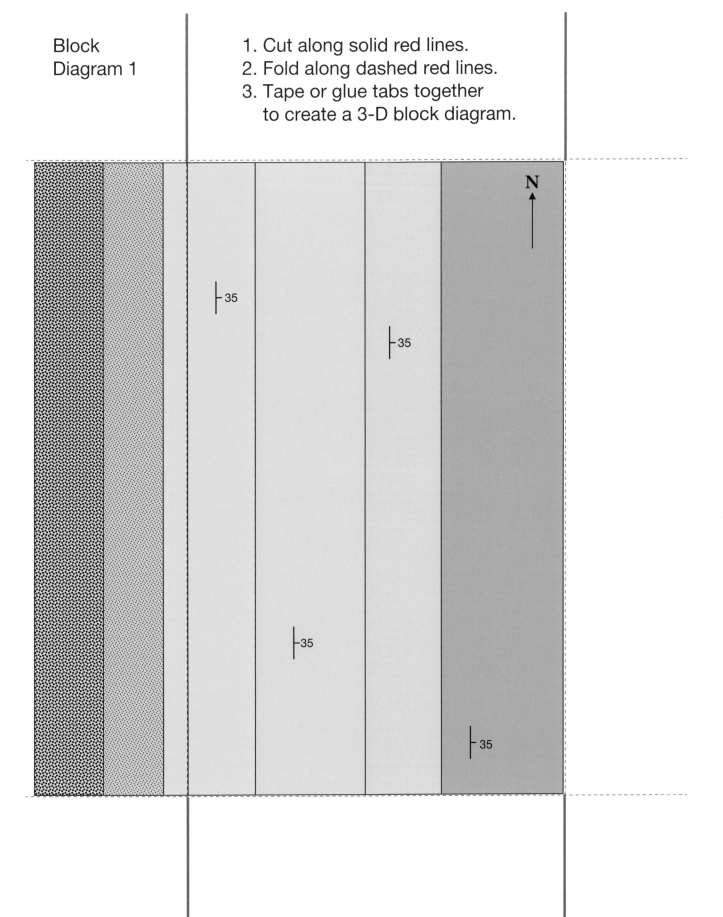

Block
Diagram 2

1. Cut along solid red lines.
2. Fold along dashed red lines.
3. Tape or glue tabs together
   to create a 3-D block diagram.

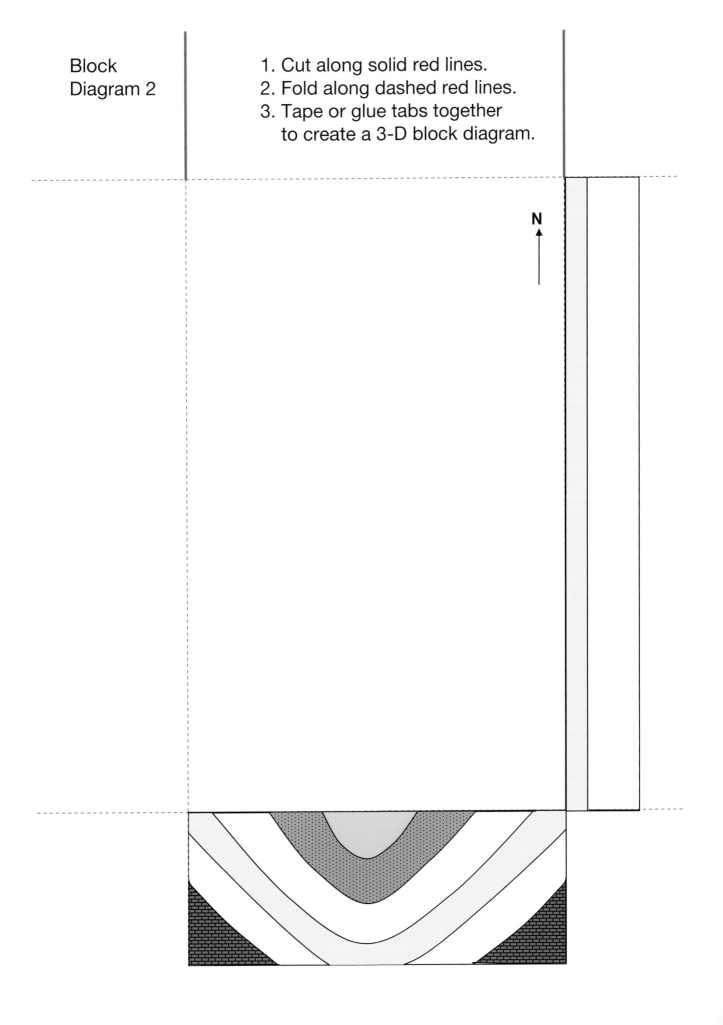

Block
Diagram 3

1. Cut along solid red lines.
2. Fold along dashed red lines.
3. Tape or glue tabs together
   to create a 3-D block diagram.

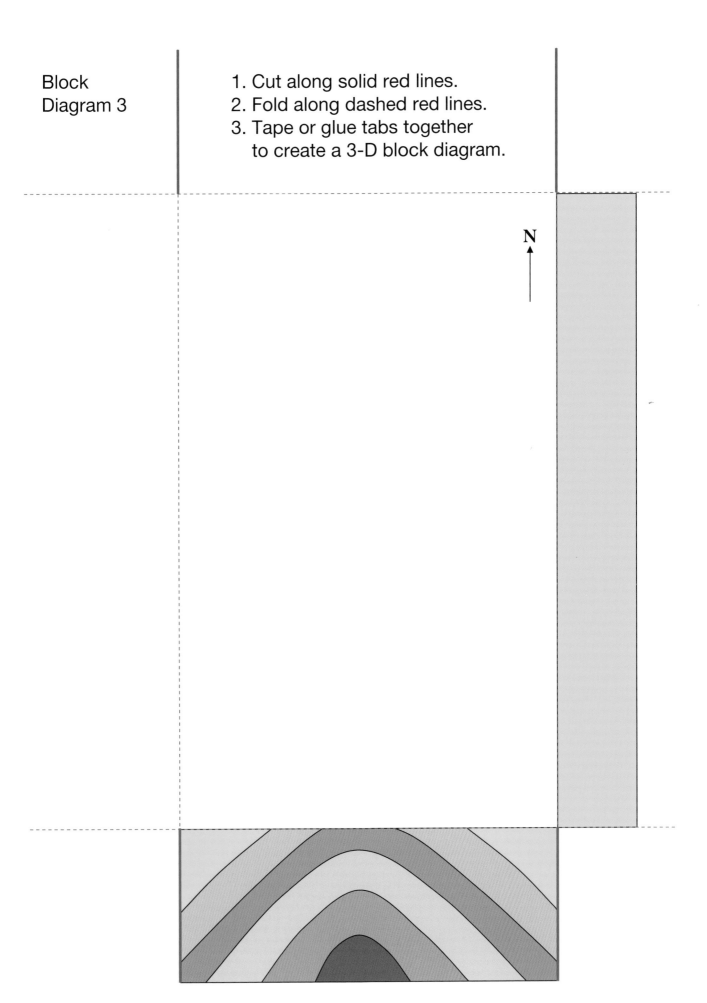

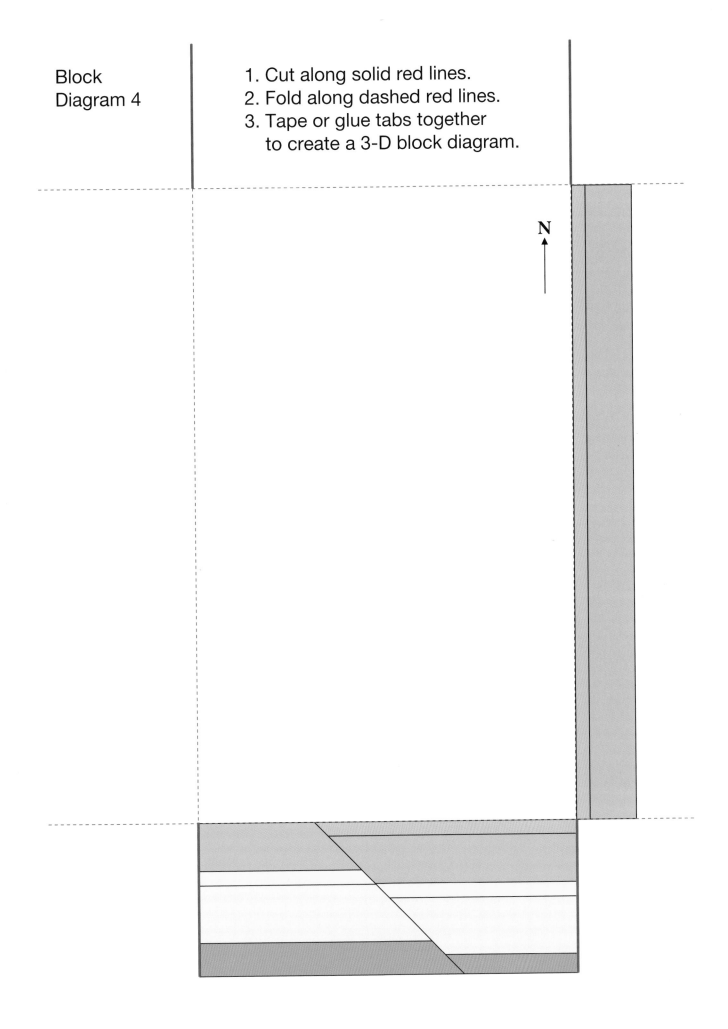

Block
Diagram 4

1. Cut along solid red lines.
2. Fold along dashed red lines.
3. Tape or glue tabs together
   to create a 3-D block diagram.

N